HPE Aruba Networking Certified Professional – Network Security

OFFICIAL CERTIFICATION STUDY GUIDE

(EXAM HPE7-A02)

Second Edition

Miriam Allred

HPE Press
660 4th Street, #802
San Francisco, CA 94107

HPE Aruba Networking Certified Professional - Network Security
Official Certification Study Guide, Second Edition (Exam HPE7-A02)
Miriam Allred

Published by:
Hewlett Packard Enterprise Press
660 4th Street, #802
San Francisco, CA 94107

ISBN: 978-1-960058-32-4

Printed in Mexico

WARNING AND DISCLAIMER

This book provides information about the topics covered in the HPE Aruba Networking Certified Professional - Network Security certification exam (HPE7-A02). Every effort has been made to make this book as complete and as accurate as possible, but no warranty or fitness is implied.

The information is provided on an "as is" basis. The author, and Hewlett Packard Enterprise Press, shall have neither liability nor responsibility to any person or entity with respect to any loss or damages arising from the information contained in this book or from the use of the discs or programs that may accompany it.

The opinions expressed in this book belong to the author and are not necessarily those of Hewlett Packard Enterprise Press.

FEEDBACK INFORMATION

At HPE Press, our goal is to create in-depth reference books of the best quality and value. Each book is crafted with care and precision, undergoing rigorous development that involves the expertise of members from the professional technical community.

Readers' feedback is a continuation of the process. If you have any comments regarding how we could improve the quality of this book, or otherwise alter it to better suit your needs, you can contact us through email at hpepress@epac.com. Please make sure to include the book title and ISBN in your message.

We appreciate your feedback.

Publisher: Hewlett Packard Enterprise Press
HPE Press Program Manager: Michael Bishop

About the Authors

Miriam Allred has many years' experience configuring, testing, and troubleshooting wired and wireless networks. Miriam combines this wide range of technical expertise with pedagogy and instructional design training, allowing her to create technical training courses for both advanced and entry-level server, storage, and networking professionals. Miriam Allred has a degree from Cleveland State University and a degree from Brigham Young University.

Introduction

This book helps you prepare for the HPE Aruba Networking Certified Professional - Network Security certification exam (HPE7-A02). This certification validates candidates' knowledge, skills, and ability to implement HPE Aruba Networking security solutions. You will learn best practices to develop and implement a secure network infrastructure utilizing many of the tools and solutions within HPE Aruba Networking's product suite.

Book topics

Topics covered in this guide include:

- HPE Aruba Networking security strategy
- Wired, wireless LAN, and WAN security
- ClearPass Policy Manager (CPPM) and ClearPass Device Insight (CPDI)
- Zero Trust Security (ZTS)
- Trusted Certificates
- Advanced firewall policies
- 802.1X and role-based access control
- Dynamic segmentation on AOS-CX Switches
- Network Analytics Engine (NAE)
- WIPS and WIDS technology
- Threat detection, analytics, and forensics
- Troubleshooting, monitoring, and more

- HPE Aruba Networking unified SASE, including HPE Aruba Networking SSE and HPE Aruba Networking EdgeConnect

To help you learn as you go, Knowledge Check questions are included in each chapter. In addition, the final chapter has a 30-question practice exam that simulates the kinds and styles of questions that you can expect to meet in the certification exam (HPE7-A02). Answers to the Knowledge Checks can be found in Appendix A.

HPE Certification and Learning

Hewlett Packard Enterprise (HPE) Certification and Learning provides end-to-end continuous learning programs and professional certifications that can help you open doors and accelerate your career. We provide:

- **Professional sales and technical training and certifications** to give you the critical skills needed to design, manage and implement the most sought-after IT disciplines; and
- **Continuous learning activities and job-role based learning plans** to help you keep pace with the demands of the dynamic, fast paced IT industry
- **Advanced training** to help you navigate and seize opportunities within the top IT transformation areas that enable business advantage today.

As a Certification and Learning certified member, your skills, knowledge, and real-world experience are recognized and valued in the marketplace. To continue your professional and career growth, you have access to our large HPE community of world-class IT professionals, trend-makers and decision-makers. Share ideas, best practices, business insights, and challenges as you gain professional connections globally.

To learn more about HPE Certification and Learning certifications and continuous learning programs, please visit: http://certification-learning.hpe.com.

Audience

This guide is designed for network engineers responsible for implementing security controls on enterprise networks.

Assumed knowledge

The candidate should be able to describe the network security stack (firewall, proxy, remote access, IDS/IPS, access control, SASE) and have a strong understanding and knowledge of network security fundamentals.

Minimum qualifications

There are no prerequisites for candidates new to this certification.

Preparing for Exam HPE7-A02

This self-study guide covers the topics that you need to prepare for the HPE Aruba Networking Certified Professional - Network Security certification exam (HPE7-A02). However, it does not guarantee that you will have all the knowledge you need to pass the exam. It is expected that you will also need to draw on real-world experience and would benefit from completing the hands-on lab activities provided in the instructor-led training course.

Recommended HPE Aruba Networking training

Recommended training to prepare for the HPE7-A02 exam is accessible from the exam's page in The Learning Center. See the exam attachment, "Supporting courses," to view and register for the courses.

Obtain hands-on experience

You are not required to take the recommended, supported courses, and completion of training does not guarantee that you will pass the exams. HPE Aruba Networking strongly recommends a combination of training, thorough review of courseware and additional study references, and sufficient on-the-job experience prior to taking a certification exam.

Exam registration

To register for the exam, go to:

https://certification-learning.hpe.com/tr/learn_more_about_exams.html

Contents

Chapter 1: HPE Aruba Networking Security Strategy and ClearPass Fundamentals......... 1

EXAM OBJECTIVES........ 1

Assumed Knowledge........ 1

Overview........ 1

Zero Trust Security........ 2

Why Perimeter Security is No Longer Enough........ 2

Common Internal Threat Vectors: Internal Users........ 3

Common Internal Threat Vectors: IoT........ 4

Common Internal Threat Vectors: Remote Workers........ 4

What is Zero Trust Security?........ 5

NIST Zero Trust Security Principles........ 6

Forrester Zero Trust Security Principles........ 6

HPE Aruba Networking Zero Trust Security........ 8

Principles of HPE Aruba Networking Zero Trust Security (ZTS)........ 8

HPE Aruba Networking ZTS Components........ 9

HPE Aruba Networking Solutions that Support ZTS........ 10

The Evolution of Device Visibility........ 11

Authentication........ 12

Role-Based Access Control........ 13

Sources of Usable Context........ 14

Continuous Monitoring & Enforcement and Response........ 15

A Sample of Our 3rd Party Integrations with CPPM........ 16

The Benefits of Integration........ 17

Example Integration with MDM Solutions........ 18

HPE Aruba Networking ZTS in Action........ 19

CPPM Services........ 20

Review Authentication, Authorization, Accounting (AAA)........ 20

Introduction to CPPM Services........ 22

Service Rules Match Requests to a Service........ 22

Service Rule Guidelines........ 23
Authentication........ 24
Options for Authentication Sources........ 25
Authorization........ 26
What Is a CPPM Role?........ 27
Overview of CPPM Role Mapping and Enforcement Policies........ 27
Why Use CPPM Roles........ 28
Overview of Role Mapping Rules........ 29
Role Mapping Rules with Multiple Conditions........ 30
Rule Evaluation Algorithm and Processing Order........ 31
Enforcement Policies........ 32
Combining Role Mapping and Policy Enforcement........ 33
More Details on Enforcement Profiles........ 34
CPPM Network Devices........ 35
Basic Network Device Entry on CPPM........ 35
CPPM Network Devices in AOS-8 Architectures........ 36
CPPM Network Desires in AOS-10 Architectures........ 37
Review RADIUS vs RadSec........ 37
Review RADIUS Dynamic Authorization........ 39
Enabling Dynamic Authorization in a CPPM Network Device Entry........ 40
RADIUS Interim Accounting........ 41
Standard Ports for Network Device and CPPM Communications........ 43
Summary........ 43
Learning checks........ 44
Chapter 2: Deploy Trusted Certificates........ 47
EXAM OBJECTIVES........ 47
Assumed Knowledge........ 47
Overview........ 47
Digital Certificates and PKI........ 47

Digital Certificate.... 47

Public/Private (Asymmetric) Encryption Algorithms for Digital Certificates.... 49

What is the Digital Certificate for?.... 49

Validating That a Certificate Is Trusted.... 50

Ways to Create Certificate Signing Requests (CSRs).... 60

Some Ways to Submit CSRs.... 64

HPE Aruba Networking ClearPass Certificates.... 67

Types of ClearPass Server Certificates.... 67

Installing Certificates on ClearPass servers.... 69

Types of Trusted CAs on ClearPass.... 70

Certificate Requirements for Forming a ClearPass Cluster.... 72

Other Certificate Considerations for ClearPass Clusters.... 73

HPE Aruba Networking Infrastructure Certificates.... 74

HPE Aruba Networking Infrastructure Certificate Options.... 74

Installing Trusted CA Certificates with HPE Aruba Networking Central.... 78

Installing CA-Signed Certificates with HPE Aruba Networking Central.... 78

Referencing Certificate Usages in HPE Aruba Networking Central.... 79

AOS-CX Trust Anchor (TA) Profiles.... 80

Enabling OCSP on an AOS-CX TA Profile.... 81

Uploading a CA-Signed Certificate on an AOS-CX Switch.... 82

Scaling Certificate Installations with HPE Aruba Networking.... 83

EST in HPE Aruba Networking Products.... 85

CA-Signed Certificates for RadSec.... 89

Summary.... 91

Learning checks.... 91

Chapter 3: Implement Certificate-Based 802.1X.... 93

EXAM OBJECTIVES.... 93

Assumed Knowledge.... 93

Overview.... 93

EAP Review.. 94

Review EAP Methods.. 94

Review the Benefits of Certificate-Based Authentication.. 94

EAP-TLS in a Windows Domain.. 95

Certificates within a Windows Domain.. 95

Windows Certificate Stores.. 96

Windows Domain Certificate Auto-Enrollment.. 96

Certificates in a Microsoft Azure AD (Entra ID) Domain.. 97

How Windows Domain Computers Authenticate the RADIUS/EAP Server.......... 98

High-Level View of EAP-TLS for Windows Domain Users/ Computers.. 102

Ensuring the Authorization Check Succeeds.. 103

Ensuring OCSP Check Works Against Windows CA.. 104

Custom Attribute to Collect AD Account Status.. 105

Using the Custom Attributes in CPPM Policies.. 106

HPE Aruba Networking ClearPass Onboard for BYOD.. 108

Why ClearPass Onboard?.. 108

Onboarding vs Guest Access.. 108

One to One Control of Clients with Onboard.. 109

Overview of a Typical Onboarding Process.. 110

Overview of Configuring ClearPass to Support Onboarding.. 113

How CPPM Authenticates and Authorizes Clients Using Onboard and AD.......... 117

TEAP.. 118

Use Case: Distinguishing Between Users on Domain Computers.. 118

Validating Domain Users Are on Domain Computers with EAP-TLS.. 119

Validating Users Are on Domain Computers with TEAP.. 120

Additional TEAP Scenarios.. 121

Configure CPPM to Use TEAP to Validate Users Are On Domain Computers....... 122

Example Setup: Edit the Enforcement Policy.. 125

Summary.. 126

Learning Checks........ 127
Chapter 4: Implement Advanced Policies on the Role-Based AOS Firewall........ 129
EXAM OBJECTIVES........ 129
Assumed Knowledge........ 129
Overview........ 129
AOS Mobility Role Derivation........ 130
Role Derivation Sequence on AOS Devices........ 130
Understanding Which Post Authentication Role Takes Precedence........ 131
What Happens if Authentication Fails?........ 132
Recommendations When Using CPPM as the Authentication Server........ 133
Example of Using CPPM to Assign Clients to Roles........ 133
Example Role Mapping Policy........ 134
Example Enforcement Policy........ 135
Applying Role-Based Firewall Policies........ 136
Role-Based AOS Firewall........ 136
AOS Role-Based Firewall Policies: Gateways or MCs........ 137
AOS Role-Based Firewall Policies: IAPs or AOS-10 APs........ 138
Access Control Rules........ 138
Review Application Visibility........ 140
Application Visibility and WebCC Use Cases........ 140
Application Rule........ 141
Example Role: Role Structure........ 142
Example Role: Blocked-Apps Policy........ 143
Example Role: Blocked-Websites Policy........ 144
Role-to-Role ACLs........ 145
Tips for Formulating and Analyzing ACLs and Policies........ 145
Summary........ 147
Learning Checks........ 147
Chapter 5: Evaluate Endpoint Posture........ 149

EXAM OBJECTIVES 149

Assumed Knowledge 149

Overview 149

ClearPass Posture Architecture 150

- Why Evaluate Posture? 150
- ClearPass Posture Tokens 151
- OnGuard Options for Posture 152
- Interaction of Information from Multiple Sources 154
- ClearPass Posture Actions 154

Evaluate Endpoint Posture with the OnGuard Agent 155

- Example of Posture Evaluation in Action 155
- Ongoing Connectivity and Temporary Tokens 156
- Overview of Posture Policies 157
- More Details on Posture Policies 158
- Posture Policy Rules 161
- Example Configuration Web-Based Health Check Service 161
- Example Configuration 802.1X Service 162
- ClearPass Clusters and CoAs 164

Microsoft Endpoint Manager Integration 165

- Third-Party Integration for Endpoint Information 165
- Microsoft Endpoint Manager 166
- CPPM Integration with Endpoint Manager 167
- How to Configure the Intune Extension 168
- Register CPPM in Microsoft Azure 169
- Install the Extension 169
- Configure the Extension 170
- Use Intune Attributes in CPPM Policies 172

Summary 172

Learning checks 173

Chapter 6: Implement a Trusted Network Infrastructure.......... 175
EXAM OBJECTIVES.......... 175
Assumed Knowledge.......... 175
Overview.......... 175
Implement Secure Management of AOS-CX Switches.......... 176
Review Management Access on AOS-CX Switches.......... 176
Adding a Control Plane ACL for Inband Management.......... 177
Example Control Plane ACL.......... 177
Review Management Authentication Options.......... 178
Review TACACS+ Server Setup.......... 179
Review Using TACACS+ for Manager Authentication.......... 180
Methods for Assigning Users to Manager Roles.......... 181
Group Privileges.......... 183
Using CPPM as the TACACS+ Authentication Server.......... 183
Defining TACACS+ Enforcement Profiles.......... 184
Defining TACACS+ Enforcement Policies.......... 185
Example TACACS+ Enforcement Services.......... 186
Using Command Authorization on AOS-CX Switches.......... 188
Adding Command Authorization to a TACACS+ Enforcement Profile.......... 188
Specifying Denied/Permitted Commands.......... 189
Enforcing Command Authorization.......... 190
SSH Public Key Authentication.......... 191
Setting up SSH Public Key Authentication.......... 192
Secure L2 and L3 Protocols.......... 193
Overview.......... 193
Review BPDU Protection.......... 194
Review DHCP Snooping.......... 195
Review ARP Inspection.......... 198
IP Source Lockdown.......... 200

Saving IP-to-MAC Bindings to External Storage 201

Secure OSPF Communications on Client VLANs 201

Secure OSPF Communications on Transit Interfaces 203

Checklist of Additional Protections 204

Summary 206

Learning checks 206

Chapter 7: Implement 802.1X and Role-Based Access Control on AOS-CX 207

EXAM OBJECTIVES 207

Assumed Knowledge 207

Overview 208

802.1X for Wired Devices 208

Why 802.1X for Wired Clients 208

Review 802.1X Setup on AOS-CX Switches 209

CPPM 802.1X Wired Service 210

Review AOS-CX Roles and Dynamic Segmentation 212

What Are AOS-CX User-Roles? 213

AOS-CX User-Role Benefits 213

Common Settings Defined in AOS-CX User-Roles 214

Methods for Creating User-Roles 214

Enabling DURs for AOS-CX Switches 216

Creating Downloadable Enforcement Profiles on CPPM 219

More Details on User-Role Assignment 220

Client vs Device Auth-Mode 223

Consideration questions 224

Methods for Handling Wired Client Traffic 225

Implementing Local Forwarding with VLAN Steering 228

Overview 228

Best Practices for Role-based VLANs 229

VLAN Steering Roles 232

Adding Port-Access Policies.......... 234
Approaches for Applying ACLs to Wired Devices.......... 234
Implementing Role-Based Port Access Policies.......... 235
Sources for Port Access Policies.......... 236
Port Access Policy Components.......... 237
Configuring Classes Locally.......... 237
Configuring Port Access Policies Locally.......... 239
Defining Classes in an Aruba Downloadable Role on CPPM.......... 240
Defining Port Access Policies in an Aruba Downloadable Role on CPPM.......... 242
802.1X Authentication of APs.......... 243
Why Authenticate APs with 802.1X.......... 243
Configuring 802.1X on HPE Aruba Networking AP Uplinks.......... 244
Example AOS-CX Roles for APs.......... 245
Example CPPM Service for Authenticating APs.......... 246
CPPM's CA Trust List.......... 248
Summary.......... 249
Learning checks.......... 249
Chapter 8: Implement Dynamic Segmentation on AOS-CX Switches.......... 251
EXAM OBJECTIVES.......... 251
Assumed Knowledge.......... 251
Overview.......... 252
Implementing UBT on AOS-CX Switches.......... 252
UBT.......... 252
Review UBT Tunnel Configuration.......... 253
Review UBT Tunnel Configuration to a Gateway Cluster.......... 254
Triggering UBT for a Client's Traffic.......... 255
Role Configuration for UBT.......... 256
Client UDG Assignment in AOS-10.......... 257
Reserved (Local) VLAN Mode.......... 258

VLAN Extend Mode 259
Enabling Jumbo Frames 260
Troubleshooting 261
Implementing VNBT on AOS-CX Switches 261
VNBT 261
VXLAN Overview (Underlay and Overlay) 262
VXLAN Virtual Network ID (VNI) 263
Centralized vs Distributed L3 Gateways 265
VNBT Configuration 267
Establish VXLAN tunnels 267
Creating User-Roles that Trigger VNBT 268
Summary 268
Learning checks 268
Chapter 9: Monitor with Network Analytics Engine (NAE) 271
EXAM OBJECTIVES 271
Assumed Knowledge 271
Overview 271
Overview of AOS-CX NAE 271
Introduction to Network Automation with AOS-CX 271
HPE Aruba Networking NAE Areas of Focus 272
NAE Agent Components 273
What NAE Agents Monitor 274
How Actions are Triggered (No Clear Condition Defined) 275
How Actions Are Triggered (Clear Condition Defined) 275
Agents That Use Baselines to Trigger Actions 276
Action Options 277
Deploying NAE Agents 279
Relationship Between NAE Scripts and Agents 279
Sources for Scripts (Standalone Switches) 280

Analytics Dashboard 282
Accessing ASE from the Analytics Dashboard 283
Obtaining NAE Scripts Using ASE 283
Creating an NAE Agent 284
Defining and Saving the New Agent 285
Managing Agents in the Analytics Dashboard 285
NAE Agent Details 286
NAE Alert Details 287
Considering NAE Maximums 288
Managing NAE Agents with HPE Aruba Networking Central 289
Monitoring NAE with NetEdit 2.0 289
Summary 290
Learning checks 291
Chapter 10: Implement WIDS/WIPS 293
EXAM OBJECTIVES 293
Assumed Knowledge 293
Overview 293
Review AOS WIDS/WIPS 293
Why Wireless Intrusion Detection/Protection? 293
Wireless Threat Protection Framework 294
Review AP Classifications 295
Review Client Classifications 296
Access Point (AP) & Air Monitor (AM) Considerations 297
Review AM Deployment Recommendations for Rogue AP Detection 299
Configure WIDS/WIPS 300
How to Configure IDS/IPS 300
The Risks of False Positives 300
Recommended Infrastructure IDS Settings 301
Recommended Client IDS Settings 305

Possible IDS Settings for Highest Security Environments 306
Review Intrusion Prevention Mechanisms 308
Recommended IPS Settings 311
Summary 315
Learning checks 315
Chapter 11: Use CPPM and Third-Party Integration to Mitigate Threats 317
EXAM OBJECTIVES 317
Assumed Knowledge 317
Overview 317
CPPM Ingress Event Engine 318
Review of HPE Aruba Networking ClearPass 360 Security Exchange 318
Categories of Third-Party Integration 319
Ingress Event Engine Overview 320
Review Syslog 321
Considering the Effects of Ingress Event Processing 322
Enabling Ingress Event Processing and Insight 323
Examples of Event Sources 324
Defining an Event Source 325
Avoiding Spoofed Event Sources 325
Ingress Event Dictionaries 326
How the Ingress Event Dictionary Works 327
Enabling Ingress Event Dictionaries 328
Using Ingress Events to Respond to Threats 329
Overview of Using Events to Mitigate Threats 329
Event Service Configuration Overview 330
Defining a Post-Authentication Enforcement Profile to Update Endpoint Attributes. 331
Gathering Additional Attributes for the Post-Authentication Profile 332
Defining an Event Enforcement Policy 334
Defining an Event Service 335

Editing the Enforcement Policies for 802.1X Services........ 335
Summary........ 336
Learning checks........ 336
Chapter 12: Implement Device Profiling with CPPM........ 339
EXAM OBJECTIVES........ 339
Assumed Knowledge........ 339
Overview........ 339
Review Endpoint Classification........ 340
Why Profile Endpoints?........ 340
Using Device Profiling to Control Devices........ 340
Consideration questions........ 341
CPPM Collectors........ 342
Set Up Device Profiling on CPPM........ 342
Review the Endpoints Repository........ 342
Endpoint Profile........ 343
Overview of How CPPM Profiles Devices........ 345
CPPM Device Fingerprints Dictionary........ 346
Example Device Fingerprint Dictionary Attributes........ 346
Updating Fingerprints Separate from the Software........ 347
Ensuring CPPM Receives DHCP Packets (Layer 3)........ 348
Ensuring CPPM Receives DHCP Packets (Layer 2)........ 349
How CPPM Receives HTTP User-Agent Strings........ 350
Use Cases for Network/Subnet Scanning........ 351
Scan Methods........ 351
Setting Up External Credentials........ 352
Setting Up Profiler Subnet Mappings........ 353
Network (Discovery) Scan........ 354
Subnet Scan........ 355
Guidelines for Running Scans........ 356

Obtaining Profiling Information from Third-Party Devices........ 357

Using Device Information........ 358

CPPM Endpoint Profiler Dashboard........ 358

ClearPass Insight........ 359

Enabling Insight........ 360

Configuring Insight Database........ 361

Using Insight Endpoint Reports........ 361

Analyzing an Active Endpoint Security Report........ 363

Creating Custom Fingerprints........ 368

Use Profiling in CPPM Services........ 369

Example L2 Authentication Process with Device Profiling........ 369

Adding the Profiler to the Service........ 370

Configure the Profiler Action........ 371

Example of Incorporating Device Profile Information in a Role Mapping Policy. 372

More Detailed Guidelines on the Rule for Devices That Need Further Profiling... 373

Example Enforcement Policy That Incorporates Device Profiles........ 374

Services for the Clients that Need to Be Profiled........ 375

Use Device Profiling to Protect Against Spoofing........ 376

MAC-Auth........ 376

How MAC-Auth and 802.1X Interact on AOS-CX Switches........ 377

MAC-Auth Vulnerabilities........ 381

Using Device Profiling to Enhance MAC-Auth........ 381

Example CPPM 802.1X MAC-Auth Service that Uses Profiling........ 382

Example Process of Denying Access to Spoofing Device........ 383

Endpoints Repository: Conflict Attribute........ 385

Adding Conflict Detection to the Enforcement Policy........ 386

Ignoring Conflict Flags on Devices Using PXE........ 386

Summary........ 387

Learning checks........ 387

Chapter 13: Device Profiling with HPE Aruba Networking.......... 389
EXAM OBJECTIVES.......... 389
Assumed Knowledge.......... 389
Overview.......... 389
Introduction to the Solutions.......... 389
Why Device Profiling from HPE Aruba Networking Cloud Solutions.......... 390
Two Main Options.......... 390
Device Profiling with CPPM vs HPE Aruba Networking Central/CPDI.......... 392
Automatic Device Classification.......... 393
What Data Are System Rules Based On?.......... 393
ML-Based MAC Range Classification.......... 394
ML Clustering.......... 395
Latent Dirichlet Allocation (LDA) Clustering.......... 396
Cluster Recommendations.......... 396
Classification Recommendations and Label Seeding.......... 397
Auto-Classification through Cluster Recommendations.......... 398
Use HPE Aruba Networking Central Client Insight.......... 399
Client Profiling Integrated in HPE Aruba Networking Central.......... 399
Data Collected for Client Profiling.......... 400
Recommendations for Collecting All Attributes on APs and Gateways.......... 401
View Client Profiles.......... 402
View Client Profile Details.......... 402
View Device Network Activity.......... 403
Tags.......... 404
Summary.......... 406
Learning check.......... 406
Chapter 14: Deploy ClearPass Device Insight.......... 407
EXAM OBJECTIVES.......... 407
Assumed Knowledge.......... 407

Overview.......... 407

CPDI Overview.......... 408

CPDI Architecture.......... 408

CPDI Licensing.......... 408

Data Collector Collection Mechanisms.......... 409

Data Collector Mechanisms Used by System Rules.......... 412

Deploy the CPDI Application and Data Collectors.......... 413

Select CPDI for the HPE Aruba Networking Central Account.......... 413

Launch CPDI in HPE Aruba Networking Central.......... 414

Deploy Data Collectors.......... 414

Data Collector Specifications.......... 415

Ports on Data Collectors.......... 415

Guidelines for Data Collector Management and Data Ports.......... 416

Summary of Network Services Required for Data Collectors.......... 417

Guidelines for Data Collector Span Ports.......... 418

Installing a Virtual Data Collector.......... 419

Obtain Registration Tokens.......... 420

Perform the Initial Setup in the Data Collector CLI.......... 420

Add Data Collectors in HPE Aruba Networking Central.......... 421

Set Up CPDI Users.......... 422

Obtain Data on the Data Collectors.......... 423

Initial Log Into CPDI.......... 427

Understand Subnet Segments.......... 427

Create Subnet Segments.......... 429

Create Augmentation Methods.......... 429

Schedule a Discovery Scan to Discover NADs.......... 431

NAD Management.......... 432

Schedule Subnet Scans for Endpoints.......... 433

Analyze CPDI Data.......... 434

CPDI Dashboard........ 434
View Device List and Access Device Details........ 435
View Device Profile Details........ 436
Device History........ 437
View Generic Clusters........ 438
Applying User Classifications and User Rules to Generic Clusters........ 438
Classifying a Cluster with a Recommendation........ 439
Creating User Rules........ 440
Defining the User Rule........ 441
Summary of the Classification Workflow With User Input........ 442
Using Filters........ 443
Creating Tags........ 445
Creating Reports........ 445
Use CPDI Security........ 446
CPDI Security Context Capabilities........ 446
CPDI Agentless Scanning........ 447
Enabling CPDI Vulnerability and Posture Assessments........ 449
When Agentless Scanning Runs........ 449
Vulnerability Assessment........ 450
Posture Assessment........ 451
Controlling Posture for Devices That Cannot Be Scanned........ 452
Understanding the Risk Score........ 453
Summary........ 454
Learning checks........ 454
Chapter 15: Integrate Device Insight with CPPM........ 457
EXAM OBJECTIVES........ 457
Assumed Knowledge........ 457
Overview........ 457
Integrate Device Insight with CPPM........ 458

Why Integrate CPPM and Device Insight? 458
CPPM and Device Insight Integration Architecture 459
Device Insight Integration in a Cluster 460
Prerequisites for Integrating CPPM and Device Insight 460
Understand Implications of Integration 461
Security Context Integration between CPPM and CPDI 461
Obtaining Registration Tokens 462
Configuring Device Insight Integration on CPPM 463
Using Device Insight Information in CPPM Services and Policies 464
Use Device Insight Integration to Detect and Mitigate Threats 465
Device Insight Tags and CPPM Integration 465
Understanding Tag Update Action Options 466
Overview of Using Tags to Apply Enforcement Profiles 469
Example Tag-Based Rule in an Enforcement Policy 469
Example Use Case Using Network Activity 470
Summary 472
Learning checks 472
Chapter 16: Use Packet Captures To Investigate Security Issues 473
EXAM OBJECTIVES 473
Assumed Knowledge 473
Overview 473
Review Packet Captures 474
Review Packet Capture 474
Considerations for Implementing the Packet Capture 474
Capturing Packets on a Client Known By HPE Aruba Networking Central 476
Implementing a Packet Capture with HPE Aruba Networking Central 477
Capturing Packets to and from Clients on a Gateway or MC 478
Implementing an MC or Gateway Datapath Packet Capture to a Remote IP 479
Implementing an MC or Gateway Datapath Packet Capture to a File 480

Capturing Packets to and from a Wired Client on an AOS-CX Switch 481
Running a Packet Analysis Directly on the Switch 482
Mirroring Traffic from a Switch Port to a Packet Analyzer 484
Use Wireshark 485
Wireshark 485
Start a Wireshark Capture 485
Open a PCAP File 486
Wireshark Output 487
Capture Filter 487
Display Filter 488
Display Filter Notation 489
Decoding GRE Packets From HPE Aruba Networking Devices 490
Decoding Aruba_ERM Packets 491
Decoding 802.11 Frames Captured in HPE Aruba Networking Central 492
Tracing a Conversation 493
Detect Port Scans with Wireshark 494
Detect Ping Sweeps 494
Other Types of Sweeps 495
TCP Syn Scan 497
Detect TCP Syn Scans 498
Another Way to Search for TCP Syn Scans 499
Further Investigating the TCP Syn Scan 500
TCP Syn Scan versus Legitimate TCP Session 501
Other Types of TCP Scans 502
UDP Port Scans 503
Detect a UDP Scan 504
Investigating ARP Poisoning with Packet Captures 505
Filter for ARP Poisoning 505
Find the Original MAC Address 506

Collecting Information about Suspicious Clients........ 507
Example of Finding Information in ClearPass Insight........ 507
Scenario: Use Wireshark to Troubleshoot........ 508
The Problem........ 508
Initial Investigation........ 509
Usual Checks Do Not Show the Problem........ 509
Capture Packets to Find the Issue........ 511
Analyze the Packet Capture to Find the Issue........ 511
Summary........ 512
Learning checks........ 513
Chapter 17: Secure Remote and Branch Access........ 515
EXAM OBJECTIVES........ 515
Assumed Knowledge........ 515
Overview........ 515
Introduction to Securing Remote Access........ 516
Traditional Option for Securing Remote Access........ 516
Problems with the Traditional Client-to-Site VPN........ 516
How HPE Aruba Networking SSE is Modernizing Remote Access........ 518
HPE Aruba Networking SSE Architecture........ 519
HPE Aruba SSE Management Console........ 520
HPE Aruba Networking SSE and Identity Providers (IdPs)........ 521
Using the Built-In IdP........ 522
Using Third-Party IdPs........ 523
Implementing HPE Aruba Networking ZTNA........ 524
Overview of the HPE Aruba Networking ZTNA Process (Option A)........ 524
Overview of the HPE Aruba Networking ZTNA Process (Option B)........ 525
Agent versus Agentless Deployments........ 526
Overview of HPE Aruba Networking ZTNA Setup Process........ 527
Creating Connector Zones........ 528

Creating Connectors.. 529

Defining Applications.. 530

Application Types.. 531

Establishing Policy Rules.. 532

Implementing HPE Aruba Networking SWG.. 536

HPE Aruba Networking SWG Overview.. 536

Overview of HPE Aruba Networking SWG Setup.. 537

Policy Rules for HPE Aruba Networking SWG.. 537

Typical SWG Policy Options.. 538

Implementing HPE Aruba Networking CASB.. 541

HPE Aruba Networking CASB Overview.. 541

Overview of HPE Aruba Networking CASB Setup.. 542

Defining SaaS Applications.. 542

Defining Policy Rules for HPE Aruba Networking CASB.. 543

HPE Aruba Networking SSE conclusion.. 544

Integrating with SD-WAN Solutions.. 544

Benefits of HPE Aruba Networking SSE Versus Legacy VPN.. 545

Securing Branch Access.. 547

Traditional Option for Securing Branch Access.. 547

IP Security (IPsec).. 547

IPsec Security Associations (SAs).. 548

Review the HPE Aruba Networking Unified SD-WAN Fabric.. 549

Review HPE Aruba Networking EdgeConnect Components.. 550

Example EdgeConnect SD-WAN Architectures.. 551

Example EdgeConnect SD-Branch Architectures.. 553

Centralized IPsec SA Establishment.. 555

Periodic Rekeying.. 556

Viewing IPsec SAs on HPE Aruba Networking Central.. 557

End-to-End, Role-Based Segmentation.. 558

Summary of EdgeConnect Benefits.......... 559
Summary.......... 561
Learning checks.......... 561
Chapter 18: Configure HPE Aruba Networking Gateway IDS/IPS.......... 563
EXAM OBJECTIVES.......... 563
Assumed Knowledge.......... 563
Overview.......... 563
HPE Aruba Networking Gateway IDS/IPS.......... 564
HPE Aruba Networking 9x00 Gateway Security Features.......... 564
IDS and IPS.......... 564
Licensing for Gateway IDS/IPS on 9x00 Gateways.......... 565
How Gateway IDS/IPS Works.......... 566
Gateway IDS/IPS Ruleset.......... 566
Gateway IDS/IPS Policies and CVSS.......... 567
Gateway IDS/IPS Configuration.......... 568
Overview of Gateway IDS/IPS Configuration.......... 568
Enable Traffic Inspection.......... 568
Update Rulesets.......... 569
Choose Whether to Use IDS or IPS.......... 570
Configure IDS.......... 571
Configure IPS.......... 571
Select the Gateway IDS/IPS Fail Strategy.......... 572
View Rules and Allow List Rules.......... 573
Send Data to a SIEM.......... 573
Security > Gateway IDS/IPS Dashboard.......... 574
Example Threat Details.......... 575
Summary.......... 576
Learning checks.......... 576
Chapter 19: Use HPE Aruba Networking Central Alerts.......... 579

EXAM OBJECTIVES.......... 579

Overview.......... 579

HPE Aruba Networking Central Security Alerts.......... 579

HPE Aruba Networking Central Alert Overview.......... 579

HPE Aruba Networking Central Alert Types.......... 580

HPE Aruba Networking Central Alert Severities.......... 581

Notification Options.......... 582

Configure HPE Aruba Networking Central Alerts.......... 583

Configure the Default Email Recipient.......... 584

Configuring Site-Specific Email Notifications.......... 584

Enable Alerts.......... 585

Configure Alerts Settings.......... 586

Useful AP Security Alerts.......... 587

Follow Up on AP Security Alerts.......... 587

Useful Gateway Security Alerts.......... 588

Follow Up on Gateway Security Alerts.......... 589

Useful Connectivity Alerts for Security Auditing.......... 590

Follow Up on Authentication Failures.......... 591

AI Insights.......... 591

Summary.......... 592

Learning checks.......... 593

Chapter 20: Practice Exam.......... 595

Introduction.......... 595

Minimum Qualifications.......... 595

Exam Details.......... 595

Testing Objectives.......... 596

Advice to Help You Take This Exam.......... 598

Practice Exam Items.......... 598

Explanations.......... 612

Appendix: Learning Check Answers.................................... 623
Chapter 1.................................... 623
Chapter 2.................................... 624
Chapter 3.................................... 625
Chapter 4.................................... 625
Chapter 5.................................... 627
Chapter 6.................................... 628
Chapter 7.................................... 629
Answers to Consideration Questions.................................... 629
Answers to Learning Checks.................................... 629
Chapter 8.................................... 630
Chapter 9.................................... 630
Chapter 10.................................... 631
Chapter 11.................................... 631
Chapter 12.................................... 632
Answers to Consideration Questions.................................... 632
Answers to Learning checks.................................... 633
Chapter 13.................................... 634
Chapter 14.................................... 634
Chapter 15.................................... 635
Chapter 16.................................... 636
Chapter 17.................................... 637
Chapter 18.................................... 638
Chapter 19.................................... 638

1 HPE Aruba Networking Security Strategy and ClearPass Fundamentals

EXAM OBJECTIVES

✓ Explain HPE Aruba Networking Zero Trust Security

✓ Explain how HPE Aruba Networking solutions apply to different security vectors

Assumed Knowledge

- Basics of deploying HPE Aruba Networking infrastructure products
- Basics of HPE Aruba Networking solution architectures
- Common types of threats
- Authentication, Authorization, and Accounting (AAA) framework and basics of protocols such as RADIUS and TACACS+
- Forms of Layer 2 authentication such as 802.1X
- Basic understanding of encryption and wireless security technologies

Overview

In the first section of this chapter, you will learn about Zero Trust Security. You will first learn about the pressures that are pushing companies toward a new security approach. You will then learn what Zero Trust Security is and how it helps companies meet their challenges.

In the second section, you will learn specifically about HPE Aruba Networking Zero Trust Security. You will learn about HPE Aruba Networking's approach to Zero Trust Security and the HPE Aruba Networking Zero Trust Security components, including visibility, authentication, role-based access control, continuous monitoring, and enforcement and response. You will see how HPE Aruba Networking solutions work together to deliver these features.

You will then examine HPE Aruba Networking ClearPass Policy Manager (CPPM) in more depth, as a foundational piece of Zero Trust Security. What you learn about creating CPPM services, as well as setting up network device entries on CPPM, in these sections will lay a foundation for the next several chapters in this study guide.

Zero Trust Security

In this section, you will review Zero Trust Security.

Why Perimeter Security is No Longer Enough

Traditionally, companies have focused on perimeter security, which uses appliances such as firewalls and Intrusion Detection Systems/Intrusion Protection Systems (IDS/IPS) to create a secure perimeter around the network. Perimeter security defines zones, including the untrusted, external zone of the Internet and the internal, trusted zone of the corporate LAN.

However, in the modern cyber security landscape, perimeter security alone is hardly enough. As shown in Figure 1-1, threats can arise from many internal vectors as well. If a company focuses all of its efforts on preventing attacks from outside, and few efforts on controlling behavior from inside, then once hackers have infiltrated the network, they can proceed to wreak havoc.

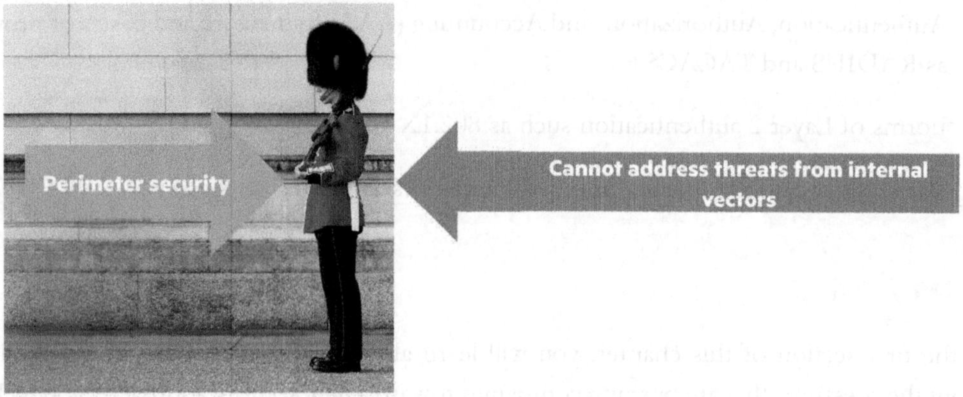

Figure 1-1: Why Perimeter Security is No Longer Enough

Common Internal Threat Vectors: Internal Users

First, companies should be aware that internal users are not always worthy of trust. As indicated in Figure 1-2, malicious employees might launch attacks. Or negligent employees might improperly secure resources, leaving those resources vulnerable to compromise. The Ponemon Institute found that organizations lose, on average per year, 11.45 million USD due to both criminal insiders and employee negligence (https://www.ibm.com/downloads/cas/LQZ4RONE).

- Malicious employees
- Negligent admins or users
- Shadow IT
- Users who fall for social engineering and other attacks

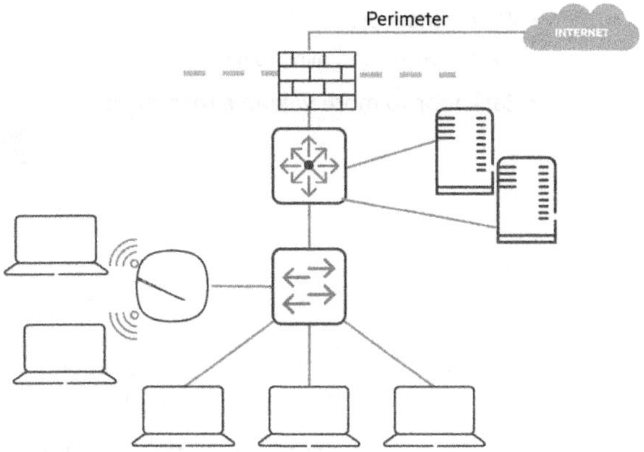

Figure 1-2: Common Internal Threat Vectors: Internal Users

Users might also introduce "shadow IT," which refers to any unauthorized IT resources deployed by users on their own. For example, users might want wireless access in an area where the company does not allow it, so they add their own AP. Shadow IT creates vulnerabilities because users rarely understand how to deploy resources in a secure way. Returning to the example of an unauthorized AP, hackers can take advantage of an insecure AP to gain access into the corporate LAN from outside the building. In any case, when users deploy shadow IT, they are circumventing policies put in place for a reason and potentially creating vulnerabilities.

Bad actors can also evade perimeter security controls by targeting employees with social engineering, which relies on manipulating individuals rather than penetrating technical defenses. Using social engineering the bad actor can gain control over internal devices. The average cost of a social engineering attack, according to SecurityInfoWatch.com, has risen to 130,000 USD (https://www.securityinfowatch.com/cybersecurity/article/21203580/social-engineering-cyberattacks-and-how-theyre-impacting-businesses#:~:text=On%20average%2C%20social%20engineering%20attacks,parties%20or%20new%20cybersecurity%20software).

Common Internal Threat Vectors: IoT

Internet of Things (IoT) devices, such as printers, cameras, smart thermostats, and many more, pose an increasingly critical vector for internal attacks, as indicated in Figure 1-3. The number of these devices is growing exponentially. "IoT Connectivity Industry Forecast by 2030," an article by 101 Blockchains, predicts that 30 billion IoT devices will be connected worldwide by 2022.

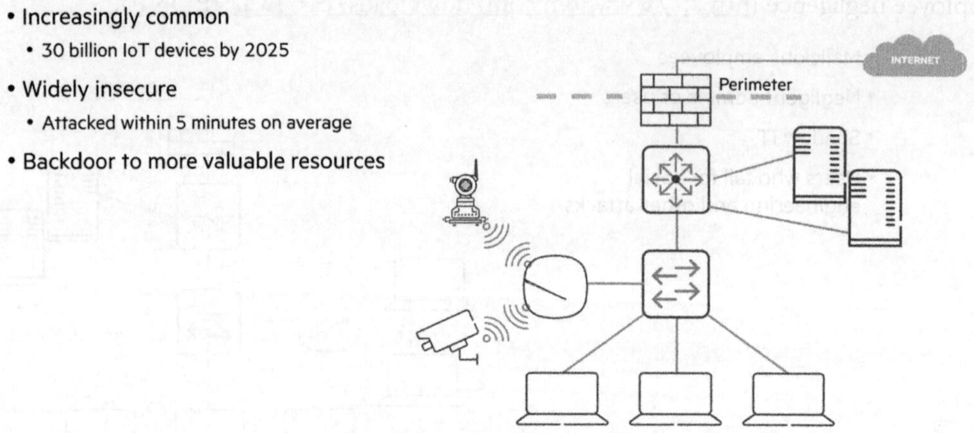

Figure 1-3: Common Internal Threat Vectors: IoT

Because these devices often lack strong security, they represent billions of weak points by which hackers can gain internal access to a corporate network. An attacker with proximity to the corporate environment—perhaps from a lobby or parking lot—can try to access and compromise wireless-enabled IoT devices. If the perimeter firewall permits external management access to IoT devices, hackers could even attempt to log into the devices from the Internet using default credentials or common passwords. IDC found that only 5 minutes pass, on average, between the time that an IoT device goes online and the time that an attack is launched on that device.

Once hackers have compromised an IoT device, they can use that device as a backdoor for reaching other internal resources, exfiltrating data or launching other attacks.

Common Internal Threat Vectors: Remote Workers

Remote workers pose another mechanism by which hackers can infiltrate corporate networks, as indicated in Figure 1-4. A February 2024 Gallup poll of US workers in remote-capable jobs revealed that 27% of the employees worked entirely remotely, and 54% worked in a hybrid manner, sometimes remote and sometimes on-site. Because the workers are accessing the Internet from outside the safety of the corporate firewall, their devices might become compromised by malware that gives hackers a backdoor into that device. The hackers could then use the remote workers' access into the corporate network as a way to infiltrate the network.

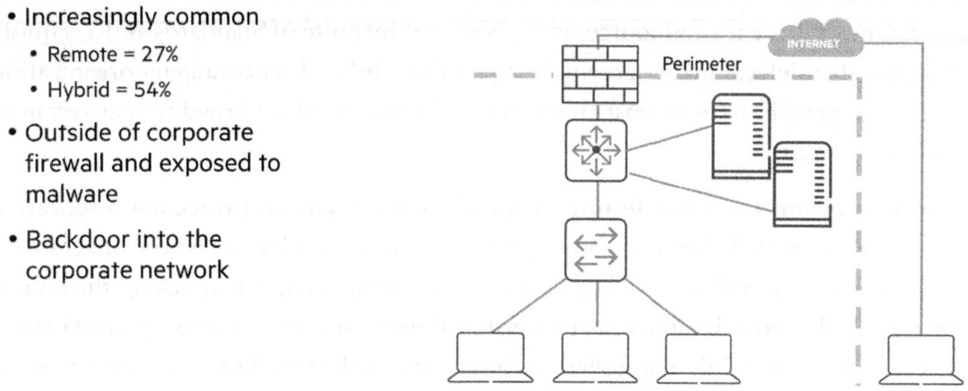

Figure 1-4: Common Internal Threat Vectors: Remote Workers

What is Zero Trust Security?

These trends in threat vectors are dissolving the line between an external "untrusted" network and an internal "trusted" network. Companies need a security approach that acknowledges this reality: simply because a user or device is internal does not mean that the device can be trusted.

As shown in Figure 1- 5, Zero Trust Security confronts this reality. It requires companies to treat all devices as inherently untrusted, regardless of their location outside or inside of the corporate network. By controlling internal devices and continuously monitoring them, the organization can limit the scope of damage that a bad actor who gains internal access can do.

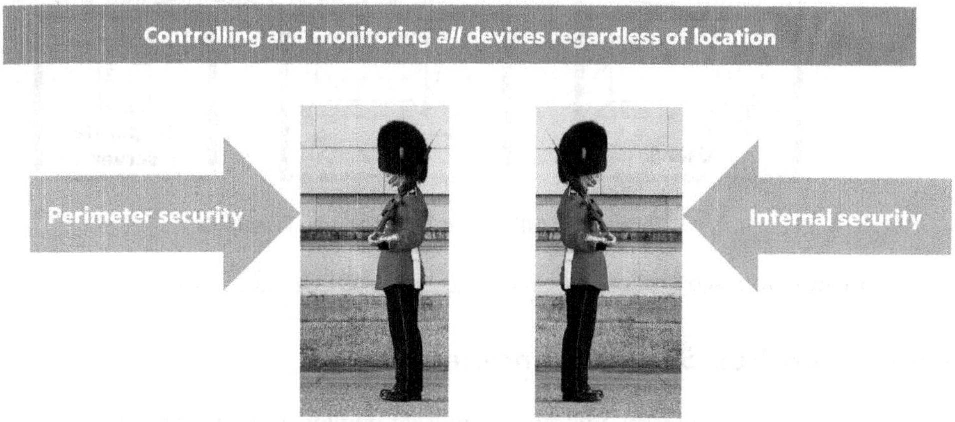

Figure 1-5: What is Zero Trust Security?

NIST Zero Trust Security Principles

Zero Trust Security is an industry-wide trend, and several trusted sources have begun to define what it means. One of those trusted sources is the National Institute of Standards and Technology (NIST), a US agency that defines many security best practices. NIST is encouraging organizations to adopt a zero trust approach. Its zero trust architecture (ZTA) provides a broad framework in which many technologies can fit.

As shown in Figure 1-6, a key feature of the ZTA is a focus on protecting resources rather than protecting boundaries. When companies protect perimeters, a hacker who manages to get past the perimeter has an easier time invading systems, exfiltrating data, and attacking the company. And a perimeter-based approach cannot protect against the internal threat vectors you just examined. With ZTA, every client is tightly controlled, including internal ones. Policy enforcement and decision points identify each user and device on the network and grant that user or device only the level of access to resources required for that particular entity to do their job. This approach is sometimes called micro-segmentation. Rather than control traffic as it passes from one large segment, such as the external network, to another large segment, such as the internal network, micro-segmentation controls all traffic between all devices, applying granular controls based on identity and device type.

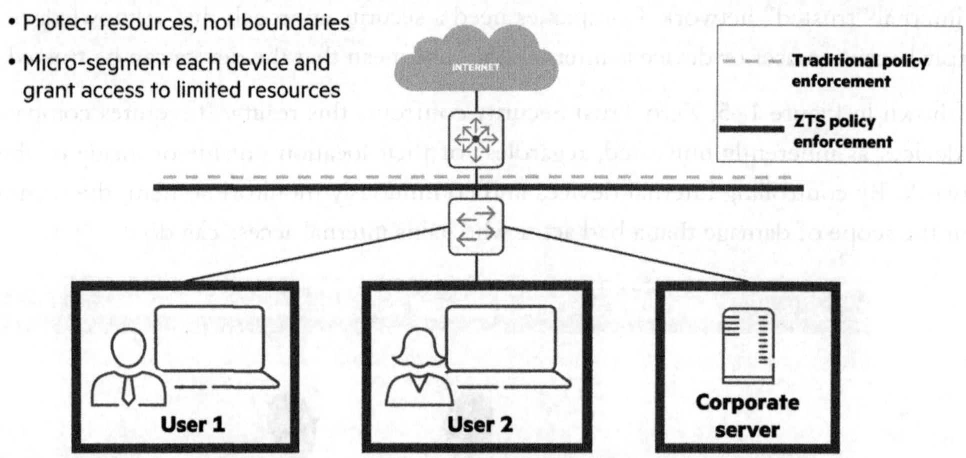

Figure 1-6: NIST Zero Trust Security Principles

For more information, see https://csrc.nist.gov/pubs/sp/800/207/final.

Forrester Zero Trust Security Principles

Forrester, a prominent industry analyst, is similarly guiding organizations towards a Zero Trust Security approach. Forrester emphasizes that in a Zero Trust Security network you do not trust any data, people, devices, or workloads absolutely. Instead, you must monitor all users, devices, and applications and take appropriate steps to control how they use the network. In this effort, visibility, analytics, and orchestration across multiple security components will be key.

As shown in Figure 1-7, Forrester recommends that customers protect their network by identifying sensitive data that needs to be isolated and protected. To achieve this protection, the company should implement micro-segmentation by applying access controls to all clients. Remember: no users are absolutely trusted. Companies must define policies to dictate which authenticated users are allowed access to which data. And these policies must apply to users no matter how and where the users access the network, although customers might choose to use this contextual information to adjust the access policies to the situation.

Finally, Forrester recommends that customers continuously monitor and analyze the environment to discover threats and address security incidents in real-time.

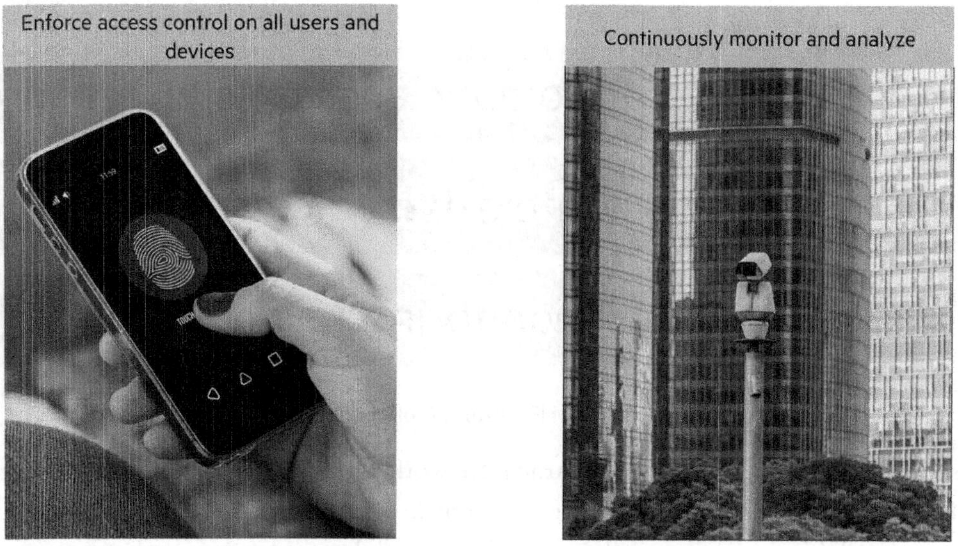

Figure 1-7: Forrester Zero Trust Security Principles

(The information about Forrester's ZTN approach is taken from "Five Steps To A Zero Trust Network," Forrester, Stephanie Balaouras, Chase Cunningham, and Peter Cerrato, October 2018.)

HPE Aruba Networking Zero Trust Security

In this section, you will learn about the HPE Aruba Networking solutions that provide Zero Trust Security. Note that this study guide covers AOS-10 APs and gateways. It covers features in HPE Aruba Networking Central 2.5.6. For switches this study guide covers features available in AOS-CX switches version 10.08 and below. The study guide focuses on HPE Aruba Networking CPPM version 6.9.

Principles of HPE Aruba Networking Zero Trust Security (ZTS)

HPE Aruba Networking Zero Trust Security (ZTS) follows industry guidelines with three key principles: control over internal access, role- and context-based control, and continuous monitoring and assessment. (These principles are summarized in Figure 1-8.)

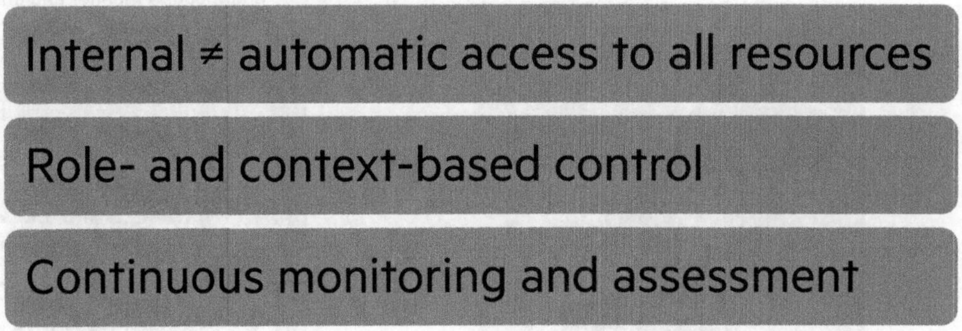

Figure 1-8: Principles of HPE Aruba Networking Zero Trust Security (ZTS)

Default deny means that the HPE Aruba Networking network never grants a device access to resources simply because that device is internal. Instead, the network uses authentication to determine who is connecting to the network and visibility to determine what is connecting to the network.

An HPE Aruba Networking network dynamically segments the traffic for every device that connects to it. This means that the network does not rely on static or location-based segments such as VLANs to determine how to control user traffic. Instead the network dynamically adapts to each device as that device connects, implementing role- and context-based access controls specific to the user role and device type. HPE Aruba Networking recommends a least access privilege approach, which means that users and devices are given access to as few resources as possible, while still enabling productivity.

Continuous monitoring and assessment entails an immediate response to threats, followed by recurring validation, so if a user is determined to be cleared from one security threat, their future access is not guaranteed without further monitoring.

HPE Aruba Networking ZTS Components

HPE Aruba Networking takes a multi-faceted approach to security, as shown in Figure 1-9.

HPE Aruba Networking ZTS solutions monitor and control every user and device that connects to the network, not just once, but continuously.

A broad range of authentication options, including both Authentication, Authorization, and Accounting (AAA) and non-AAA options, ensure that only authorized devices receive network access. With HPE Aruba Networking, a user's network experience is defined not by their location or access type, but by their role. The "one role, one network" approach means that users receive consistent and appropriate access rights based on their role, whether they are connected from a wired device at their desks, a wireless device in a conference room, or a remote device from home.

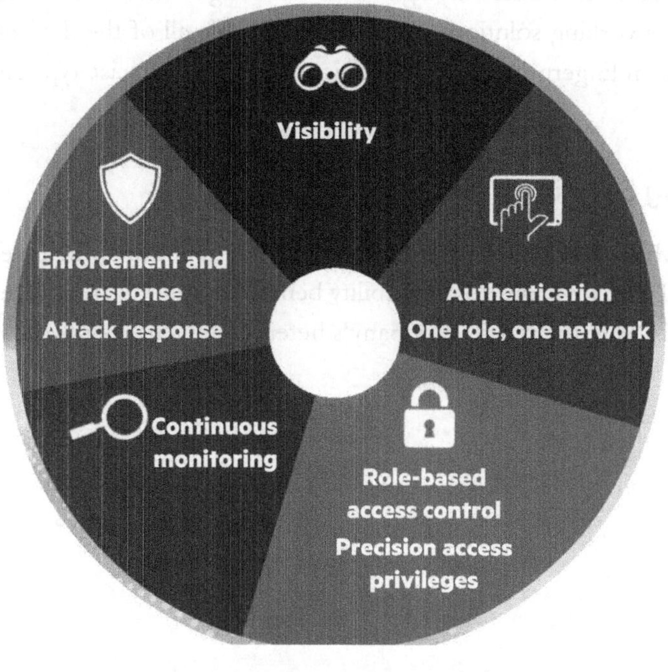

Figure 1-9: HPE Aruba Networking ZTS Components

Role-based access controls power HPE Aruba Networking's approach. After authenticating every client and assigning it a role, the solutions enforce precise access privileges to micro-segment every client. While based on identity, these privileges also take context into account. For example, the solutions can enforce different privileges for a user on a laptop and for the same user on a smartphone, in keeping with the different security profiles of those devices.

Because HPE Aruba Networking understands that a ZTS solution can never authorize a client and then forget about it, connected devices are continuously monitored. HPE Aruba Networking solutions collect real-time telemetry information to detect intrusions and other threats. Companies can monitor a broad range of clients, traffic, and behaviors because not only do HPE Aruba Networking solutions have embedded capabilities, but they also integrate with more than 150 third-party products.

Every threat detected through monitoring requires a response. HPE Aruba Networking solutions make those responses faster and more powerful. When a triggering event occurs, such as the detection of a malware-infected device, HPE Aruba Networking solutions can take a variety actions, including locking an infected device out of the network or placing it in a quarantine segment for remediation.

Finally, because you can only secure something if you know it's there and understand what it is, HPE Aruba Networking grants customers a high level of insight into the devices in their environment. HPE Aruba Networking solutions discover and profile all of the devices connected to a network. Advanced, custom fingerprinting technologies identify the precise type and operating system of each client.

HPE Aruba Networking Solutions that Support ZTS

In Figure 1-10, you can see which HPE Aruba Networking solutions support which pieces of ZTS. ClearPass Policy Manager (CPPM) provides some visibility benefits, while ClearPass Device Insight (CPDI) provides complete visibility for all of a company's heterogenous network.

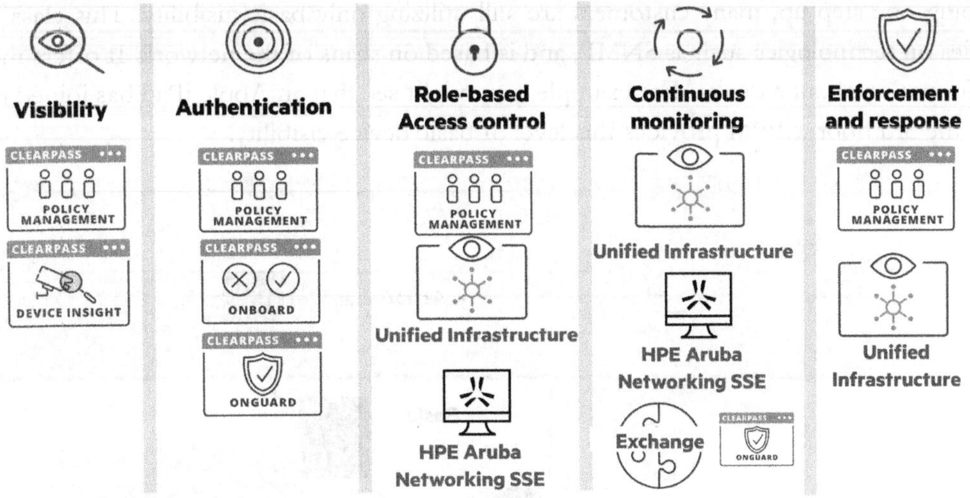

Figure 1-10: HPE Aruba Networking Solutions that Support ZTS

CPPM provides a broad range of sophisticated authentication methods, and HPE Aruba Networking ClearPass Onboard and ClearPass OnGuard add further capabilities. Building on its authentication capabilities, CPPM delivers role- and context-based access control based on highly flexible policies. The HPE Aruba Networking Unified Infrastructure receives the instructions from CPPM and enforces them with a powerful firewall. HPE Aruba Networking SSE secures access for clients anywhere to applications anywhere.

Both HPE Aruba Networking Unified Infrastructure and HPE Aruba Networking SSE also helps to deliver continuous monitoring. ClearPass OnGuard can also continuously assess clients' posture. And companies can extend monitoring with a broad range of third-party solutions that are part of ClearPass Exchange.

Based on monitoring results, CPPM can make new decisions to respond to threats, and the Unified Infrastructure can enforce those decisions and mitigate threats.

You will next dive deeper into how exactly each of these products helps businesses achieve ZTS.

The Evolution of Device Visibility

Good security begins with visibility—since security is only as good as the information it's based on.

As shown in Figure 1-11, visibility has naturally evolved as time has passed, but you may find that some customers are still relying on antiquated visibility. This level of visibility is often based on MAC address and DHCP. It might, for example, only inform you that an Apple iPad has joined the network. With the rising sophistication of exploits, this level of visibility is simply not enough.

Going one step up, many customers are still utilizing only basic visibility. This class of visibility relies on technologies such as SNMP, and is based on scans of the network. It offers more specific information about a device. For example, you might see that an Apple iPad has joined the network on the 3rd floor. CPPM provides this level of basic device visibility.

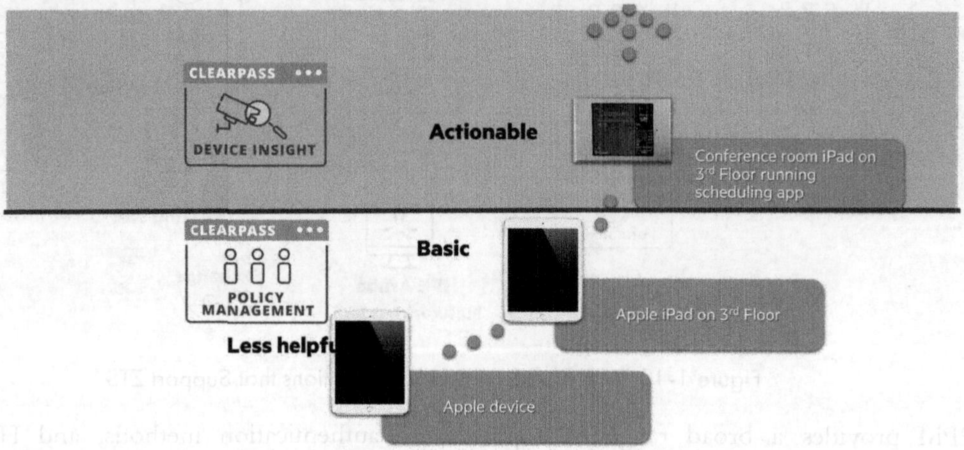

Figure 1-11: The Evolution of Device Visibility

With CPDI, HPE Aruba Networking leads customers to actionable visibility. This level of visibility is based on the behavior of a client, and provides granular insights about each device on the network. For example, you would see that an Apple iPad has joined the network in the conference room on the 3rd floor and is now running a scheduling app. Actionable visibility provides the kind of detailed information that organizations need to protect themselves from today's attacks.

Authentication

CPPM acts as the central policy engine for controlling all forms of access. It supports multiple authentication methods, including integration with Microsoft Active Directory, a portal for guests, certificate authentication, and multi-factor authentication. CPPM allows network device admins to define custom policies based on a user's identity or device type. It also enables them to define policies based on context, such as a user's location or the time of day. And CPPM integrates with third-party solutions to enhance security orchestration.

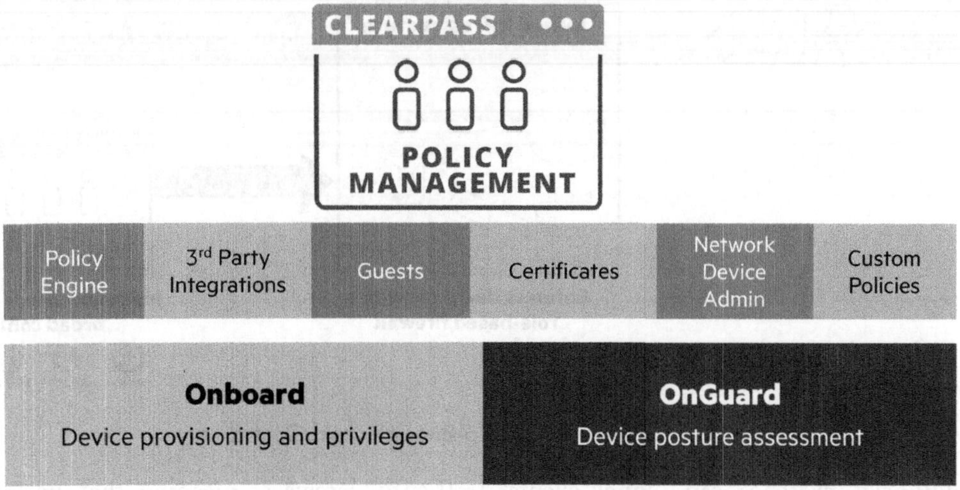

Figure 1-12: Authentication

As shown in Figure 1-12, authentication with CPPM can be enhanced with ClearPass Onboard and OnGuard.

ClearPass Onboard simplifies the process of provisioning devices with the settings and certificates required for a secure authentication. Onboard provides a self-service app that enables users to provision their own devices, rather than making this another task for IT.

ClearPass OnGuard scans endpoints to determine their security posture. For example, it can verify firewall settings and ensure that a device's patches are up-to-date. OnGuard communicates information about an endpoint's security posture to CPPM. CPPM can then use that information to affect its access control decisions, deciding, for example, whether to quarantine the device.

Role-Based Access Control

To implement role-based access control, customers rely on HPE Aruba Networking infrastructure to take action based on the roles that CPPM assigns, as shown in Figure 1-13. HPE Aruba Networking APs and gateways provide a role-based Policy Enforcement Firewall (PEF), which filters all devices' traffic based on the CPPM-assigned role. For example, the PEF might allow a user with the HR role to access a subnet of servers that hold academic records while preventing other users on the same network from such access. The PEF might further restrict IoT devices to sending and receiving traffic from a very limited set of IP addresses.

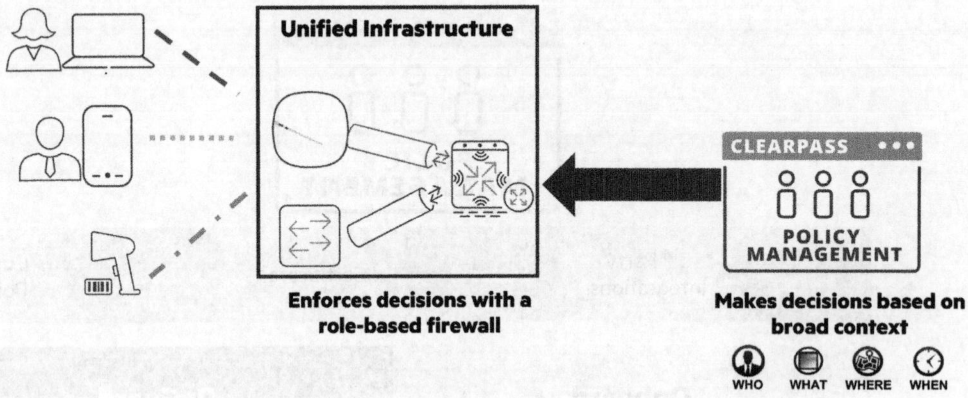

Figure 1-13: Role-Based Access Control

Additionally, switches can tunnel clients' traffic to gateways, meaning that no matter what connection method users choose, wireless or wired, the same role-based policies will be applied to them.

An HPE Aruba Networking network can furthermore implement context-aware policies, which allows for the application of different policies to the same user, depending on the user's device. A user on a personal, insecure smartphone can be restricted from accessing certain corporate resources that the user can otherwise access on a company-issued laptop.

Sources of Usable Context

As shown in Figure 1-14, CPPM has several sources of information on which to draw for "context" for its policies. First, CPPM gains information from the devices themselves. CPPM can implement device profiling, or it can integrate with CPDI for more advanced profiling. Device profiling yields information such as the device OS and category. For example, HPE Aruba Networking solutions can distinguish between a laptop and a smartphone, and they can determine the hardware vendor for the smartphone. As you learned, CPDI distinguishes itself in the depth of visibility that it provides for IoT devices. Using this information, CPPM can assign a role with less trust to more vulnerable devices, helping to prevent these devices from becoming backdoors for attackers.

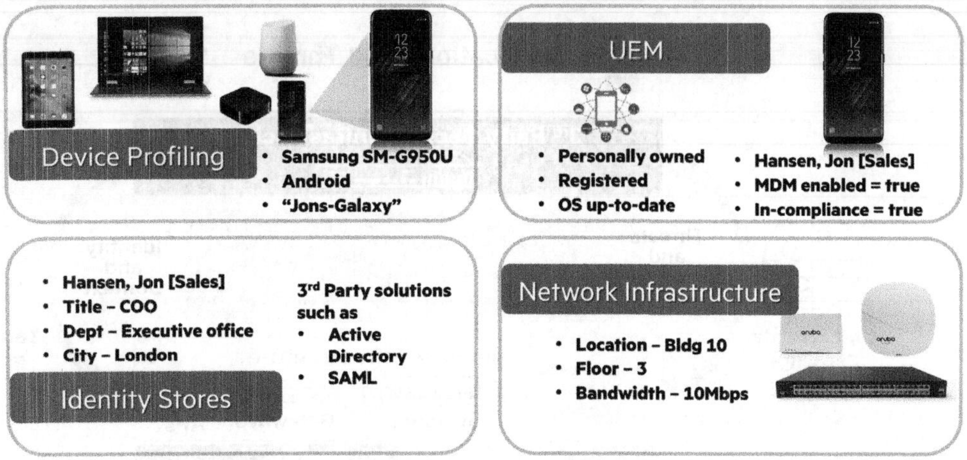

Figure 1-14: Sources of Usable Context

CPPM can also integrate with unified endpoint management (UEM) tools to gather context. With this data, CPPM can determine if a device's OS is up-to-date or if the device is overall in compliance with policies. If the OS is not up-to-date or the device is not in compliance, the HPE Aruba Networking network can react accordingly to protect the network from this vulnerability.

CPPM also uses identity stores to obtain more user-centric context. For example, the name, title, department, and location of an employee will be available through sources like Active Directory, which allows CPPM to precisely limit users' access based on their actual needs.

Finally, CPPM can also obtain contextual information from network infrastructure devices. Based on information included by the network access device (NAD) in its RADIUS request, CPPM can determine the client's location and apply location-based policies. It can also use accounting information from NADs to track clients' bandwidth consumption.

Continuous Monitoring & Enforcement and Response

The last two elements of HPE Aruba Networking's Zero Trust Security approach are closely related, as shown in Figure 1-15.

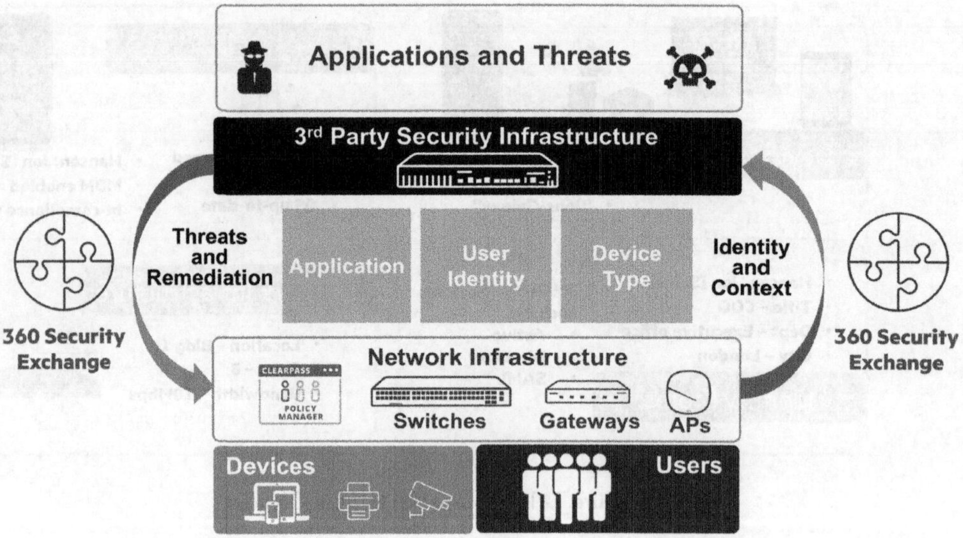

Figure 1-15: Continuous Monitoring & Enforcement and Response

HPE Aruba Networking infrastructure conducts continuous monitoring on all connected endpoints to ensure that they are behaving appropriately. Built-in functionality such as intrusion detection, deep packet inspection (DPI), and web-traffic filtering based on site reputation make this possible. HPE Aruba Networking solutions also integrate with more than one hundred and fifty third-party solutions in the HPE Aruba Networking 360 Security Exchange.

Through 360 Security Exchange, HPE Aruba Networking can feed third-party tools important contextual information such as user identity and device category. In return, those third-party solutions can detect endpoints that are infected with malware, identify the source of an attack, and so on, before subsequently remediating the threat.

Finally, HPE Aruba Networking supports enforcement and response primarily by adjusting an endpoint's access level. For example, if CPPM receives information indicating that an endpoint has been infected with malware, HPE Aruba Networking infrastructure solutions will lock out that endpoint, or they will place it in quarantine and direct the user's browser to a remediation server.

A Sample of Our 3rd Party Integrations with CPPM

In Figure 1-16, you see just some of the many third-party solutions that can integrate with CPPM. These solutions span many categories, including security and authentication. Companies can even choose to let guests login with credentials from popular social media platforms.

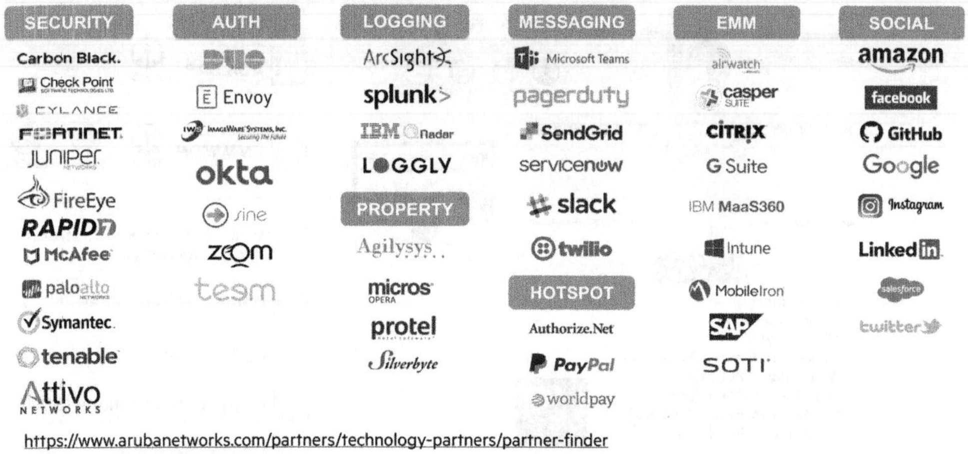

Figure 1-16: A Sample of Our 3rd Party Integrations with CPPM

By integrating with logging solutions, CPPM helps to create audit trails and support companies' organization-wide security efforts. Integration with Enterprise Mobility Management (EMM) solutions can help to provide important context, as you learned earlier, and CPPM integrates with many commonly used ones.

The Benefits of Integration

By integrating with third-party solutions, CPPM is able to assign policies with much greater accuracy based on who the user is, what device is in use, where the client is located, and the time at which the device is trying to access network resources.

With third-party integrations, CPPM is able to derive users' identity primarily from apps that users regularly interact with, like Active Directory. CPPM relies on other third-party solutions, especially mobility management solutions, and its integrations with them to further derive user identity, and all the other relevant information about that client.

Consider a company that is using AD for its directory service and MobileIron MDM for security and management. CPPM can utilize integrations with these solutions to build a profile of the user, who, for example, tries to access files stored on a corporate server using a mobile device. As shown in Figure 1-7, rather than being an anonymous user on a mobile device, CPPM now identifies this user as Robert from Marketing, using his Pixel 3 in Conference Room 103 at 9 AM on Monday morning to access corporate resources. The device is determined to be compliant with security policies. Its MAC and IP address are naturally known as well.

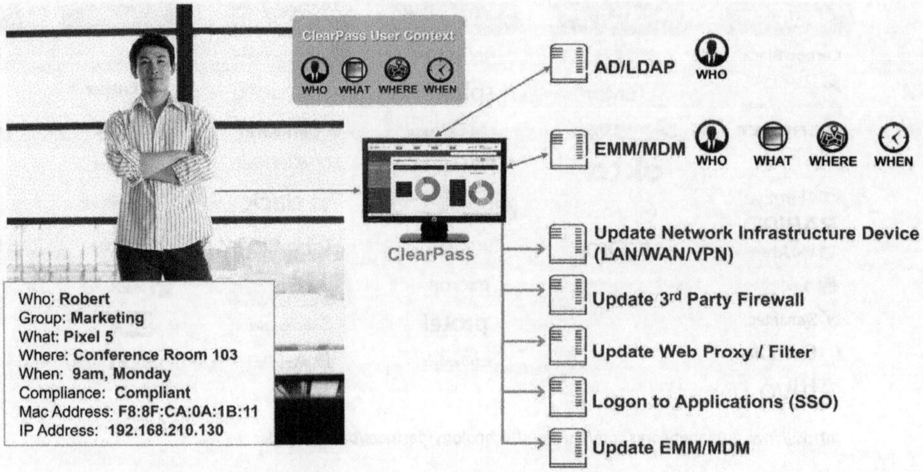

Figure 1-17: The Benefits of Integration

Now, CPPM can update enforcement devices, the firewall, and the web proxy so they can apply policies that are appropriate for Robert. CPPM can adjust Robert's ability to logon to applications, and even update MobileIron to ensure Robert and his device are still secured effectively.

Example Integration with MDM Solutions

As another example of strengthening security with third-party integration, consider what might happen if a jailbroken device joins the network (see Figure 1-18). A mobile device management (MDM) solution would be capable of recognizing that the device is jailbroken—and represents a security threat to an enterprise environment. Through integration, CPPM could then be alerted about the jailbroken device, and subsequently determine that access should be denied. A helpdesk ticket would be autogenerated by CPPM. And CPPM can request the MDM solution to notify the user why they did not receive a connection, and what action the user can take.

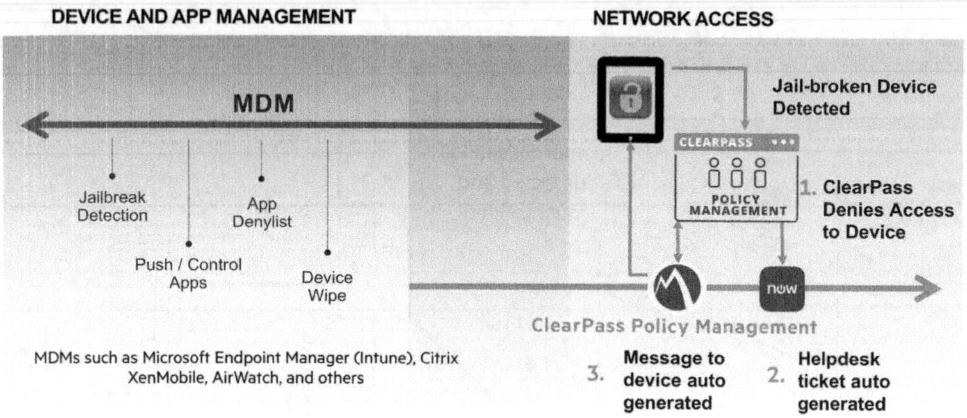

Figure 1-18: Example Integration with MDM Solutions

HPE Aruba Networking ZTS in Action

Now consider some more examples of how HPE Aruba Networking ZTS can secure a network against internal threat vectors.

Perhaps, as shown in Figure 1-19, hackers succeed in compromising an IoT device. Because HPE Aruba Networking ZTS enabled the company to tightly control the types of traffic that IoT devices can send, the hackers will find it much more difficult to use the devices to reach sensitive resources. Similarly, perhaps a bad actor uses phishing and social engineering to install malware on an employee's device and covertly take control of that device. However, role-based access controls have denied this user from using SSH and also limits which data center networks the user can access. Again, the hacker is limited in ability to expand the attack to other devices in the system.

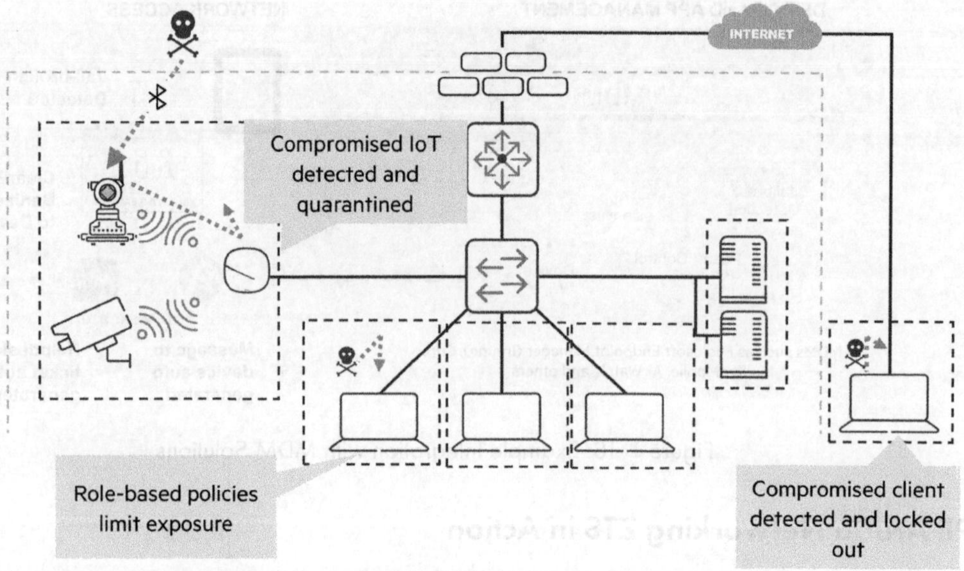

Figure 1-19: HPE Aruba Networking ZTS in Action

And by continuously monitoring internal devices, HPE Aruba Networking ZTS can detect suspicious behavior such as an IoT device that suddenly begins trying to establish SSH sessions with a broad range of IP addresses. The network can then take action to lock out the device behaving suspiciously.

This is the substance of Zero Trust Security. The network never lets a device in and then forgets about it. By restricting each device to the lowest level of appropriate resources, and continually monitoring whether devices continue to merit that access, ZTS minimizes real damage to the network.

CPPM Services

In this section, you will learn about the main components of CPPM services.

Review Authentication, Authorization, Accounting (AAA)

CPPM provides the foundation for HPE Aruba Networking Zero Trust Security, so you will now learn some fundamental concepts that will help you use this powerful tool to secure company's networks.

Among other capabilities, CPPM acts as an authentication server within the Authentication, Authorization, Accounting (AAA) framework. RADIUS is a common AAA protocol, and the one on which you will focus for the first part of this study guide.

In AAA, a user or device—often called a client—requests access to a service, as shown in Figure 1-20. The client can request any service, but most often in this study guide, the client is requesting network access by connecting to the network. In that case, the AP, switch, gateway, or Mobility Controller (MC) to which the client connects is the Network Access Device (NAD). (In some cases, the NAD is referred to as the Network Access Server (NAS). The NAD sends a RADIUS request to its authentication server to ask whether the client is permitted access.

As authentication server, CPPM must first authenticate that the client is who and what it claims to be and, second, authorize the client for the correct level of access. It matches the RADIUS request to a service and uses that service to process the request. First, it implements the authentication method listed in the service. Often that method calls for CPPM to exchange several more RADIUS messages with the NAD. If the NAD is enforcing 802.1X, the NAD actually acts as a go-between for the client and CPPM, which exchange Extensible Authentication Protocol (EAP) messages.

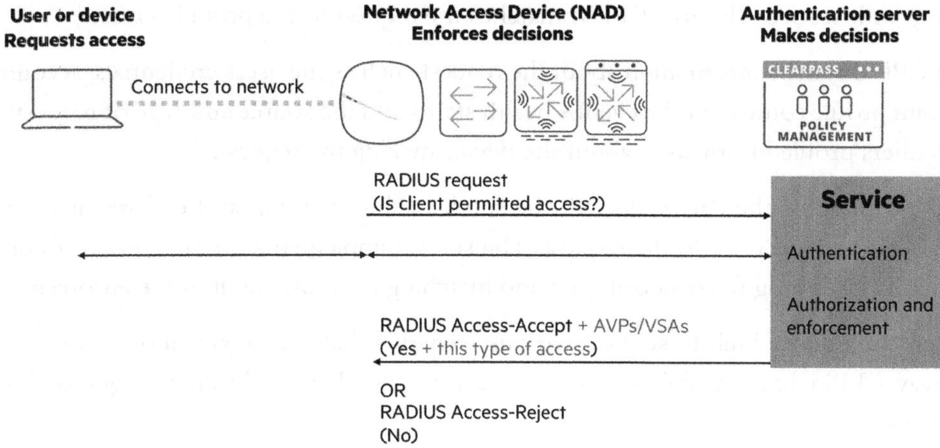

Figure 1-20: Review Authentication, Authorization, Accounting (AAA)

As an end result of these communications, the client has passed or failed authentication. The CPPM service further processes the request and determines whether an authenticated client is allowed access and at what level. If CPPM decides the client is allowed, it sends the NAD a RADIUS Access-Accept. This message often includes attribute value pairs (AVPs) or vendor specific attributes (VSAs), which further define the level of access. If CPPM determines that the client is not authorized, it sends a RADIUS Access-Reject.

Introduction to CPPM Services

When CPPM processes a request it follows a set pattern, as shown in Figure 1-21.

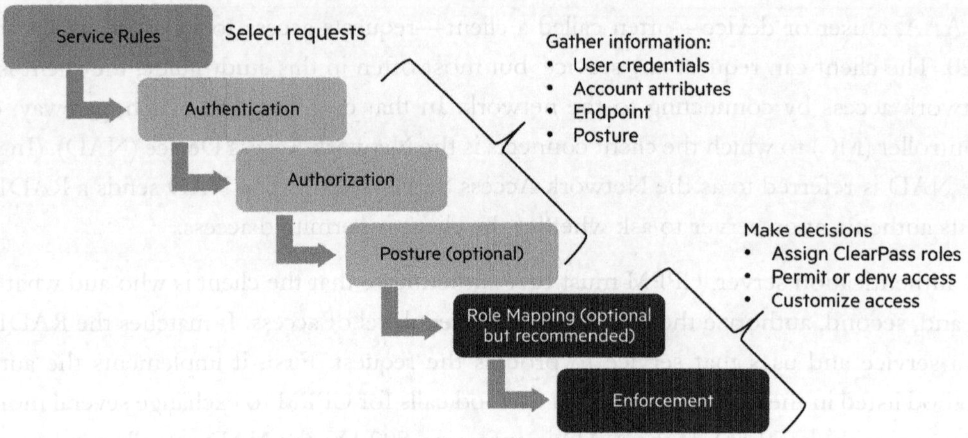

Figure 1-21: Introduction to CPPM Services

Starting with service selection, CPPM matches the request to the proper service.

Then CPPM gathers information about the request such as the users credentials, account attributes, endpoint profile context, or the client's health status and the source and type of request. The service also gathers profile information about the device making the request.

Finally, with all of the data collected, the service roles mapping and enforcement processes can make decisions and reply to the request. The service implements the organization's desired access policies by evaluating the data collected and matching it to rules built in the enforcement policies.

Services are custom built to support various scenarios that the organization wants to support. In, this way CPPM has the ability to process a large number of different requests from different networks.

Service Rules Match Requests to a Service

The service selection rules (or service rules) are the beginning of service processing. As shown in Figure 1-22, the service selection rules filter incoming requests and select the proper service for the request.

This process is important because different types of network access requires different considerations when processing. Even though both APs and switches might send RADIUS requests to CPPM, CPPM will most likely need to process a RADIUS request associated with a WLAN differently from a RADIUS request associated with a wired connection. Similarly, it needs to process a RADIUS request for a corporate WLAN, which typically uses 802.1X, differently from a RADIUS request for a guest WLAN. This means that a different service needs to process each of these requests.

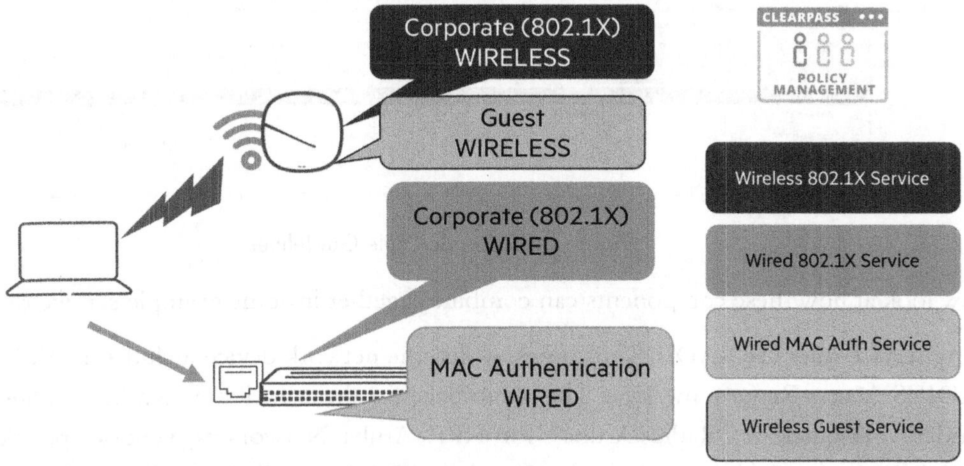

Figure 1-22: Service Rules Match Requests to a Service

Service Rule Guidelines

Even when you create a service from scratch, you can often select a service type, and CPPM will automatically create the correct service rules for that use case. However, it can still be useful for you to understand a bit about how a service rule is put together.

As shown in Figure 1-22, each rule consists of four components. The type defines a "namespace," or dictionary of attributes, while the name specifies the precise attribute. You can think of the attribute as a parameter. The value is a specific value for that parameter. The operator specifies the desired relationship between the value in the rule and the value in the actual request, in order for CPPM to say that a match has occurred. For example, does the value in the actual request need to equal the specified value exactly? Equal the value without regard to case? Start with a value? CPPM supports all these operators and many more.

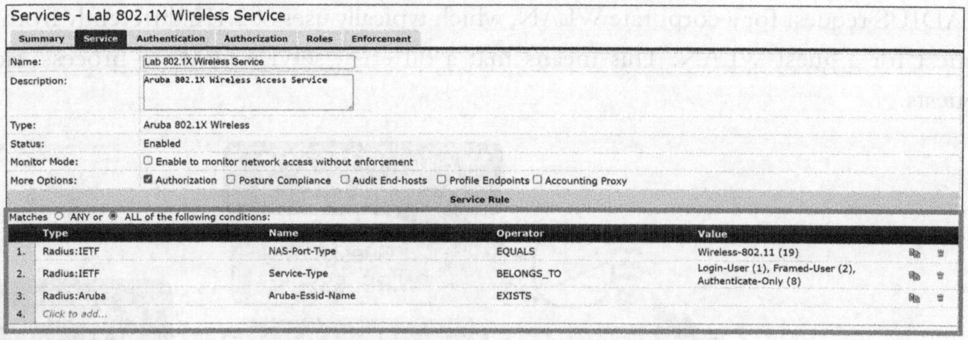

Figure 1-23: Service Rule Guidelines

Now look at how these components can combine together in some example service rules.

Many service rules use RADIUS attributes, which the network device includes as AVPs within the RADIUS Access-Request message. The namespace for these attributes can be "Radius:IETF" for standard attributes or "Radius:Aruba" for HPE Aruba Networking vendor specific attributes (VSAs). For example, in a wireless 802.1X policy, CPPM typically looks for the RADIUS Access-Request to specify the NAS-Port-Type attribute with a value of 19 for Wireless-802.11.

CPPM can also use AVPs within the access request to derive attributes in another namespace. For example, the RADIUS NAS-IP attribute indicates the network device's IP address. CPPM stores information about all network devices, so it can use this IP address to determine the network device's name or location. You can create rules that use these derived attributes as a more intuitive way to set up rules.

Authentication

The next component of the service is authentication, and you specify the authentication settings in the **Authentication** tab, as shown in Figure 1-24. You need to select the valid authentication methods and authentication sources.

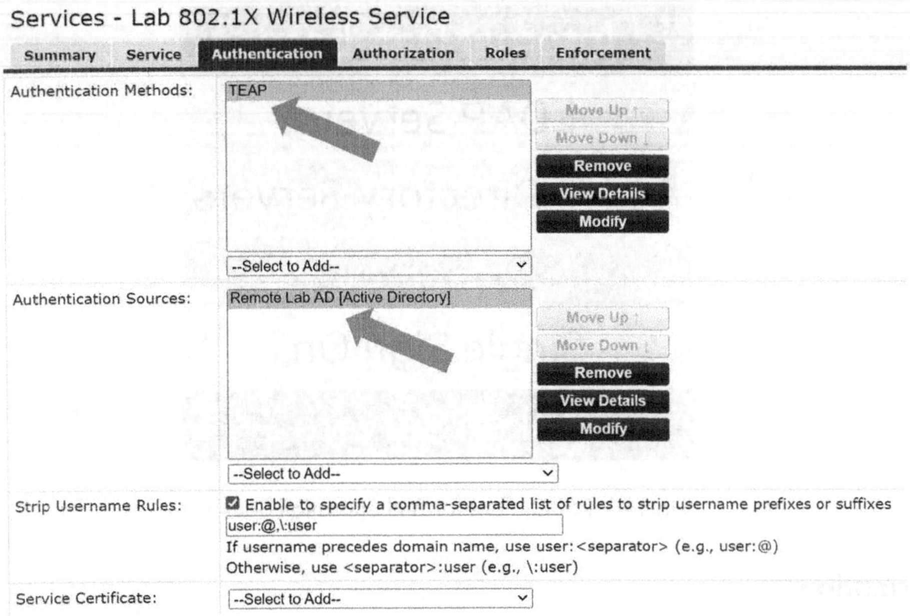

Figure 1-24: Authentication

CPPM proposes each authentication method to the client in turn until the client accepts one of the methods. If the client does not support any of the methods, the authentication fails. HPE Aruba Networking recommends you clean up the Authentication Methods by reducing the list to only what is valid for the clients being authenticated by this service.

Assuming that the client does accept a method, CPPM uses the selected authentication method to collect the user's credentials. It then submits those credentials to the authentication source for verification.

You can select more than one authentication source. CPPM attempts to authenticate the client credentials against the sources in the list from top down.

Options for Authentication Sources

CPPM uses authentication sources to validate the user's identity and credentials. The most basic authentication source is the internal database, which should be avoided except for use as a last resort. You may use it for a very small installation if necessary. LDAP servers, and specifically Active Directory servers, provide easy tools for managing accounts as well as rich context about the user. CPPM also supports SQL servers, single sign-on situations, and token servers, giving CPPM great versatility in authenticating users.

These options are summarized in Figure 1-25.

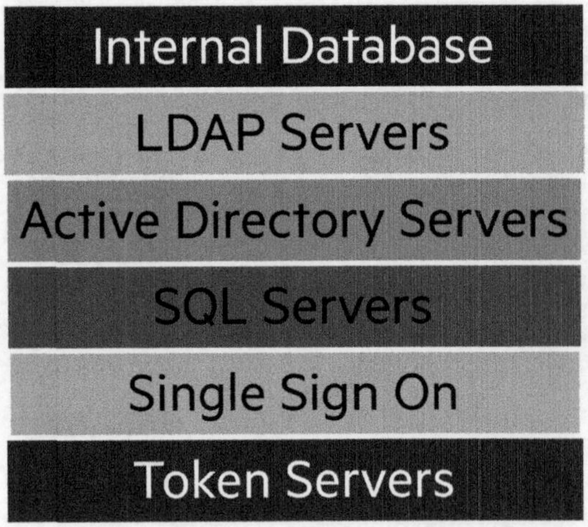

Figure 1-25: Options for Authentication Sources

Authorization

As shown in Figure 1-26, the **Authorization** tab for a service specifies the information sources from which CPPM will draw when it applies role mapping and policy enforcement policies.

The first thing you'll notice in this tab is that the sources from the Authentication tab are added by default. This is because CPPM automatically authorizes against the authentication sources. In fact it collects authorization attributes from the authentication source even if you do not select the **Authorization** check box in the Service tab.

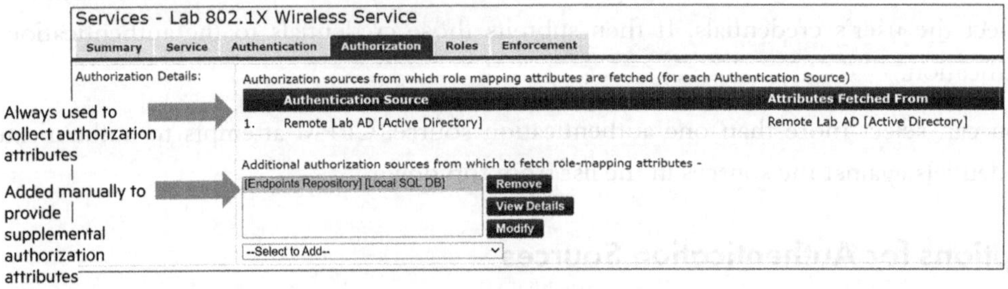

Figure 1-26: Authorization

Sometimes the service requires input from sources other than the Authentication source. This includes things like endpoint context, timestamps, and even queries against a SQL database. To add these supplemental sources, you must first select the **Authorization** check box in the Service tab. The **Authorization** tab then appears. You can then go to that tab and add the desired sources.

What Is a CPPM Role?

Having authenticated the user and collected information from authorization sources, the service is ready to start determining whether the client is permitted access to a network service and with what rights. This process generally starts with role mapping.

Before you look at role mapping in detail, you should take a moment to learn what "role" means in CPPM and how CPPM roles are different from the "roles" that you might be familiar with on HPE Aruba Networking gateways, APs, and switches. As you know, a role on an AOS device is a group for assigning firewall policies and other settings to the client. Each client is assigned to only one role at a time.

In CPPM, on the other hand, a "role" is simply a descriptive tag, which gives the service context about the client, including characteristics of both the user and the device. As shown in Figure 1-27, a CPPM role can describe the user's identity, such as indicating that the user belongs to the loan officer department. But the role can indicate other information as well. It can convey the device type. It can identify the client's location. You can create any CPPM roles that you want to convey the information that is important for your organization's policies. And CPPM can assign multiple roles to the same client in order to build up a complete picture of the user and device.

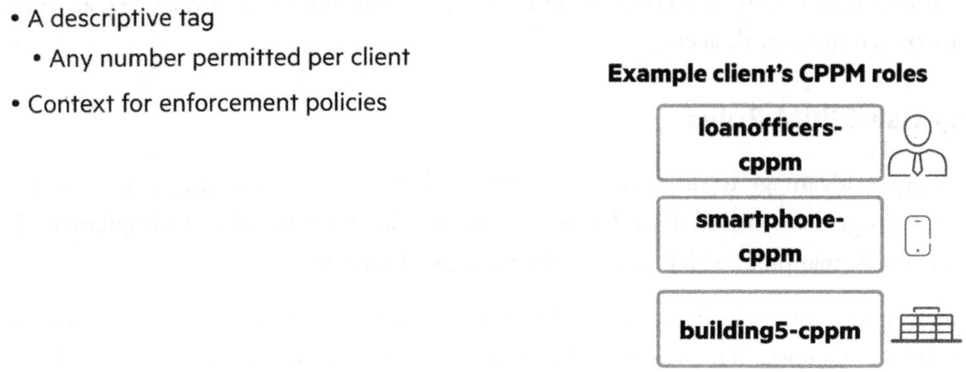

Figure 1-27: What Is a CPPM Role?

Overview of CPPM Role Mapping and Enforcement Policies

CPPM primarily assigns roles to clients based on rules within role mapping policies. Those rules draw on authorization attributes collected during the authorization stage of the service processing. Two examples are shown in Figure 1-28, but many other examples exist. After the role mapping policy finishes processing, the client has one or more roles assigned to it.

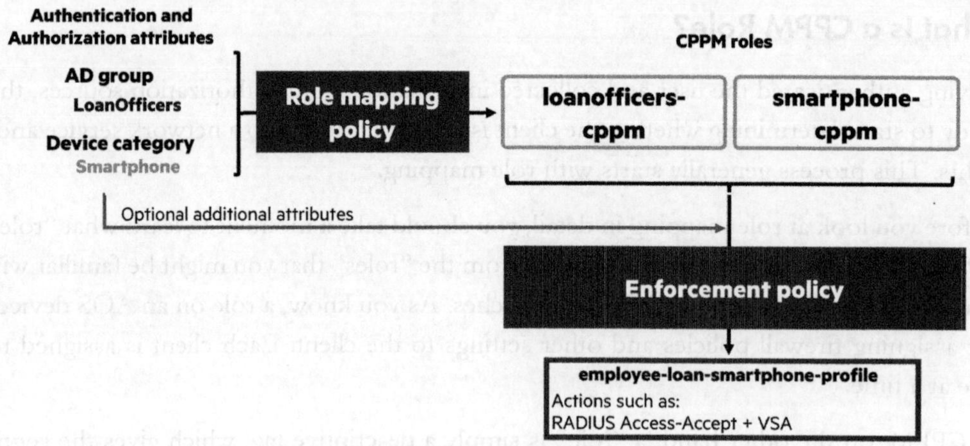

Figure 1-28: Overview of CPPM Role Mapping and Enforcement Policies

You can then use the roles as conditions in rules within an enforcement policy. You can also optionally include other authorization attributes in the enforcement policy rules. Each rule applies one or more enforcement profiles to clients that match the rule.

The enforcement profile defines actions to apply to the client. For example, the profile might tell CPPM to send a RADIUS Access-Accept message to the authenticator, as well as to include VSAs to customize the client's access.

Why Use CPPM Roles

The biggest advantage to using roles in CPPM is in bringing consistency across multiple services. Once the organization is able to define how to describe a client, you can implement the rules in a universal role mapping, which you can then use in all services.

In the example shown in Figure 1-29, the same client might attempt to connect to either the wired or wireless network. The client authentication needs to be processed by a different service depending on the connection type because a different enforcement profile is required for enforcing access on the wireless versus the wired network infrastructure device. However, by using the same role mappings, the client will always be categorized in the same way. The difference comes in how CPPM communicates this categorization to the network infrastructure device. This approach makes enforcement more consistent.

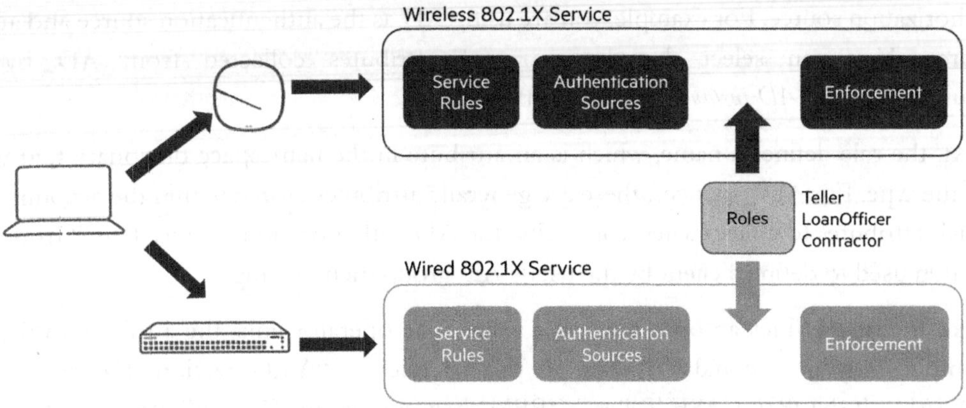

Figure 1-29: Why Use CPPM Roles

In other words, the role defines the client while the enforcement policy defines how to treat the client.

CPPM can assign multiple roles to a client, indicating that the client meets multiple conditions.

Overview of Role Mapping Rules

CPPM primarily determines which roles to assign to a client based on rules within the service's role mapping policy.

Before you look at role mapping and enforcement policies in more detail, it will be useful to examine the general format for these policies' rules, as shown in Figure 1-30.

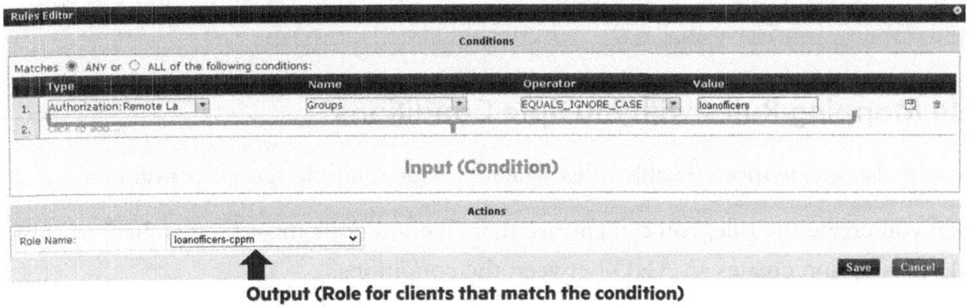

Figure 1-30: Overview of Role Mapping Rules

Every rule has a type. The type defines the namespace dictionary from which information about the user or client is retrieved. CPPM supports many namespaces. In this study guide, though, you will focus on just a few commonly used namespaces. For both role mapping and policy enforcement rules, you can use the authorization namespace, which is further defined by a particular authorization source. For example, you are using AD as the authentication source and authorization source. You can select the dictionary of attributes collected from AD by specifying "Authorization:<*AD instance name*>" for the type.

Next, the rule defines a name, which is an attribute in the namespace dictionary that you selected for the type. For AD instances, these are generally attributes stored within the account. You define which attributes to collect when you define the AD authentication source. The "Groups" attribute is often used to define a client by the user's AD group membership.

Next, the rule defines an operator and a value. The operator links the attribute to the value. For example, you could create this rule: "Authorization:*AD_instance* Groups EQUALS_IGNORE_CASE Teller." CPPM then checks the "Group" attributes that it collected for each user, and it applies this rule if the attribute equals "Teller" or "teller." In other words, it applies the rule if the user is in the Teller group. Similarly, you could create the rule shown in Figure 1-30 to apply a role to clients with a Group that equals "Loanofficers," "loanofficers," "LoanOfficers," and so on.

CPPM supports many other operators to enable you to create flexible rules. For example, you could use "STARTS_WITH" to match multiple attribute values, all of which start with a specific string.

Those four components create the condition.

Finally, the rule specifies an output—a decision for CPPM to support. For a role mapping rule, the output is a CPPM role. For an enforcement policy rule, the output is an enforcement profile.

For a complete list of namespaces supported in CPPM v6.9, refer here: https://www.arubanetworks.com/techdocs/ClearPass/6.9/PolicyManager/Content/CPPM_UserGuide/Rules/Namespaces_RADIUS.htm?Highlight=namespace

Role Mapping Rules with Multiple Conditions

You can also create more flexible rules which include multiple sets of conditions.

When you create the rule, you can choose that all conditions must be matched, as shown in Figure 1-31. This option creates an AND between the conditions.

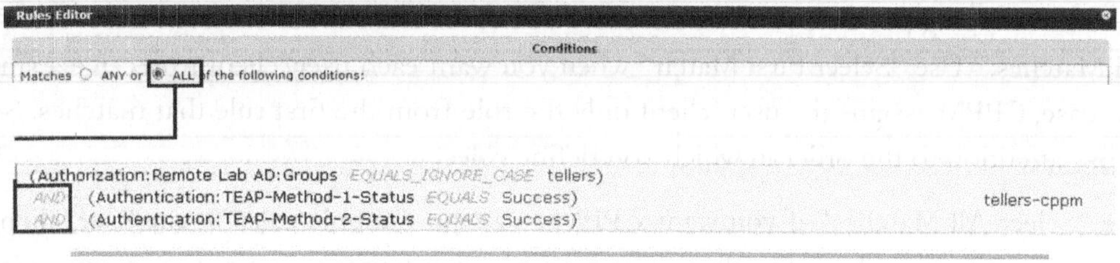

Figure 1-31: Role Mapping Rules with Multiple Conditions (All)

Or, as shown in Figure 1-32, you can choose that the client can match any of the conditions. In this case, conditions are separated by OR. In other words, the client receives the role if it matches any one (or more) or the conditions.

The rule in Figure 1-32 shows a condition in which a client receives a role if it belongs to either one of two AD groups. In the example shown here, you could achieve the same effect with one condition that uses the operator CONTAINS. However, this rule serves to provide an example of how the OR operator works.

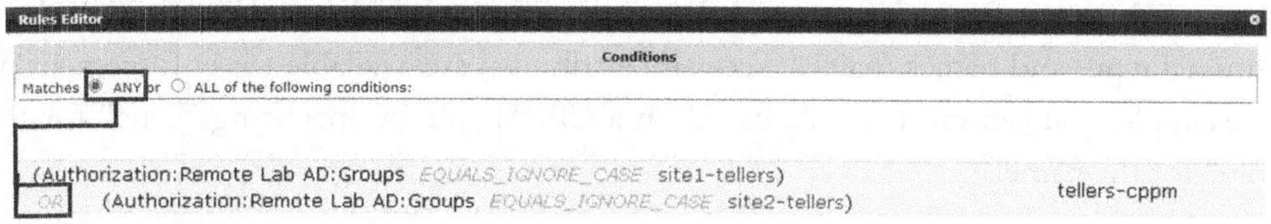

Figure 1-32: Role Mapping Rules with Multiple Conditions (All)

Rule Evaluation Algorithm and Processing Order

CPPM processes policies in order from the top rule to the bottom, as shown in Figure 1-33.

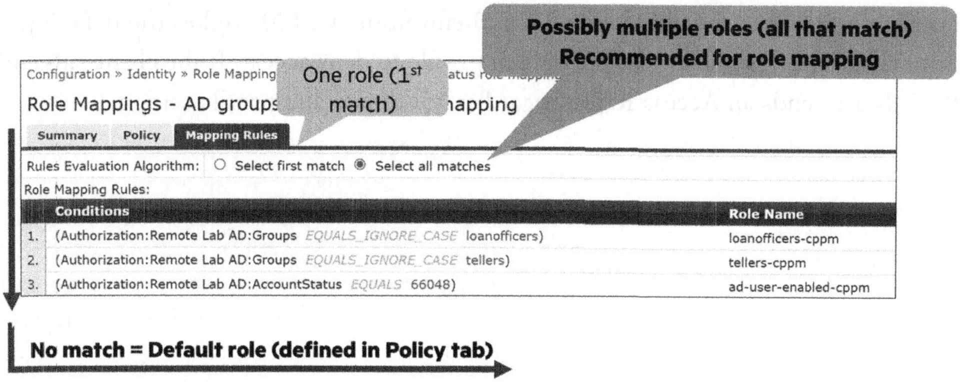

Figure 1-33: Rule Evaluation Algorithm and Processing Order

For role mapping policies, you can choose whether CPPM should “Select First Match” or “Select All Matches.” Use “Select First Match” when you want each user/client to receive a single role. In this case, CPPM assigns the user/client only the role from the first rule that matches. So you need to pay attention to the order in which you define roles.

Use “Select All Matches” if you want CPPM to assign multiple roles to the user/client based on various information about the client. For example, you might assign a user to two roles: one role based on AD group membership and one general role for all users with enabled accounts in AD. HPE Aruba Networking generally recommends using the “Select All Matches” option for role mapping policies. Then you can assign as many roles as might be required to each client to fully characterize it. The roles will not have any effect until you use them in enforcement policies.

If CPPM does not find a matching rule at all, it applies the default role, which you define in the Policy tab.

Enforcement Policies

The general format for enforcement policy rules is very similar to role mapping rules. However, additional types and names (namespaces and attributes) are available for enforcement policy rules. For example, you can create a rule based on a CPPM role by specifying “Tips” for the type and “Role” for the name.

The output, or action, for an enforcement policy rule is an enforcement profile or multiple enforcement profiles.

For enforcement policies, HPE Aruba Networking recommends setting the Rules Evaluation Algorithm to First Match. CPPM then evaluates the rules from top to bottom and applies only the first rule that matches the client, as shown in Figure 1-34.

If no rules match, but the client did pass authentication, CPPM applies the default profile. Often, but not always, you should set up the default profile to deny access. If the client fails authentication, CPPM always sends an Access-Reject regardless of the default profile setting.

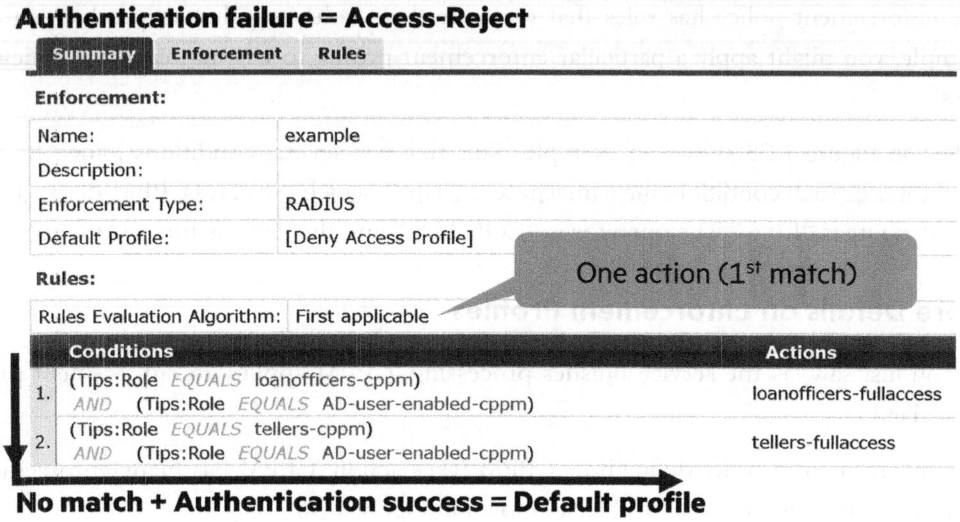

Figure 1-34: Enforcement Policies

Combining Role Mapping and Policy Enforcement

In Figure 1-35, you see an example of how you can combine role mapping and policy enforcement policies.

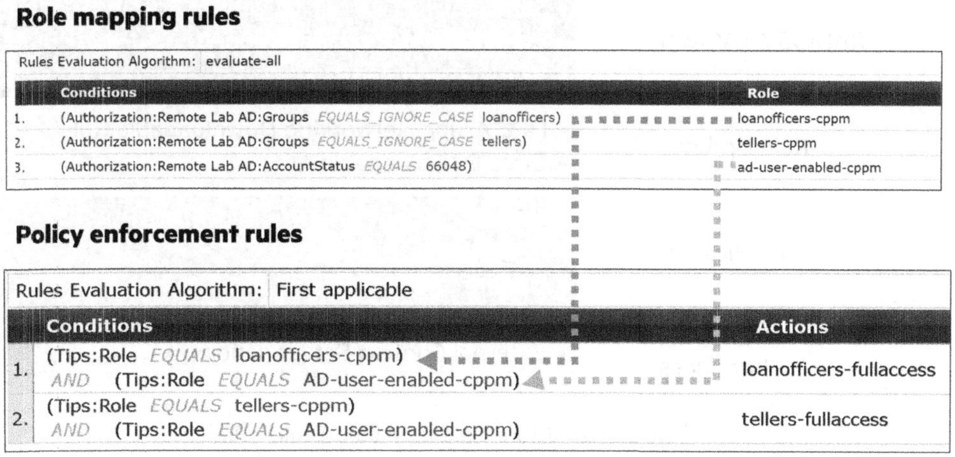

Figure 1-35: Combining Role Mapping and Policy Enforcement

The role mapping policy takes an "evaluate-all" approach and assigns clients to every role that applies. In this example, users receive either the "loanofficers-cppm" or "tellers-cppm" role based on AD group. They might also receive the "ad-user-enabled-cppm" role based on their account status.

The enforcement policy has rules that combine these roles together to produce an output. For example, you might apply a particular enforcement profile to clients that have achieved multiple roles.

Rule 1 in Figure 1-35 shows an example. This rule has several conditions joined by the "AND" operator. For each condition, the namespace is "Tips," which indicates CPPM-derived information. The attribute is "Role." The operator is "EQUALS," and the value is the role in question.

More Details on Enforcement Profiles

As you just saw, as the service finishes processing, CPPM applies an enforcement profile to the user/client.

The enforcement profiles define how CPPM takes action. CPPM has many enforcement actions, making it very powerful and flexible, as shown in Figure 1-36.

The example in Figure 1-35 showed a common enforcement profile type: RADIUS. With this type, CPPM sends a RADIUS message that tells the authenticator to accept or reject access. An Access-Accept message can further customize the access with settings such as VLAN or role assignments. TACACS profiles similarly tell CPPM to send a TACACS message to permit or deny user access and specify the authorization level.

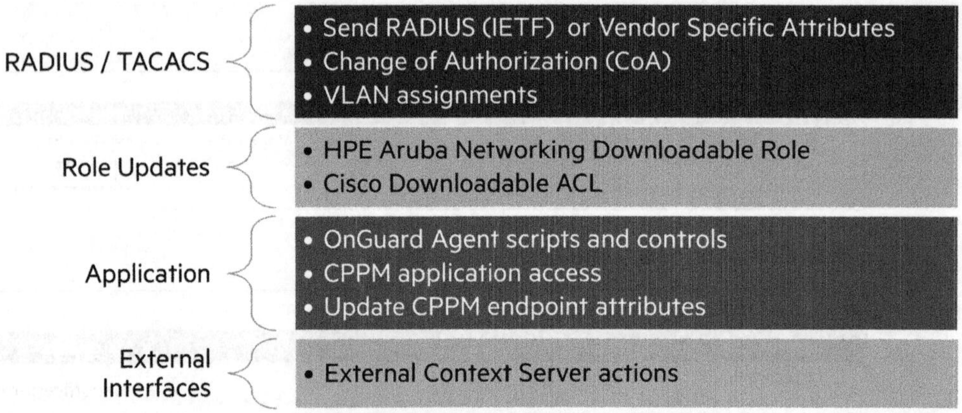

Figure 1-36: More Details on Enforcement Profiles

Enforcement may also include situations in which CPPM formulates a User Role or ACL and then sends it to the network access device.

You can also use SNMP or CLI commands to control various aspects of the network devices. For example, if a wired IoT client fails MAC authentication, CPPM could execute the commands for a port shutdown to remove that client from the network.

Similarly, CPPM can control clients by imposing session restrictions, such as a session-timeout or re-authentication interval. With some Network Access Devices (NADs), CPPM can execute a captive portal or external web redirect to the client. This is helpful with network access devices that do not natively support external web authentication or redirection when the client has already authenticated.

Often you will need to build services to control which clients have access to CPPM applications such as Onboard or OnGuard. CPPM can also interact with the OnGuard agent to execute controls on the clients. CPPM can update endpoint information, giving context to previous endpoint activity. For example, what if a company does not want client devices on the corporate network to access the guest network? The company can use a CPPM Entity Update Enforcement action to tag a device as a corporate client. In this way, when a corporate client attempts to access the guest network, the guest access service reads the attribute and denies access to the guest network.

CPPM can also interface with external services and servers to exchange context information about a client accessing the network.

CPPM Network Devices

This section explains how to define network devices on CPPM.

Basic Network Device Entry on CPPM

For CPPM to accept RADIUS requests from a NAD and process those requests with a service, you must create a network device entry for that NAD on CPPM.

In Figure 1-37, you see a basic network device entry. The entry specifies a single IP address or a complete IP subnet. When CPPM receives a RADIUS request, it checks the source IP address for the request and matches it to a network device entry with that address. In this example, the request arrives from 10.1.5.5, so it matches a network device entry with Subnet Address 10.1.5.0/24. If no entry exists for the IP address, CPPM drops the request (and logs the event in the Event Viewer). Therefore, you must make sure that you understand the IP addresses that your network infrastructure devices will use to reach CPPM and create entries for them.

The network device entry also includes a RADIUS shared secret. You must match the secret configured here in the RADIUS server configuration on the NAD.

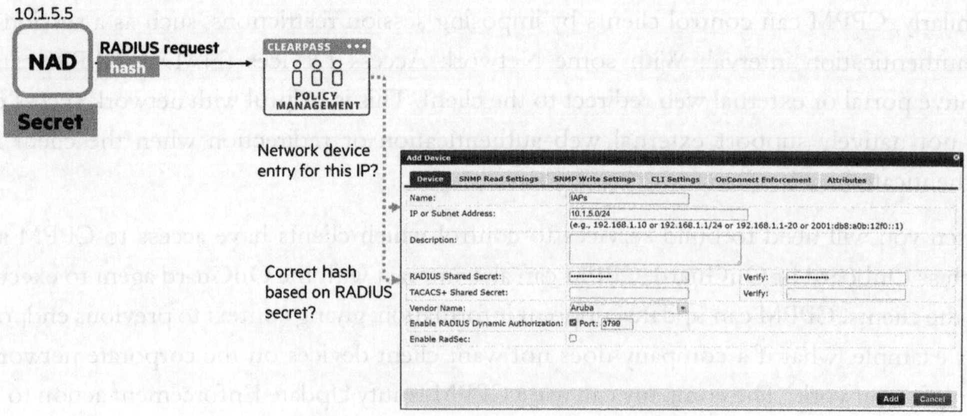

Figure 1-37: Basic Network Device Entry on CPPM

CPPM Network Devices in AOS-8 Architectures

To create the correct network device entries, you must understand which network infrastructure devices take the NAD role in your architecture.

In an AOS-8 architecture, MCs act as NADs and authenticate wireless clients to CPPM. Therefore, as shown in Figure 1-38, you must create network device entries for the MCs on CPPM.

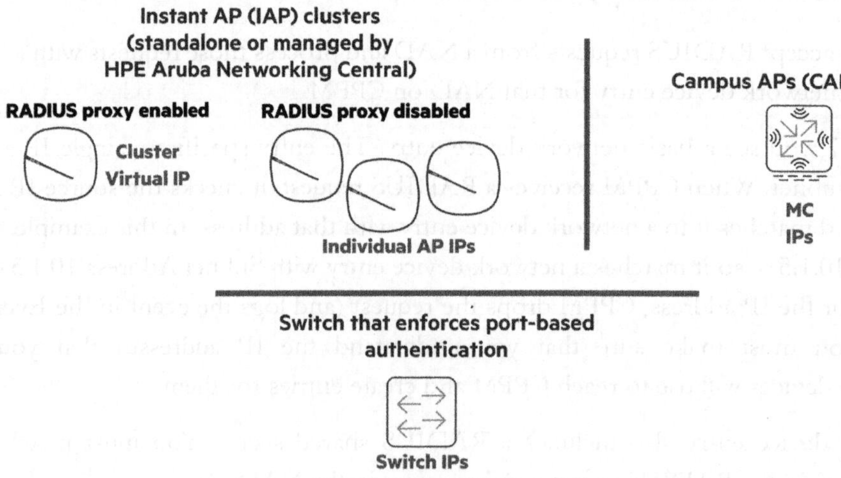

Figure 1-38: CPPM Network Devices in AOS-8 Architectures

In an Instant AP (IAP) cluster, which might be standalone or managed by HPE Aruba Networking Central, the APs within the cluster might act as NADs. If RADIUS proxy is enabled on the cluster, though, the APs send their RADIUS requests through the virtual conductor. CPPM only needs a network device entry for the cluster's virtual IP address. However, if RADIUS proxy is disabled, it needs an entry that applies to all of the APs.

AOS-CX switches can also act as NADs in any of these architectures. When you configure port-based authentication on the switch, it is the NAD, and CPPM requires a network device entry for it.

CPPM Network Desires in AOS-10 Architectures

APs act as NADs for wireless clients in AOS-10. When the WLAN operates in bridged mode, the APs use their IP addresses for the NAS IP address, as shown in Figure 1-39. When the WLAN operates in mixed mode or tunneled mode, the gateway acts as the RADIUS proxy. You then need to add a network device entry for the gateway in CPPM.

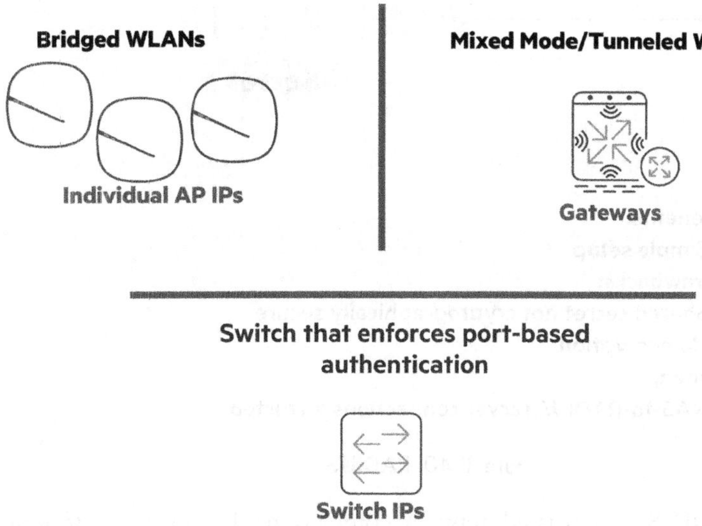

Figure 1-39: CPPM Network Desires in AOS-10 Architectures

In AOS-10 architectures, as well, AOS-CX switches might implement port-based authentication, in which case CPPM must have network device entries for them.

Review RADIUS vs RadSec

As you learned, network devices and CPPM communicate with RADIUS during the process of authenticating clients for network access. You can implement RADIUS alone or add a layer of security with RadSec.

RADIUS

RADIUS uses a shared secret to authenticate the RADIUS client (NAD) and RADIUS server to each other, as shown in Figure 1-40. You must configure a matching shared secret on both devices. The NAD and server include a hash of the message and shared secret within all RADIUS messages. If the hash does not check out, indicating that the secret is incorrect or that the message was tampered with, the receiving device drops the message. However, the shared secret does not provide rigorous security. You can make it more secure by setting a very long, random key. However, RADIUS itself is not considered a secure protocol. It also transmits data in plaintext.

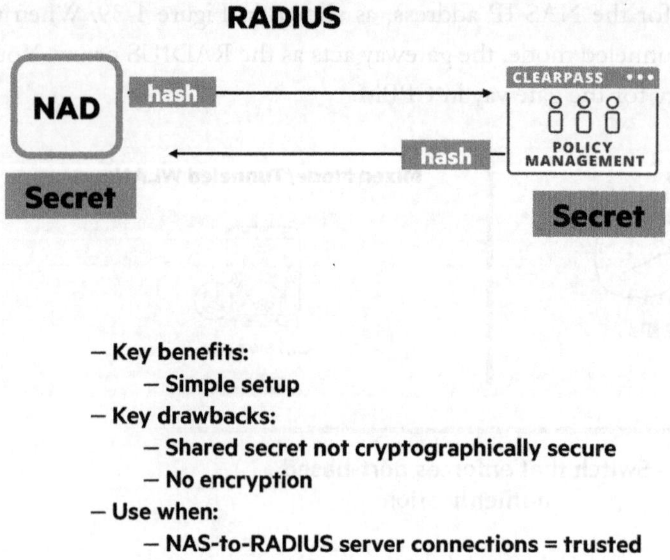

Figure 1-40: RADIUS

You should only run RADIUS over trusted network connections. For example, if network devices and CPPM are deployed at the same site, you can generally use RADIUS between them. You can also use RADIUS between network devices and CPPM servers at another site if the site-to-site connection is secured with a protocol like IPsec.

RadSec

As shown in Figure 1-41, RadSec establishes a TLS tunnel between the RADIUS client (NAD) and RADIUS server. During the TLS tunnel establishment, the NAD and RADIUS server authenticate each other with certificates. After the TLS tunnel comes up, the NAD and RADIUS server exchange all RADIUS messages over the tunnel, including authentication and accounting. The tunnel provides encryption and data integrity for the messages. Because the tunnel uses TCP, rather than RADIUS's typical UDP, the NAD can also detect when it loses contact with the RADIUS server more quickly.

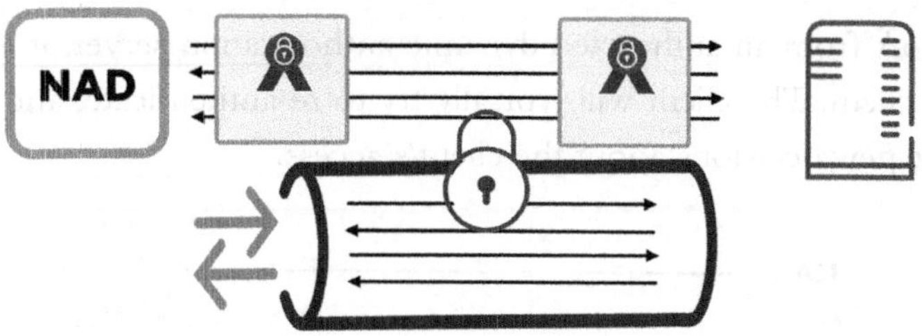

Figure 1-41: RadSec

When you enable RadSec on the NAD or on CPPM, the RADIUS secret is automatically set to "radsec." Instead, you set up certificates to provide the authentication.

RadSec provides greater security than RADIUS, but the certificate requirements can make it harder to deploy and maintain. Issues can occur such as a NAD's certificate expiring. HPE Aruba Networking recommends RadSec when network devices and CPPMs must communicate over untrusted network connections. Some companies also require RadSec as part of their security policies.

Review RADIUS Dynamic Authorization

A RADIUS server like CPPM is capable of making decisions about a user's or device's authorization beyond the initial authentication. For example, ClearPass OnGuard can detect a client's security posture and tell CPPM to disconnect a client that does not have the proper firewall and antivirus settings. Or CPPM might find out after the initial authentication that the client is an IoT device, which has particular security policies associated with it. Many other examples of such events exist.

So that CPPM can enforce the new policies, it must force the client to disconnect and re-authenticate. As shown in Figure 1-4, CPPM sends a Change of Authorization (CoA) message to the NAD such as an AP, switch, gateway, or MC. By default, CPPM sends the message on standard port UDP 3799. But you can change that port, as long as you match the setting on CPPM and the NAD. If CPPM and the NAD are using RadSec, CPPM sends the CoA messages in the RadSec tunnel.

When the NAD receives a CoA from an authorized dynamic authorization server, it enforces the message and disconnects the client. The client will typically try to re-authenticate, and CPPM can then apply its policies to make new decisions about the client's access.

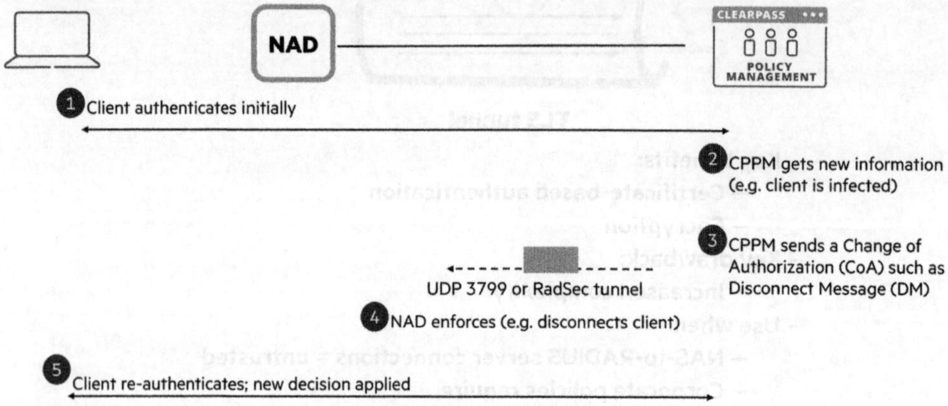

Figure 1-42: Review RADIUS Dynamic Authorization

Enabling Dynamic Authorization in a CPPM Network Device Entry

To enable the CoAs, you must make sure that CPPM's network device entry for the device in question has **Enable RADIUS Dynamic Authorization** enabled, as shown in Figure 1-43.

Also note the Vendor Name. CPPM uses CoA enforcement profiles to trigger the CoAs. It has different enforcement profiles for devices from different vendors. The vendor name within the network device entry helps CPPM determine whether a profile is valid for a particular network device, so make sure to set the name correctly. The vendor name also enables the correct VSAs for the network device. Use **Aruba** for HPE Aruba Networking APs, MCs, gateways, and AOS-CX switches.

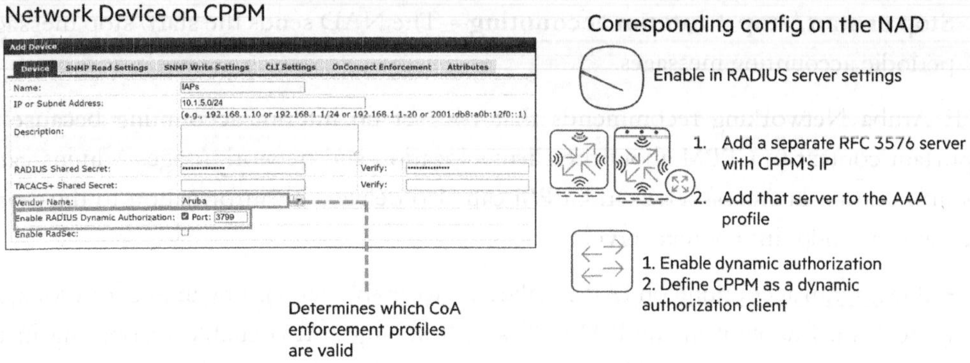

Figure 1-43: Enabling Dynamic Authorization in a CPPM Network Device Entry

Similarly, you must make sure that the network device is configured to accept CoA messages. On IAPs or AOS-10 APs managed within HPE Aruba Networking Central, you enable **Dynamic Authorization** in the RADIUS server configuration.

The setup on gateways (or MCs) has a few more steps. Instead of enabling dynamic authorization on the RADIUS server, you set up a separate Dynamic Authorization server with CPPM's IP address. You set a key for the Dynamic Authorization server that matches the RADIUS key in CPPM's Network Device entry for the gateway or MC. (If you are using RadSec, the key is "radsec.") You then activate the Dynamic Authorization server in the AAA profiles that the device is using to authenticate clients to CPPM.

On AOS-CX switches, you also define the CoA settings separately from the RADIUS server. You must enable dynamic authorization with this command: **radius dyn-authorization enable**. Specify CPPM's IP address as a dynamic authorization client on the switch with this command: **radius dyn-authorization client <*CPPM IP or FQDN*>.**

RADIUS Interim Accounting

As an AAA protocol, RADIUS not only provides authentication and authorization, but also accounting. RADIUS accounting logs a broad range of information such as a client's username, NAD port, and bandwidth consumption in packets and bytes. A NAS can implement RADIUS accounting in these modes:

- **Stop**—When the client's session terminates, the NAD sends a message that summarizes the session.
- **Start/stop**—The NAD sends an accounting message at the beginning of the session and then a session summary at the end.

- **Stop or start/stop + interim accounting**—The NAD sends the start/stop messages and also periodic accounting messages.

HPE Aruba Networking recommends that you set up interim accounting because it provides important context to CPPM about the client's location and network usage. Admins can then view this information in Access Tracker. CPPM can also deliver this information to third-party devices integrated as endpoint context servers.

To enable interim accounting on IAPs, make sure to enable accounting in two locations, as indicated in Figure 1-44. Enable it in the RADIUS server settings. Also enable accounting in the WLAN security settings. Choose to send accounting information to the authentication server defined for the WLAN. Then set the interval for the interim updates.

On AOS-CX switches, you enable interim accounting for authenticated clients with this command: **aaa accounting port-access start-stop interim <*minutes*> group <*group_name*>**

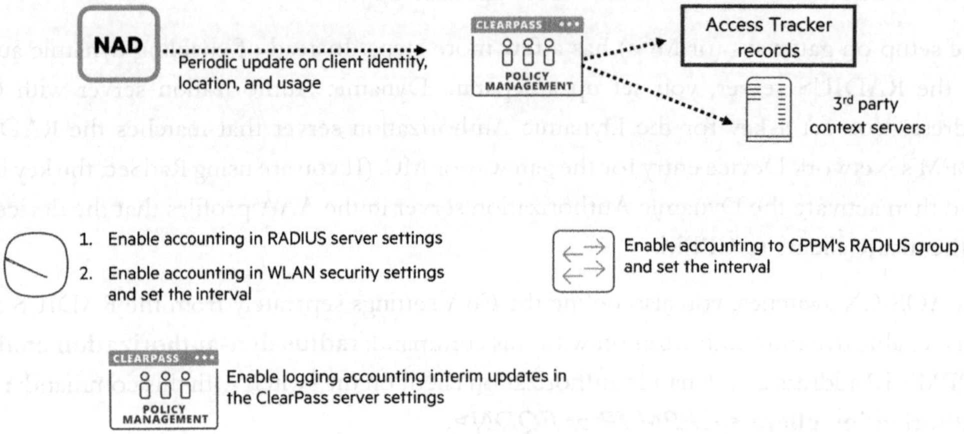

Figure 1-44: RADIUS Interim Accounting

As indicated in Figure 1-44, you must also enable processing of interim updates in the ClearPass server configuration (go to **Administration > Server Manager > Server Configuration** and Click the server). In the **Service Parameters** tab, for the **RADIUS server** settings, set **Log-Accounting-Interim-Update-Packets** to **True**.

Standard Ports for Network Device and CPPM Communications

If a firewall exists between NADs and CPPM, you must ensure that the firewall permits the required ports. When NADs and CPPM communicate with RADIUS, these ports must be open (see also Figure 1-45):

- UDP 1812 for RADIUS authentication
- UDP 1813 for RADIUS accounting
- UDP 3799 for CoAs

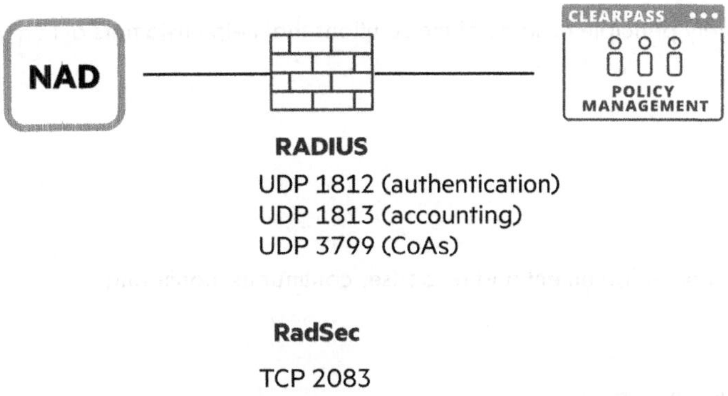

Figure 1-45: Standard Ports for Network Device and CPPM Communications

For RadSec, only one port is required: TCP 2083. The TCP session carries all RADIUS authentication, accounting, and CoA messages.

Note that these are the standard ports. If you change your configuration to use non-standard ports, the firewall must permit those instead.

Summary

Congratulations! You should now be able to explain why companies require a Zero Trust Security approach and how HPE Aruba Networking provides superior ZTS. You also know the basics of setting up CPPM to act as the foundation for a ZTS solution.

Learning checks

1. Which correctly describes Zero Trust Security?
   a. External clients are untrusted, while internal clients are trusted.
   b. The perimeter is enhanced by a demilitarized zone (DMZ).
   c. All users and devices are identified and controlled with granular, identity and context-based policies.
   d. All users are considered equally untrustworthy and are given the same privileges regardless of identity.
2. Match each Zero Trust Security principle to some of the solutions that help customers act on that principle.

   ZTS principles
   a. Visibility
   b. Authentication
   c. Role-based access control, enforcement and response, continuous monitoring

   Solutions

   __ Unified Infrastructure, ClearPass Exchange

   __ ClearPass Onboard

   __ CPDI
3. What role do service rules play within a CPPM service?
   a. Determining whether a user is permitted access
   b. Selecting the preferred authentication method
   c. Matching the incoming request to the correct service
   d. Determining whether the NAD is authorized

4. What is a reason to use role mapping policies?
   a. Because role mapping policies simplify configuration by eliminating the need for enforcement profiles and policies
   b. Because you want to characterize clients with a global policy that you can use in multiple services
   c. Because you want to use more complex conditions which are not supported in enforcement policies
   d. Because you want to define role-based firewall policies directly within CPPM

Note

For answers to all learning checks, refer to the Appendix.

---

2 Deploy Trusted Certificates

EXAM OBJECTIVES

- ✓ Describe PKI dependencies
- ✓ Set up appropriate certificates and trusted root CAs on HPE Aruba Networking solutions

Assumed Knowledge

- Basic understanding of asymmetric encryption
- Basics of managing HPE Aruba Networking products

Overview

To protect their valuable resources and implement a Zero Trust Security network, companies must secure wireless access. Best practices dictate securing WLANs with certificate-based authentication. However, many companies avoid certificates because they do not understand them or believe that they are too hard to manage.

In this chapter, you will review and expand on your knowledge of digital certificates and public key infrastructure (PKI). You will also learn how to deploy certificates on HPE Aruba Networking ClearPass servers and HPE Aruba Networking infrastructure devices.

Digital Certificates and PKI

In this section, you will review digital certificates and PKI.

Digital Certificate

You will now review certificates and how they exist within a PKI. You will also review ways that devices can obtain and use certificates.

As shown in Figure 2-1, a digital certificate enables the certificate owner to tie its identity to a public key. This public key is also tied to a private key, held by the owner. The entity can then use the private key for various purposes such as "signing" data and authenticating itself.

In addition to the owner's identity, public key, and authorized usages for that key, the certificate includes the name of the entity that issued the certificate and that issuer's signature. The issuer name and signature essentially provide other entities a reason to trust that the identity within the certificate is real. Think of a certificate like a governmental ID. Anyone can print out a page with a picture of their face and a name. You only trust that the name is legitimate if a trusted government issued the ID.

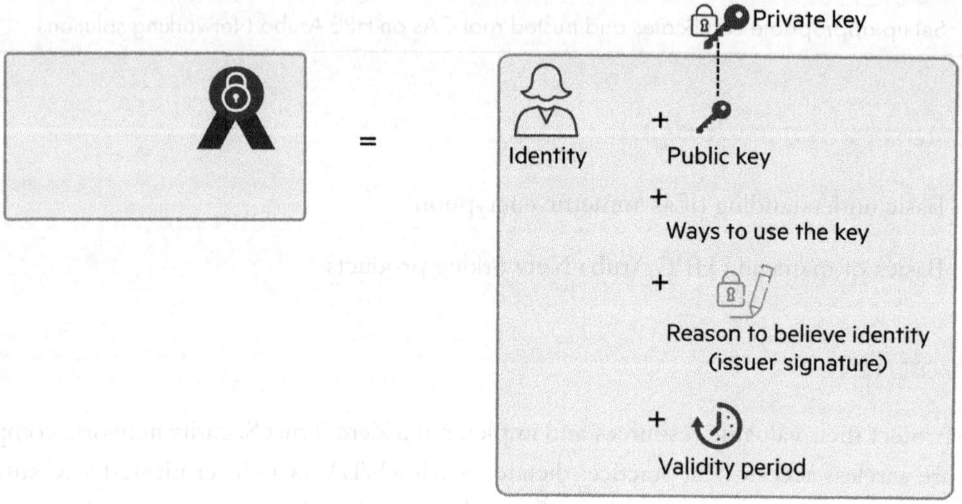

Figure 2-1: Digital Certificate

Finally, a certificate also has a validity period associated with it. Like a passport or other ID, the ID is not good forever; it must be renewed.

(As a side note, you might wonder how a device authenticates itself with a digital certificate. Typically, the device that needs to authenticate hashes a conversation with the peer that wants to authenticate it. The device then encrypts that hash with its private key. The peer decrypts the hash with the public key in the certificate and validates that the result matches its own hash of the conversation. This proves that the original device truly owns the certificate—and has the identity within the certificate.)

Public/Private (Asymmetric) Encryption Algorithms for Digital Certificates

The public/private keypair within the certificate use asymmetric encryption. Most certificate keypairs use one of two algorithms: Rivest Shamir Adleman (RSA) and Elliptic Curve Digital Signature Algorithm (EC-DSA).

RSA was the original asymmetric encryption algorithm. It involves the use of large prime numbers and an operation called modulus. As shown in Figure 2-2, RSA remains secure, but requires long keys to provide adequate security. NIST recommends RSA keys of at least 3072 bit to resist brute force attacks.

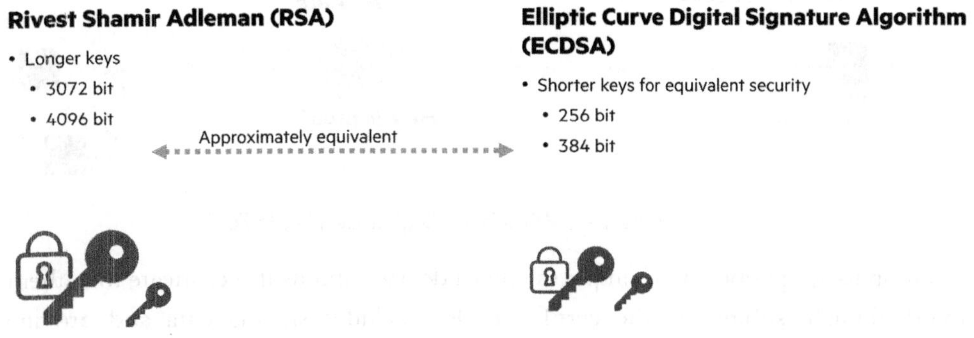

Figure 2-2: Public/Private (Asymmetric) Encryption Algorithms for Digital Certificates

A more modern asymmetric encryption algorithm, ECDSA involves calculating positions across an elliptic curve (EC). The EC approach has the advantage of providing relatively higher security with much shorter keys. Shorter keys lead to faster encryption and decryption and better performance. (However, devices rarely use asymmetric encryption to encrypt large amounts of data in transit.) A 256-bit ECDSA key provides roughly the same security as a 3072-bit RSA key. A 384-bit ECDSA key delivers the same security as a 7680-bit RSA key, which is much longer than RSA keys used in the real world.

What is the Digital Certificate for?

Certificates also often have extended key usages (EKUs), sometimes called enhanced key usages (EKUs). An EKU specifies a particular purpose for the certificate, building on the underlying key usages.

Of most interest to you in this study guide are the client authentication and server authentication EKUs. These EKUs permit the certificate owner to use the certificate to prove its identity. Most authentication frameworks define a client and server. The client needs a certificate with the client authentication EKU while the server needs a certificate with the server authentication EKU.

These EKUs rely on the digitalSignature key usage, illustrated in Figure 2-3. The certificate owner submits the certificate to prove its identity. But, because anyone can intercept a certificate and try to claim it as their own, the owner further offers proof that it truly owns the certificate. The certificate owner signs some data, such as a hash of the preceding conversation, with the private key. The authenticating device can then use the certificate's public key to validate the signature.

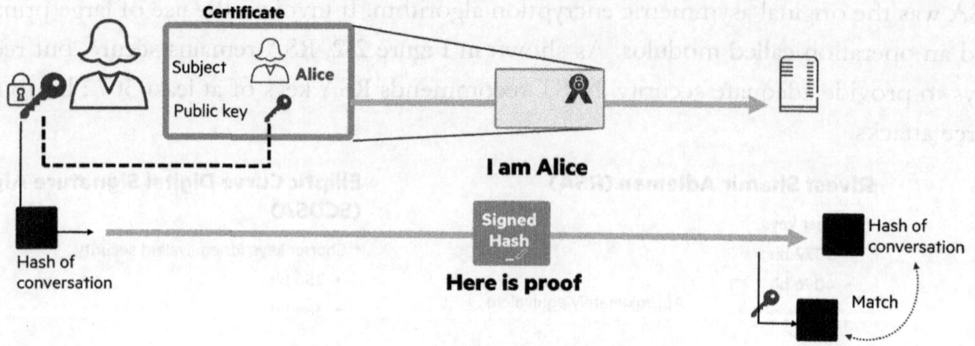

Figure 2-3: What is the Digital Certificate For?

In common usage, people often simply say that a device submits its certificate to authenticate. Keep in mind, though, submitting the certificate also includes signing data and proving certificate ownership.

Validating That a Certificate Is Trusted

You now have a sense of how users and devices can use their certificates. In this section, you will explore how entities validate that certificates are trusted. If, for example, a client submits a certificate to authenticate, how does the server verify that the certificate is valid?

First, the certificate has to have a valid usage. In other words, if the certificate owner is using the certificate to authenticate as a client, the certificate must have the clientAuth EKU specified within it.

The certificate must also be within its validity period. In addition, the certificate must be issued by a trusted source, must not be revoked by that source, and must have an identifier that checks out with the party with which the receiver believes itself to be communicating.

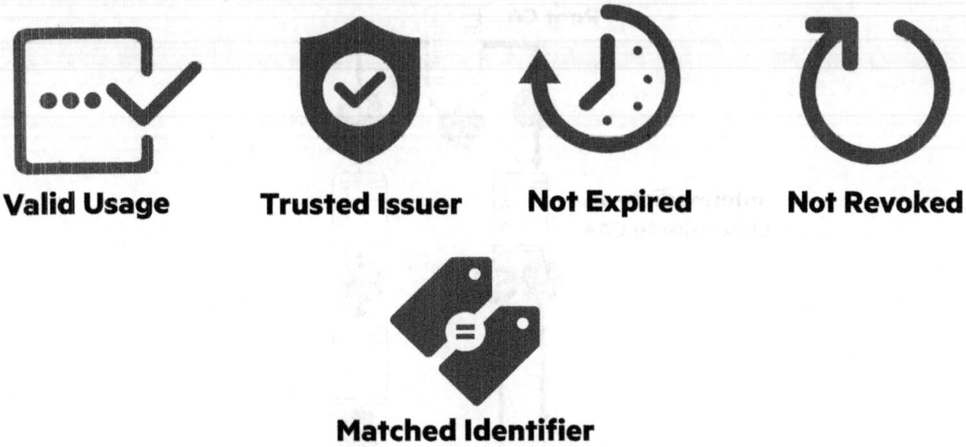

Figure 2-4: Validating That a Certificate Is Trusted

Several of these criteria, summarized in Figure 2-4, call for a deeper look.

Trusted Issuer—PKI

Every certificate has an "issuer," which is the entity that vouches for the identity specified in the certificate. It also has a signature, which proves that the issuer truly issued the certificate.

The issuer forms a crucial component of the certificate's trustworthiness. Anyone can create a certificate with any identity that they want. However, other devices only "trust" certificates that are issued by authorities that they trust.

As shown in Figure 2-5, certificates exist within a public key infrastructure (PKI) hierarchy, which enables devices to determine which certificates they should trust.

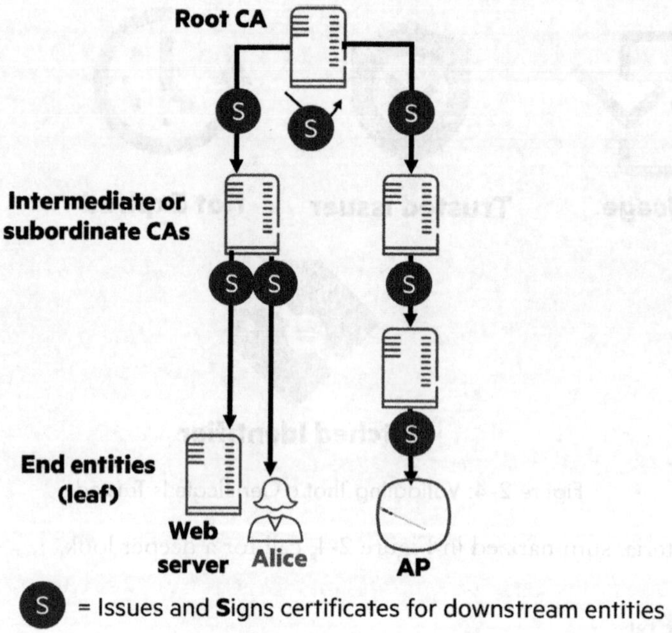

Figure 2-5: Trusted Issuer—PKI

Root Certification Authorities (CAs) are at the top of the hierarchy. A root CA signs its own certificates, called root CA certificates. Root CAs are well-known entities that are widely believed to be trustworthy. For example, DigiCert and Comodo are well-known CAs. A company can also set up its own internal or private root CA, which is believed to be trustworthy within the company.

As the ultimate source of trust, root CAs must be rigorously secured. The root CA issues just a few certificates itself, issuing and signing certificates for intermediate CAs. This ensures minimal contact between the root CA and other entities. In fact, the root CA is typically kept offline for security purposes.

The intermediate CAs, sometimes also called subordinate or signing CAs, are responsible for issuing and signing certificates for end entities such as Web servers, network infrastructure devices, and users. The end entity certificates are sometimes called server certificates, user certificates, or client certificates, depending on the type of entity to which they are assigned. They are also sometimes called "leaf" certificates because they are at the other side of the PKI "tree" from the root.

A PKI could also have multiple tiers of intermediate CAs, as shown on the right of Figure 2-5.

Trusted Issuer—Chain of Trust

When a device attempts to validate a certificate, it follows a chain of trust through the PKI until it reaches the root CA.

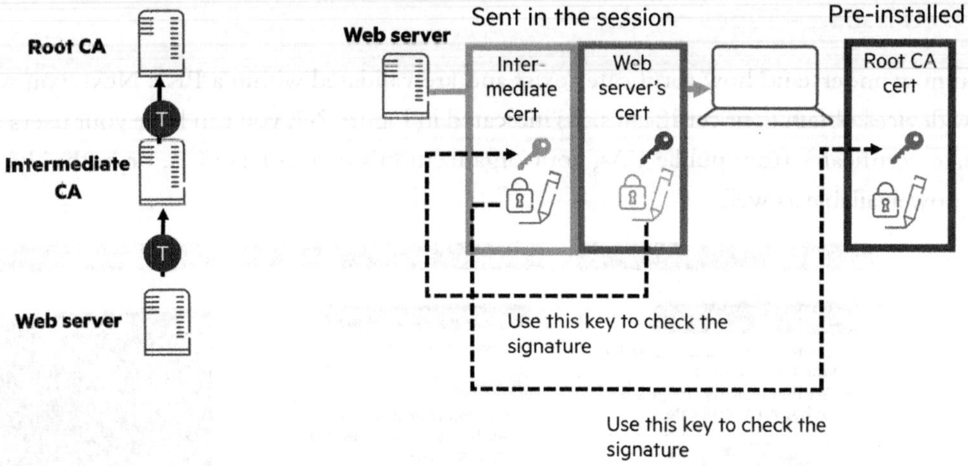

Figure 2-6: Trusted Issuer—Chain of Trust

In Figure 2-6, you see an example of how a client can use a chain of trust to validate a Web server's certificate. The chain of trust has two links: intermediate CA and root CA.

The server sends its own certificate and the certificate for the intermediate CA that signed that certificate. The server does *not* send the root CA certificate above the intermediate CA, though.

The client receives these certificates and validates that it can link the Web server's certificate back to a root CA that it trusts. First, the client looks for the issuer of the server's certificate—in this case, the Intermediate CA. Then the client looks in the Intermediate CA's certificate for the public key, and it uses that key to check the signature in the server's certificate.

The client continues to follow the chain of trust by checking the issuer of the Intermediate CA certificate. In a three-tier PKI, this is the root CA certificate.

As mentioned earlier, the Web server does *not* send this certificate. Instead the root CA certificate must be pre-installed as a trusted root CA on the client. The client can then use the public key in that pre-installed certificate to check the Intermediate CA certificate's signature. If the signature checks out, the client has completed the chain of trust and can validate the server's certificate as legitimate.

It is important for the client to have the root CA certificate pre-installed, rather than the Web server send that certificate. This is because users should not be prompted to trust root CA certificates on the fly; they will often make insecure choices. Instead security experts should install the root CA certificates in advance. OS typically come with trusted root CA stores populated with well-known trusted root CA certificates.

Trusted Issuers—Types of CAs

You now understand how certificates exist and are validated within a PKI. Next, you will consider how devices obtain their certificates. As indicated in Figure 2-7, you can have your users and devices obtain certificates from public CAs, your organization's own private CA, or both. Hybrid options are now available as well.

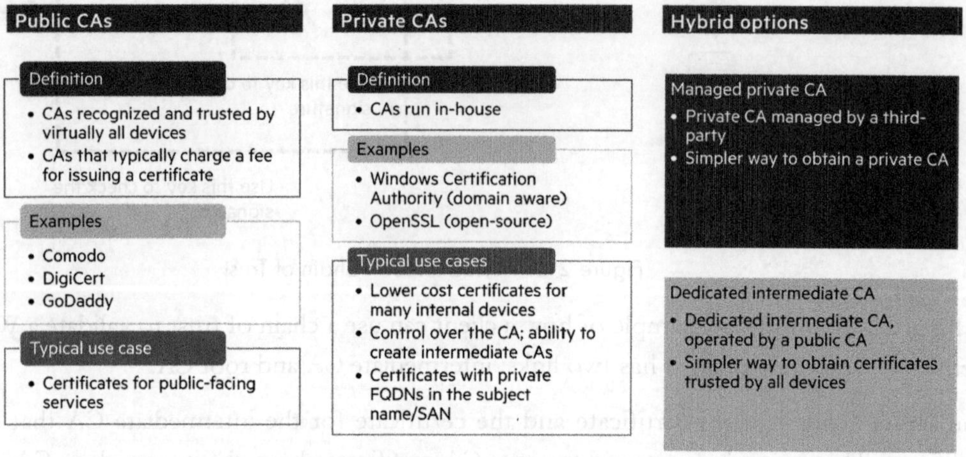

Figure 2-7: Trusted Issuers—Types of CAs

Public CAs are recognized and trusted by virtually all devices. Windows, Apple, Android, and other vendors pre-install these CAs' root certificates in their devices' operating systems. Examples of prominent public CAs include Comodo, DigiCert, and GoDaddy, but many other public CAs exist. You typically install CA-signed certificates on public-facing services so that you know that any devices that interact with those services will trust the certificates automatically. However, you must pay a fee to a public CA to obtain the certificate, which is why organizations are selective in which services use these certificates. In addition, public CAs now only issue certificates with 13-month validity periods.

Companies can set up their own private CAs that are specific to their own organization. They often use the Windows Certification Authority (CA) because it is domain aware. However, open-source options such as OpenSSL also exist. A private CA can be a cost effective choice for issuing certificates to many end-users and endpoints. In addition, the company can maintain more control over exactly which users and devices receive certificates, which can establish greater security for internal communications. With a private CA, the company has control over the PKI and can create the intermediate CAs that it requires. In addition, network infrastructure devices and non-public-facing services often use private IP addresses and fully qualified domain names (FQDNs). A public CA cannot issue certificates with a private IP address in the subject name or subject alternative name (SAN), based on RFC 1918. Similarly, public CAs do not issue certificates with subject names or SANs that specify private domain names, such as mydevice.example.local. Therefore, a company must use a private CA to obtain certificates for devices and services with private IP addresses or FQDNs. Companies might also use a private CA to issue certificates with longer validity periods.

A company that does not want to manage its own PKI could also obtain a managed private CA; for example, Globalsign offers such an option. In this way, the company can obtain the benefits of a private CA without needing the expertise to establish a CA in-house. Some public CAs such as DigiCert also offer a dedicated intermediate CA option. The public CA creates an intermediate CA, which it dedicates to the organization and operates for it. The organization can then obtain all the certificates that it needs from that intermediate CA. Because the certificates are ultimately backed by the public CA, they will be trusted by nearly all devices inside *and* outside the organization.

Not Expired—Ways to Avoid Issues

Every certificate has a start date and end date, and the receiver will only trust the certificate if the current date falls between those two dates. This requirement is fairly straight-forward, but it creates some considerations, as shown in Figure 2-8.

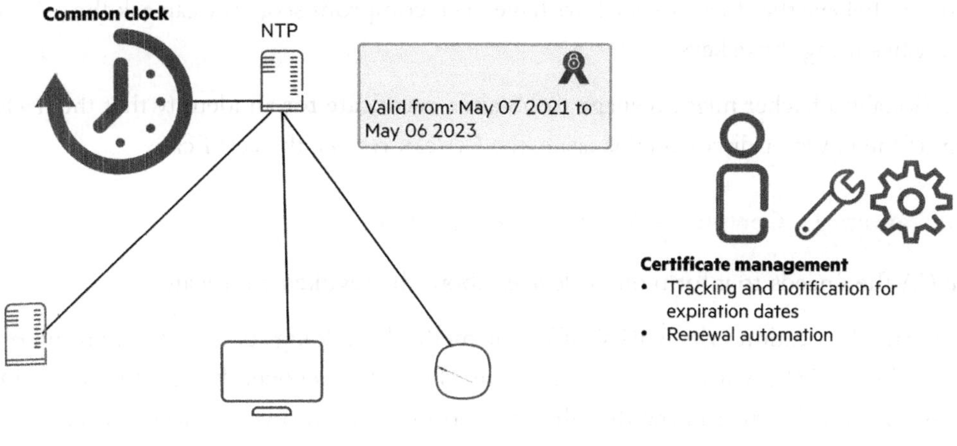

Figure 2-8: Not Expired—Ways to Avoid Issues

First, you need to ensure that all of the devices in the network, from clients to servers to network infrastructure devices, agree on the time. Otherwise, one device might reject another's certificate. Setting up Network Time Protocol (NTP) is the best way to meet this requirement on your network infrastructure devices.

Second, you need to be aware of when each certificate that you have installed will expire so that you can obtain and install a new certificate before that occurs. While updating one certificate is fairly simple, updating many certificates can be time-consuming and complex. And if admins miss even one certificate and fail to update it, the functions based on that certificate fail too. For this reason, companies often seek to automate this process. You will look at a few options for automation throughout this study guide.

Not Revoked—Why Revocation Is Important

An important part of maintaining a trusted PKI is revoking certificates that are no longer valid.

Say, for example, that a user has installed a certificate on a laptop. Then the user loses the laptop, or someone steals the laptop. Hopefully the certificate was installed with a measure of security, including a password to protect the private key. However, you cannot rely on that security alone. The certificate on the laptop is now compromised, and the CA that issued it should revoke it.

Companies might also retire devices before the certificates on those devices expire. It is best practice to revoke the retired device's certificates so that they can no longer be used.

Devices typically renew their certificates before they expire. The old certificate should be revoked so that a single valid certificate remains in use.

A certificate owner might also fail to properly secure the private key. If certificate owners have reason to believe that their private keys have been compromised, they can ask the CA to revoke any certificates using those keys.

Occasionally a hacker might manage to obtain a certificate for an identity that the hacker does not own. If the CA later discovers this issue, the CA can revoke the certificate.

Not Revoked—Certificate Revocation Lists (CRLs)

The CA then needs to inform other devices about the revoked certificates.

Certificate Revocation Lists (CRLs) offer one method for doing so. CAs (typically intermediate CAs) can publish a CRL, which lists any certificates that it has revoked by serial number. Devices with access to the CRL then know that they should not trust those certificates, even though the CA signature and chain of trust checks out. The CRL is signed by the intermediate CA that issues it to prevent tampering.

Admins can install CRLs on devices manually. However, it is most useful for devices to be able to obtain the latest CRL dynamically to prevent the CRL from becoming out of date.

The CA publishes its CRL at a CRL Distribution Point (CDP), which is a location on an LDAP server or Web server. Intermediate CAs often include the CDP within their certificates, so a device that trusts an intermediate CA can find the location where that CA publishes its CRL. CRLs have a validity period, and the CA periodically updates and republishes the CRL at the CDP, typically on a weekly basis.

As shown in Figure 2-9, when a device needs to validate a certificate signed by a CA, it checks whether it has a currently valid CRL associated with that CA3. If it does not, it contacts the CDP and downloads a new CRL. It then looks for the certificate's serial number in that list. If it finds a match, the device rejects the certificate. Otherwise, the certificate passes the "not revoked" check.

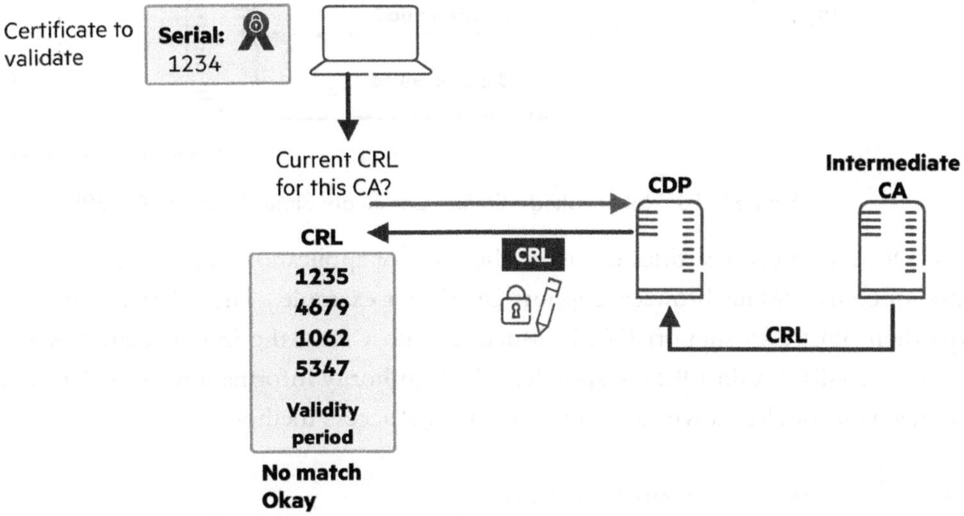

Figure 2-9: Not Revoked—Certificate Revocation Lists (CRLs)

CRLs do have some drawbacks. If the validity period is too long, a device might miss an update to the CRL and trust a certificate that has been revoked. In addition, a device has to download and check the *entire* CRL to find out whether a particular certificate has been revoked. Because CRLs can become long, this process can cause latency.

Not Revoked—Online Certificate Status Protocol (OCSP)

Online Certificate Status Protocol (OCSP) is designed to address the shortcomings of CRLs. OCSP enables a device to check on-demand the revocation status of *just* the certificate in which it is currently interested.

OCSP uses a client-server framework, and it uses HTTPS for the transport. As shown in Figure 2-10, when a client needs to validate a certificate, it sends a query that contains the certificate's serial number to an OCSP server, or OCSP responder. The OCSP responder is maintained by the CA. When the OCSP responder receives a query, it checks the certificate's status in the CA database, which is stored securely. The OCSP responder then sends a signed reply, indicating whether the certificate is still valid or not. Another advantage of OCSP versus CRL is that OCSP provides positive validation that the certificate is valid. A CRL, on the other hand, only gives negative information; it shows whether a CA has actually revoked the certificate or not, but the CA does not attest as to whether that specific certificate is valid.

Figure 2-10: Not Revoked—Online Certificate Status Protocol (OCSP)

To use OCSP to check a certificate's status, the client or application running the check must support OCSP, of course. (Many browsers support OCSP, for example.) The CA that issued the certificate in question must also support OCSP. Much as with CRLs, the intermediate CA certificate can advertise the URL for the OCSP responder. The "Authority Information Access" field includes this URL in a name or alternative name under the OCSP access method.

Matched Identifier—Making the Match

As part of validating the certificate, the validator also checks the subject name.

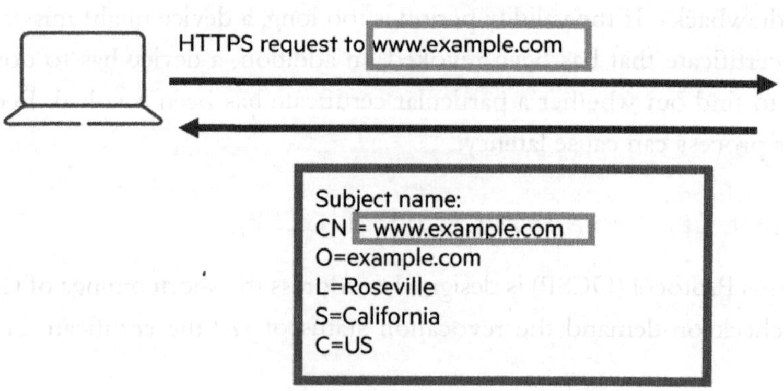

Figure 2-11: Matched Identifier—Making the Match

In the example shown in Figure 2-11, the client makes an HTTPS request to www.example.com. The server responds with its certificate as part of setting up the HTTPS session. The client validates that the subject name in the certificate matches the hostname with which the client believes it is communicating. If the certificate includes a different subject name from the one expected, this could indicate a man-in-the-middle (MitM) attack in which a hacker is attempting to insert itself and its own certificate in between the client and legitimate server.

Sometimes certificates are exchanged as part of a Layer 2 process such as 802.1X. In this case, the device validating the certificate cannot specifically match the subject name to the owner's FQDN or IP address. But it can check it against a list of trusted names; you will learn more about this process later.

Matched Identifier—Subject Names and SANs

Now look at that subject name in more detail. As shown in Figure 2-12, the subject name uses ASN.1 format. The subject name always includes a common name (CN) but might or might not include additional components in the ASN.1 hierarchy. As for what the common name specifies, X.509 makes no requirements. It could be a user name, an FQDN, an IP address, an email address, or any string.

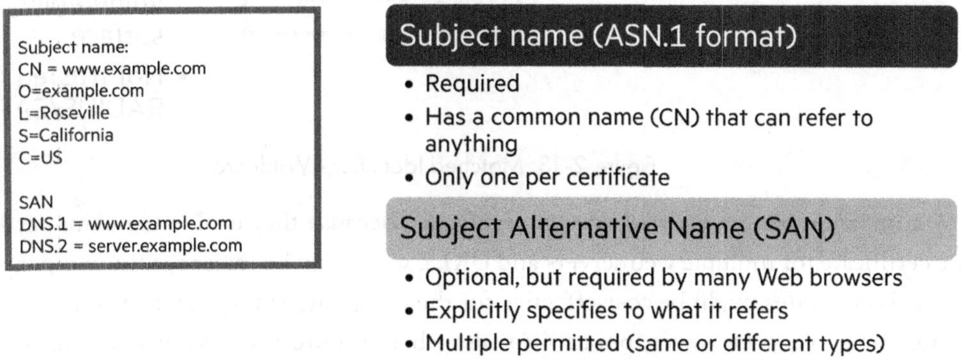

Figure 2-12: Matched Identifier—Subject Names and SANs

Subject alternative names (SANs) offer a more precise way for identifying the certificate owner. The SAN specifies a particular type, such as "DNS" for an FQDN or "IP" for an IP address. SANs also make it possible for a certificate to include multiple identities. For example, a single certificate could include two DNS SANs and be used by servers with different hostnames. A certificate could also have SANs of various types such as DNS and email.

Even though SANs are technically optional, some applications will only validate a certificate if it has a valid SAN. For example, Chrome and Mozilla require Web servers to present certificates with valid DNS-type SANs.

Matched Identifier—Wildcard

A subject name or DNS SAN can also include a wildcard, as shown in Figure 2-13. The wildcard is an asterisk, which matches any value for one level of an FQDN. For example, *.example.com matches www.example.com, mail.example.com, and server.example.com. However, it does not match my.server.example.com.

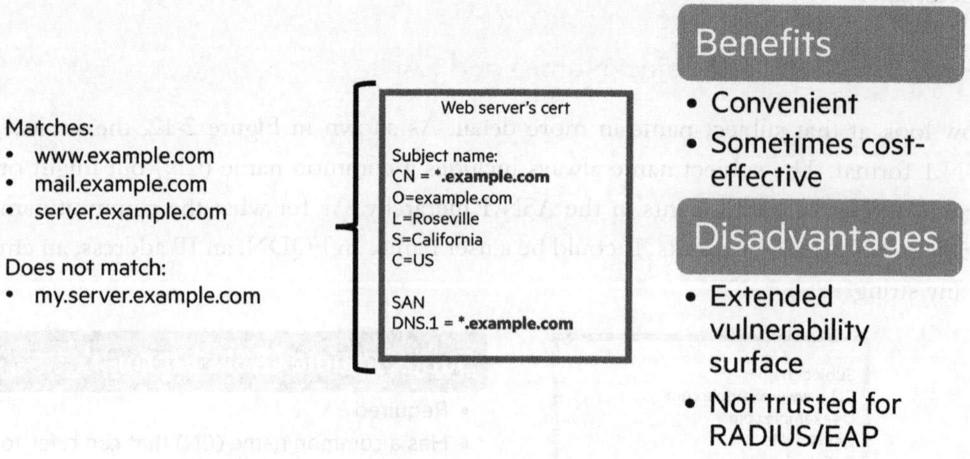

Figure 2-13: Matched Identifier—Wildcard

Many companies like to use wildcards in certificates because they enable the company to reuse the same certificate for multiple web servers and URLs, which can be easier for the company to manage. Wildcard certificates might be cost-effective for the company, as they can install the same certificate on multiple servers. However, public CAs also charge more for wildcard certificates, off-setting these advantages.

Certificates that use wildcards can also be less secure. Distributing the same certificate and associated private key across multiple servers inherently increases the vulnerability surface for that certificate and key. If one server is compromised, the security for communications on *all* servers is compromised. Some security audits will flag wildcard certificates. And some clients do not trust wildcard certificates for some purposes such as for a RADIUS/EAP server authenticating itself.

Ways to Create Certificate Signing Requests (CSRs)

Now that you have reviewed how certificates are used and validated, review how you can obtain a certificate for a device. You need to generate a certificate signing request (CSR) and a public/private keypair. You must also get the CSR signed by a valid CA for your purposes.

As shown in Figure 2-14, you often have two choices for how you can create the public/private keypair and CSR: online, on the device that will own the certificate, or offline using another application such as OpenSSL.

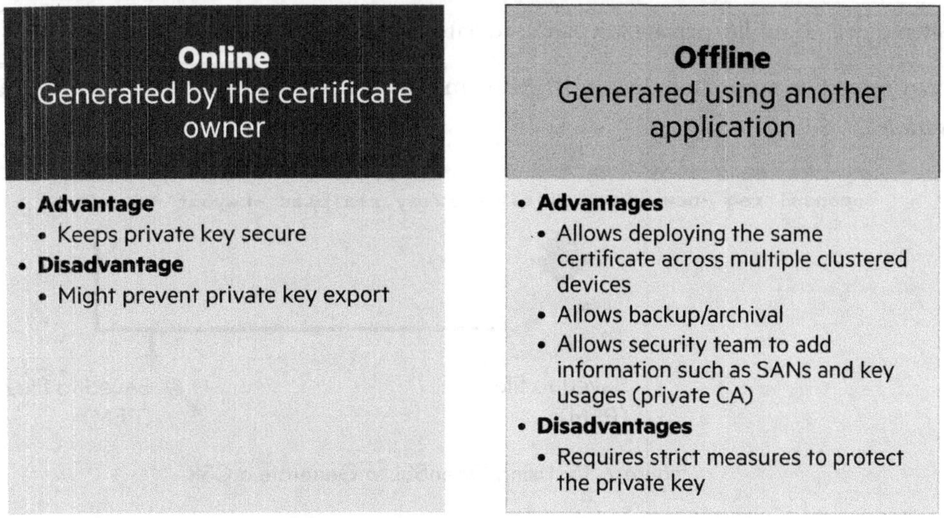

Figure 2-14: Ways to Create Certificate Signing Requests (CSRs)

When you create the keypair and CSR online, the private key remains on the device—perhaps permanently depending on the device. This can be good for security, but might interfere with certain applications. For example, the company might want to back up the private key so that admins can install the same certificate on new hardware if the original device fails. Or the company might want to deploy the same certificate across multiple devices in a cluster, which requires installing the private key on all of the devices so that the devices can use the certificate.

When you create the CSR offline, though, remember that you are creating the private key offline too, which makes it more likely to be compromised. You must take appropriate measures to secure the private key. This could mean encrypting the private key, storing it in a physically secure location, or both.

Using OpenSSL to Generate Offline CSRs

You will now look at using OpenSSL to generate offline CSRs.

OpenSSL is an open-source toolkit for generating certificates and the associated keys. OpenSSL has distributions for Windows, Linux, and macOS. With it you can perform tasks such as generating public/private key pairs, generating CSRs, and exporting certificates in a new file. You can even set up your own CA with OpenSSL and use it to sign certificates and create CRLs.

This study guide will focus on just a few common tasks that you can complete with OpenSSL.

Illustrated in Figure 2-15, the **openssl req -new** command generates a CSR. Indicate the file to which you want to save the CSR with this option: **-out <*filename*>**.

A CSR requires a public/private keypair associated with it.

You can reference an existing key in PEM format by adding this option to the command: **-keyin <*keyfile*>**.

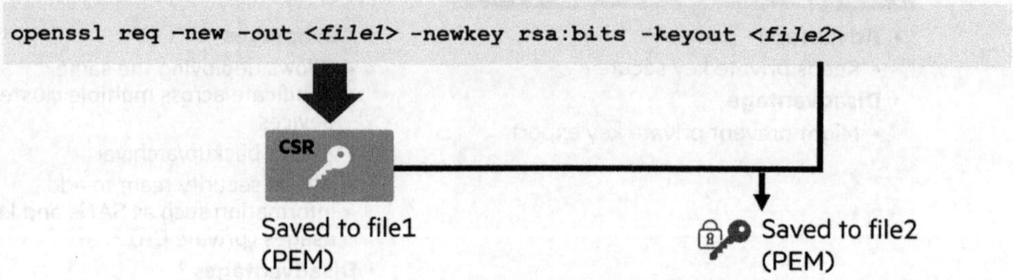

Figure 2-15: Using OpenSSL to Generate a CSR

You can alternatively create a new keypair along with the CSR by adding this option: **-newkey rsa:bits –keyout <*filename*>**. When you use RSA for the algorithm, you specify bits after the colon. (You can also specify another algorithm such as **ec** for elliptic curve, but in that case you must specify an algorithm parameter file in advance to define the parameters. See Openssl documentation for more information.)

The key and CSR are both saved in PEM format, which is sometimes called base64.

As shown in Figure 2-16, when you run this command, you will be prompted to set a passphrase, which encrypts the private key. Now no one can read or use the private key file without first entering this passphrase. You will also be prompted to enter fields for creating the ASN.1 subject name. Usually the CN is the most important part of this name.

```
.........................................................................
writing new private key to <file2>
Enter PEM pass phrase: **********
Verifying - Enter PEM pass phrase: **********
-----
<output omitted>
-----
Country Name (2 letter code) [AU]:US
State or Province Name (full name) [Some-State]:California
Locality Name (eg, city) []:Sunnyvale
Organization Name (eg, company) [Internet Widgits Pty Ltd]:example.com
Organizational Unit Name (eg, section) []:Infrastructure
Common Name (e.g. server FQDN or YOUR name) []:clearpass.example.com
```

Figure 2-16: Output from OpenSSL CSR Generation

Using OpenSSL to Create a CSR with Additional Options

The basic **openssl req –new** command gives you prompts for inputting the ASN.1 DN, and it generates the CSR based on some default settings in the openssl.conf file. However, you often want to customize the CSR settings. For example, you might want to include one or more SANs or EKUs.

To add these custom settings, edit the openssl.conf file, as shown in Figure 2-17. Create a section with a name within brackets such as [myDevice]. Under that heading you can add your desired options.

Add a SAN with this format:

subjectAltName = *Type : value.*

For example:

subjectAltName = DNS:clearpass.example.com.

To specify extended key usages, use this format:

extendedKeyUsage = *usage1*, *usage2.*

For example:

extendedKeyUsage = serverAuth

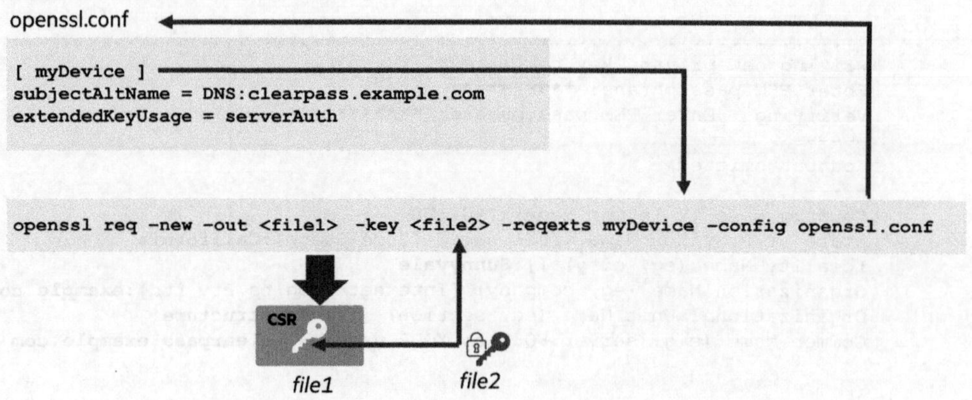

Figure 2-17: Using OpenSSL to Create a CSR with Additional Options

After you have created the section, save and close the openssl.conf file. As shown in Figure 2-17, you can then reference the section and config file when you run the openssl req -new command with these options: **-reqexts <*sectionname*> -config openssl.conf**.

The example command in Figure 2-17 also illustrates that you can reference an existing key file when you generate the CSR. This option is useful when you want to create a new certificate for a device, but using an existing public/private keypair, which other hosts already trust.

If you need to specify multiple SANs of the same type, you can use this format:

subjectAltName = *Type.#: value, Type.#: value*.

For example:

subjectAltName = DNS.1:cppm1.example.com,DNS.2:cppm2.example.com

Some Ways to Submit CSRs

If you need to get your CSR signed by a public CA, you will need to review that specific CA's procedures for submitting CSRs. The CA will tell you the correct format for the request. OpenSSL outputs CSRs in base 64 PEM, by default, but you can also use OpenSSL to convert CSRs to another format if necessary. Keep in mind that some CAs do not accept SANs within CSRs for security purposes. Instead, as shown in Figure 2-18, you must submit the SANs in another form along with the CSR. The CA then vets the SANs and adds them to the certificate during the signing procedure. Also keep in mind that public CAs might charge different prices for certificates with various numbers of SANs, as well as for wildcard certificates.

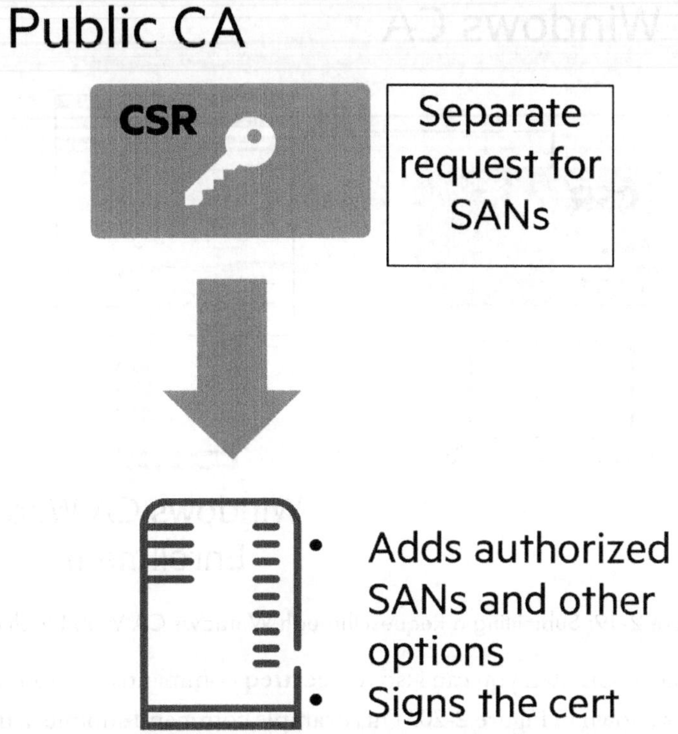

Figure 2-18: Submitting a Request to a Public CA

There are many ways to submit CSRs to a private Windows CA. The CA might maintain a Web Enrollment server, as shown in Figure 2-19. If so, you can log into this server with your domain credentials and submit the CSR as an advanced certificate request. Open up the PEM certificate with a text editor and paste the complete text into the request window. You will also have a chance to choose the certificate template. Certificate templates control factors such as key usages, EKUs, how the subject name is generated, and validity period. The certificate templates available for you depend on your group memberships.

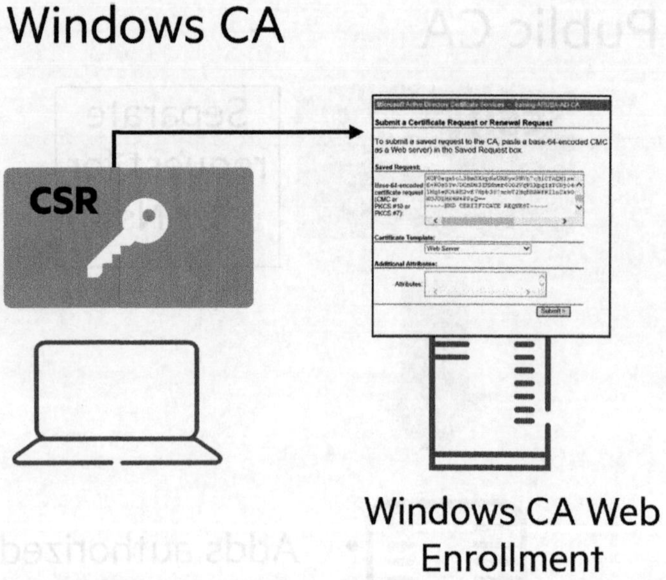

Figure 2-19: Submitting a Request through Windows CA Web Enrollment

If you are on a domain computer, you can also use **certreq** commands to submit a certificate request to the domain CA, as shown in Figure 2-20. This example command submits a file named mycsr.req, requests the Web server template, and outputs the signed certificate to a file named mycert.cer:

certreq -submit -attrib “CertificateTemplate:WebServer” mycsr.req mycert.cer

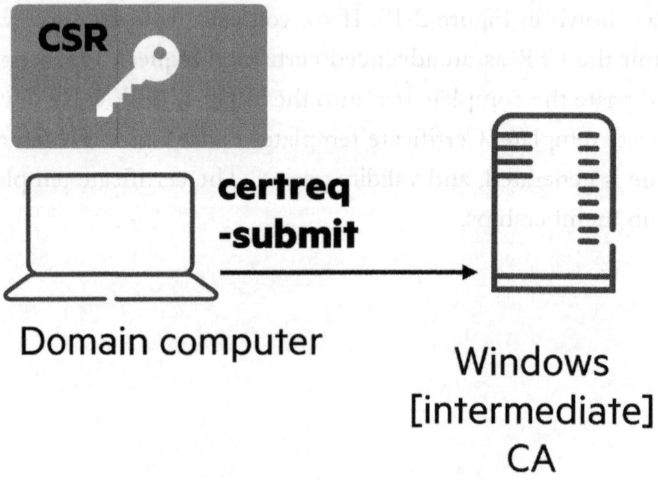

Figure 2-20: Submitting a Request to a Window CA with Certreq

HPE Aruba Networking ClearPass Certificates

This section explains how to manage certificates on HPE Aruba Networking ClearPass.

Types of ClearPass Server Certificates

When you install a certificate on ClearPass, you can choose the usage. You can also choose an existing certificate and assign additional usages to it. In this way, you can ensure that ClearPass has a certificate signed by the correct CA for each use case. ClearPass uses server certificates for four purposes. Note that if you are using a Windows CA for any of the certificates, the WebServer template is recommended.

Type	Usage	CN in subject name	SAN	Typical issuing CA
RADIUS/EAP	Authenticating to 802.1X supplicants	FQDN wildcard prohibited	DNS (match CN)	Private
RadSec	Authenticating to RadSec clients	FQDN (or IP address)	Optional (match RADIUS address on RadSec clients)	Private or public
HTTPS	Establishing secure Web sessions to the UI, guest portal, Onboard, etc.	FQDN	DNS (match CN plus optional extra)	Public
Database	Establishing secure sessions to the Insight Database	FQDN or IP address	DNS, but must be IP address	Private

Figure 2-21: Types of ClearPass Server Certificates

The ClearPass certificate types are summarized in Figure 2-21. Read the section below for a more detailed description of each type.

RADIUS/EAP

HPE Aruba Networking ClearPass Policy Manager (CPPM) uses this certificate during the 802.1X process to authenticate itself to 802.1X supplicants (users and clients). This certificate must not use a wildcard for its CN, as some clients will reject RADIUS/EAP certificates with wildcards. The certificate should have a DNS SAN that matches the CN. The CN and SAN should not specify IP addresses to avoid exposing the IP address to users, as well as to permit the certificate to remain valid if the IP address changes. The supplicants must trust this certificate. Generally, you should obtain the certificate from a private CA. 802.1X authentication typically applies to corporate devices, which are either managed or onboarded, so the organization can ensure that the clients trust the private CA. Using a private CA gives the organization more control over the certificate. Public CAs typically only issue certificates that are valid for 13 months, but a company can issue its own certificates with longer validity periods, making them easier to maintain.

You can install multiple RADIUS/EAP certificates on CPPM and associate a specific RADIUS/EAP certificate with a specific service. This feature is useful if CPPM is authenticating users from multiple organizations, and clients in different organizations trust different CAs. You simply install certificates signed by different CAs on CPPM and then match the correct certificate with the service used by each domain.

RadSec

Because RADIUS itself does not provide rigorous security, CPPM (a RADIUS server) and network devices (RADIUS clients such as APs, gateways, or switches) can use RADIUS Secure (RadSec). RadSec carries RADIUS messages inside a secure TLS session and is useful for cases in which CPPM and the network devices are separated by an untrusted network. CPPM uses the RadSec certificate to authenticate itself to network devices that use RadSec. The network access devices (NADs) must be manually configured by admins to trust the root CA associated with this certificate.

Often a private CA works well for issuing the RadSec certificate. The company has control over its own devices and can ensure they trust the CA. However, some companies have particular requirements that call for public CA-signed certificates, and those work for RadSec as well.

The RadSec certificate generally uses an FQDN for the CN. However, the CN can alternatively be an IP address. A SAN is optional. The SAN, if specified, or the CN must match the RadSec server address specified on the RadSec clients. For example, if the SAN or CN is an FQDN, the RadSec clients must also specify the RadSec server address with that FQDN, not an IP address. (If you are using AOS-Switches, not covered in this study guide, using an IP address works better.)

HTTPS

ClearPass uses this certificate to authenticate itself to clients that access its web interfaces, including its management interface, Guest portal, and Onboard. Any clients that might need to access these interfaces must trust this certificate. If you are providing guest authentication services for users outside of the corporation, those users' devices will not trust the corporate CA. Therefore, the HTTPS certificate should typically be signed by a well-known public CA.

This certificate should also follow other best practices for HTTPS certificates. It is allowed to have a wildcard for the CN, but you may not necessarily have one, as wildcards can be less secure. The certificate *must* have a SAN with a DNS name that matches the CN; it can also have additional DNS names in the SAN. The CN and SAN should not specify IP addresses to avoid exposing the IP address to users, as well as to permit the certificate to remain valid if the IP address changes.

Database

ClearPass uses the database certificate to authenticate to clients that access its Insight Database. When you deploy a cluster, the database certificates of cluster members must have a DNS type SAN, but with the server's IP address as the value. Because public CAs might not issue a certificate that meets these criteria, you will often obtain the database certificate from a private CA.

Installing Certificates on ClearPass servers

You have two choices for installing certificates on ClearPass servers.

As shown in Figure 2-22, you can create the private key and CSR online using the Web UI. You can then copy the Base-64 CSR into a file and submit that file to the proper CA for signing. When you receive the signed certificate, you upload that certificate on ClearPass.

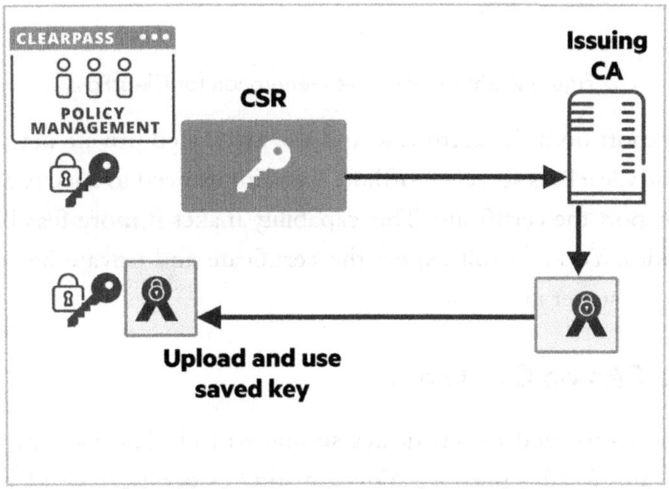

Figure 2-22: Online CSR Generation for ClearPass

Or, as shown in Figure 2-23, you can create the private key and CSR offline using a tool like OpenSSL. Some admins prefer this method so that they can script the generation of many CSRs. You then submit the CSR to the proper CA for signing. When you receive the signed certificate, you combine it with the private key into a PKCS12 file using a tool like OpenSSL. You can then upload the PKCS12 file on CPPM. Or you can even upload the private key and signed certificate as separate files.

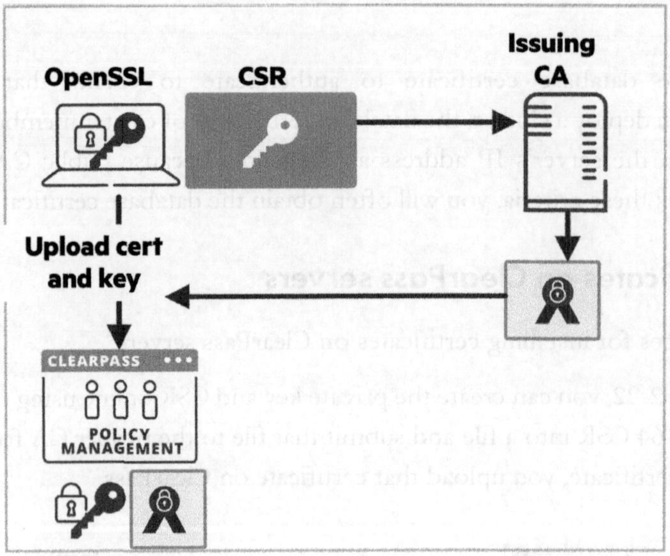

Figure 2-23: Offline CSR Generation for ClearPass

Note that you can export both the certificate and the associated private key, whether you created the CSR online on the ClearPass server or offline. You simply need to specify a password to encrypt that key when you export the certificate. This capability makes it more feasible for you to use the online CSR capabilities. You can still export the certificate and private key and archive them or install them on another server in a cluster.

Types of Trusted CAs on ClearPass

CPPM sometimes needs to validate certificates submitted to it. Just like any other device, CPPM uses root CA certificates for this purpose. You can add CA certificates to CPPM's trusted list for several usages, shown in the table in Figure 2-24. Note that CPPM also supports several other types of CA certificates not listed in this table, which are less relevant for this study guide.

Type	Usage
AD/LDAP	Validating an AD/LDAP authentication source using secure LDAP
Database	Validating database certificates Providing the root of trust for the Database cert issuer
EAP	Validating 802.1X supplicant certificates Providing the root of trust for the RADIUS/EAP cert issuer
Endpoint context server	Validating the certs submitted by Endpoint Context servers
RadSec	Validating certificates submitted by RadSec-enabled network devices
Other	Providing the root of trust for the HTTPS cert issuer

Figure 2-24: Types of Trusted CAs on ClearPass

You can upload multiple trusted CA certificates. Each certificate can have one or more usages. You set the usage when you upload the certificate, and you can also add usages at a later point.

Multiple trusted CAs can also be permitted for the same usage. For example, different endpoint context servers might have certificates issued by different CAs, and ClearPass can have multiple trusted CA certificates with the Endpoint context server usage.

AD/LDAP

CPPM can integrate with AD or another LDAP server to authenticate clients. CPPM and the AD/LDAP server can secure their communications with a protocol such as Start TLS or LDAP over SSL. In this case, AD or the LDAP server authenticates itself to CPPM with a certificate. CPPM verifies that certificate using the root CA certificate that you assign the AD/LDAP usage.

Database

This is the root CA certificate for the CA that issues the ClearPass server's Database certificate. If acting in a cluster, a server also uses CA certificates with the database usage to validate database certificates on other members of the cluster.

EAP

CPPM uses the certificate with the EAP usage type to validate certificates submitted by supplicants during an 802.1X EAP authentication process. CPPM must also have an EAP-type root CA certificate for the CA that signs CPPM's own RADIUS/EAP certificate. (In practice, CPPM and supplicants might have the same root CA for certificates that they use during RADIUS/EAP.)

Endpoint Context Servers

CPPM can integrate with endpoint context servers, such as a Palo Alto firewall or mobile device management (MDM), to exchange contextual information with those servers. To secure these communications, the servers can authenticate to CPPM with certificates. CPPM uses this type of CA certificate to validate these servers' certificates.

RadSec

RadSec also requires NADs to submit certificates to authenticate themselves. CPPM uses the RadSec-type CA certificate to validate those certificates. CPPM must also have a RadSec-type root CA certificate for the CA that signs CPPM's own RadSec certificate.

Other

You should upload the root CA certificate for the CA that signs CPPM's HTTPS certificate and assign it the Other usage.

Supplemental Information about Default Root CA Certificates

Starting with Policy Manager 6.9.1, the Comodo RSA Certification Authority root CA is included by default in the certificate trust list. This CA is enabled by default and is used by ClearPass Device Insight integrations to connect to HPE Aruba Networking Central. Policy Manager 6.9.2 also introduces a new SHA-2 root certificate authority CA USERTrust RSA Certification Authority. This certificate is disabled by default.

The trust list might also include Department of Defense (DoD) certificates: A DoD certificate allows a browser to trust web sites whose secure communications are authenticated by a DoD agency. These certificates are disabled by default. It might also include an Alcatel root certificate, which allows Alcatel Lucent IP phones to authenticate using EAP-TLS. These certificates are disabled by default.

Certificate Requirements for Forming a ClearPass Cluster

Many enterprises choose to deploy ClearPass clusters to provide better redundancy and scale out services. A ClearPass cluster consists of a Publisher and multiple Subscribers; one Subscriber can also act as a standby Publisher. The Publisher acts as the central point for configuration and maintains a central database to which Subscribers synchronize. In a production cluster, the Publisher is typically dedicated to managing and controlling the cluster while Subscribers respond to service requests and apply policies.

Every ClearPass server starts as a Publisher. To join the future Subscribers to the cluster, you convert them to Subscribers and point them to the Publisher.

There are many considerations for cluster design, but this study guide focuses on the certificate requirements, illustrated in Figure 2-25. To join the cluster, Subscribers must trust the HTTPS certificate on the Publisher. The best way to meet this requirement is to obtain a certificate signed by a well-known public CA on the Publisher. Then check that the Subscribers trust the *complete* chain for that certificate, including the root CA and one or more intermediate CAs. The certificates might be pre-installed in the server trust list if they are well-known. In this case, enable the certificates and set the Others usage. If the certificates are not already present, add them to the trust list and set the Others usage.

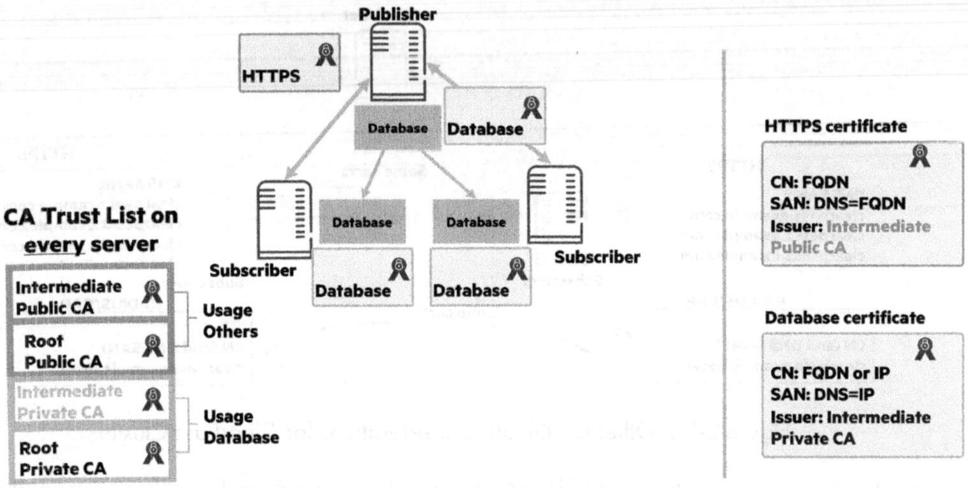

Figure 2-25: Certificate Requirements for Forming a ClearPass Cluster

The Publisher must also have a database certificate that has its own IP address listed as the *DNS* SAN. Similarly, each Subscriber must have its own database certificate with its own IP address listed as the DNS SAN. Typically, you obtain these certificates from a private CA. On each server, add the private CA root certificate and any intermediate certificates to the trust list with the Database usage. Obtain a certificate signed by a CA that meets the requirements listed above and install that as the server's Database certificate.

The certificates are then ready, and you can convert each server to a Subscriber and connect it to the Publisher.

Note that these are the requirements for ClearPass 6.9. In 6.10, admins receive a prompt to trust the Publisher certificate when they add the Subscriber.

Other Certificate Considerations for ClearPass Clusters

Each member of a ClearPass cluster continues to have its own certificates installed on it. However, you can choose to install the same certificate for the same function across multiple members, as shown in Figure 2-26. In that case, the certificate's subject name or SAN must be valid for all of the members using it. For example, if you install the same HTTPS certificate on multiple members of a cluster, the certificate needs a SAN for the hostname of each member. Or it could have a wildcard name.

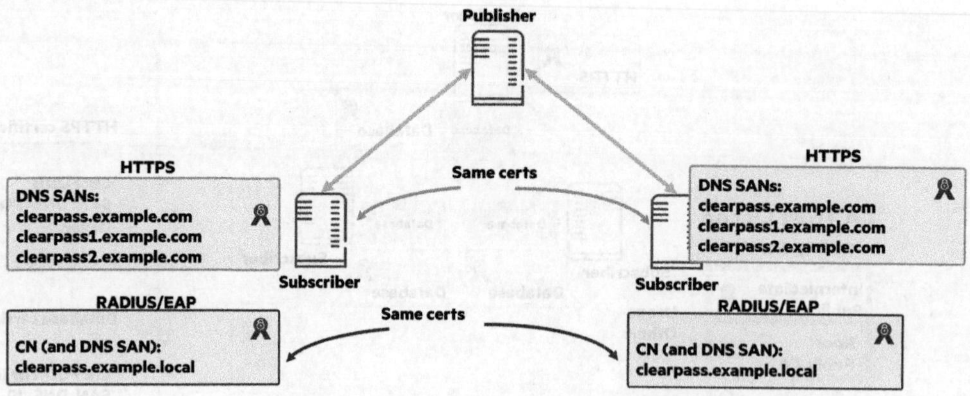

Figure 2-26: Other Certificate Considerations for ClearPass Clusters

Alternatively, clusters can use virtual IP groups, in which one member takes responsibility for the virtual IP address while the standby member takes over the virtual IP address if the first member fails. It is generally best practice to use virtual IP addresses to simplify failover when the company has not deployed a load balancer for the CPPM cluster. If you are using the virtual IP approach, the members can effectively share the hostname to which the virtual IP address resolves. You must define that hostname as a DNS SAN in the certificate. You might also assign the devices' individual hostnames as extra SANs in case admins want to sometimes contact the ClearPass servers on their individual IP addresses.

Clients only check the name in the RADIUS/EAP certificate, so multiple members can share it without an issue, whether or not the members are using a virtual IP.

HPE Aruba Networking Infrastructure Certificates

In this section, you review managing certificates on HPE Aruba Networking infrastructure.

HPE Aruba Networking Infrastructure Certificate Options

Like ClearPass, HPE Aruba Networking infrastructure devices can use certificates for many usages, depending on the device type and its role in the architecture.

Self-Signed Certificates

The HPE Aruba Networking devices come with self-signed certificates, which let admins obtain initial access to the web UI using HTTPS. As shown in Figure 2-27, a self-signed certificate is issued by the device itself. Self-signed certificates are not considered secure, and should not be used in production.

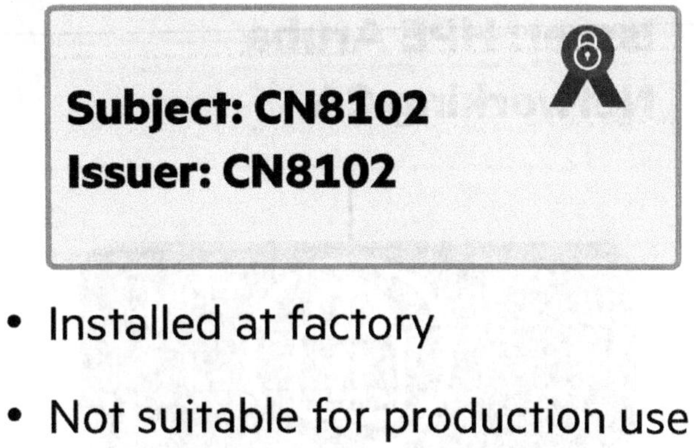

Figure 2-27: Self-Signed Certificates

Device (TPM) Certificates

Current models of HPE Aruba Networking APs, Mobility Controllers (MCs), gateways, and AOS-CX switches also come with factory-installed device certificates, as shown in Figure 2-28. This certificate is issued by an HPE Aruba Networking or HPE CA, and it is installed at the factory in a tamper-proof hardware Trusted Platform Module (TPM), making its private key highly secure. Admins find the certificates convenient to use as they are pre-installed and have a long validity period, eliminating the need for renewals. Because the certificate is issued by a secure HPE Aruba Networking or HPE CA, rather than self-signed, it can be used in production.

Device certificates are trusted by many HPE Aruba Networking products, so they can be useful for setting up secure communications between HPE Aruba Networking devices. In the HPE Aruba Networking Wireless Operating System 8 (AOS-8) architecture, APs can discover MCs and set up CPsec tunnels (as long as the MC is configured to auto-accept them or has an allow list). In an AOS-10 architecture, devices can establish secure connectivity with HPE Aruba Networking Central. HPE Aruba Networking Central typically learns the serial numbers, included in the device certificates, from Activate.

Figure 2-28: Device (TPM) Certificates

You do not need to set anything up to let HPE Aruba Networking devices use these certificates in these ways; they do so out of the box.

As you will learn later, you can also set up APs to use their TPM certificates when they act as 802.1X supplicants.

Note that virtual Mobility Controllers (vMCs) do *not* have TPM certificates.

CA-Signed Certificates

As shown in Figure 2-29, you can also obtain a CA-signed certificate and install it in your devices when the devices need to establish trusted communications with non-HPE Aruba Networking devices.

- Must be installed by admins
  - Public CA-signed certificate from HPE Aruba Networking Central
- Secure and trusted by all devices that trust that CA
- Typical use cases: Web UI, captive portal

Figure 2-29: CA-Signed Certificates

The Web UI is one example of this use case. While admins might be used to skipping most certificate errors to reach devices' web interfaces, it is more secure to install a trusted certificate on devices. Doing so minimizes the chances of a MitM attack, in which hackers insert their own devices between an admin and the managed device. You can use a private CA to issue these certificates to lower costs. However, if you are managing devices through HPE Aruba Networking Central, this step might be less important to you.

The HPE Aruba Networking device also requires a public CA-signed certificate if it forms part of a captive portal solution, even if the captive portal is hosted on an external server such as ClearPass Guest. The HPE Aruba Networking device redirects HTTPS traffic, and it needs a certificate that clients trust to do so successfully. Note that HPE Aruba Networking Central comes with a public CA-signed certificate, which it can push to devices that it manages, eliminating the need for you to acquire a certificate for this purpose.

HPE Aruba Networking recommends using ClearPass to provide RADIUS server and captive portal capabilities in most cases. However, sometimes a company will want to host these services on their HPE Aruba Networking mobility devices. In this case, the device should use a CA-signed certificate for these functions. The RADIUS/EAP certificate can be signed by a private CA. But the devices will need public CA-signed certificates for the captive portal so that guests who access the portal do not see errors.

Note that all of these functions require ServerAuth for the EKU.

Installing Trusted CA Certificates with HPE Aruba Networking Central

To install a trusted CA certificate on HPE Aruba Networking APs and other devices managed by HPE Aruba Networking Central, first obtain the PEM-formatted CA certificate file on the local machine with which you will access the cloud-based management solution.

In HPE Aruba Networking Central, access certificate settings in the **Global** view (go to the **Organizations** page and click the **Certificates** card). Add a certificate, specifying an intuitive name and browsing to the PEM-formatted file in which the certificate is saved on your local machine. Set the certificate type to **CA Certificate**.

This process is summarized in Figure 2-30.

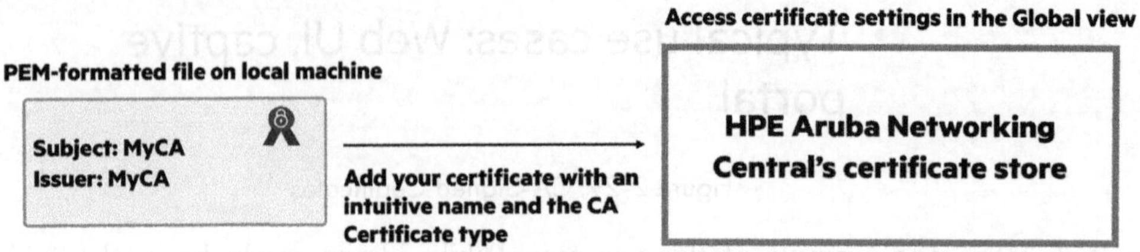

Figure 2-30: Installing Trusted CA Certificates with HPE Aruba Networking Central

Note that the reference to the location for the settings are based on HPE Aruba Networking Central 2.5.5. This location might change in later updates; refer to the latest documentation for details. The same holds true throughout this study guide.

Installing CA-Signed Certificates with HPE Aruba Networking Central

APs, including ones managed by HPE Aruba Networking Central, do not provide an online mechanism for generating the CSR from the UI. You can use a tool like OpenSSL to generate a CSR and private key and submit the certificate to a CA for signing.

As shown in Figure 2-31, you should combine the certificate and its private key in the same PEM-formatted file. When the certificate was signed by an intermediate CA, which is typical, you should also place the intermediate CA certificate within that file.

1. Add the root CA certificate for the certificate to HPE Aruba Networking Central.
2. Upload the CA-signed certificate in HPE Aruba Networking Central's Global certificate settings

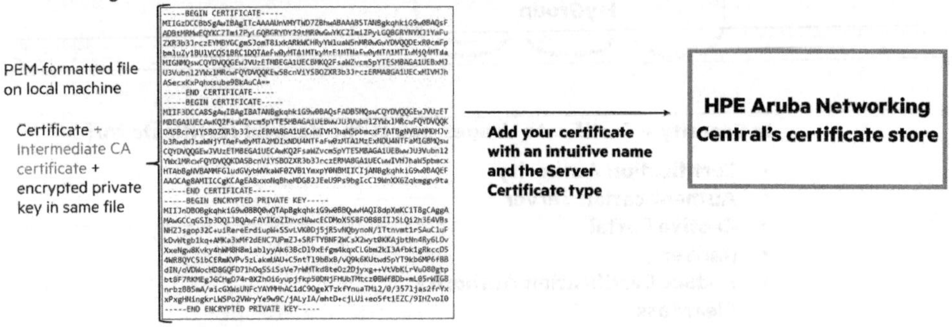

Figure 2-31: Installing CA-Signed Certificates with HPE Aruba Networking Central

You are then ready to install the certificate on HPE Aruba Networking Central. First make sure that you have installed the root CA certificate associated with the CA that signed your certificate. Then install the signed certificate on HPE Aruba Networking Central from the same location where you install CA certificates. You give the certificate a name, set the type to **Server Certificate**, and choose the format. Set the passphrase to the phrase that encrypted the private key. Browse to your PEM file and add the certificate.

Referencing Certificate Usages in HPE Aruba Networking Central

To actually deploy the CA and server certificates on APs and other devices managed by HPE Aruba Networking Central, you select usages for those certificates in the devices' group-level device configuration.

As indicated in Figure 2-32, In HPE Aruba Networking Central 2.5.5, you go to the group view. You then select **Devices** and the desired device type, such as Access Points. You then click the **Config** icon. (You can also configure the usages at the device level, but group-level configuration is typically preferred.)

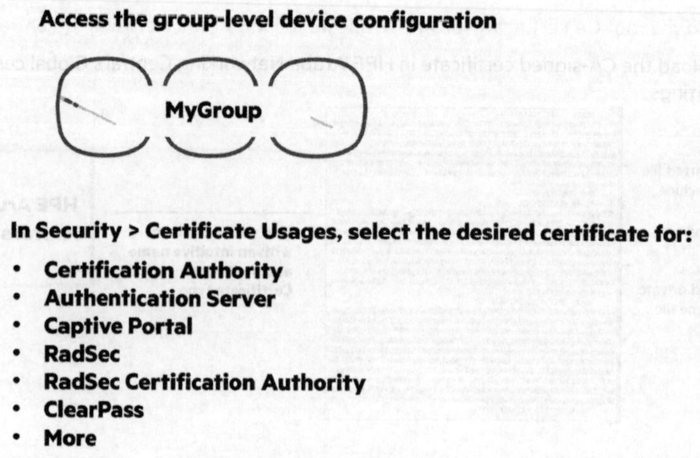

Figure 2-32: Referencing Certificate Usages in HPE Aruba Networking Central

For APs you set usages in the **Security > Certificate Usage** settings. Reference a trusted CA certificate for the AP's general CA. (The AP uses that CA certificate to validate client certificates when the AP acts as RADIUS server, for example). You can also reference a specific CA certificate for RadSec and for ClearPass (validating CPPM's HTTPS certificate for downloading roles).

You can also choose a CA-signed server certificate for functions such as the Web UI, captive portal, and RadSec. Additional usages are also supported; you can reference the documentation to learn more about them.

The concept is similar for HPE Aruba Networking MCs. You upload a certificate file and specify its type (such as Trusted CA or Server). You can then select the certificate for various usages.

AOS-CX Trust Anchor (TA) Profiles

You will now learn how to manage certificates on AOS-CX Switches. Installing a trusted root CA serves two purposes. It enables you to install a certificate signed by that CA on the switch. It also enables the switch to trust certificates that were issued by that CA or an intermediate CA below it. For example, to implement downloadable user roles (DURs), about which you will learn more later, AOS-CX Switches must trust CPPM's HTTPS certificate. So, they need the root CA for CPPM's HTTPS certificate installed in a TA profile. If you want to use RadSec, you must also ensure that the AOS-CX Switches have a TA profile for the root CA for CPPM's RadSec certificate.

You need one TA profile for each CA that you want the switch to trust. AOS-CX switches come with several pre-installed TA profiles, and you can add your own for the CAs that your company uses.

Copying the root CA certificate into the TA profile activates the profile. As shown in Figure 2-33, you can issue a command and copy a Base64 formatted certificate directly into the CLI, or you can use SFTP.

```
crypto pki ta-profile myCA
    ta-certificate import <terminal | sftp username@location >
```

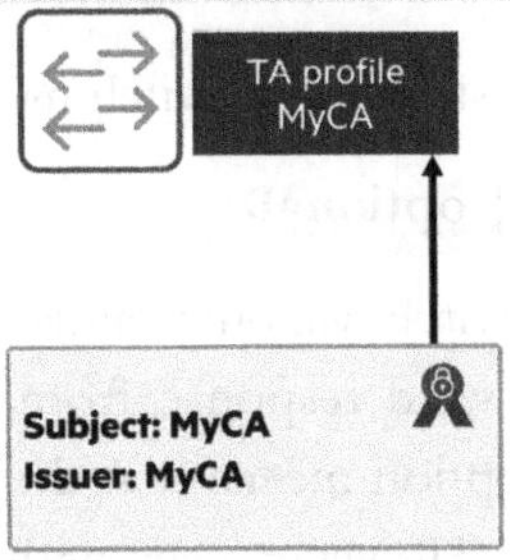

Figure 2-33: AOS-CX Trust Anchor (TA) Profiles

Enabling OCSP on an AOS-CX TA Profile

As you learned, OCSP enables a device to dynamically check whether a certificate is still valid or has been revoked. If it supports OCSP, the CA maintains a database.

As shown in Figure 2-34, to enable OCSP on an AOS-CX Switch TA profile, enter this command: **revocation-check ocsp**.

```
crypto pki ta-profile myCA
    ta-certificate import <terminal | sftp <username@location>
    revocation-check ocsp
    ocsp url primary http://ocsp.digicert.com
```

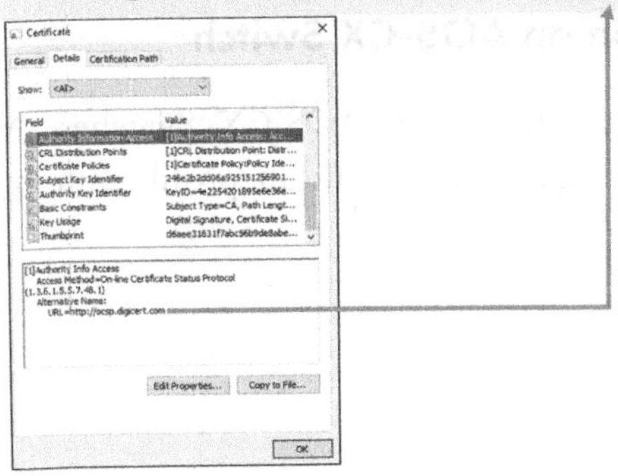

Figure 2-34: Enabling OCSP on an AOS-CX TA Profile

You can use the **ocsp** command to specify the OCSP responder's URL. You can ask your CA about what URL to use. Or you can look in the intermediate CA certificate. In the certificate details, find the "Authority Info Access" field and look for URL under the OCSP access method.

Supplemental Information about OCSP Settings

You can regulate a few more settings for OCSP.

Use this command to set a secondary URL in case the primary one is inaccessible:

ocsp url secondary <URL>

Use this command to specify how strictly the switch uses OCSP to validate the certificate:

ocsp enforcement-level <strict | optional>

The **strict** option means that the switch will only validate a certificate if it passes all checks; in other words, the switch *must* receive a valid response from the OCSP responder, indicating that the certificate is valid. The **optional** option means that the switch will still validate the certificate if it cannot reach the OCSP responder. However, even with the **optional** option, the switch will reject the certificate if the OCSP responder actually sends a response that the certificate is invalid, or if the OCSP response has an invalid signature or nonce.

The default option is **strict**.

Use this command to disable inclusion of a nonce:

ocsp disable-nonce

The switch includes a nonce (a random value) in the query; it then expects the responder to include the same response in replies. This mechanism helps to protect against replay attacks. However, some CAs' OCSP responders do not support using nonces, so you would need to turn off the feature if this is the case with your CA.

Uploading a CA-Signed Certificate on an AOS-CX Switch

Now you will look at installing CA-signed certificates on AOS-CX Switches. As shown in Figure 2-35, you start the configuration by creating a named certificate profile with this command: **crypto pki certificate <*name*>**.

```
AOS-CX(config)# crypto pki certificate mycert
AOS-CX(config-cert-mycert)# import terminal ta-profile myCA password
<key_password>
Paste the certificate in PEM format below, then hit enter and ctrl-D:
AOS-CX(config-cert-import)#
```

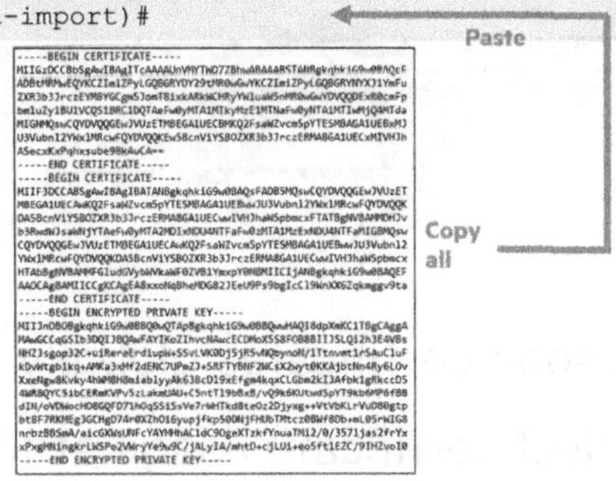

Figure 2-35: Uploading a CA-Signed Certificate on an AOS-CX Switch

If you have created the certificate offline, you import the file through the terminal. First make sure that the certificate file is formatted properly in PEM (Base64) format. The file needs to include the certificate, the intermediate certificate that signed it, and the encrypted key. Copy the full contents.

Enter this command in the switch certificate profile context: **import terminal ta-profile <*name*> password <*key_password*>**. The key password is the password used to encrypt the private key file. Paste the contents of the PEM file into the command line. Press **[Ctrl+D]** to accept the certificate.

You can alternatively create a CSR directly on the AOS-CX switch and submit that to the CA. Refer to the switch documentation if you want to use that option.

Scaling Certificate Installations with HPE Aruba Networking

You can easily use the processes about which you just learned to install a certificate on one network infrastructure device. However, many customers need to install certificates across many devices. And they need to repeat that process to install new certificates before the old ones expire. HPE Aruba Networking offers several ways to simplify the deployment of certificates at scale.

HPE Aruba Networking Central

As you have learned, if you are using HPE Aruba Networking Central to manage your devices, you upload any certificates that you want to use in the global settings (see Figure 2-36). If you use features like wildcards, you can upload the same certificate for multiple devices. You can then specify how you want to use each certificate in the device configuration settings at group level or individual device level.

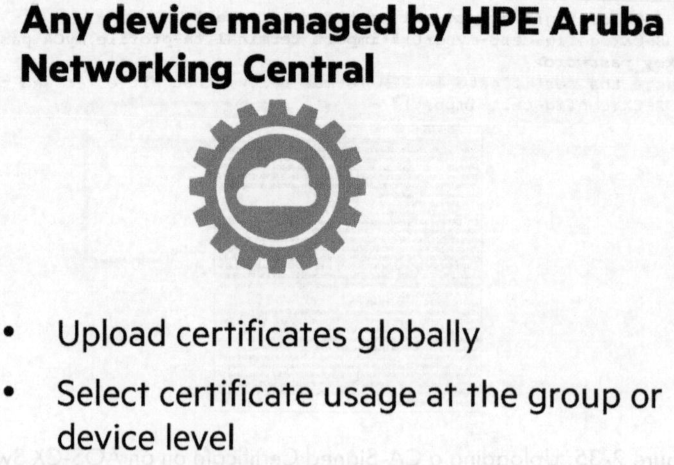

Figure 2-36: Scaling Certificate Installations with HPE Aruba Networking Central

NetEdit

NetEdit provides centralized management for AOS-CX switches, as shown in Figure 2-37. You can easily discover the AOS-CX switches within NetEdit. You can then create CLI-based configuration templates and apply them to your switches. In this way, you can install certificates on multiple switches at once. NetEdit also supports features such as configuration validation and compliance checks.

To learn more about using NetEdit, attend other HPE Aruba Networking training.

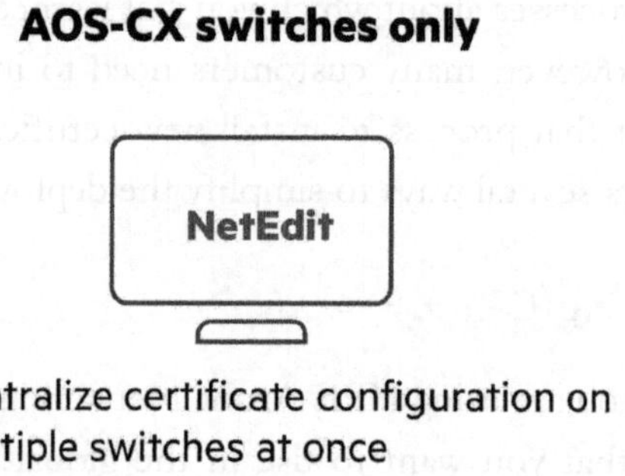

Figure 2-37: Scaling Certificate Installations with NetEdit

EST in HPE Aruba Networking Products

You can also use Enrollment over Secure Transport (EST) to manage certificates on HPE Aruba Networking devices more automatically. HPE Aruba Networking devices that support ESP include:

- HPE Aruba Networking gateways and MCs
- HPE Aruba Networking APs (for Instant APs, in Instant 8.7 and later)
- AOS-CX switches (AOS-CX 10.06 and later)

The device is the EST client. As shown in Figure 2-38, it creates an HTTPS session with an EST server, authenticating the EST server by its HTTPS certificate. The EST server also authenticates the client (it might not require authentication for all functions but does for enrolling for certificates). If the device passes authentication, the EST server helps it to enroll or re-enroll for certificates with a CA.

The EST client appends specific paths to the EST URL to make requests. For example, it uses /cacerts to download the CA chain for the PKI in which it is enrolling. It uses /simpleenroll to enroll for a certificate. In this message, the client sends its CSR, as well as proof that it owns the private key associated with the CSR public key. The EST server passes the CSR to the issuing CA and also passes the signed certificate back to the client.

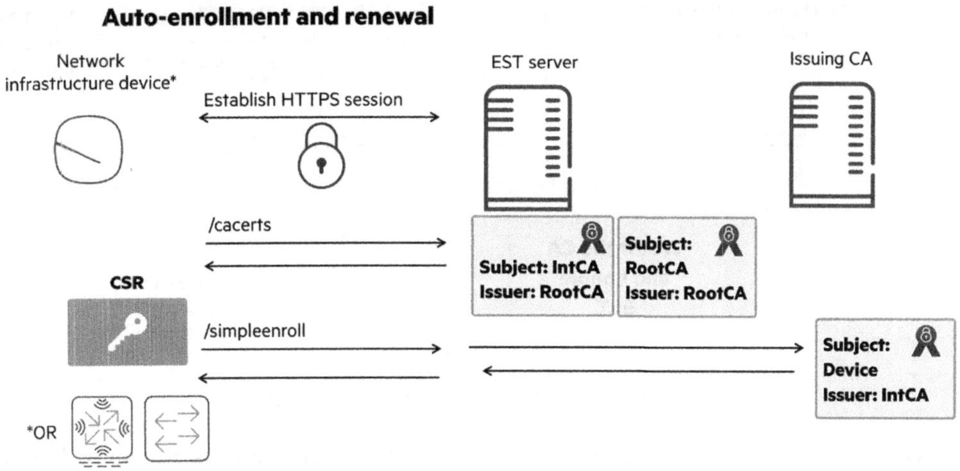

Figure 2-38: EST in HPE Aruba Networking Products

When the client needs to renew its certificate, it follows much the same process, but uses the /simplereenroll process.

All of these functions occur transparently after you have configured the network infrastructure device with the proper EST profile for contacting an EST server.

This option is preferred for wider scalability because it simplifies the deployment of certificates. Automatic re-enrollment before a certificate expires also decreases the risks that admins will forget to renew a device's certificate, which would cause functions that use that certificate to fail. Using EST does require the deployment of an EST server, which can communicate with the issuing CA.

ClearPass Onboard can act as an EST server for the company's private CA, if the company chooses.

Tips for Using Onboard EST

If the company wants to use EST to automate the deployment of private CA-signed certificates, you can set up HPE Aruba Networking ClearPass Onboard as the EST server.

First, you create the CA on Onboard. As you will learn in more detail later in this study guide, Onboard can act as a root CA or an intermediate CA. You should choose the root CA option if you want Onboard to act as an EST server for certificates that it issues itself. These certificates will only be valid for use within the ClearPass cluster. You should choose the intermediate CA option if you want Onboard to act as an EST server for a PKI that the company has already established.

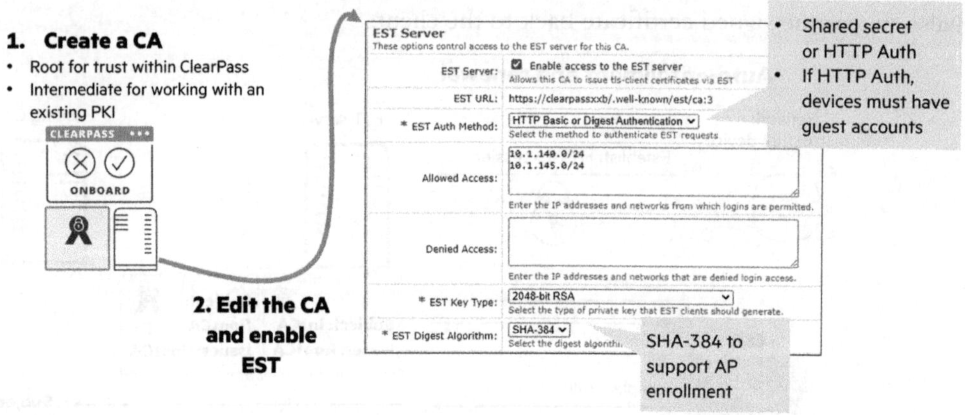

Figure 2-39: Tips for Using Onboard EST

After you create the CA, you can edit the settings and enable EST, as shown in Figure 2-39. The EST Auth Method can be **Shared Secret** or **HTTP Basic or Digest Authentication**. If you choose the **HTTP Basic or Digest Authentication** option, you must set up ClearPass guest accounts for the devices that will act as EST clients.

You can control which devices can access the EST services by adding subnets to the Allowed Access and Denied Access settings.

The EST Key Type and EST Digest Algorithm settings dictate the settings that EST clients must use in the private keys and CSRs that they generate. Make sure to select **SHA-384** for the EST Digest Algorithm because HPE Aruba Networking APs do not support longer SHA digests.

EST Prerequisites on HPE Aruba Networking Devices

All HPE Aruba Networking devices have similar prerequisites for using EST (see Figure 2-40).

You must install the root CA certificate for the EST server's HTTPS certificate as a trusted CA certificate. Also ensure that the HPE Aruba Networking device and the EST have synchronized time by setting up NTP on the device. Typically you need to specify the EST server by FQDN because the name that you specify must match the subject name and DNS SAN within the EST server's HTTPS certificate. Therefore, you must specify a DNS server on the device so that the device can resolve the FQDN.

If a firewall lies between the HPE Aruba Networking device and the EST server, make sure that it permits communications on the correct port. Typically the communications use TCP port 443, but check with the EST server admins to make sure.

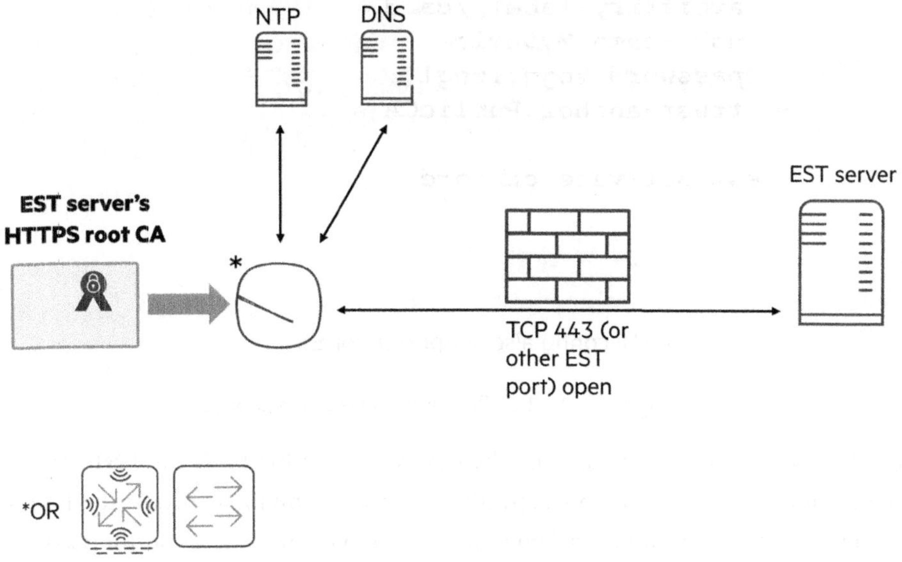

Figure 2-40: EST Prerequisites on HPE Aruba Networking Devices

EST Profile Configuration on HPE Aruba Networking Devices

You can then configure the EST profile.

AOS-8 Devices

See Figure 2-41 for an example of an EST profile.

You must specify the **server-host**, which is the FQDN at which the device reaches the EST server. Note that the device automatically appends this standard path after the specified FQDN: **/.wellknown/est**. You must also specify the server-port, at which the device reaches the EST server.

The arbitrary-label helps the EST server select the correct CA and settings for this device. The device appends the arbitrary-label at the end of the URL. For example, this device will contact the following URL: clearpassxxb.example.com/.wellknown/est/ca:3.

```
est profile onboard
 server-host clearpassxxb.example.com
 server-port 443
 arbitrary-label /ca:3
 user-name MyDevice
 password %ojg;lzagl
 trust-anchor PublicCA

est-activate onboard
```

*Web UI config also supported for EST

Figure 2-41: EST Profile on an AOS-8 Device

You must also specify the credentials that the device uses to log in. If the EST server uses HTTP basic or digest authentication, you must specify a username and password. If the EST server uses a shared secret, you must specify a challenge password. You must also tie the EST profile to the trusted root CA certificate for the EST server's HTTPS certificate.

You can then activate the EST profile. The AOS device then contacts the EST server, downloads the root CA certificate, and enrolls for its own certificate signed by a CA within that PKI.

Note that MCs support Web UI configuration for the EST profile. Follow the same guidelines for determining the correct settings.

AOS-CX Switches

As shown in Figure 2-42, the EST profile settings on AOS-CX switches are similar to those on AOS devices. Note, however, that you specify the EST server by URL. You must include the "https://" and "/.wellknown/est" portions of the URL in the command, as well as the port. Note that, even though, the display does not fit on one line in the example shown in Figure 2-42, this is one command: **url https://clearpassxxb.example.com:443/.wellknown/est**.

AOS-CX switches support HTTP basic or digest authentication, not shared secrets. You specify the username and password in one command: **username <*username*> password plaintext <*password*>**.

```
crypto pki est-profile onboard
  url https://clearpassxxb.example.com:443
  /.wellknown/est
  arbitrary-label /ca:3
  username MyDevice password plaintext
    %ojg;lzagl


crypto pki certificate mycert
  key-type rsa key-size 2048
  subject common-name P46-T14-6300
  enroll est-profile onboard
```

Figure 2-42: EST Profile on an AOS-CX Switch

You can also use the **reenrollment-lead-time** command within the EST profile to configure how long before the certificate expires the switch attempts to reenroll.

To activate the EST profile, you create a certificate profile on the switch with this command: **crypto pki certificate <*name*>**. Set the key algorithm and key size (or curve size for ECDSA), making sure to specify settings that match with the EST server requirements. Also set the subject name for the CSR. The example in Figure 2-42 shows just a common name for the subject name, but you can add additional fields. You can then activate the enrollment; enter this command: **enroll est-profile <*EST_profile_name3*>**.

CA-Signed Certificates for RadSec

If you want to implement RadSec between HPE Aruba Networking devices and CPPM, you need to use a CA-signed certificate.

You must obtain certificates from your choice of CA. To minimize costs, you should usually use a private CA to issue these certificates. This option also lets you choose longer validity periods or wildcards for CNs or SANs.

You then install the certificates on the infrastructure devices, such as APs or switches, and select the RadSec usage for the certificates. If you have IAPs, you can install a single certificate for the entire cluster. If you are using HPE Aruba Networking Central, you might use the same certificate across a group of APs or other devices.

As shown in Figure 2-43, on CPPM, you must complete two steps to ensure that CPPM trusts the certificate. You must add the CA that signed the devices' certificates to CPPM's trust list with the RadSec usage. You must also create one or more network device entries that match your devices' IP addresses.

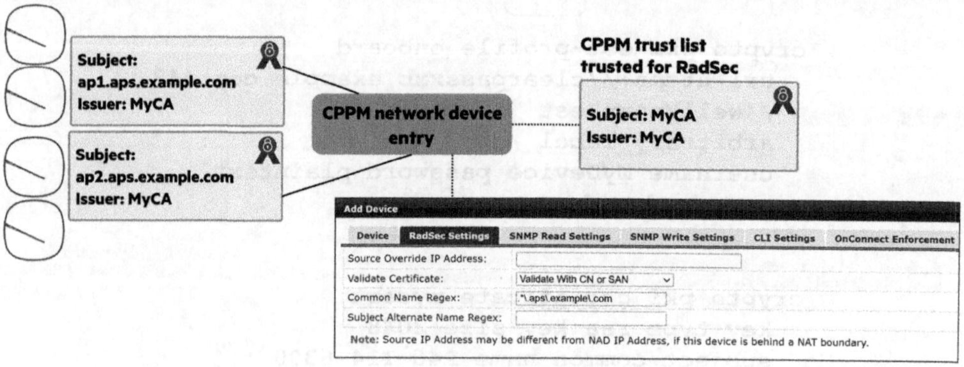

Figure 2-43: CA-Signed Certificates for RadSec

You then use the **RadSec Settings** tab to specify which certificates CPPM trusts. You have three choices:

- **No Authorization Checks**—CPPM only checks that the certificate is signed by a trusted CA, not expired, and not revoked. HPE Aruba Networking generally recommends against using this option as it is less secure.

- **Validate With CN or SAN**—You must specify the common name or SAN within the certificate or certificates on your devices. You can use regular expressions to match multiple subject names or SANs. In this example, **.*\.aps\.example\.com** matches any CN that ends with **aps.example.com**. This option lets you use a single network device entry that matches the CNs or SANs for multiple valid certificates.

- **RFC Compliant (Serial + Issuer)**—You must specify the Issuer's DN from a drop-down menu which includes the DNs for all RadSec usage trusted CAs on CPPM. You must also specify the serial number for that certificate. You can also put in a regular expression for the certificate's CN or SAN. Some companies' security policies require this option. When you use this option, each network device entry can match only one certificate. However, you might be able to install the same certificate on multiple devices.

Summary

Congratulations! You should now be able to explain how certificates exist within a PKI. You understand how devices validate certificates and what issues might cause validations to fail. And you can use OpenSSL to create CSRs and perform other tasks.

You also know how to apply these principles and set up certificates properly on your HPE Aruba Networking solutions, including ClearPass, APs, MCs, gateways, and AOS-CX switches.

Learning checks

1. What is required for a client to trust a web server's certificate? (Select two)
   a. The client has the certificate's public key installed on it in advance
   b. The certificate was issued within the previous six months
   c. The client can trace the certificate back to a trusted root
   d. The certificate DNS SAN matches the hostname accessed by the client
2. What is one requirement for joining a Subscriber to a ClearPass cluster?
   a. The Subscriber trusts the entire chain for the Publisher's RADIUS/EAP certificate.
   b. The Subscriber trusts the entire chain for the Publisher's HTTPS certificate.
   c. The Subscriber has an HTTPS certificate with its IP address listed as a DNS SAN.
   d. The Publisher has an HTTPS certificate with its IP address listed as a DNS SAN.
3. What correctly describes the Device Certificate on HPE Aruba Networking APs?
   a. It is signed by an HPE Aruba Networking CA and installed in a TPM.
   b. It is a self-signed certificate, which should be replaced in production.
   c. It uses the AP's default IP address as its CN in the subject name.
   d. It is signed by a CA of the company's choice and must be installed manually.

Note

For answers to all learning checks, refer to the Appendix.

3 Implement Certificate-Based 802.1X

EXAM OBJECTIVES

✓ Use HPE Aruba Networking ClearPass Policy Manager (CPPM) to provide AAA for WLANs

✓ Deploy certificate-based authentication for users and devices

Assumed Knowledge

- Authentication, Authorization, and Accounting (AAA) framework and basics of protocols such as RADIUS and TACACS+
- Forms of Layer 2 authentication such as 802.1X, including common EAP methods such as PEAP and EAP-TLS
- Fundamental ClearPass Policy Manager (CPPM) components, such as services, roles, enforcement profiles, and enforcement policies

Overview

HPE Aruba Networking recommends that companies use certificates to authenticate all users and clients, as far as is possible. In this chapter, you will review the Extensible Authentication Protocol (EAP) methods that support certificate-based authentication, as well as the benefits of certificate-based authentication. You will then learn about implementing EAP-TLS to authenticate Windows domain computers.

If a company needs to support a Bring Your Own Device (BYOD) environment with unmanaged devices, you need to onboard these devices to install certificates and configure the proper authentication methods on them. You will learn how to use HPE Aruba Networking ClearPass Onboard and other HPE Aruba Networking solutions to support this use case.

Finally, you will learn how to use Tunneled EAP (TEAP) to support machine and user authentication within a single EAP session. You will see how this method better enables you to distinguish between users on domain computers and users on other types of devices. And you will learn how to set up TEAP on CPPM.

EAP Review

The first section in this chapter briefly reviews EAP authentication.

Review EAP Methods

While WPA3-Enterprise security requires 802.1X, 802.1X itself supports multiple EAP methods, and not all EAP methods deliver the same level of security.

In Figure 3-1, you see a comparison between PEAP MSCHAPv2 and EAP-TLS. Both of these EAP methods require the server to authenticate with a certificate. However, with PEAP MSCHAPv2, the supplicant uses a username and password to authenticate. EAP-TLS, on the other hand, requires the supplicant to authenticate with a certificate.

EAP method	Credentials
Protected EAP (PEAP) MSCHAPv2	Server = Server Certificate Supplicant = Username and password
EAP-TLS	Server = Server Certificate Supplicant = Client Certificate

Figure 3-1: Review EAP Methods

Review the Benefits of Certificate-Based Authentication

Why does it matter than EAP-TLS requires the supplicant to authenticate with a certificate? Take a moment to review the benefits of certificate-based authentication.

Users often choose insecure passwords, which hackers can target with dictionary attacks in which the hacker simply tries many common passwords. If the password is short enough, the hacker might be able to crack it with a brute force attack. Certificates, on the other hand, use public/private keypairs. It is cryptographically infeasible for hackers to derive the private key from the public key. So, as long as private keys are stored securely, the certificate is protected against hackers.

Users often share passwords or leak them in other ways. Certificates are stored on a specific machine, making it much harder for users to extract them and use them inappropriately.

Certificates provide greater control for admins. It becomes easier for admins to control exactly which users and devices can authenticate.

Users often use the same passwords for many accounts. If one account is compromised, the user might not even know that hackers have learned the password. However, the password is now compromised. With certificates, the certificate is stored on a machine. If the machine is lost or stolen, the user is aware of the fact. Admins can revoke the certificate so hackers cannot use it to obtain unauthorized access.

Certificates also provide ease of use for users. Once the certificate is properly deployed to the user's device, the device can automatically use it to authenticate. The user does not have to worry about forgetting the password or having the periodically renew the password. Later in this chapter you will learn about options for automatically deploying and updating certificates.

EAP-TLS in a Windows Domain

This section explains how to implement EAP-TLS to authenticate clients in a Windows domain.

Certificates within a Windows Domain

You have been examining how certificates work in general. You will now focus in on a particular certificate use case that you will often encounter in the real world. The company has a Windows domain. As shown in Figure 3-2, employees are Domain Users. They connect to the network on desktops and laptops that are also joined to the domain.

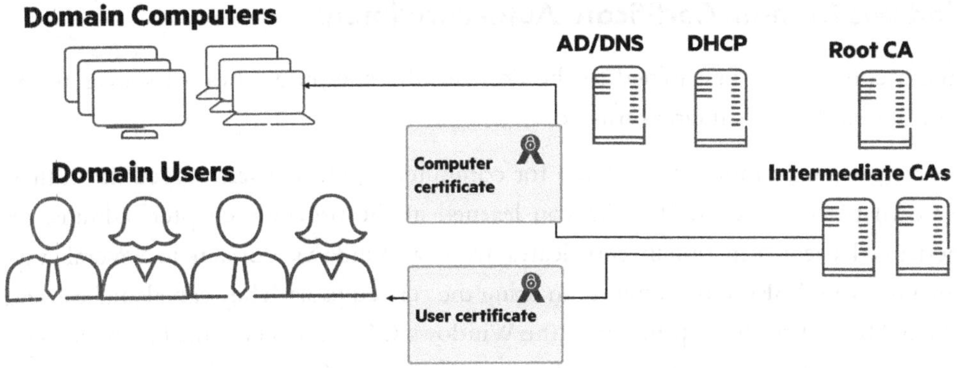

Figure 3-2: Certificates within a Windows Domain

The company has a Windows certificate authority (CA) service, which consists of a root CA and several intermediate CAs.

The company uses its CAs to issue certificates to internally-facing servers and network infrastructure devices. The company now wants to deploy certificates to domain computers and users so that they can authenticate. The next few sections take you through how this process could work.

Windows Certificate Stores

First it can be helpful to review how Windows operating systems store certificates. Each Windows OS has two main certificate stores, as shown in Figure 3-3.

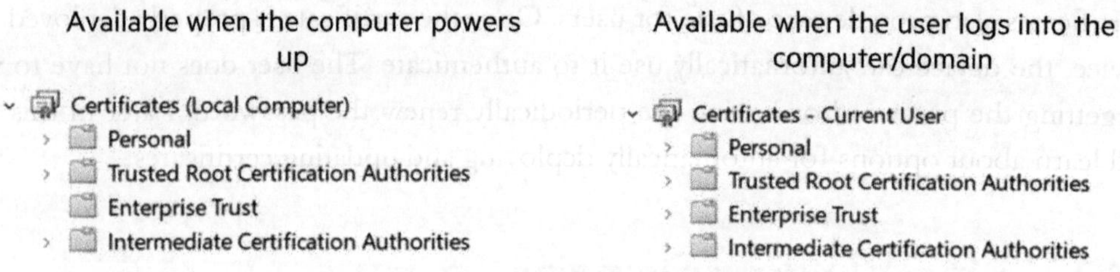

Figure 3-3: Windows Certificate Stores

The Local Machine store includes certificates that belong to the computer itself, as well as CA root certificates that are trusted by the computer. The computer can always use certificates within the Local Machine store.

The Current User store includes certificates that belong to the user who is currently logged into the computer, as well as CA root certificates trusted by that user. This store only becomes available for the computer to use after the user has logged in.

Windows Domain Certificate Auto-Enrollment

When a computer is first joined to the domain, the computer always receives the Windows CA certificates and installs them as trusted.

However, it can be more complicated for computers, and the users logged in on them, to receive their own personal certificates. As you learned in the previous chapter, admins, or even more experienced users, can obtain certificates from a Windows domain in several ways. They can manually create CSRs and submit them using the certreq tools. They can also use a wizard available in the MMC certificate snap-in. Or, if the Windows CA has a Web Enrollment server, they can log into those web pages and obtain certificates in that way. However, these methods can be complex for users, and, if admins need to install all certificates, the process is not scalable.

Auto-enrollment provides the simplest and most scalable ways for computers and users to receive certificates from a Windows CA. With auto-enrollment, Windows CA admins create certificate templates that support auto-enrollment for computer and user certificates. Domain admins also create group policy objects (GPOs) to enable auto-enrollment for Domain Computers and for user groups.

With these policies in place, computers and users receive certificates automatically, as shown in Figure 3-4. Computers joining the domain or powering up check for domain computer GPOs. When the computer applies the auto-enrollment GPO, it creates a private key and CSR. It submits the CSR to the Windows signing CA, which signs the certificate and returns it to the computer. The computer then installs the signed certificate in the Local Machine store. This process completes automatically.

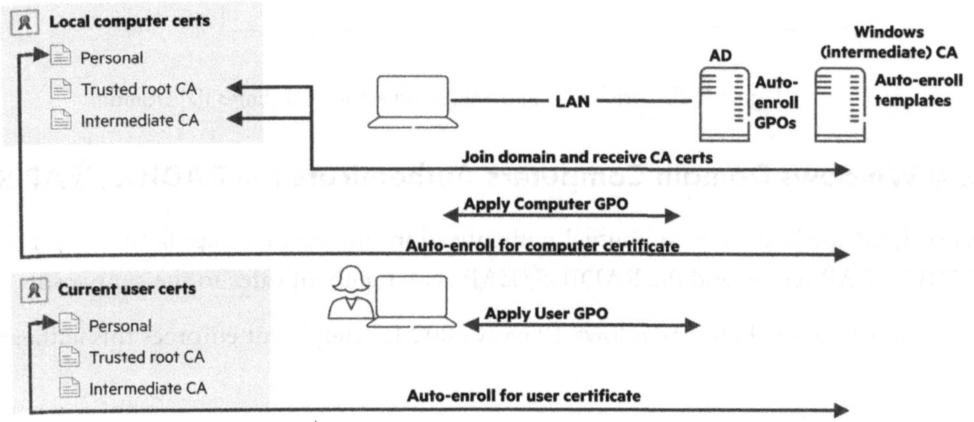

Figure 3-4: Windows Domain Certificate Auto-Enrollment

When a user first logs in on a computer, the user group GPO manages a similar process, as shown in Figure 3-4. The computer creates a private key and CSR for the user certificate. It submits the CSR to the signing CA. And, when it receives the signed certificate, the computer installs it in the Current User store.

Certificates in a Microsoft Azure AD (Entra ID) Domain

Some companies are now using Azure Active Directory (AD), a cloud-based version of AD, recently renamed Microsoft Entra ID. Like on-prem AD, Entra ID enables companies to manage their users and devices, organizing them into groups with different privileges. However, unlike a traditional Windows domain, Entra ID can manage mobile devices, as well as Windows PCs and servers. It does so by integrating with Microsoft Endpoint Manager, a Mobility Device Management (MDM) solution (which used to be called Intune).

To support 802.1X with EAP-TLS in an Entra ID environment, Endpoint Manager should deploy certificates to all the devices within the domain, as shown in Figure 3-5. Endpoint Manager can work with a Windows CA to deploy the certificates, and it supports Simple Certificate Enrollment Protocol (SCEP) and Public Key Cryptography Standards (PKCS) for deploying the certificates.

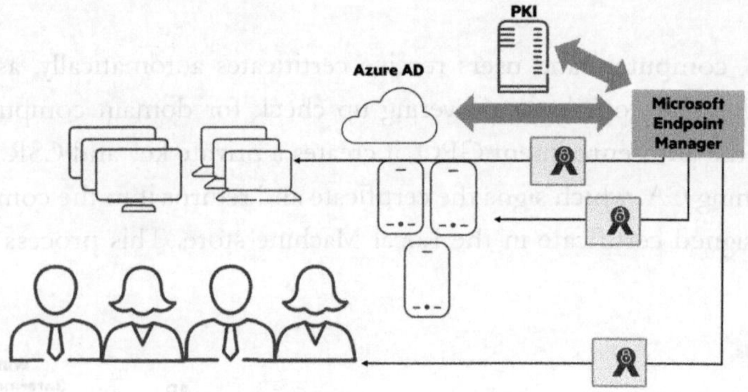

Figure 3-5: Certificates in a Microsoft Azure AD (Entra ID) Domain

How Windows Domain Computers Authenticate the RADIUS/EAP Server

Secure EAP methods impose mutual authentication: the 802.1X supplicant authenticates to the RADIUS/EAP server, and the RADIUS/EAP server authenticates to the supplicant.

Next you will review how a Windows 10 native 802.1X supplicant enforces this authentication.

Default settings

At default settings in Windows 10, the supplicant attempts to validate the RADIUS/EAP server's certificate. As shown in Figure 3-6, it has a list of potential trusted root CAs for validating the RADIUS/EAP server certificate, which comes from the Current User and Local Machine trusted root CA store. As you learned earlier, this store includes many well-known public CAs. If the machine is a part of a domain, and the domain has a private CA, the store also includes the domain CA certificate.

When the Windows supplicant using default settings tries to connect to a network that enforces 802.1X, it checks the RADIUS/EAP server certificate and prompts the user to trust the root CA that signed the RADIUS/EAP server certificate. The behavior is the same for a root CA certificate that exists in the machine's trusted root CA stores and for a root CA certificate that does not. If the user chooses to trust the certificate, the associated root CA will then be trusted for future authentications on this network. However, if the supplicant later receives a RADIUS/EAP server certificate with a chain of trust back to a different root CA, the supplicant will again be prompted to trust that root CA.

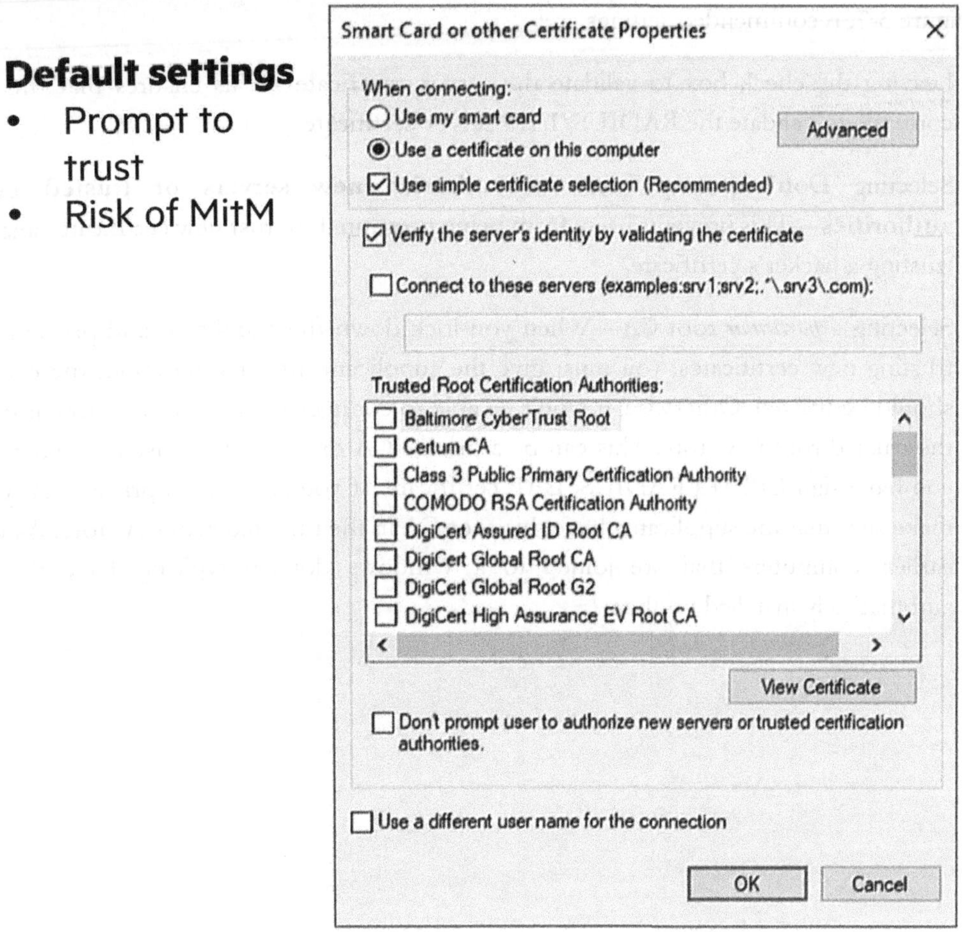

Figure 3-6: Default Settings for EAP on Windows Clients

This default behavior can introduce security vulnerabilities. You should avoid training users that it is normal to trust a new root CA in response to a prompt, and it is best practice to avoid situations in which users are ever prompted to do so. Often users do not have the information or expertise that they need to make the correct decisions. Some might call IT and report a problem when no problem truly exists. But, even worse, many will simply trust any certificate. They might end up trusting certificates submitted by hackers in a man-in-the middle attack.

Secure Settings

To close the vulnerability, the supplicants must be configured to trust specific certificates. As shown in Figure 3-7, recommended settings are:

- Leaving the check box to validate the server certificate—This ensures that the supplicants continue to validate the RADIUS/EAP server certificate.
- Selecting **Don't prompt user to authorize new servers or trusted certification authorities**—This prevents users from being prompted to trust new certificates and potentially trusting a hacker's certificate.
- Selecting a *particular* root CA—When you lock down the supplicant and prevent users from trusting new certificates, you must give the supplicant information about the certificate that should be trusted. Otherwise, it won't be able to connect at all. You can select any root CA in the trusted root CA store. This can be a private CA or public CA based on what type of CA you have sign CPPM's RADIUS/EAP certificate. If you are using a private CA, you need to make sure that the supplicants have this root CA in their trusted root CA store. As you learned earlier, computers that are joined to a Windows domain typically have this certificate automatically installed on them.

Figure 3-7: Recommended Settings for EAP on Windows Clients

- Adding a valid server name or names—When you select this setting, the supplicant will check the CN in the RADIUS/EAP certificate and only validate certificates that have the correct CN or SAN. Note that the supplicant cannot validate anything beyond the CN. With HTTPS, the client can further check that the CN or SAN matches with the website the client is accessing, but 802.1X operates at Layer 2, so the supplicant has no awareness of CPPM's IP address or FQDN. Therefore, the security of this setting comes *in conjunction with* the specification of the trusted root CA. In other words, a hacker could theoretically create a self-signed certificate or even a CA-signed certificate with the valid server CN. So the supplicant should check both that the certificate has the correct CN and comes from the correct CA. Similarly, a hacker could theoretically obtain a valid certificate from the correct CA, *but* not with the correct CN. By specifying both the server name and the trusted root CA, you ensure greater security than by specifying either one of the settings alone.

Configuring Secure Settings

You can configure the recommended settings manually. However, doing so for all devices would impose a large burden on IT staff. The best approach for automating the deployment of these settings depends on the type of device. For the domain joined devices that you have been examining in this section, domain admins can use GPOs to push the 802.1X configuration settings to devices, much as they use GPOs to auto-enroll the devices for their certificates.

You will learn how to use ClearPass Onboard to set up certificates and 802.1X settings for unmanaged devices later in this chapter.

High-Level View of EAP-TLS for Windows Domain Users/ Computers

Now you will consider the other side of the authentication: how CPPM authenticates the Windows native 802.1X supplicant using EAP-TLS and AD as the authentication source.

The Windows supplicant looks for a personal certificate that it can use for EAP-TLS authentication. If no user is logged in, it uses the computer certificate from the Local Machine store. However, as long as a user is logged in, the supplicant looks for a personal certificate in the Current User store.

When CPPM uses EAP-TLS as the authentication method, it can actually run two checks, as shown in Figure 3-8. First, it authenticates the client by validating the client's certificate. It makes sure that it can trace the root of trust for the certificate to its EAP root CA certificate. In this example, because the domain uses a Windows CA, you should upload the Windows root CA certificate and set it to the EAP type. This step validates that the user has the identity that they are claiming.

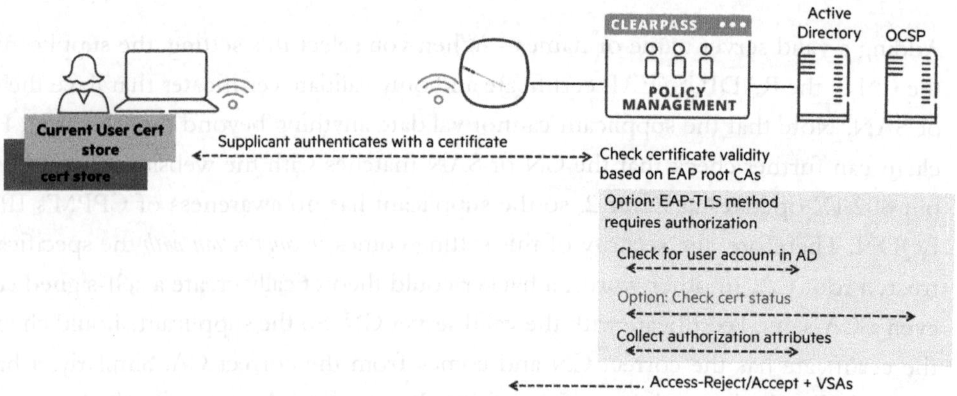

Figure 3-8: High-Level View of EAP-TLS for Windows Domain Users/ Computers

CPPM performs the next check if the EAP-TLS method has the **Authorization Required** check box selected. The default [EAP-TLS] method uses this setting. CPPM checks the authentication/authorization source defined in the service, here an AD server, for an account that matches the client's username, which is derived from the certificate. This step validates that the user is truly authorized to access the network.

CPPM also collects authorization attributes from the authentication source, as you learned in an earlier chapter. It then returns an Access-Reject or Access-Accept plus VSAs to the authenticator, based on the results of the enforcement policy.

Ensuring the Authorization Check Succeeds

As shown in Figure 3-9, a user certificate received from a Windows CA typically includes the user account's userPrincipalName. When the Windows client uses that certificate as part of EAP-TLS authentication, it takes the EAP-TLS username from the userPrincipalName. This name includes the domain name.

By default, when CPPM checks for an account in an AD authentication source, it checks only the sAMAccountName. The sAMAccountName does not include the domain name. Therefore, unless you take additional steps, the authorization check will fail. You can take one of two approaches to prevent such failures, as shown in Figure 3-9.

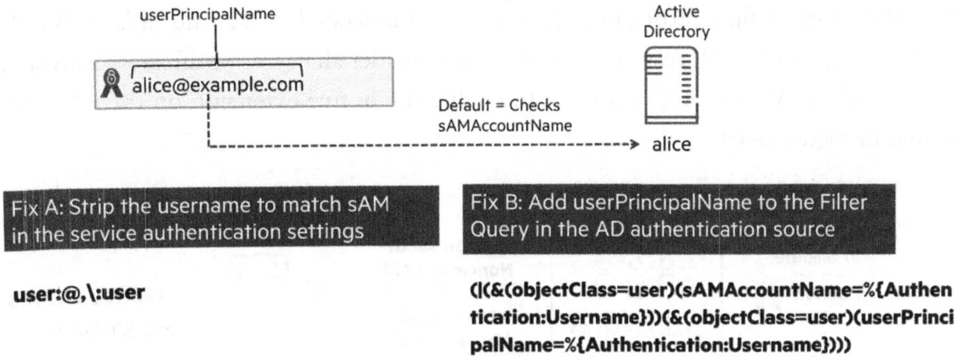

Figure 3-9: Ensuring the Authorization Check Succeeds

Fix A: Strip the Username

At the service Authentication tab, you can configure rules for stripping the user name. This rule works for most user certificates: user:@. This rule tells CPPM that it should use the characters in front of @ as the username. Sometimes the username includes the domain name after a backslash. This rule tells CPPM to find the name after the backslash: \:user.

Separate the rules with a comma (,).

Fix B: Add userPrincipalName to the Filter Query

Rather than creating stripping rules, which might not work in all scenarios, you can configure CPPM to query AD for both sAMAccountNames and userPrincipalNames. You configure this setting within the authentication source for this AD instance. Within the **Configuration > Authentication > Sources** page, edit the authentication source. Then go to the **Attributes** tab and edit the authentication attributes. Change the Filter Query to:

(|(&(objectClass=user)(sAMAccountName=%{Authentication:Username}))(&(objectClass=user)(userPrincipalName=%{Authentication:Username})))

Ensuring OCSP Check Works Against Windows CA

When CPPM sends an OCSP request, it includes the nonce extension. In other words, CPPM includes a pseudo-randomly selected value, a nonce. The OCSP responder must include the same nonce in its response. The nonce extension helps to prevent replay attacks, in which hackers capture and replay a valid OCSP response to trick the OCSP requester into accepting a revoked certificate.

By default, the Windows OCSP responder for the Windows CA does not support the nonce extension. This fact can cause issues when you configure CPPM's EAP-TLS method to use OCSP and CPPM is validating certificates signed by a Windows CA. At the default Windows settings CPPM will reject all OCSP responses, and it will consider all users' certificates as invalid. To prevent this issue, ask the Windows CA admins to enable the nonce extension on the OCSP responder, as indicated in Figure 3-10.

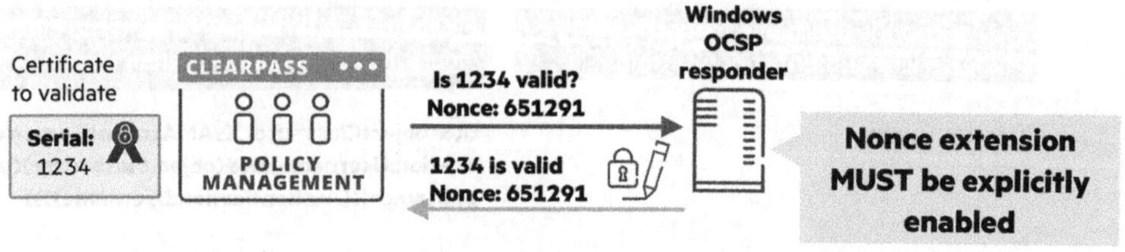

Figure 3-10: Ensuring OCSP Check Works Against Windows CA

Alternatively, you can configure the CPPM server not to require the nonce. You configure that setting in the individual server configuration (click the server in **Administration > Server Manager > Server Configuration**). In the **Service Parameters** tab, select **RADIUS server**. Set **Reject if OCSP response does not have Nonce** to **False**. You might also need to configure CPPM to not send the nonce at all in those same settings.

For your reference, you enable the nonce extension as follows:

1. On the OCSP server, open the Online Responder Management console.
2. Right-click **Revocation Configuration** and select **Edit Properties**.
3. Click the **Signing** tab.
4. Select the **Enable NONCE extension support** check box.
5. Click **OK**.

Custom Attribute to Collect AD Account Status

As you saw, when the EAP-TLS method enables authorization, CPPM checks that an account with the client's username exists on AD. However, this check does not distinguish between enabled and disabled accounts. For true security, you need to ensure that CPPM only authorizes a user or machine for access if its account is enabled.

First you must ensure that CPPM collects the account status as an authorization attribute from AD. You must edit the authentication source to add it. As shown in Figure 3-11, in the CPPM UI, go to **Configuration > Authentication > Sources**. Find your Active Directory source and click it to edit it. Then go to the Attributes tab and edit the authentication attributes. There you can add the attribute. The correct Name is userAccountControl. For Alias Name, you can use any intuitive name such as AccountStatus. You reference the Alias Name when you create conditions. The attribute data type is String. Select the Attribute check box to indicate that this is an authorization attribute that CPPM should collect from AD.

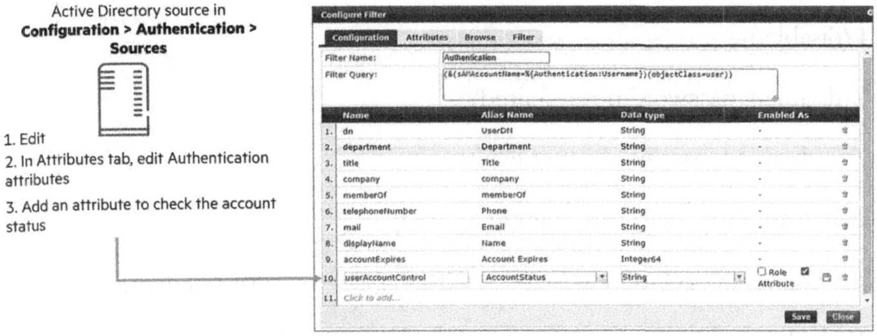

Figure 3-11: Custom Attribute to Collect AD Account Status

Using the Custom Attributes in CPPM Policies

AD returns the attribute value as a string. The string is a number based on flags that are enabled or disabled on the account. The strings typically most relevant to CPPM are:

- For enabled machine accounts
  - 4096
- For disabled machine accounts
  - 4098
- For enabled user accounts
  - 512 (enabled)
  - 544 (enabled, password not required)
  - 66048 (enabled, password never expires)
  - 262656 (enabled, smart card logon required)
  - 262688 (enabled, smart card logon required, password not required)
  - 328192 (enabled, smart card logon required, password never expires)
  - 328224 (enabled, smart card logon required, password never expires, password not required)
- For disabled user accounts:
  - 514 (disabled)
  - 546 (disabled, password not required)
  - 66050 (disabled, password never expires)
  - 262658 (disabled, smart card logon required)
  - 262690 (disabled, smart card logon required, password not required)
  - 328194 (disabled, smart card logon required, password never expires)
  - 328226 (disabled, smart card logon required, password never expires, password not required)

You can then specify these attributes and values within role mapping policies or enforcement policies.

Consider an example with two roles:

- "AD-user-enabled-cppm" for users whose accounts are enabled
- "AD-machine-enabled-cppm" for computers with enabled accounts

As shown in Figure 3-12, the role mapping policy has rules to assign those roles.

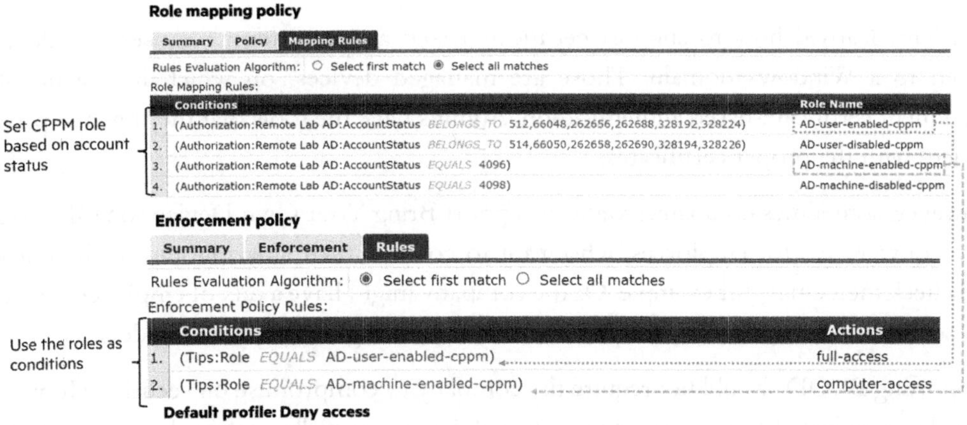

Figure 3-12: Using the Custom Attributes in CPPM Policies

You should be careful when creating those rules to include the correct statuses based on the company's domain and requirements. Some companies will want to permit access to any enabled user account, regardless of other flags. The role mapping rule shown in this example does just that by using the "BELONGS_TO" operator and a list of many values in the rule that assigns the "AD-user-enabled-cppm" role. You then avoid locking users out of the network if domain admins enable or disable a particular, unrelated setting on their account. Other companies only set the password to never expire for special accounts such as the accounts that solutions like CPPM use to bind to the domain. Those companies might want to deny access to users with a 66048 or 66050 account status to prevent unauthorized users from connecting with those accounts. In that case, you might create an additional CPPM role with a name like "AD-user-password-never-expires" and deny access to that role. Communicate with domain admins and the security team to formulate rules that work for your company.

You can then reference roles within the enforcement policy. Here you see a very simplified example in which clients with the "AD-user-enabled-cppm" role receive full access and clients with the "AD-machine-enabled-cppm" role receive computer access. This policy does not have rules for clients with disabled accounts, so they will be denied access. However, it can still be useful to assign those roles so that you can see in Access Tracker if a user attempts to authenticate with a disabled account.

HPE Aruba Networking ClearPass Onboard for BYOD

Now you will learn how to use HPE Aruba Networking ClearPass Onboard to provision certificates to Bring Your Own Device (BYOD) clients.

Why ClearPass Onboard?

You have learned how to support certificate-based authentication for users on devices that are joined to a Windows domain. Those are managed devices, on which it is generally easier to implement certificate-based authentication. Admins have the tools that they need to provision those devices with the proper certificates.

However, sometimes companies want to support Bring Your Own Device (BYOD). BYOD might be a convenience for employees, who want to connect their smartphones to the corporate LAN and check their email, for example. Or the company might have a mix of employees and contractors; the contractors bring their own laptops, rather than using company-issued devices.

Supporting BYOD should not require the company to compromise on security. However, it can be a challenge to install certificates on unmanaged devices, as well as set up the devices with the proper settings to trust the RADIUS server certificate. The company might also struggle to provision BYOD clients with software that the company requires.

ClearPass Onboard helps companies overcome these challenges by making it easy for users to self-provision their BYOD devices with all the required certificates, settings, and applications. Onboard can provision Windows, Android, iOS, and MacOS clients.

Onboarding vs Guest Access

Before you begin learning about the onboard process, you need to make sure that you understand the difference between users who require onboarding and guests. In both cases, you might have users bringing smart devices and trying to get access to the network.

However, in onboarding, the typical users are people known to the organization, like company employees or university staff and faculty members. These individuals are bringing their personal devices to use on the secure network. Guests are typically users who are unknown to the organization and who do not exist in a corporate identity store, like Active Directory. For example, guest users include guests who check in and out of hotels, customers visiting a company's office, or visitors at a hospital. These users may bring their own personal devices.

Another difference is that in onboarding, you are trying to get employee mobile devices connected to a secure network like an 802.1X network; you want to control what devices they can use on this network. A guest network typically uses an open SSID, and anyone can connect to the network.

These differences are summarized in Figure 3-13.

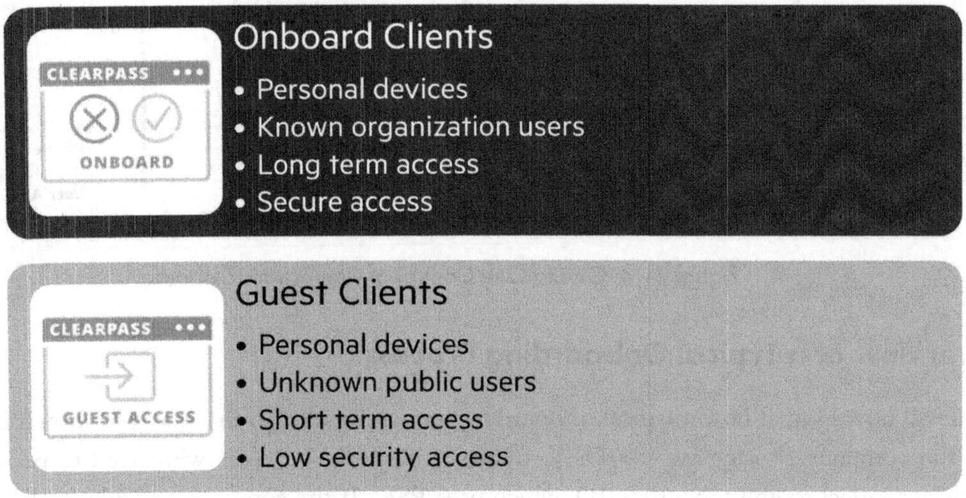

Figure 3-13: Onboarding vs Guest Access

One to One Control of Clients with Onboard

Now that you understand the difference between onboarding and guest access, you will explore Onboard in more detail.

Remember that clients are really a combination of the user identity and the device connecting to the network. In a conventional network that authenticates users with usernames and passwords, the same user owns all of the clients and completes the authentication.

Conventional network access says if the user is good, then all the users devices are good. What about a situation when the user loses one of his devices? If the client credentials are only the user account, then the only control you have is to block the user, but this will block all of the user's devices. As shown in Figure 3-14, this solution is not very functional.

Onboard, on the other hand, establishes a unique identity for each client, as shown at the right of Figure 3-14. The client will be issued a certificate that references the user plus the hardware. If the need arises, individual certificates can be revoked, and only one device is blocked from the network. Of course, because the certificate references back to the owner's user account, if the account gets disabled, then ClearPass will block all the devices.

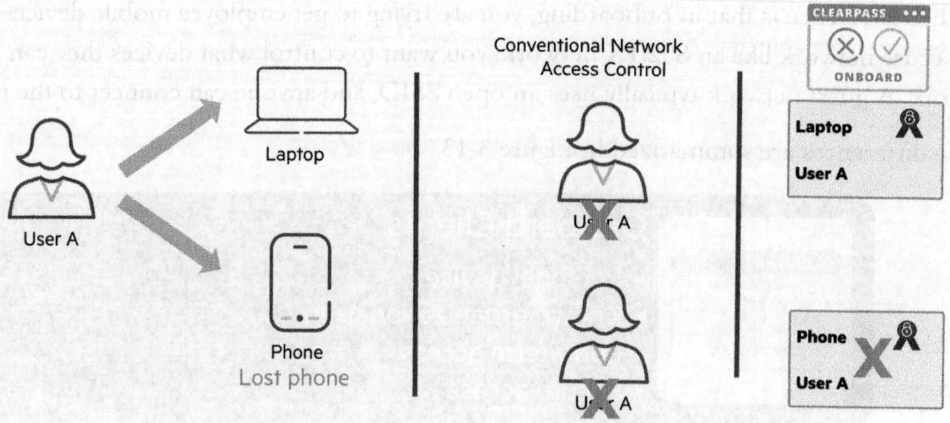

Figure 3-14: One to One Control of Clients with Onboard

Overview of a Typical Onboarding Process

You will now look at how a typical onboarding process works. In the example shown in Figure 3-15, the company is using two SSIDs to manage the process. Users who need to provision their devices connect to an Onboard SSID, which uses PSK. If the company provides guest access, you might also use the same SSID for guests and for onboarding devices. Users who already have certificates on their devices connect to a secure corporate SSID.

When a user needs to onboard a new device, such as a smartphone, this process occurs (see also Figure 3-15):

- The user connects the client to the Onboard or Onboard/Guest SSID. Because this SSID is open or uses PSK, the user can connect the device without any problems. If you alternatively had users connect to an SSID that implements 802.1X with PEAP, the clients would need to trust CPPM's certificate. That can cause issues for some smartphones.
- The AP is set up to implement captive portal on the open SSID. It redirects the client's HTTP/HTTPS traffic to the portal indicated in the captive portal settings. If the SSID is only for onboarding clients, it can redirect the clients directly to the Onboard portal. Otherwise, it can redirect clients to a generic guest portal, which has a link to the Onboarding portal.
- The user logs into the Onboard Portal.
- Onboard checks with CPPM to authenticate the user; CPPM has an Onboard Authentication service for this purpose.

- As long as the user and device type are authorized for access, the user enters the portal. The user follows simple instructions there, and the client is automatically provisioned for access. This process typically involves installing a certificate on the client and configuring its network settings.

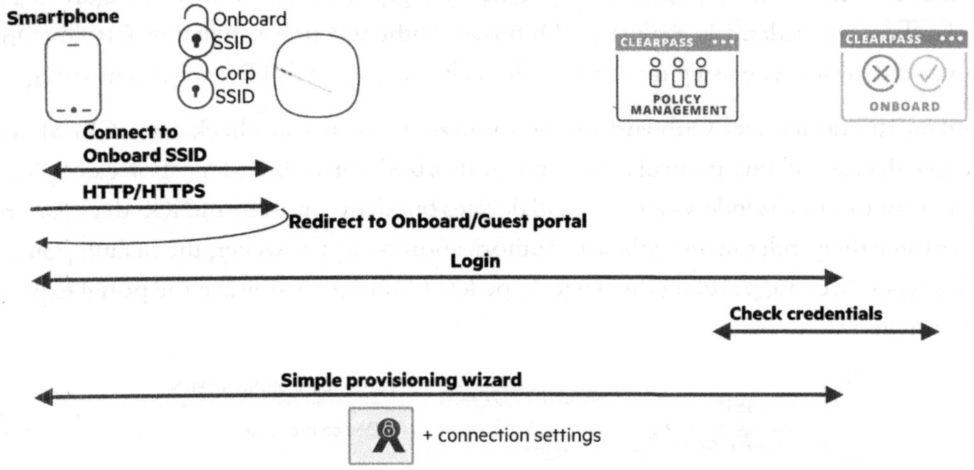

Figure 3-15: Steps in a Typical Onboarding Process

- As shown in Figure 3-16, at the end of the process, the user clicks **Connect**. The client now connects to the Corporate SSID for which the settings were configured on the client during the onboarding process. Using different SSIDs for onboarding clients and for the secure corporate WLAN also helps the reconnection process run more smoothly.
- The client can now authenticate with EAP-TLS using its certificate.
- CPPM assigns the client to the correct role, based on policies in the service. In this example, that role is Tellers.

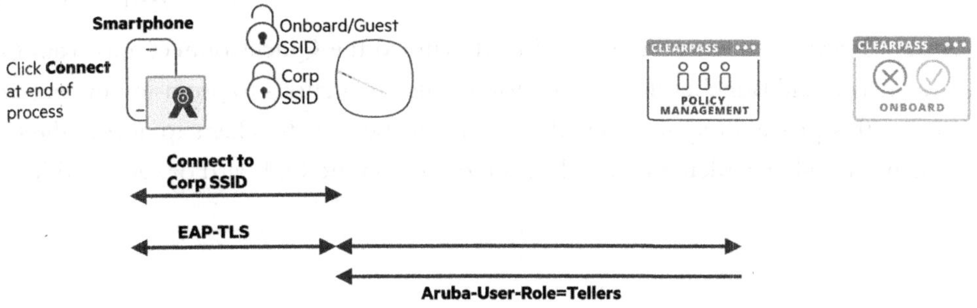

Figure 3-16: Connection after Onboarding

Onboard Experience Based on Device Type

You will now look in a bit more detail at the part of the process in which the user is interacting with the Onboard portal.

No matter what device type, the user begins by logging in. As shown in Figure 3-17, Onboard checks the user credentials against an Onboard Authentication service on CPPM. Onboard next profiles the device type using methods such as checking the HTTP User-Agent string.

Profiling the device lets Onboard do two things. First, it can check with CPPM to determine whether devices of this particular type are authorized for onboarding. For example, a company might want to allow Windows and MacOS devices but deny Android and iOS devices. The company can enforce those rules in an Onboard Authorization policy; however, the default policy permits all device types. Second, profiling the device type lets Onboard customize the portal experience based on that type.

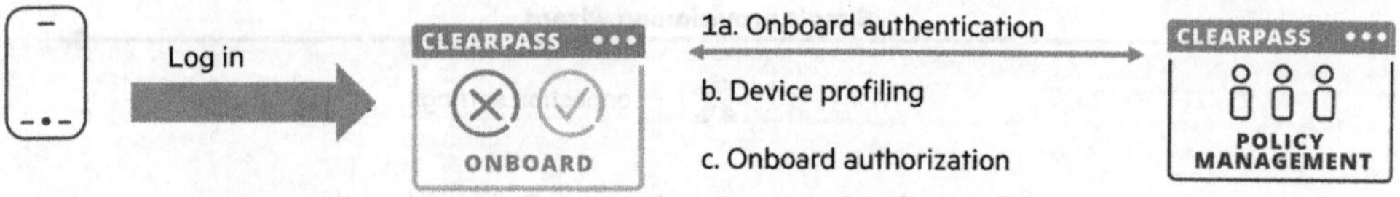

Figure 3-17: Onboard Authentication and Authorization

These three steps occur in the background; the user simply experiences logging into the portal (assuming that the user and device type are authorized for access).

What the user sees next depends on the device type, as shown in Figure 3-18.

On iOS devices, the user is prompted to accept a certificate. Onboard then automatically provisions the device.

On Windows devices, the user is prompted to download the QuickConnect client. The user then launches the client and follow instructions within the client to complete the provisioning process.

On Android devices, the user is directed to download the QuickConnect app from Google Play. The user installs and launches the app and follows its instructions to provision the device. Note that supporting this process requires the AOS device to permit the client to access the Google Play. Some form of walled garden, or allow list, is required on the Onboard or Onboard/Guest SSID.

Figure 3-18: Onboard Experience Based on Device Type

Overview of Configuring ClearPass to Support Onboarding

You will now look at the process for configuring ClearPass Onboard and CPPM to support onboarding. This study guide does not have the time to cover every detail of the configuration, but you will get a basic sense of the process, which is outlined in Figure 3-19.

Figure 3-19: Overview of Configuring ClearPass to Support Onboarding

First you need to determine what option you will use for the ClearPass Onboard Certificate Authority and set up that option.

Then you set up the ClearPass Onboard Portal, including Network Settings, a Configuration Profile, and a Provisioning Profile.

You then set up CPPM services to support the onboarding process.

Options for Onboard Certificate Authorities

You can configure ClearPass Onboard as one of three types of certificate authorities, as shown in Figure 3-20. In every case, ClearPass Onboard is the authority that issues the certificate to the client. The difference is the location of the certificate's root.

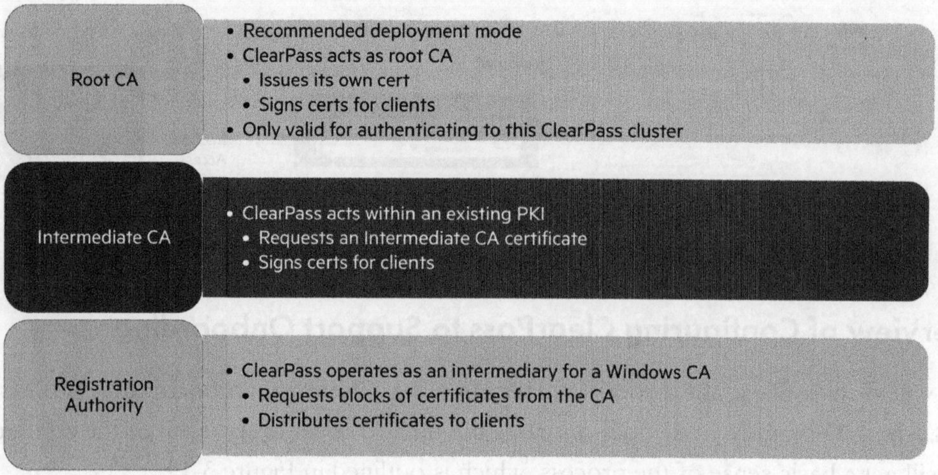

Figure 3-20: Options for Onboard Certificate Authorities

Root CA

You can set up ClearPass Onboard as its own root CA. It issues its own certificate and also issues certificates to clients. It is generally recommended that you use this straightforward option. Many companies want to limit how users can use the certificates deployed to them through the onboarding process. They should be able to use the certificates to connect to the network but not to gain any other privileges. When Onboard issues the certificates from its own PKI, the certificates are only valid for authenticating to that ClearPass cluster.

Intermediate CA

However, if a company wants to use its existing PKI to issue certificates, you can set up ClearPass Onboard as an intermediate CA within that PKI. ClearPass Onboard requests an Intermediate CA certificate, which the root CA must sign for it. Onboard can then sign and issue certificates to clients. These certificates will be trusted throughout the existing domain.

Registration Authority

ClearPass Onboard can also act as a registration server for a Windows CA. Basically, ClearPass Onboard requests a block of certificates from the Windows CA. These certificates are already signed by that CA. Onboard then distributes the certificates to clients.

Onboard Portal Components

You will now look in a bit more detail at how the remaining settings fit together. (See also Figure 3-21.)

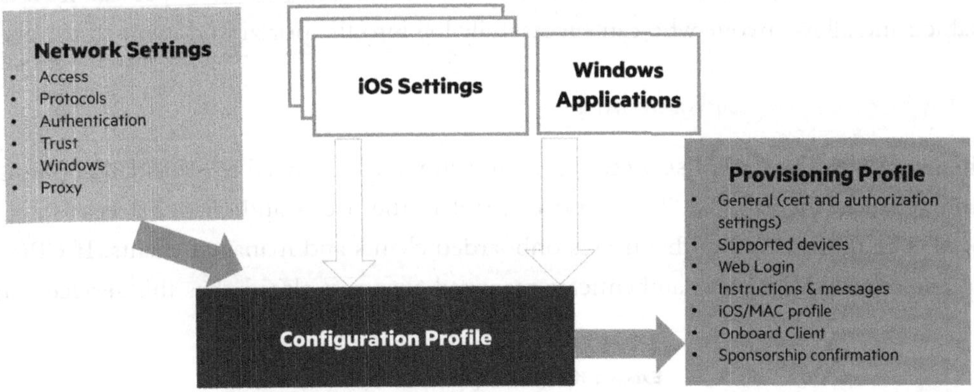

Figure 3-21: Onboard Portal Components

You must create at least one Network Settings definition. Network Settings define all the parameters for provisioning the clients' network connections. This object defines whether wired, wireless, or both types of connections will be provisioned, as well as settings for the wireless network. It defines which EAP authentication methods are supported (protocols), and where the certificates that Onboard sends will be stored on the client (authentication). It also configures the client to trust the subject name and root CA certificate for the proper RADIUS server certificate, typically the RADIUS/EAP certificate on CPPM. Finally, it provides some additional settings that might be required for configuring networking on Windows devices, and in an environment that uses a proxy server.

You can optionally create iOS Settings and Window Applications objects. These objects provision iOS or Windows devices with applications that the company wants to deploy to BYOD clients.

You then bundle the Network Settings, iOS Settings, and Windows Applications into a Configuration Profile. The Configuration Profile tells the Onboard portal what it should push to the client.

Finally, you create the Provisioning Profile. This profile references the Configuration Profile and the certificate authority that you established earlier. It further specifies support devices that can use this profile, and all the settings for the portal such as the Web Login page, customized instructions and messages, and settings specific to iOS and MAC clients. The Web Login page defines the page name to which HPE Aruba Networking infrastructure devices should redirect devices that need onboarding. Another important setting can be Prevent CNA. Captive Network Assistant (CNAs), often provided in smartphones, help users log into captive portals. However, *CNAs* cannot support the onboarding process; the Prevent CNA setting allows ClearPass Onboard to bypass the CNA. You can enable sponsorship in the Provisioning Profile, which would require a user to receive permission from a sponsor before downloading the Onboard Client. Or you can leave sponsorship disabled and allow anyone who can successfully log into the portal to download the client.

CPPM Services for Onboarding

When analyzing the CPPM services related to Onboard, you need to consider all phases of access, summarized in Figure 3-22. Two services regulate the user's and client's access to the Onboard Portal. The third service authenticates onboarded clients and managed clients. If CPPM is already implementing EAP-TLS to authenticate managed clients, it already has this service configured on it.

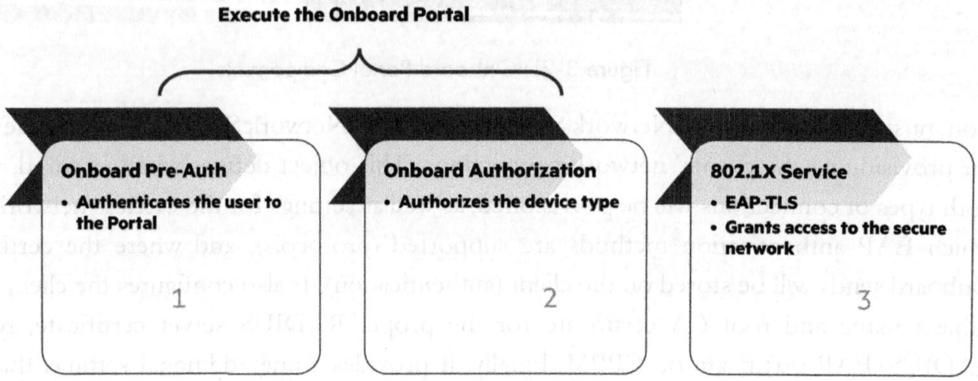

Figure 3-22: CPPM Services for Onboarding

When the user logs into the portal and the portal begins to execute, Onboard triggers the two Onboard services. The first is the Onboard pre-auth, which controls which users may execute the portal. The second service is the Onboard authorization service. Enforcement in this service filters what types of client devices you can onboard.

After the client onboards and reconnects to the secure network, the authentication triggers the 802.1 X service. The client will execute an EAP-TLS authentication, presenting the TLS certificate received from Onboard as the credentials. CPPM simply needs to trust the Onboard CA certificate with the RADIUS/EAP usage. This trust is added automatically on CPPM when you set up the Onboard CA.

Onboard Service Only Template

You can use the Onboard Service Only template to create the Onboard Pre-Auth and Onboard Authorization services that you need for the two-SSID scenario.

The two new services added by this template are shown in Figure 3-23. You should edit the Onboard Service Pre-Auth service to permit access only for the desired users. For example, you might edit this service to use your AD authentication source. Then only users who have valid domain credentials can log into the Onboard portal. You can also reference role mapping and enforcement policies within the Onboard Service Pre-Auth and Onboard Service Authorization services to enforce more sophisticated policies based on user or device attributes.

You can then use an existing 802.1X wireless service, which you should validate will work for onboarded and non-onboarded clients.

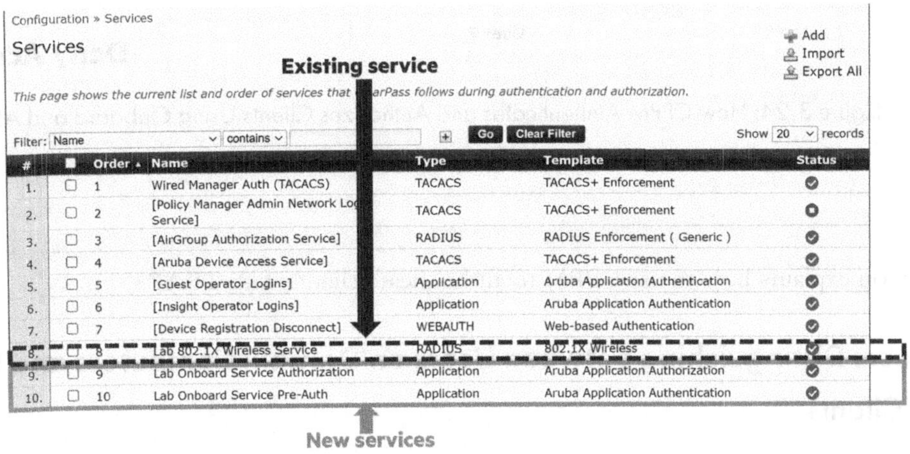

Figure 3-23: Onboard Service Only Template

How CPPM Authenticates and Authorizes Clients Using Onboard and AD

As you learned earlier, when CPPM uses EAP-TLS to authenticate a client and AD is the authentication source, CPPM checks the certificate validity and it also checks that the user has an account in AD.

This process continues to hold true when CPPM is validating certificates issued by an Onboard CA. CPPM checks the validity of the certificate with Onboard during the authentication cycle. It also checks with the AD server to determine if the user account on the certificate exists in AD. During the authorization cycle, ClearPass will also gather the AD attributes for the user account, so you can use enforcement rules based on AD attributes for your Onboard clients, just as you do for managed clients. This includes checking whether the user account is enabled.

If the user has been deleted from AD, then, even though the certificate is valid, CPPM will block the client's access, as shown in Figure 3-24.

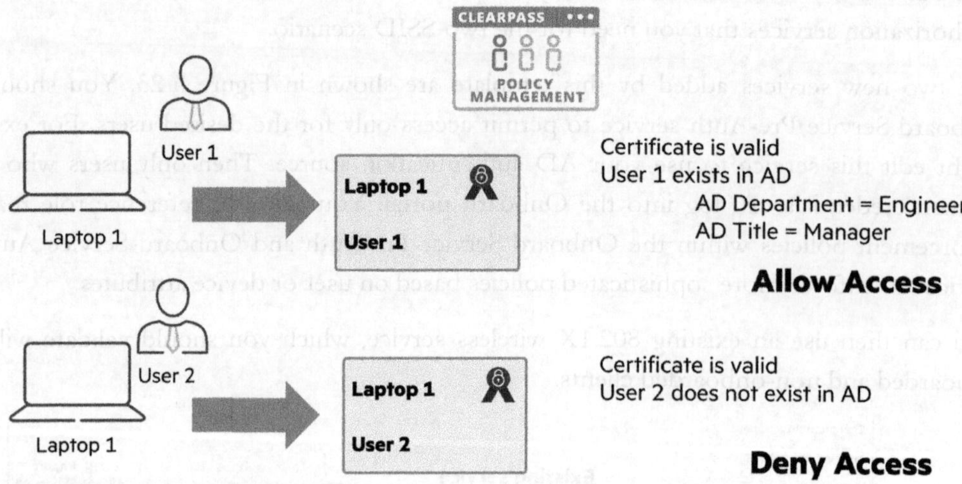

Figure 3-24: How CPPM Authenticates and Authorizes Clients Using Onboard and AD

TEAP

This section explains how to use CPPM to authenticate clients with TEAP.

Use Case: Distinguishing Between Users on Domain Computers and Users on Other Clients

Often companies want to distinguish between users who access the network on domain computers and users who access the network on unmanaged devices. The company manages the domain computers to minimize their vulnerability to malware and exploits. However, the company has no idea whether the unmanaged devices are configured securely or not.

You learned earlier about CPPM's [Machine Authenticated] and [User Authenticated] roles. You can use those roles to distinguish between several types of client. As shown in Figure 3-25, a domain computer with no user logged in has the [Machine Authenticated] role. CPPM assigns a domain user on an unmanaged device to the [User Authenticated] role, on the other hand.

You want a domain user logged in through a domain computer to receive both the [User Authenticated] and [Machine Authenticated] roles. Then you can use that combination of roles to provide the user a higher privilege level than a user who is only [User Authenticated].

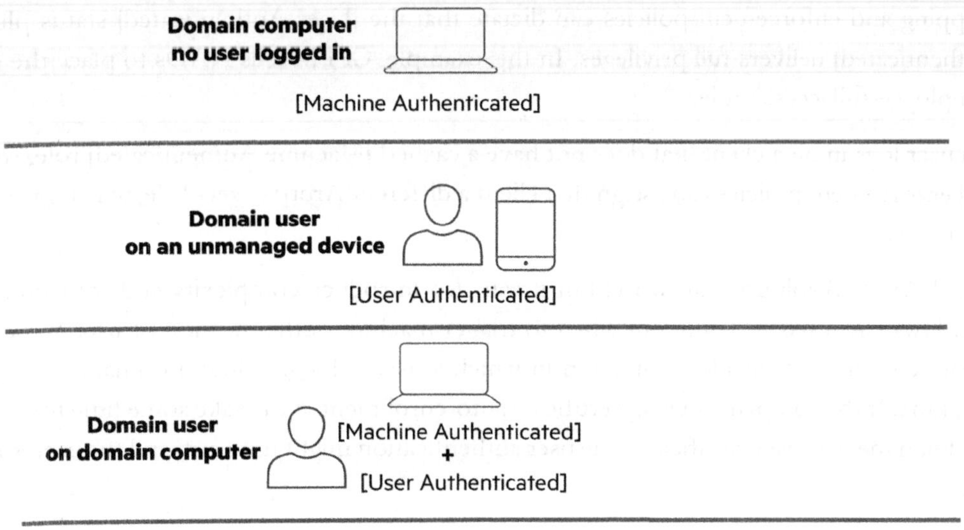

Figure 3-25: Use Case: Distinguishing Between Users on Domain Computers and Users on Other Clients

Validating Domain Users Are on Domain Computers with EAP-TLS

However, with EAP-TLS (or PEAP), the computer authentication and user authentication occur separately, as shown in Figure 3-26. CPPM has to cache the [Machine Authenticated] role for the client across sessions.

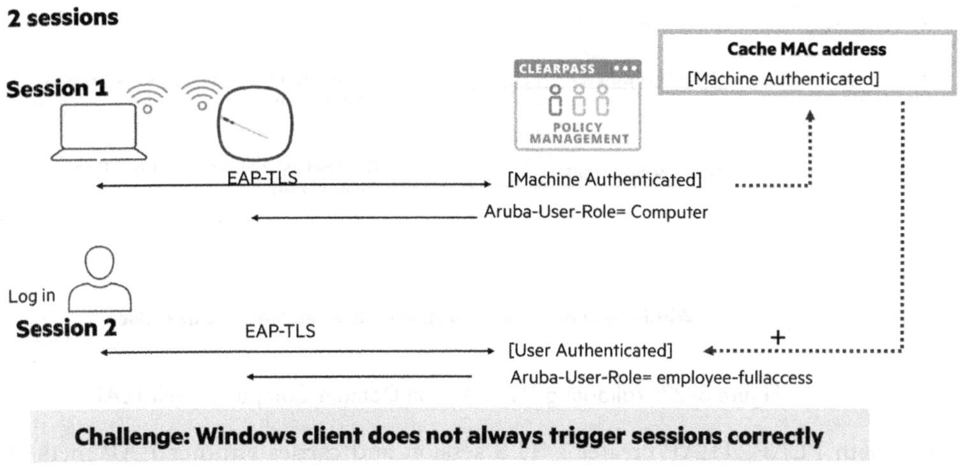

Figure 3-26: Validating Domain Users Are on Domain Computers with EAP-TLS

When a client authenticates as a computer, CPPM marks the client as Machine Authenticated and caches the client's MAC address with this status. When a user later logs in, CPPM marks the client as User Authenticated and also matches the client with the cached Machine Authentication. Role mapping and enforcement policies can dictate that the [User Authenticated] status plus [Machine Authenticated] delivers full privileges. In this example, CPPM tells NADs to place the client in the "employee-fullaccess" role.

If a user logs in on a client that does not have a cached [Machine Authenticated] role, role mapping and enforcement policies can assign that client a different Aruba-User-Role, which provides limited privileges.

This EAP-TLS solution can work, but it can also introduce complexity and sometimes fails. The Windows client has to figure out when to trigger machine authentication or user authentication as separate sessions. Consider a situation in which a user is logging into a domain computer for the first time. If the company is using certificate auto-enrollment, it can take some time for the computer to obtain the new user certificate. The user authentication might time out, and the user is left without full access.

Validating Users Are on Domain Computers with TEAP

Windows 10 2004 and CPPM 6.9 and later support EAP-TEAP, or TEAP, which provides a more graceful solution to validating whether domain users connect on domain computers. Whenever the Windows client authenticates, it always does so as both a machine and a user, as shown in Figure 3-27.

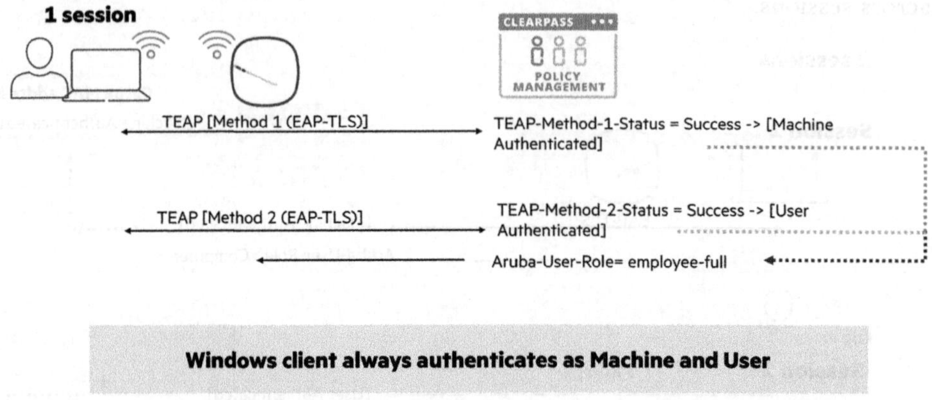

Figure 3-27: Validating Users Are on Domain Computers with TEAP

Much as with PEAP, TEAP creates a TLS session and carries another EAP method within the session. However, TEAP supports the chaining of multiple authentication processes. The supplicant "chains" a token from the earlier authentication to the subsequent authentication. In this way, the machine and user authentication can take place within the *same* session. Whenever a user logs in, the Windows OS using TEAP runs two methods. The first method authenticates the computer. When a client succeeds at TEAP Method 1, CPPM assigns that client the [Machine Authenticated] role. The second method authenticates the user, and when a client succeeds at Method 2, CPPM assigns the [User Authenticated] role.

TEAP supports multiple types of inner methods. However, HPE Aruba Networking recommends EAP-TLS for both Method 1 and Method 2. The computer authenticates with its computer certificate, and the user authenticates with the user certificate.

TEAP makes it easier for the Windows domain client to authenticate as both a computer and user in a consistent fashion. The client is less likely to fail to authenticate as a user and get stuck in a computer role. On CPPM's side, CPPM no longer needs to cache the [Machine Authenticated] role.

Additional TEAP Scenarios

Scenarios can also occur in which a client succeeds at one TEAP method but not at another. CPPM counts this client as authenticated, and you can set up enforcement policy rules to assign appropriate profiles to these clients.

A domain computer that is set up to use TEAP still implements both TEAP methods when no user is logged in. The computer succeeds at Method 1, and CPPM assigns it the [Machine Authenticated] role. The client fails at Method 2. CPPM still counts this client as authenticated, but only with the [Machine Authenticated] role. You can create an enforcement policy rule that assigns clients with the [Machine Authenticated] role to an enforcement profile that applies an Aruba-User-Role VSA. In the example shown in Figure 3-28, that VSA value is "domaincomputers," which matches up with an AOS role with appropriate firewall rules for domain computers.

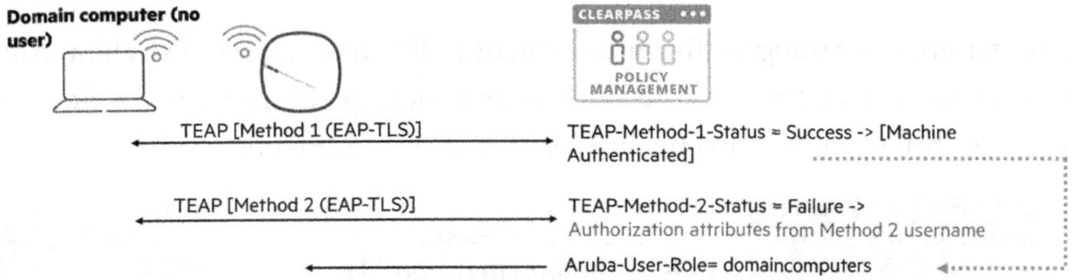

Figure 3-28: Additional TEAP Scenarios

Keep in mind that, even though TEAP Method 2 failed, CPPM still collects authorization attributes for the session because TEAP Method 1 succeeded. However, CPPM uses the *Method 2* username to collect those authorization attributes. You will consider the implications of this fact as you explore an example setup.

As another example, shown in Figure 3-29, a BYOD client that is using TEAP will fail TEAP Method 1 but pass TEAP Method 2. In this case, CPPM assigns the client to [User Authenticated] role. You can use that role to assign the client to an appropriate enforcement profile. In this example, that enforcement profile applies "Aruba-User-Role=internet-only."

Note that the BYOD client might be provisioned simply to use EAP-TLS.

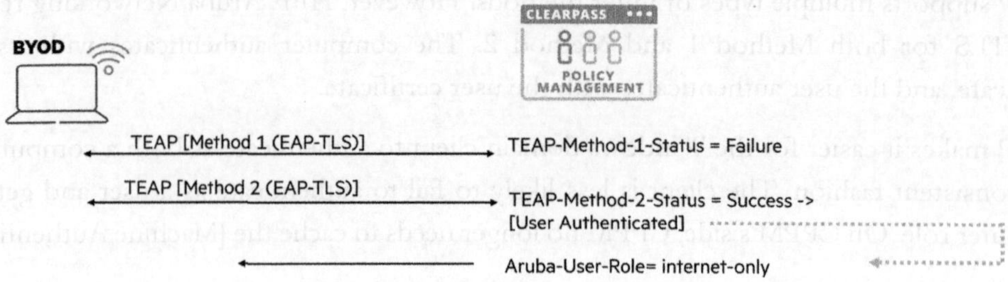

Figure 3-29: Additional TEAP Scenarios

Configure CPPM to Use TEAP to Validate Users Are On Domain Computers

You will now look at how to configure CPPM to use TEAP to validate that users are on domain computers and assign clients to roles based on the answer to this question. You can easily edit an existing 802.1X wireless service, such as the one that you examined in the previous scenario, for this purpose. In Figure 3-30, the additional steps are summarized.

CPPM supports TEAP as of version 6.9, but you must create the TEAP method manually. You can then add that method to your 802.1X services.

You should also create a special enforcement profile, which informs the NAS of the client's username. This profile is necessary because TEAP often obscures the username in the outer method.

To finish the set up, you configure the enforcement policy rules to take [Machine Authenticated] and [User Authenticated] statuses into account when assigning enforcement profiles. You also add the enforcement profile that informs the NAS of the client's username.

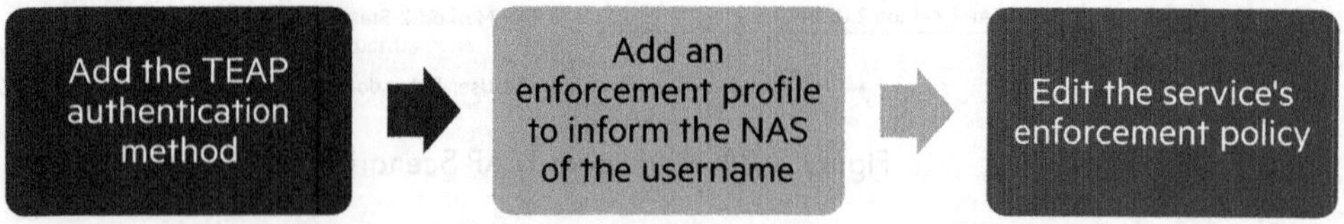

Figure 3-30: Configure CPPM to Use TEAP to Validate Users Are On Domain Computers

Add the TEAP Authentication Method

As shown in Figure 3-31, add a new method and name it an appropriate name such as TEAP. Choose **TEAP** for type.

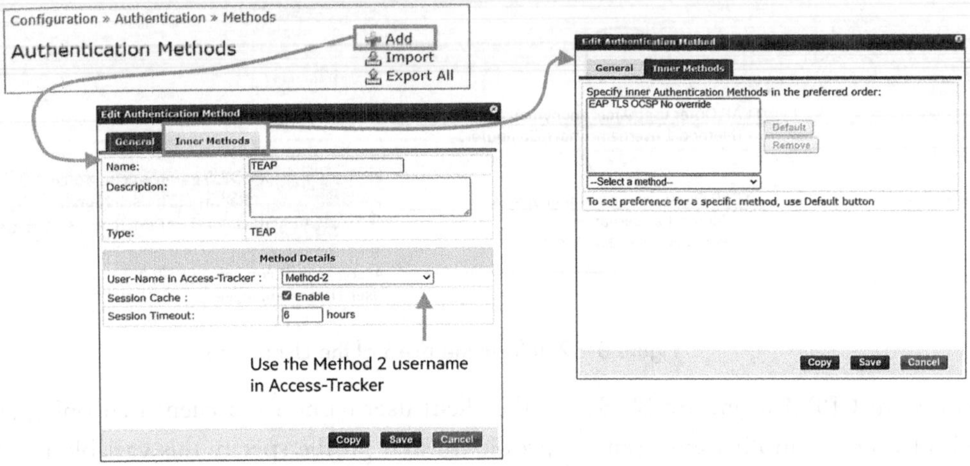

Figure 3-31: Add the TEAP Authentication Method

As you learned, TEAP supports multiple inner methods. Its Method 1 is the machine authentication because the computer authenticates first. Its Method 2 is the user authentication. You should typically choose to use Method 2 as the User-Name in Access Tracker because the username is more relevant to track than the computer name.

For this scenario, you want to use certificate-based authentication for both the computer and the user. Therefore, you should select Inner method and specify an EAP-TLS method. Here you see an EAP-TLs method that uses OCSP with no override of the OCSP URL specified in the certificate.

When you are finished configuring the method, click **Save**. You can now add this authentication method to the service.

Inform the NAS of the Username

Often clients' TEAP settings dictate that the client sets the username for the outer method to "anonymous." Only the inner methods reveal the client's actual name, which is the computer name for Method 1 and the username for Method 2. The NAS, such as the AP or switch, only sees the outer method name. You want your NASes to know client's actual user names so that you can properly monitor, troubleshoot, and audit.

You can create an enforcement profile that CPPM applies together with the enforcement profile that assigns the client's Aruba-User-Role. As shown in Figure 3-32, this second enforcement profile specifies a standard Radius-IETF attribute called "User-Name." The value for the attribute is this variable: **%{Authentication-TEAP-Method-2-Username}** CPPM then replaces that variable with the username submitted as part of TEAP Method 2.

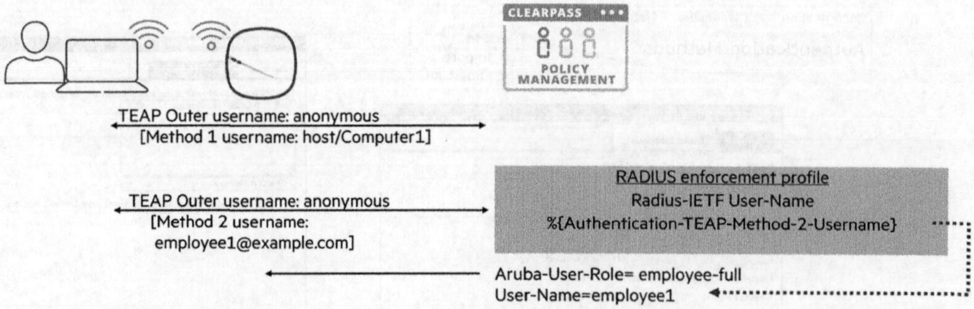

Figure 3-32: Inform the NAS of the Username

If you want CPPM to inform NASes of the client user name for a client that only passes TEAP Method 1, create another enforcement profile. In that profile specify this variable for the Radius-IETF User-Name attribute: **%{Authentication-TEAP-Method-1-Username}**

Example Role Mapping Policy

If you have already set up a role mapping policy that works for 802.1X authentication with EAP-TLS to your AD server, you just need to adjust that policy slightly to work with TEAP, as outlined in Figure 3-33.

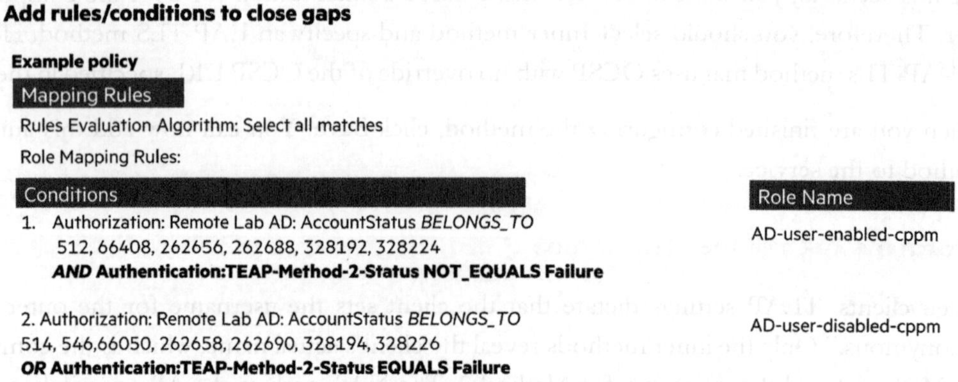

Figure 3-33: Example Setup: Example Role Mapping Policy

Remember that when a client passes *either* or both TEAP Methods, CPPM collects authorization attributes using the Method 2 username. If Method 2 fails, often CPPM does not obtain any authorization attributes because, for example, AD does not have a user account that matches the TEAP Method 2 username. But sometimes CPPM might obtain attributes for the user account even if Method 2 fails; for example, perhaps a user has an expired certificate but an active AD account. To prevent the client receiving higher privileges than it should in this case, you should add rules or conditions to the role mapping policy.

In this example, you want to treat a user with an expired certificate like a user with a disabled account. So, you could adjust the rule that assigns the "AD-user-disabled-cppm" role to also assign that role if TEAP Method 2 fails (Authentication:TEAP-Method-2-Status EQUALS Failure). (You could change the assigned role name to something more descriptive or keep it as is for simplicity). You set the rule to match on any condition ("OR") so that the client receives the role if it the authentication is deemed not truly valid for either reason.

If you are only using the "AD-user-account-enabled" role in enforcement policies, you might add a condition to that rule that TEAP Method 2 has *not* failed. Checking for lack of failure, rather than success, ensures that users can connect with EAP-TLS also if necessary. You set the rule to match all conditions ("AND") so that the client must meet both criteria to receive the role that indicates a valid account.

Also keep in mind that, with default AD authentication source filters, you cannot collect authorization attributes about the computer account. Therefore, you should not attempt to use those attributes in rules. You can configure *custom* filters in the CPPM authentication source to collect authorization attributes based on the TEAP Method 1 username. This topic is covered at the expert level of this security certification track.

Example Setup: Edit the Enforcement Policy

You need to edit the enforcement policy to change the level of access based on whether users are connecting on authenticated domain computers or not.

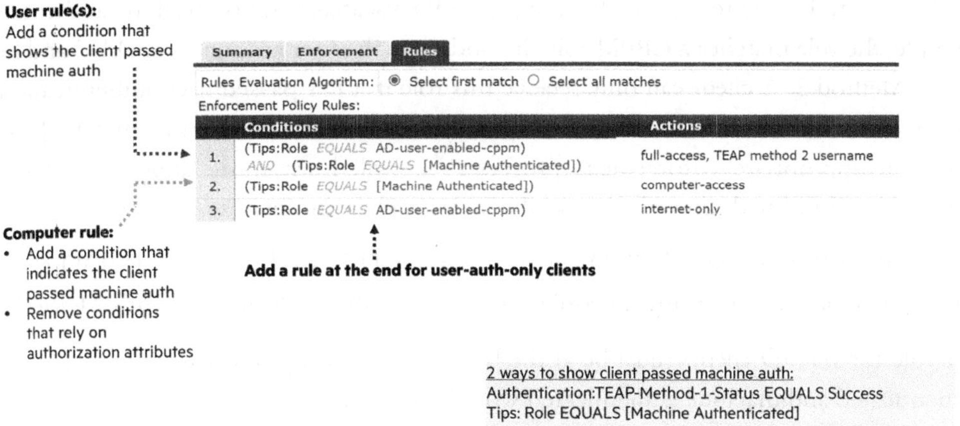

Figure 3-34: Example Setup: Edit the Enforcement Policy

In the real world, the enforcement policy might have many rules. But in the example shown in Figure 3-34, you are focusing on a very minimal policy. In this way, you can focus on the main points related to TEAP.

This policy used to have a single rule that provided full-access to users with enabled accounts in AD. You now want to add a further condition. The user client must have authenticated as a computer. As long as all domain computers support and use TEAP, it is most secure to explicitly require the client to have succeeded at TEAP Method 1 with this condition:

Authentication:TEAP-Method-1-Status EQUALS Success

Alternatively, you could use the built-in [Machine Authenticated] role for this purpose. You could also add a condition that requires the users to have the [User Authenticated] role to this first rule. However, in the example that you have been following, the first condition implicitly vouches for the fact that the user authenticated, based on the role mapping rules. For this rule, you continue to use the enforcement profile associated with full access. You also add the new enforcement profile that informs the NAS of the client's Method 2 username.

This example policy also has a second rule, which applies to domain computers without users logged in. Again, you should use the computer's success in passing TEAP Method 1 as a condition. Or, if some domain computers do not support TEAP, you could use the "Tips: Role EQUALS [Machine Authenticated]" condition. If you were using different conditions, such as a role assigned based on the machine account's status, you must remove those conditions. Remember that, with default authentication filters, CPPM collects authorization attributes using the Method 2 username, and this username does not exist in this scenario. Although not shown here, you could also add a second enforcement profile to this rule. This profile would inform the NAS of the client's Method 1 username.

Finally, you add a rule for clients that are user authenticated, but not machine authenticated. In this example, the rule matches a CPPM role that indicates the user account is enabled and has not failed TEAP Method 2. A client can only receive this role if it has passed user authentication, so you do not have to add a separate condition to explicitly check for the [User Authenticated] role. You then assign an appropriate enforcement profile. (You will learn more about coordinating profiles assigned in CPPM with firewall roles in the infrastructure in the next chapter.)

Note that, in this example, BYOD clients are using TLS, so the rule for BYOD clients does not need to include the enforcement profile for informing the NAS of the client's Method 2 username.

The rule for BYOD clients must be at the bottom of the list. Otherwise, clients that are both user authenticated and machine authenticated would match it.

Summary

Congratulations! You should now be able set up certificate-based 802.1X authentication for wireless clients in a variety of scenarios.

You can now follow best security practices to authenticate clients that are part of a domain, as well as unmanaged clients. You can also create policies to distinguish between users on domain clients versus unmanaged clients and control each type of client appropriately.

Learning Checks

1. A CPPM service uses the default [EAP-TLS] method for the authentication method and AD for the authentication source. What is required for a client to pass authentication? (Select two.)
   a. The client certificate is signed by a different CA from the one that signed CPPM's RADIUS/EAP certificate.
   b. The client certificate is signed by a CA in the CPPM trust list with the EAP usage.
   c. The serial number for the client certificate is listed on a CRL.
   d. The client username, derived from the client certificate, matches an account in AD.
2. What function does the Onboard Pre-Auth service provide in the onboarding solution?
   a. It authenticates users who are using 802.1X to authenticate with certificates issued by Onboard.
   b. It checks whether the user is attempting to onboard an authorized device type.
   c. It checks the credentials that the user submits to the Onboard portal.
   d. It authenticates users who are attempting to connect to an Onboard SSID.
3. A client has completed TEAP. It succeeded at TEAP Method 1 and failed at TEAP Method 2. What role does CPPM assign it?
   a. [Unauthenticated]
   b. [Partial Authenticated]
   c. [Machine Authenticated]
   d. [User Authenticated]

Note

For answers to all learning checks, refer to the Appendix.

---

4 Implement Advanced Policies on the Role-Based AOS Firewall

EXAM OBJECTIVES

✓ Deploy AAA for WLANs using HPE Aruba Networking ClearPass Policy Manager (CPPM)

✓ Define and apply advanced firewall policies

Assumed Knowledge

- AOS firewall and how to create firewall policies and rules that use basic actions
- Fundamental ClearPass Policy Manager (CPPM) components, such as services, roles, enforcement profiles, and enforcement policies

Overview

To continue implementing Zero Trust Security (ZTS), an organization must assign each client to a role. The network infrastructure can then apply access control appropriate to the client.

In this chapter, you will review how AOS devices derive a role for every client that connects to a WLAN. You will also learn how to use CPPM to assign a role and communicate that assignment to the AOS device.

You will then review how AOS devices take that role and enforce controls on it. You will review the role-based firewall and access control rules. You will also look at application policies in more detail and see how they enhance organizations' security. You will next learn about role-to-role ACLs and receive some general tips for formulating and analyzing tips.

AOS Mobility Role Derivation

In this section, you will review how AOS mobility solutions derive roles for connected clients.

Role Derivation Sequence on AOS Devices

Because the power of the AOS firewall partially derives from its role-based nature, you will now briefly review how an AOS controller or gateway assigns a client to a role.

As shown in Figure 4-1, the role deviation sequence begins when a user associates to the network, and the AOS device assigns an initial role. Any user-derived role can override the initial role. User-derived roles are based on criteria such as SSID, location, MAC address, and encryption types. Be aware, though, that user-derived roles are less commonly used now; as you will see, HPE Aruba Networking solutions now have better ways to derive the role.

If the user connects to an SSID that requires authentication, the role returned based on the authentication process takes precedence over both the initial role and the user-role derivation.

After authentication, the authentication server, such as CPPM can return attributes, which will assign a role to the user. HPE Aruba Networking has a set of vendor-specific attributes (VSA) that you can integrate into a RADIUS server. If the server returns an Aruba-User-Role VSA, the AOS device will automatically place the user into the role or specified VLAN.

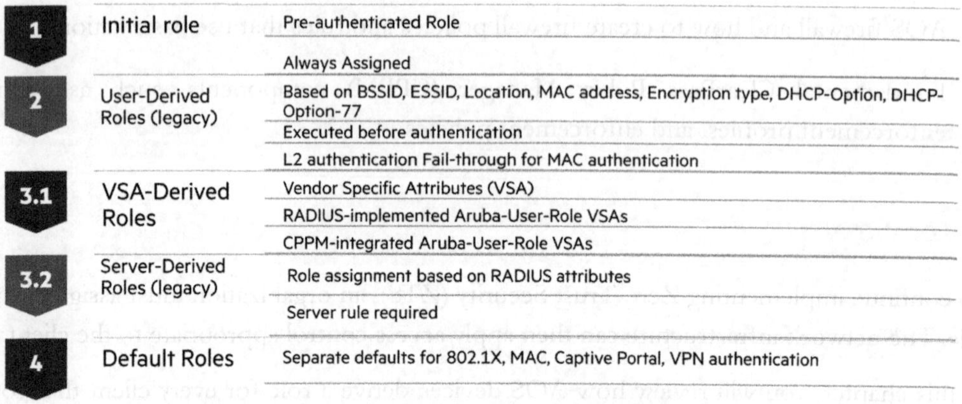

Figure 4-1: Role Derivation Sequence on AOS Devices

If you are using a third-party authentication server and cannot configure or integrate the VSAs in that server, the server will return a typical RADIUS Access-Accept with standard attributes. In that case, you can define rules to specify the server-derived roles. These rules tell the AOS device to look for particular standard attributes returned by the RADIUS server and then to assign a specific role (or VLAN). However, HPE Aruba Networking recommends using the Aruba-User-Role VSA in most situations.

If the AOS device does not receive a role from a VSA and cannot derive a server-derived role based on a rule, it assigns a default authentication role. Each authentication method has a default role. The AOS device will assign a role depending on which authentication method you configure on the SSID.

Understanding Which Post Authentication Role Takes Precedence

After successful authentication, the client is reassigned to a new, post-auth role. As shown in Figure 4-2, the AOS device can derive this role in one of three ways.

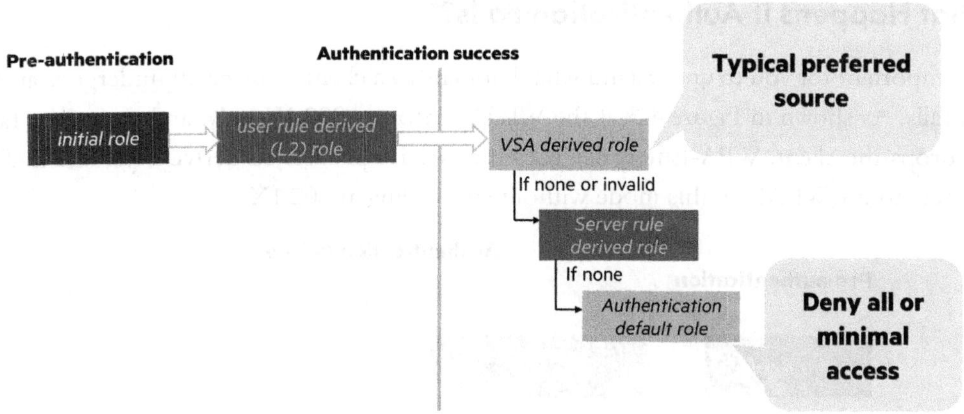

Figure 4-2: Understanding Which Post Authentication Role Takes Precedence

First it looks for an Aruba-User-Role VSA in the Access-Accept from the authentication server. If this VSA is present, and if the VSA specifies the valid role name—in other words, a role that exists on the AOS device—the device assigns this role to the client. (CPPM could also send a downloadable role in a VSA, but this feature is less often used with AOS devices.)

Only if there is no valid VSA does the AOS device move on to the next type of role derivation, which are server rules. These rules allow the AOS device to make the ultimate decision about the authenticated client's role, but under advisement from the authentication server, based on attribute value pairs (AVPs) or VSAs sent by that server.

Typically, HPE Aruba Networking recommends that you use VSA-derived roles. This approach lets CPPM make decisions about the role to which clients belong, drawing on all the contextual information that it has. You usually use the server rule-derived roles only if you are using a third-party authentication server, which is not the focus for this study guide.

Finally, only if no valid VSA- or server-derived role applies, the AOS device assigns the client to the default role for the authentication method that the client passed. This is sometimes called the role of last resort because you typically want to assign the client to a granular role based on information received from the authentication server. For security, choose a role with the lowest access level appropriate to authenticated users or even "denyall."

In most cases, role assignment ends at a VSA-derived role, server rule-derived role, or authentication default role. However, if an Authentication, Authorization, and Accounting (AAA) Profile's user rule set includes rules with L3 criteria, such as DHCP fingerprints, this rule can take effect after authentication and reassign the client to a new role. But L3 user rules are not often used; preferably, you would use CPPM or ClearPass Device Insight (CPDI) to fingerprint and classify devices instead.

What Happens if Authentication Fails?

It is important for you to understand what happens to a client if the client undergoes authentication and fails. As shown in Figure 4-3, if the WLAN enforces 802.1X only, authentication fails, and the AP drops the client. WPA-Enterprise uses the 802.1X process to derive keys, and a client cannot connect to the WLAN in this mode without succeeding at 802.1X.

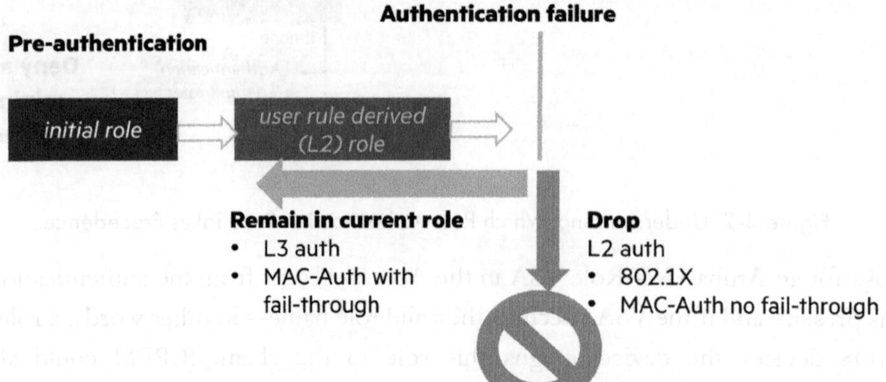

Figure 4-3: What Happens if Authentication Fails?

If the WLAN enforces MAC-Auth, you have choices. If you do not enable fail-through, and a client fails authentication, the client is denied access. But if you enable fail-through, a client that fails authentication remains connected with its current role. That could be the initial role or a user rule-derived role, if a user rule applied to the client.

If the WLAN enforces Layer 3 authentication, such as captive portal, a client that fails authentication also remains connected in the current role.

Recommendations When Using CPPM as the Authentication Server

HPE Aruba Networking recommends that, whenever possible, you leverage the capabilities of CPPM and other solutions in the ClearPass suite to assign devices to roles.

Consider an example, summarized in Figure 4-4.

You need to support connectivity for some headless devices that do not support 802.1X authentication. You want to sort these devices into different roles based on their type. You might consider setting up an open WLAN and creating user rules to assign various lists of MAC addresses to various roles. However, a more powerful solution might leverage CPPM. You can create a WLAN that uses MAC-Auth and authenticates clients to CPPM. CPPM could then handle assigning the clients to their roles based on profiling information. It sends the correct role for each device in a VSA to the AOS device as usual.

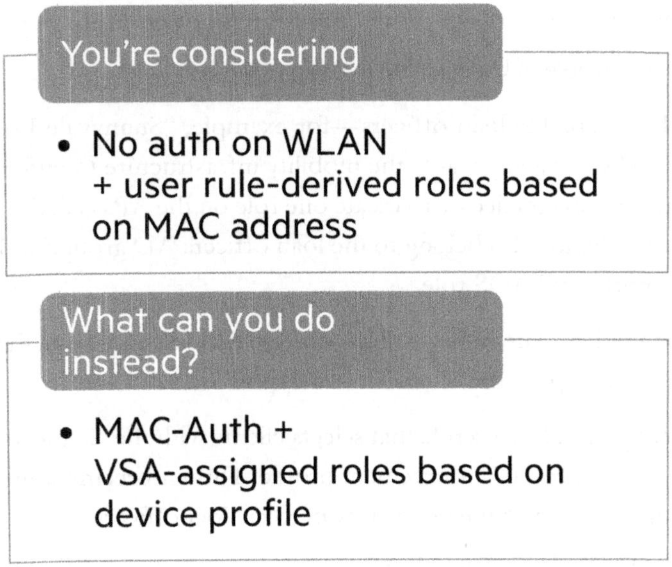

Figure 4-4: Recommendations When Using CPPM as the Authentication Server

Example of Using CPPM to Assign Clients to Roles

You will now learn how to set up CPPM to assign authenticated clients to roles by following an example scenario, illustrated in Figure 4-5.

A company has a Windows domain, HPE Aruba Networking IAPs and CPPM. The company assigns users to several AD groups, and it wants to apply different firewall policies to users in different groups. You need to plan how to use CPPM to communicate the correct AOS role for each client to APs. You will start by looking at a single AOS role for simplicity, the role for loan officers.

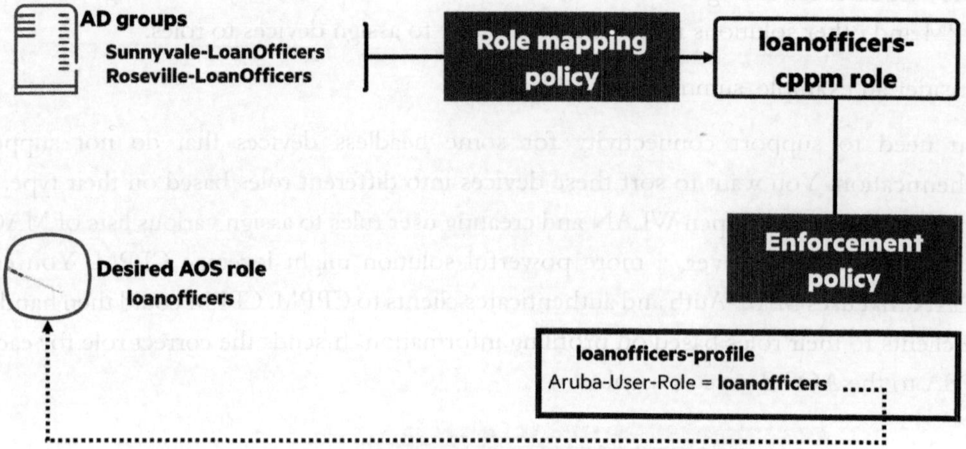

Figure 4-5: Example of Using CPPM to Assign Clients to Roles

This company has several AD groups for loan officers—for example, "Sunnyvale-LoanOfficers" and "Roseville-LoanOfficers." The company wants the mobility infrastructure to enforce the same firewall policies for all loan officers, so you decide to create one role on the APs called "loanofficers." It will be up to CPPM to identify clients who belong to the loan officers AD groups and tell the AP to put those clients in the "loanofficers" AOS role.

First, you create a role mapping policy. The policy will have a rule that selects clients that belong to any of the loan officers groups and applies the "loanofficers-cppm" role.

Next you create an enforcement policy. It has a rule that selects clients with the "loanofficers-cppm" role and applies the "employee-loan-profile" enforcement profile. This enforcement profile defines an Aruba-User-Role VSA with the name of the desired AOS role: "loanofficers."

At this point, creating both a role mapping policy and an enforcement policy might seem like extra steps. However, you will see the benefits of this approach as you create more complex policies, which draw on additional context. You will also later see that you can add other services that reuse the same role mapping policy.

Example Role Mapping Policy

In Figure 4-6, you see the role mapping policy for this example scenario. It has two rules to assign clients to ClearPass roles based on AD group. The example rules use the EQUALS_IGNORE_CASE operator to match the precise AD group name, except for capitalization. As a variation, you could specify that the Group must contain a value, or begin or end with a value. In that way, the rule can match multiple similar group names. In the real world, you would likely have many more rules for other AD groups.

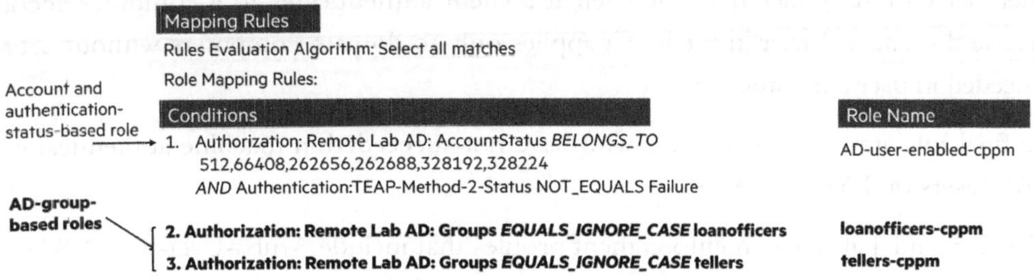

Figure 4-6: Example Role Mapping Policy

As you see, this example policy still has a rule for assigning a CPPM role based on the AD account status, which is important for solutions that use EAP-TLS with authorization to AD. And you will learn about more ways to assign CPPM roles throughout this study guide.

If a client does not match any of these rules, it is assigned the default [Other] role.

Example Enforcement Policy

The enforcement policy, shown in Figure 4-7, features several rules which combine together multiple CPPM roles and output an enforcement profile. That enforcement profile is configured to send an Aruba-User-Role value to the authenticator. For example, users in the AD security group named Loan Officers (CPPM role, "loanofficers-cppm") who have active AD accounts (CPPM role, "AD-user-enabled-cppm") and who are on domain computers (CPPM role, "[Machine Authenticated]") receive the "loanofficers" enforcement profile. That profile has an Aruba-User-Role VSA set to "loanofficers," which matches a firewall role on the AOS devices.

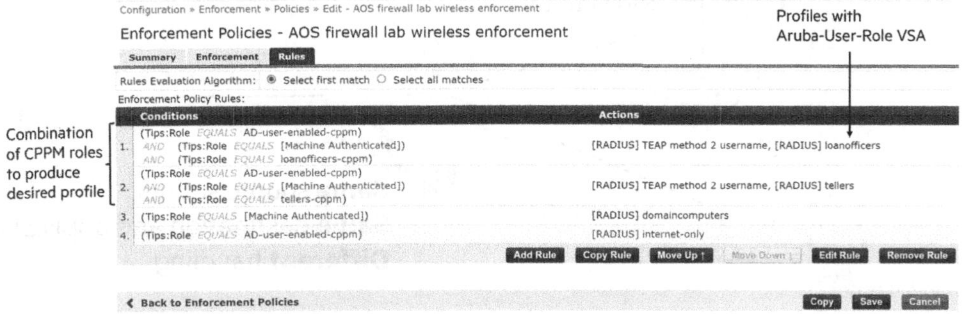

Figure 4-7: Example Enforcement Policy

The second rule is similar, but for users in a different AD security group.

The third rule uses the [Machine Authenticated] role to assign clients to a "domaincomputers" enforcement profile. Your role mapping policy did not assign the [Machine Authenticated] role; remember CPPM applies that role itself if a client authenticates to a computer account on AD. Because this rule is lower in the list, it applies only to domain computers without users who have succeeded in user authentication.

The final rule applies only to users who have not succeeded in machine authentication—in other words, users on BYOD devices.

Rules 2, 3, and 4 also assign enforcement profiles that include Aruba-User-Role VSAs.

Clients that fail authentication and clients with the [Other] role are denied access.

As a side note, remember that if all domain computers use TEAP, you can replace the "Tips:Role EQUALS [Machine Authenticated]" conditions with "Authentication:TEAP-Method-1-Status EQUALS Success."

Applying Role-Based Firewall Policies

This section explains how to apply role-based firewall policies.

Role-Based AOS Firewall

As shown in Figure 4-8, the role-based AOS firewall can enforce different rules based on user identity and assigned role, rather than simply on features in packet headers. For example, the firewall can restrict contractors to sending Web traffic to limited ranges of IP addresses while it allows managers' traffic to flow more or less unrestricted.

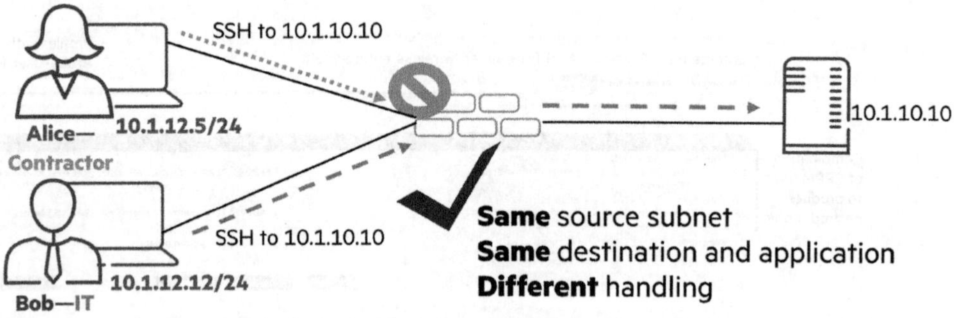

Figure 4-8: Role-Based AOS Firewall

With less sophisticated firewalls, you would try to create such behavior by ensuring that contractors and managers receive IP addresses in different ranges. You would then create different rules based on source IP address. However, such an approach makes planning user connections more complex—you need more VLANs/subnets and a way to get users in the right VLAN every time. With the AOS firewall, you can place users in the same subnets, but easily create different rules for different users.

By sending all internal user traffic through the role-based AOS firewall, you can change from a model in which you protect a single perimeter to a model in which you place every user within their own unique and properly customized firewall.

AOS Role-Based Firewall Policies: Gateways or MCs

You are now ready to look at the role-based firewall policies in more detail. First you will look at how rules are organized on HPE Aruba Networking Mobility Controllers (MCs) or gateways. (See Figure 4-9.)

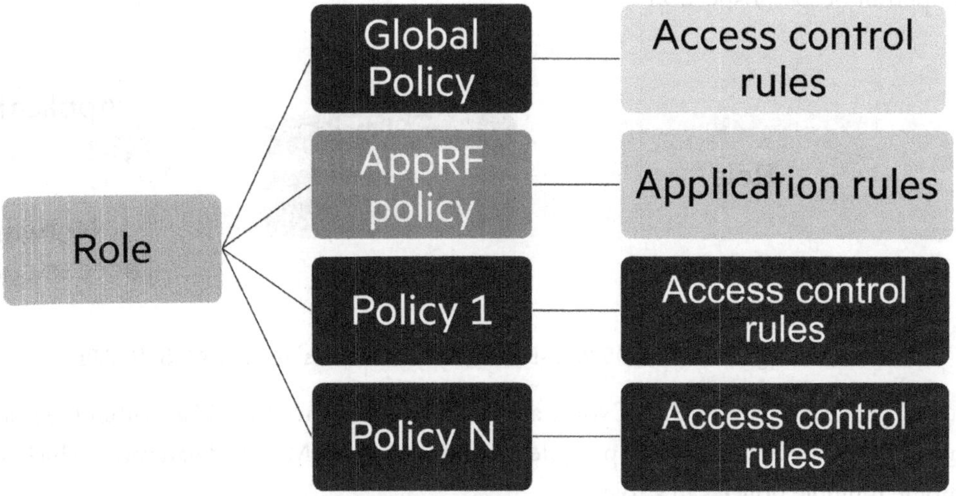

Figure 4-9: AOS Role-Based Firewall Policies: Gateways or MCs

The AOS firewall implements a different set of policies for each role. Each policy contains a series of rules or access control entries (ACEs), which filter traffic for users assigned to that role.

At the top of every role's list of policies is the "global-sacl" policy, in which you can place any rules that you want to apply to every user and device, no matter what the role. As just one example, you might include a rule for allowing DHCP traffic in this policy.

The AppRF policy comes next in every role; it is related to application visibility. It includes application rules that are specific to the role. (More precisely, application rules are categorized as application, application category, Web category, and Web reputation.)

When you create a role in the Web UI, the role has a policy with the same name as the role automatically created for it. This is the default policy for access control rules that are specific to the role.

In addition to adding rules to the default role-specific policy, you can also add other policies to the role. An AOS device comes with many pre-configured policies, and you can also create policies of your own.

AOS Role-Based Firewall Policies: IAPs or AOS-10 APs

Instant APs (IAPs) and AOS-10 APs controlled by HPE Aruba Networking Central also take a role-based approach to access control rules. However, you cannot create separate policies and apply the policies to multiple roles. Instead, as shown in Figure 4-10, you create access control rules within the role itself. Network access control rules control traffic based on the IP header, while application rules permit deeper inspection.

Figure 4-10: AOS Role-Based Firewall Policies: IAPs or AOS-10 APs

You can create a separate ACL, if you want, and then reference that ACL within the role. However, you can only reference one ACL per role, as opposed to on MCs and gateways, which allow you to combine multiple policies in a role.

Access Control Rules

You will now review access control rules (called network access control rules on IAPs). As shown in Figure 4-11, you can create rules that select IPv4 or IPv6 traffic. On gateways and MCs, the source and destination can be user, any, host, network, Local IP or an alias. You can select the service from pre-defined services like DHCP or a specific port like TCP 21.

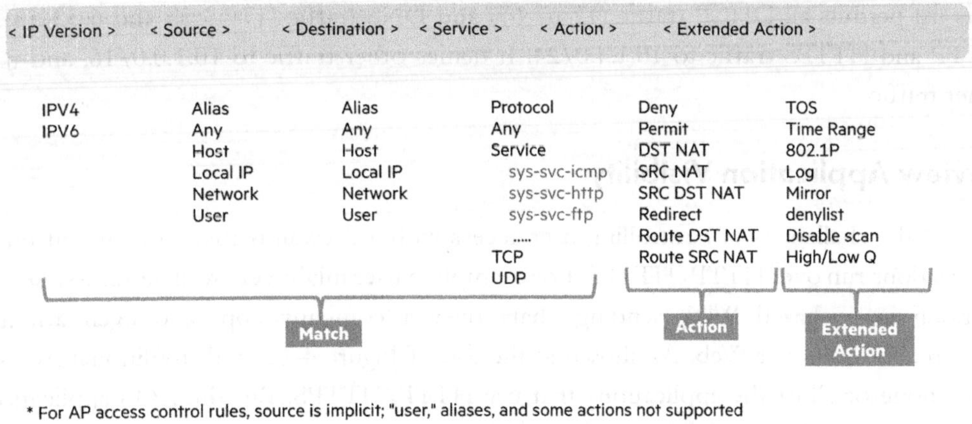

Figure 4-11: Access Control Rules

The various actions include permit, drop, NAT, and route. When a packet matches a rule, the AOS device executes the main action, as well as any extended actions such as log, mirror, or time range. You can add multiple extended actions to a single rule.

Note that HPE Aruba Networking APs (IAPs or APs managed by HPE Aruba Networking Central) support similar rules to MCs or gateways. However, these rules do not specify a source. Instead, the rule always implicitly matches clients assigned to the role (called "wlan access-rule" in the CLI). In addition, "alias" and "user" are not options for destinations. Also, the APs do not support the redirect or routing actions. They also might not support the mirroring extended action.

Format and Examples

For APs, if you configure the rules through HPE Aruba Networking Central WebUI, you will be able to select the destination and services from a drop-down list. If you configure the rules as commands, the format is **rule *destination_ip destination_mask* match *protocol_number first_port_in_TCP/UDP_range last_port_in_TCP/UDP_range*.** For example:

```
wlan access-rule contractors
 rule any any match 17 67 67 permit
 rule any any match 17 53 53 permit
 rule any any match 6 53 53 permit
 rule 10.1.4.0 255.255.255.0 match 6 80 80 permit
 rule 10.1.4.0 255.255.255.0 match 6 443 443 permit
 rule 10.1.0.0 255.255.0.0 match any any any deny
 rule any any match any any any permit
```

This list permits all DHCP traffic (17 67 67) and DNS traffic (17 53 53 and 6 53 53). It permits HTTP and HTTPS traffic to 10.1.4.0/24. It denies other traffic to 10.1.0.0/16, and it permits all other traffic.

Review Application Visibility

One of the challenges of controlling user access with a firewall is that so many different modern applications run over HTTP/HTTPS. For example, a user might very well be accessing social media, working in a shared Wiki, sending chats over a company app, and even attending video conferences—all over Web. As shown at the top of Figure 4-12, with traditional ACLs you must allow none or all of the applications that use HTTP/HTTPS. But the AOS application visibility feature provides deep packet inspection (DPI), which gives visibility down to the application layer, as shown at the bottom of Figure 4-12.

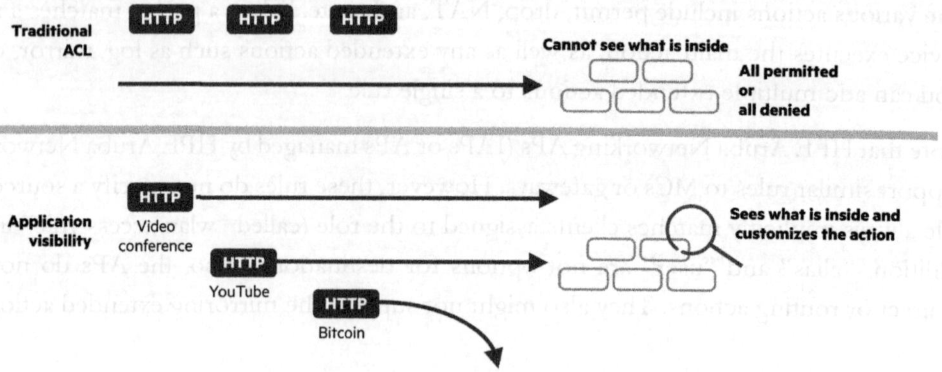

Figure 4-12: Review Application Visibility

The first step to using application visibility is enabling Deep Packet Inspection (DPI) globally, which requires a reboot. You must also ensure that DPI is enabled on the role (it is by default). If you want to filter by Web category, as well as application, you must also enable WebCC globally and on the role (it is enabled on the role by default).

Also make sure that you have the proper licenses for these features. DPI is supported with the PEF license, but WebCC requires the PEF and WebCC license.

Application Visibility and WebCC Use Cases

By enabling admins to create rules that filter traffic based on application or Web site category, rather than IP header alone, application visibility and WebCC make it much easier for admins to create useful policies. They also enable the company to give admins much greater control over how users connected to the network use the network.

By filtering traffic based on application or web content, the organization can better control the types of tasks that users perform on the network. This can help eliminate resource-intensive applications, that are not important to the company's mission. For example, bitcoin mining can consume many resources and is usually an unauthorized activity. With application rules, you can deny this activity entirely.

You can also protect your users from untrustworthy sites, which are often sources of malware or phishing attempts. In this way, you can help keep the internal network free of malware and eliminate some of the most common ways that hackers compromise devices.

Many companies also want to limit their legal liability for content that users access on their network. Application and web content filtering rules make it much easier for the company to implement policies that meet the approval of the legal team. These use cases are summarized in Figure 4-13.

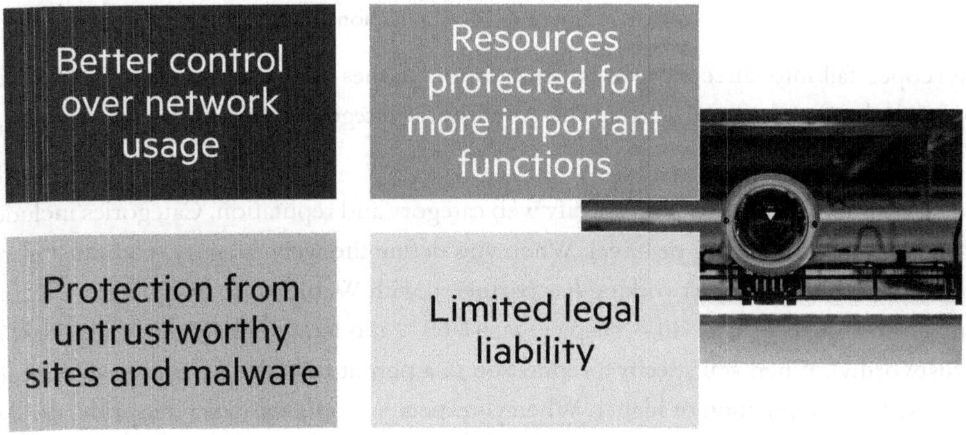

Figure 4-13: Application Visibility and WebCC Use Cases

Application Rule

A rule can be made up of application scopes. In an application rule, the source and destination options are the same as the services' rules. However, instead of selecting services, you can select scopes, as shown in Figure 4-14.

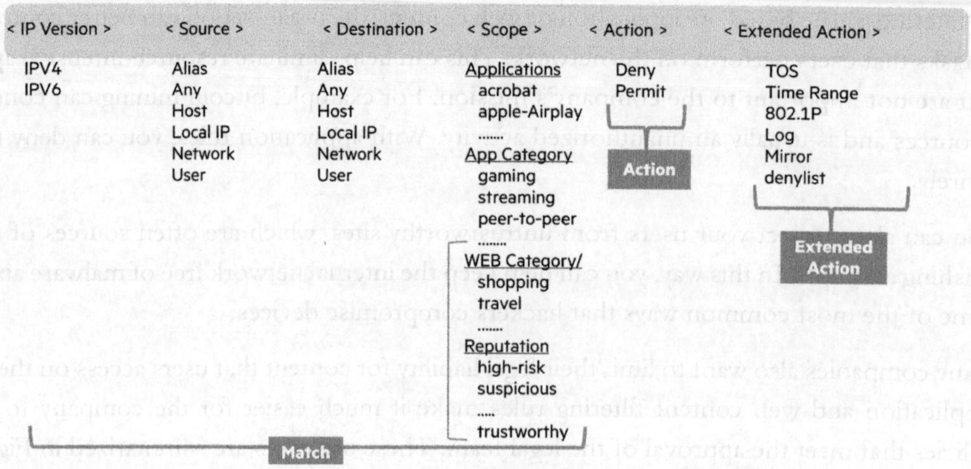

Figure 4-14: Application Rule

The scopes fall into three categories. Application scopes deal with applications such as Acrobat, Apple Airplay® and several others. The Application category breaks down into gaming, streaming, and other selections.

If you are using WebCC, you can specify web category and reputation. Categories include categories for web sites like shopping or travel. When you define the web category, you must also define the reputation. HPE Aruba Networking has partnered with WebRoot to assign each URL a reputation score (1-20 = high-risk, 21-40 = suspicious, 41-60 = moderate-risk, 61-80 = low-risk, and 81-100 = trustworthy). When you specify a reputation in a permit rule, the AOS firewall permits matching traffic with that reputation or higher. When you specify a reputation in a deny rule, the AOS firewall denies matching traffic with that reputation or lower.

You can only configure two actions: deny and permit. You can also apply extended actions if you want.

Example Role: Role Structure

You will now look at an example of some application and WebCC rules that many corporations apply.

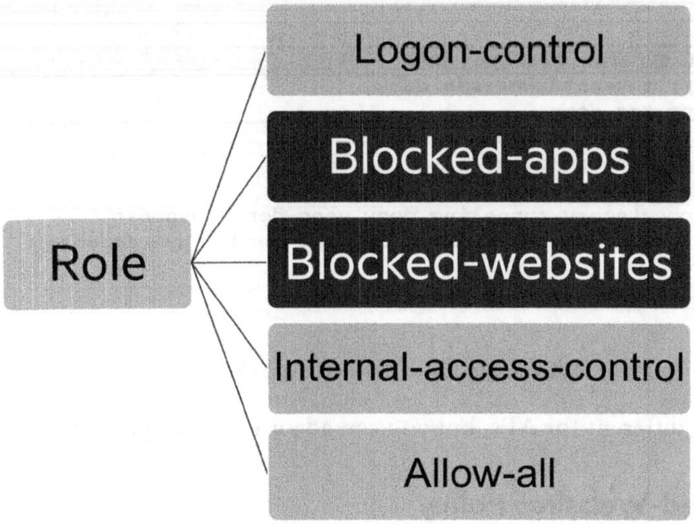

Figure 4-15: Example Role: Role Structure

In Figure 4-15, you see a basic structure for the role.

It begins with a standard "logon-control" policy, which permits DHCP, DNS, and denies IPv6 router advertisements.

It then has policies for "blocked-apps" and "blocked-websites."

Next the role could have a policy or policies for controlling the users' access to the corporate LAN. The rules in these policies depend on the corporation's particular needs, and they are not the focus for this example.

Finally, the policy permits all other traffic, which will be traffic destined to the Internet. The total effect is that users can access the Internet, but the applications and sites that they access will be restricted by the "blocked-apps" and "blocked-websites" policies.

You can create a similar role for IAPs, including ones managed by HPE Aruba Networking Central. However, rather than defining rules in separate policies, you would define them all within the role.

Example Role: Blocked-Apps Policy

In Figure 4-16, you see an example of the applications that many corporations block in a blocked-apps policy. You can adjust this policy based on your organization's particular requirements. For example, a school offering access to students might want to prohibit Youtube, or prohibit most Youtube channels, but permit educational Youtube channels. Application policies give you the power to meet specialized requirements like these.

```
ip access-list session blocked-apps
    user any app torrentdownloads deny
    user any app tor deny
    user any app snapchat deny
    user any app bitcoin deny
    any app gmail-chat deny
    user any appcategory tunneling deny send-deny-response
    ipv6 user any appcategory tunneling deny send-deny-response
```

Figure 4-16: Example Role: Blocked-Apps Policy

Again, the syntax is slightly different for APs, but you can adapt this example to them.

Example Role: Blocked-Websites Policy

In Figure 4-17, you see an example of a blocked-websites policy. Note that this policy is particularly restrictive, making it appropriate for a guest role when guests might be minors. You would, of course, communicate with your organization's stakeholders to customize a policy appropriate for each role.

```
ip access-list session blocked-websites
    user any web-cc-category abortion deny
    user any web-cc-category abused-drugs deny
    user any web-cc-category adult/pornography deny
    user any web-cc-category alcohol/tobacco deny
    user any web-cc-category cheating deny
    user any web-cc-category cult-and-occult deny
    user any web-cc-category dating deny
    user any web-cc-category dead-sites deny
    user any web-cc-category gambling deny
    user any web-cc-category gross deny
    user any web-cc-category hacking deny
    user any web-cc-category hate/racism deny
    user any web-cc-category illegal deny
    user any web-cc-category malware-sites deny
    user any web-cc-category marijuana deny
    user any web-cc-category nudity deny
    user any web-cc-category keyloggers/monitoring deny
    user any web-cc-category peer-to-peer deny
    user any web-cc-category spyware/adware deny
    user any web-cc-category spam-urls deny
    user any web-cc-category proxy-avoidance/anonymizers deny
    user any web-cc-category phishing deny
    user any web-cc-category violence deny
```

Figure 4-17: Example Role: Blocked-Websites Policy

Again, the syntax is slightly different for APs, but you can adapt this example to them.

Role-to-Role ACLs

On MCs using an AOS-8 architecture under a Mobility Conductor (MCR), you can also implement role-to-role ACLS. A role-to-role ACL allows or denies communications between clients in different roles on a role-by-role basis. For example, you might want to permit users in the Finance role to access clients in the printer role, but not to access other user clients.

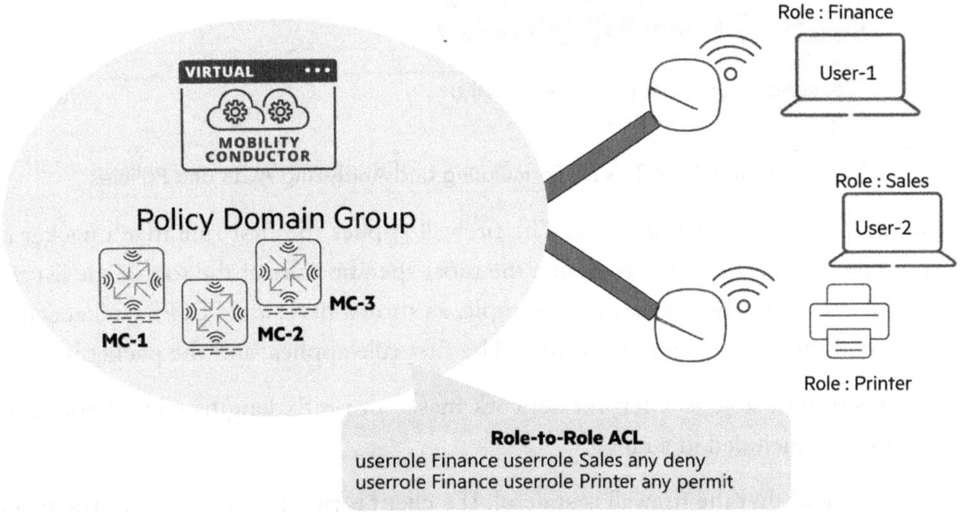

Figure 4-18: Role-to-Role ACLs

Starting from AOS 8.6.0.0, multi-MC deployments support role-to-role ACLs between two clients terminating on different MCs. (Earlier versions only permitted role-to-role ACLs between clients on the same MC.) In the example scenario shown in Figure 4-18, the MCs are all part of the same Policy domain group. Finance users are anchored on MC-1, Sales users are anchored on MC-2, and the printer has anchored on MC-3. Based on the sample Role-Role-ACL which is configured on MC-1, user traffic from the Finance user to the Sales user is denied, but traffic from the Finance user to the Printer client is allowed.

Tips for Formulating and Analyzing ACLs and Policies

As you formulate ACLs and policies within a rule, you need to continuously assess it to make sure that the policy does what you expect. In a moment you will practice doing just that. First consider a few tips, summarized in Figure 4-19.

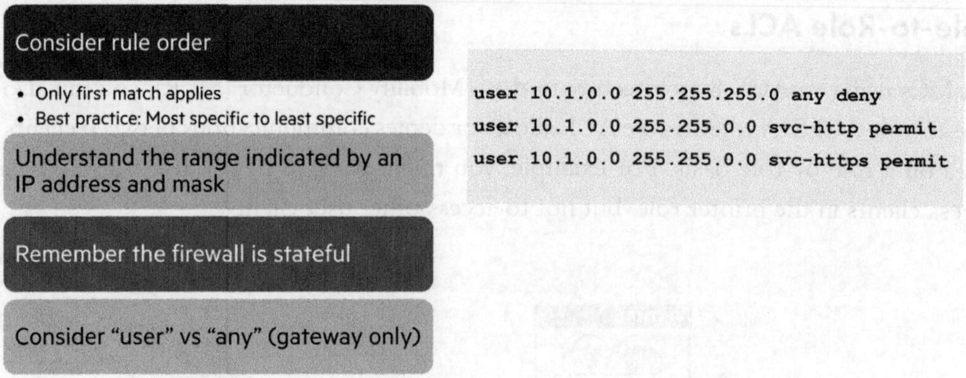

Figure 4-19: Tips for Formulating and Analyzing ACLs and Policies

It is important to consider rule order. The firewall applies the first rule that a packet matches and only that rule. It is best practice to define the most specific rules at the top of the list to ensure that packets will always hit those rules. For example, as shown here, an HTTPS packet sent to 10.1.0.24 would match both the first and third rules. The first rule applies, and the packet is dropped.

You also need to be able to interpret network masks or prefix lengths so that you can understand the IP addresses included in a range.

Also keep in mind that the firewall is stateful. If a client is permitted to send traffic to a destination, the destination can send traffic in return.

For MCs and gateways, remember the difference between "user" (IP addresses of clients assigned to the role) and "any" (all IP addresses). Note that IAPs' access control rules, configured within roles, do not specify a source; the source is always clients assigned to the role.

Subnet Mask Review

To understand which rules are more specific than others, you need to understand a bit about subnet masks. Remember: the more 1s in the binary mask, the more specific the match because each 1 indicates a bit that must match. For example, 10.1.0.0 255.255.0.0, or 10.1.0.0/16, matches all IP addresses from 10.1.0.0 to 10.1.255.255 (the first two octets must match the network address, and the second two octets do not need to). 10.1.0.0 255.255.255.0, or 10.1.0.0/24, matches all IP addresses from 10.1.0.0-10.1.0.255.

User vs Any Review

On the subject of "user" vs "any," "user" matches all IP addresses that belong to users assigned to this role, while "any" matches all IP addresses, including on devices on the other side of the firewall (connected to trusted ports). A rule that permits users to send any traffic to any destination allows users to initiate sessions and send requests to devices on the other side of the firewall. But those devices on the other side of the firewall cannot initiate sessions or send requests to clients assigned to the role. A rule that permits any source to send any traffic to any destination allows both users and devices on the other side of the firewall to initiate.

Keep in mind that the MC or gateway discovers the user IP address when the user is assigned that address by DHCP; the rule that permits DHCP traffic must use "any" as the source.

Summary

Congratulations! You can now use CPPM to authenticate clients, classify them, and tell AOS authenticators to assign the clients to roles. You can also implement role-based policies with the AOS firewall. You can take instructions from your security team and translate them into effective rules.

Learning Checks

1. An AOS device uses 802.1X to authenticate clients to ClearPass Policy Manager (CPPM). How does HPE Aruba Networking generally recommend you assign roles to clients that successfully authenticate?

   a. Using the authentication default method

   b. Using server-derived rules

   c. Using user-derived rules

   d. Using the Aruba-User-Role VSA sent by CPPM

2. If a client attempts to access a particular subnet, the company wants to lock that client out of the network for a period. How can you enforce this action?

   a. Enable DPI and create an application rule that denies traffic sent to that subnet.

   b. Enable WebCC and create an application rule that denies traffic sent to that subnet.

   c. Create a firewall rule that denies traffic sent to that subnet and also applies denylisting (blacklisting).

   d. Create a firewall rule that denies traffic sent to that subnet and also applies NAT.

3. A customer has APs managed by HPE Aruba Networking Central. Based on Aruba-User-Role VSA sent by CPPM, an AP has assigned a client to a role named "loan-officers." Refer to Figure 4-20 to see the rules associated with this role. Then consider the following questions. Take notes on your answers.

   a. To which IP address are loan officers permitted to send DNS requests?

b. The domain controller is at 10.254.1.31. What services (list by name or protocol/port) are loan officers permitted to access on the domain controller?

c. Are loan officers permitted to access web services on 10.254.1.15?

d. Are loan officers permitted to access web services on 10.254.1.24?

e. Are loan officers permitted to access web services on 10.1.19.1?

f. Are loan officers permitted to access the Internet?

```
allow udp on port 67 to all destinations
allow dns on server 10.254.1.31/255.255.255.255
allow tcp on port 53 on server 10.254.1.31/255.255.255.255
allow http on server 10.254.1.31/255.255.255.255
allow kerberos on server 10.254.1.31/255.255.255.255
allow tcp on port 88 on server 10.254.1.31/255.255.255.255
allow tcp on port 135 on server 10.254.1.31/255.255.255.255
allow netbios-ssn on server 10.254.1.31/255.255.255.255
allow https on server 10.254.1.31/255.255.255.255
allow smb-tcp on server 10.254.1.31/255.255.255.255
allow tcp on port 389 on server 10.254.1.31/255.255.255.255
allow udp on port 389 on server 10.254.1.31/255.255.255.255
allow tcp on port 1023-65434 on server 10.254.1.31/255.255.255.255
deny any to network 10.254.1.0/255.255.255.240
allow http to network 10.254.1.0/255.255.255.0
allow https to network 10.254.1.0/255.255.255.0
allow msrpc-tcp to network 10.254.1.0/255.255.255.0
allow msrpc-udp to network 10.254.1.0/255.255.255.0
allow https on server 10.1.19.11/255.255.255.255
allow tcp on port 6658 on server 10.1.19.11/255.255.255.255
deny any to network 10.0.0.0/255.0.0.0
allow any to all destinations
```

Figure 4-20: ACL Rules

Note

For answers to all learning checks, refer to the Appendix.

---

5 Evaluate Endpoint Posture

EXAM OBJECTIVES

✓ Evaluate endpoint posture

Assumed Knowledge

- Authentication, Authorization, and Accounting (AAA) framework and basics of protocols such as RADIUS and TACACS+
- Forms of Layer 2 authentication such as 802.1X
- PKI and digital certificates
- Fundamental ClearPass Policy Manager (CPPM) components, such as services, roles, enforcement profiles, and enforcement policies

Overview

A user might be trusted, but the user's device might have insecure settings, leaving it vulnerable to compromise. To obtain a Zero Trust Security (ZTS) network, companies require a way to evaluate devices' "posture," or compliance with best security practices. They further need a way to take action based on what is found—for example, granting full access to compliant endpoints but quarantining non-compliant or insecure endpoints.

This chapter will begin with an overview of the HPE Aruba Networking ClearPass posture architecture and key concepts related to posture assessments. ClearPass supports many options for posture assessment. This study guide has time to focus on just a couple examples. You will learn how to use the ClearPass OnGuard agent to run posture checks on managed endpoints, as well as how to create policies with CPPM to use the information collected.

You will be introduced to how ClearPass integrates with Microsoft Endpoint Manager (also called Intune) to evaluate endpoints' posture.

ClearPass Posture Architecture

In this section, you will learn about the ClearPass posture architecture.

Why Evaluate Posture?

Up to now, you have been primarily looking at how CPPM can use information about clients' identity to control those devices' access. Identity is a crucial aspect to understand and on which to base access decisions. However, for true zero trust security, companies must go beyond identity to include other forms of context as well.

Posture provides another crucial piece of context on which to base access control decisions. Posture refers to the state of settings on the client OS and whether those settings comply with the company's security policies. For example, is the client's firewall enabled? Is the browser set up securely? Does the client have all required patches installed? Posture could also refer to whether the client is infected with malware. These are just some of the examples of settings that can be assessed to determine a client's posture.

Many companies find it difficult to control all of these settings on every client. Companies need a tool like ClearPass, which can dynamically assess every client's posture and then use that assessment to adjust the client's rights on the network. For example, many companies now have employees who work both on- and off-site at various times. When employees take their clients off-site, those clients might be subjected to more threats, which the employees could then introduce back into the corporate environment when they return on-site.

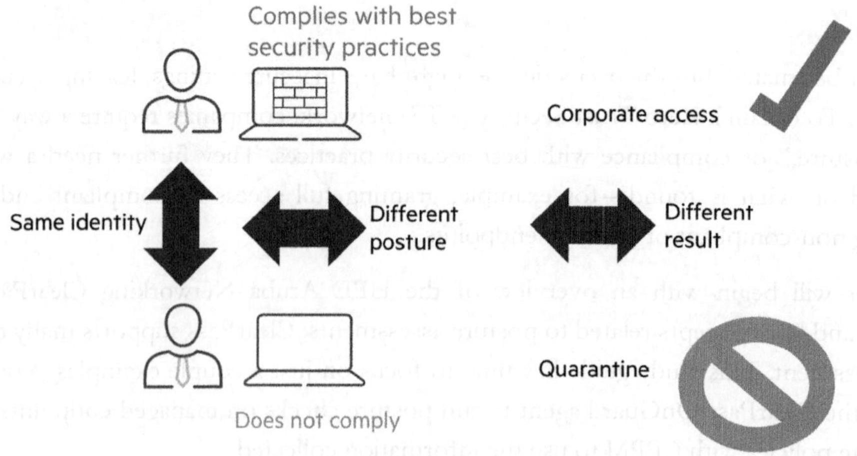

Figure 5-1: Why Evaluate Posture?

It makes sense to grant a different level of access to a client that is set up securely and a client that is vulnerable to attack, even if exactly the same user is on the two different clients, as shown in Figure 5-1. Often the company wants to "quarantine" the insecure client. In other words, the insecure client is granted very limited access, only sufficient to reach a remediation server which gives the user guidelines for bringing the client into compliance and correcting its posture.

ClearPass Posture Tokens

ClearPass tracks clients' posture using application posture tokens, which represent the clients' health, as shown in Figure 5-2.

Figure 5-2: ClearPass Posture Tokens

You may configure these tokens to take on different meanings, but they are generally used in this way:

- A Healthy token means that the client is compliant, and no restrictions should apply to network access.
- A Checkup token means that the client is compliant, but there is an update available for the client. A client can proactively remediate to a healthy state.
- When a client evaluation is in progress, the client receives a Transient token. Typically, this occurs when auditing a client. The network access granted in this case is interim.
- A Quarantine token means that the client is out of compliance and should get restricted network access. This means the client should only have access to the remediation servers.
- An Infected token returns when the client is infected and threatens other systems in the network. ClearPass should deny or severely restrict network access.

- When ClearPass cannot obtain the client's posture token, it returns an Unknown token. This might occur if there is no posture agent assigned to the client, if the client agent is unable to communicate with ClearPass, or the client agent hasn't performed a posture check.

Whenever a client first connects to the network, their posture token will always be unknown. After the agent checks in, ClearPass policies can update the token to any one of the other five depending on conditions.

Tokens are information about the endpoint. As the ClearPass administrator, you have the ability to interpret the tokens in any way that you wish. As you will see, you do this by writing enforcement rules to evaluate the token. For example, a rule might state:

IF Tips:Posture EQUALS Healthy THEN employee-full-access profile.

OnGuard Options for Posture

CPPM can receive information about clients' posture using ClearPass OnGuard.

The OnGuard agent is available as a persistent agent or dissolvable agent. The persistent agent installs an application on the user's device. Windows, Mac OSX, and Linux clients are supported. The dissolvable agent is web-based and does not require the user to install a permanent OnGuard application on their device. There are some advantages to using the persistent agent, such as auto-remediation capabilities, which are not possible with the dissolvable agent. Auto-remediation helps to resolve the client's issues by automatically enabling or disabling a function like a firewall. Furthermore, the persistent OnGuard agent performs a large range of posture checks, including checking for peer-to-peer applications, virtual machines, USB device presence, and many others. In light of this, most users choose to implement the OnGuard agent for posture analysis.

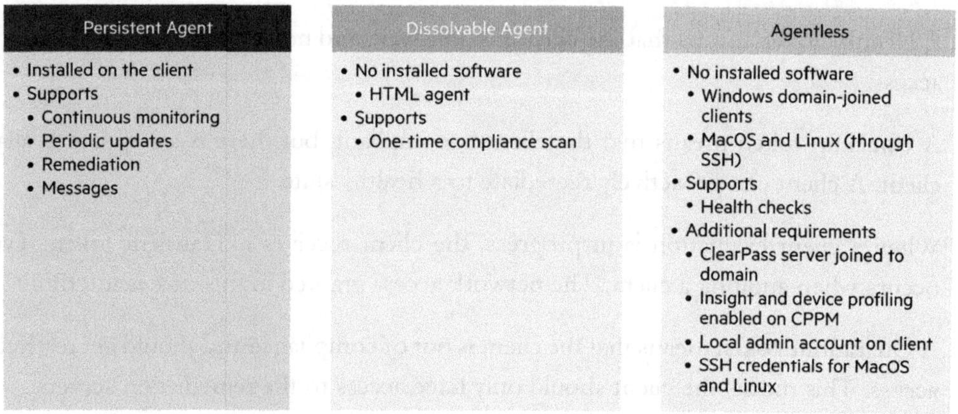

Figure 5-3: OnGuard Options for Posture

OnGuard also supports an agentless option. Agentless OnGuard supports health checks and posture analysis on Windows clients without installing the persistent OnGuard agent. It supports Windows clients as well as servers joined to the domain. CPPM can also provide agentless scanning of MacOS and Linux clients using SSH. To support agentless OnGuard, ClearPass servers themselves must be joined to the domain. You must enable Insight and device profiling on CPPM. The devices that you want to scan must also be joined to the domain, and they require a local admin account. If you want CPPM to scan MacOS or Linux devices, you must configure CPPM with the SSH credentials for their devices under **Administration > External Accounts**. CPPM uses its device profiling capabilities for agentless scanning. You will learn more about device profiling in a later chapter.

These options are summarized in Figure 5-3.

Prerequisites for Agentless OnGuard

More detailed prerequisites for agentless OnGuard, in ClearPass v6.9, are as follows:

- On CPPM:
    - Policy Manager Insight must be enabled.
    - Endpoint device profiling must be enabled.
    - The CPPM server must be joined to the domain.
    - The server requires OnGuard licenses or Compliance Suite Licenses.
- On Windows Client:
    - Windows endpoints should be joined to a domain.
    - Administrative Shares must be enabled on Windows endpoints.
    - Server Message Block (SMB) v2 should be enabled on Windows endpoints.
    - C:\Windows\winexesvc.exe and other Agentless OnGuard files should be whitelisted in Antivirus and Firewall applications.
    - In Policy Manager 6.9.3 and earlier releases, PowerShell scripts must be allowed on Windows endpoints. Starting with Policy Manager 6.9.4, Agentless OnGuard for Windows uses an .exe file instead of a PowerShell script to launch Agentless OnGuard on Windows endpoints.
- On the Network:

- Access to TCP Ports 135, 139, and 445 should be allowed.
- For DHCP profiling, the CPPM server IP address should be configured as an IP helper on the network devices.

Interaction of Information from Multiple Sources

To determine the client's system posture, CPPM primarily uses posture policies. These policies define how OnGuard agents check clients' posture and how the results of those checks translate to an application posture token for the client.

CPPM can also integrate with audit servers like Nessus. The Nessus audit server settings include a scan profile, which defines how the server's checks translate to an application posture token for the client. Note that audit servers might also have rules that determine a role for a client.

As shown in Figure 5-4, if CPPM receives application posture tokens for the same client from multiple sources, CPPM assigns the client a system posture token based on the lowest-rated application posture token. It then uses this posture token when applying enforcement policy rules.

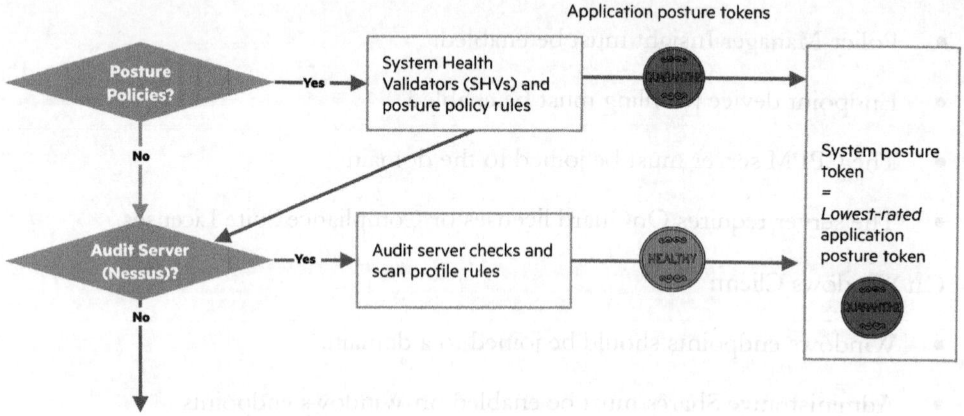

Figure 5-4: Interaction of Information from Multiple Sources

ClearPass Posture Actions

As shown in Figure 5-5, ClearPass can take a number of actions based on the posture tokens. Often you make the change in token trigger a change of authorization (CoA). Then the client re-authenticates and is assigned a new enforcement profile based on posture-based rules within the authentication service's enforcement policy. You could alternatively use a posture action to place the client in a quarantine VLAN directly.

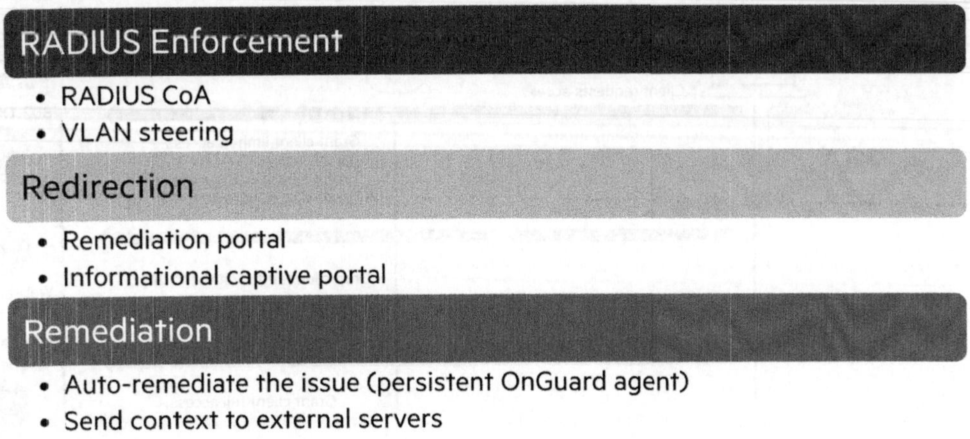

Figure 5-5: ClearPass Posture Actions

If the user's Windows security patches are out of date, you could redirect the device to a remediation portal to download the requisite application, patches, and hotfixes. Optionally, you could display a simple message to contact support or IT.

If you use a persistent OnGuard agent, you can auto-remediate the client to automatically enable a firewall, update the antivirus software, or disable peer-to-peer applications.

Evaluate Endpoint Posture with the OnGuard Agent

This section explains how to evaluate endpoint posture with the OnGuard agent.

Example of Posture Evaluation in Action

As you learned in the previous section, ClearPass provides many options for enforcing posture assessment. You will now examine a basic configuration for implementing the OnGuard persistent agent and for using posture as part of an 802.1X service.

First, examine the process by which a client authenticates, receives a posture token, and receives network access in this scenario.

When the client first requests access, the Network Access Device (NAD), such as the AP, blocks everything except Layer 2 authentication traffic, as shown in Figure 5-6. This allows the client the ability to send authentication traffic to CPPM but does not allow the agent to send an SHV report. The authentication request matches the 802.1X service on CPPM. This CPPM has an enforcement policy with a rule that matches "unknown" posture tokens.

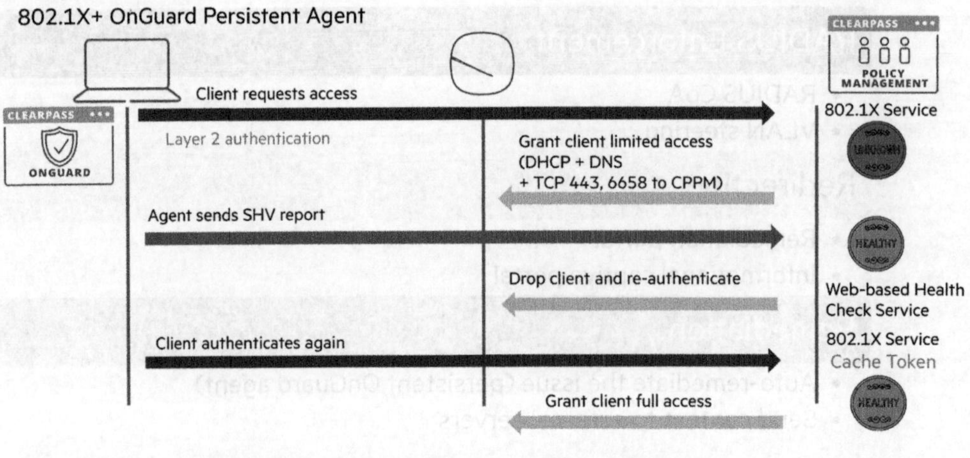

Figure 5-6: Example of Posture Evaluation in Action

This rule applies an enforcement profile, which instructs the Network Access Device to grant limited access only. With a limited role on the Network Access Device, the client can get an IP address with DHCP. It also needs DNS services. It must also be able to make an HTTPS (TCP 443) connection with CPPM and also send traffic to CPPM on TCP 6658.

This level of access allows the agent to send an SHV report to CPPM. A Web-based Health Check Service on CPPM reads the SHV report from the agent and updates the posture token.

You must configure this service to also send a terminate session message to the Network Access Device, forcing the client to re-authenticate.

It is important to note that the posture token must be cached in the service, so the updated token is available for the second authentication. If you do not select the Cached Policies and Roles option on the 802.1X service Enforcement tab, then the new posture status will disappear as the client re-authenticates.

When the client authenticates for the second time, the 802.1X service in CPPM will reevaluate the enforcement rules and, this time, because of the cached posture token, it will assign the client a new role and grant full access.

You can also use posture evaluation with Layer 3 authentication methods such as captive portal. However, this topic is beyond the scope of this study guide.

Ongoing Connectivity and Temporary Tokens

The posture tokens assigned to the endpoints are temporary and expire in five minutes if there is not an update from the agent, as shown in Figure 5-7. If the agent is continually sending updates to

ClearPass, the token will continually get updated. If the client leaves the network for more than five minutes, the token expires and returns to an unknown token.

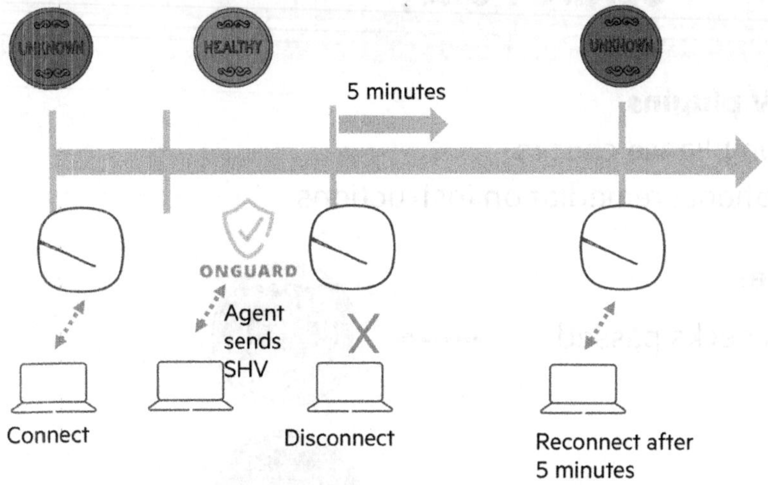

Figure 5-7: Ongoing Connectivity and Temporary Tokens

This ensures that when any client connects to the network the status is always unknown until the agent has time to check-in and update the status.

Overview of Posture Policies

To complete this configuration, you must create at least one posture policy. As shown in Figure 5-8, the posture policy defines one or more System Health Validator (SHV) plugins, each of which consists of a set of compliance checks, as well as instructions for remediation.

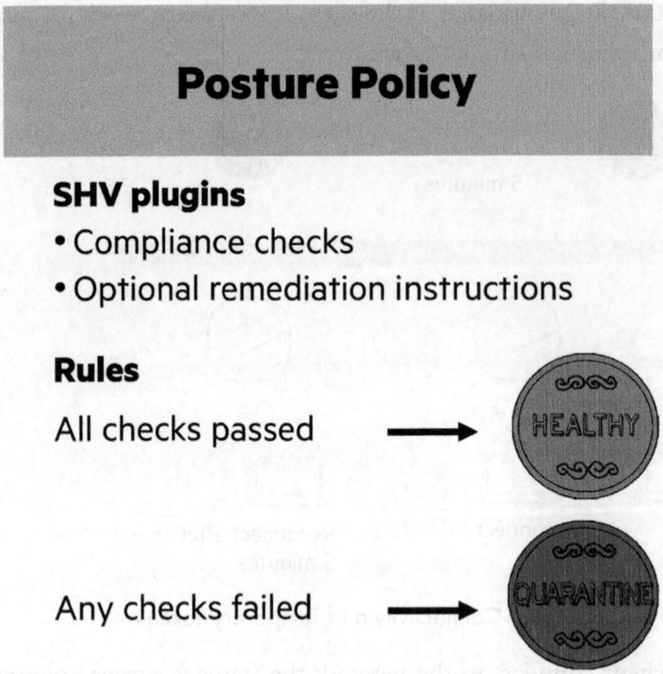

Figure 5-8: Overview of Posture Policies

The posture policy further includes rules, which define the posture token assigned to each client based on the results of the SHV checks. These example rules apply the Healthy posture to clients that pass all checks and the Quarantine posture to clients that fail any checks.

More Details on Posture Policies

Now that you have an overview of the configuration, you will look at some of the new components in a bit more detail.

Add a New Posture Policy

You can configure the posture policies in the sidebar menu under **Configuration > Posture > Posture Policy**. Click **Add**.

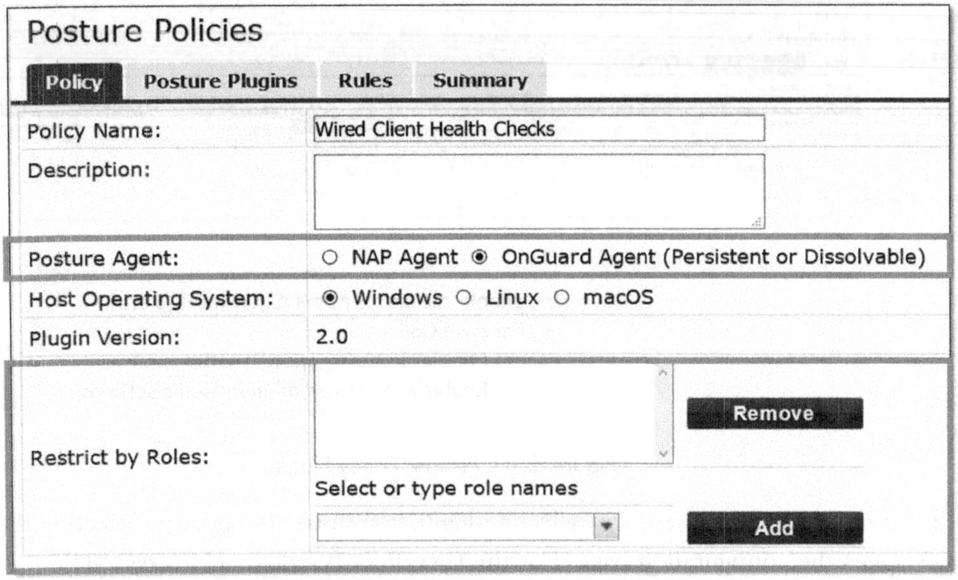

Figure 5-9: Add a New Posture Policy

When adding a new posture policy, you will need to give the policy a name and select OnGuard Agent as the Posture Agent. (See Figure 5-9.) (The NAP Agent is a legacy option.) You also select on which OS the agent will run.

You can also restrict the policy to only certain ClearPass roles. You might use this option when the company wants to apply one set of restrictions to general users and another set of restrictions to a specialized user group, for example. Based on identifications and role assignments, CPPM could instruct the agent differently depending on the user type.

You should configure a different posture policy for each family of OS that you want to assess, such as one posture policy for Windows operating systems (including multiple client and server OS versions) and one posture policy for Linux operating systems.

Posture Policy Plugins

On the Posture Plugins tab, you will need to configure one of the plugins. This study guide focuses on the ClearPass Windows Universal System Health Validator, which is supported by the OnGuard agent. (The Windows NAP agent supports the Windows System SHV and Windows Security SHV plugins.)

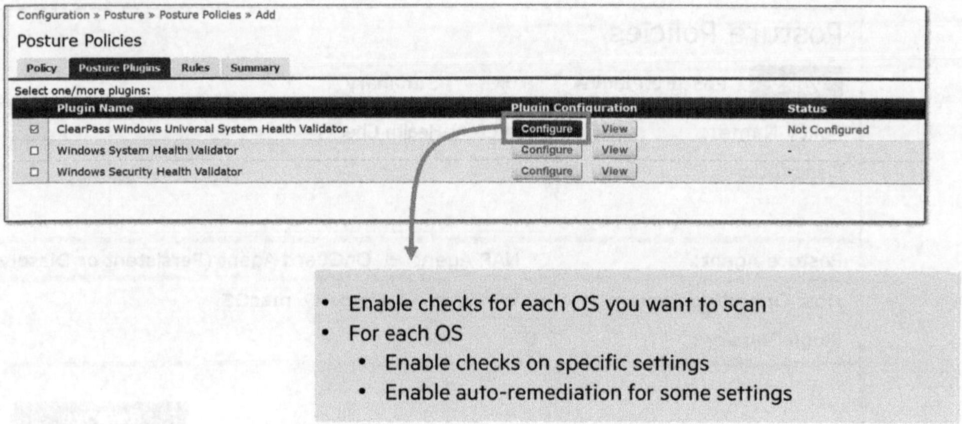

Figure 5-10: Posture Policy Plugins

As shown in Figure 5-10, when you select a plugin, you must configure it. Selecting the Configure button opens the configuration page. For the ClearPass Windows Universal SHV, you can select each OS that you want to check. You can expand various settings menus under the OS and select options. For example, you can check whether the firewall is on, whether specific services are running or not running, whether various hotfixes are applied, and many other settings.

For some settings, you can select an auto remediation option, which enables the OnGuard agent to try to fix the problem itself. If you want the agent to auto-remediate different features, you will need to select auto-remediation in the settings as well as configure auto-remediation in the service. The auto-remediation setting in the service executes the settings you have configured.

The OnGuard agent has an option for messaging where you can configure a user notification message. This message pops up on the user's screen telling them what they are in violation of—in this example, the firewall. You can also configure a quarantine message that sends when ClearPass applies the quarantine token.

After you configure the desired checks for one OS, you can then select other operating systems and configure the checks for them. When you include multiple operating systems within the same posture profile, ClearPass applies the correct checks, based on the current client OS, during the Web-Based Service process.

Many of the settings have the option to allow the system to check for a specific product and version, or you can check for any valid product.

Posture Policy Rules

You can apply OnGuard tokens through a set of rules. These rules are relatively simple in scope. You have the option for "Passes all SHV checks," "Passes one or more SHV checks," "Fails all SHV checks," or "Fails one or more SHV checks." For each of these options, you select the corresponding posture token. HPE Aruba Networking recommends you do not assign the transition token and the unknown token as the system uses these. When you configure the rule, you must also select the plugin for which the rule applies.

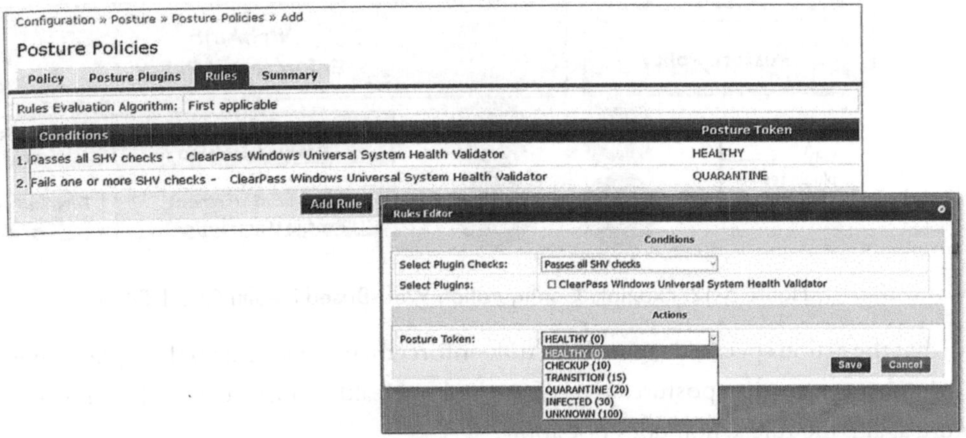

Figure 5-11: Posture Policy Plugins

The example in Figure 5-11 shows a go/no-go set up. The client either passes all checks and receives the healthy posture, or it fails one check and is assigned to the quarantine posture.

These rules do not discern which metrics the client health check passed or failed. For example, for the "Passes one or more SHV checks" condition, the client could pass any one of the checks and the result is the same. If you want to control clients based on whether they pass or fail particular metrics, you must create policy enforcement rules that reference the Posture namespace and specific checks. However, this configuration is beyond the scope of this study guide.

Example Configuration Web-Based Health Check Service

In addition to creating a posture policy, you must create a WebAuth enforcement policy. As shown in Figure 5-12, this policy's rules use posture tokens as conditions and define the actions that should be taken on clients, when they first receive that posture. For example, a rule might say "If a client has a healthy posture, apply the Agent Enforcement Profile (Healthy)." Agent enforcement profiles define the posture actions that you examined earlier. In this scenario, though, you are not using the agent enforcement profiles for their enforcement and remediation capabilities. You will simply use them to send a message to the OnGuard agent. Instead you are also attaching a second profile to each rule that tells the Network Access Device (NAD) to terminate the session. The NAD then forces the client to log in again, at which point the 802.1X service can take over.

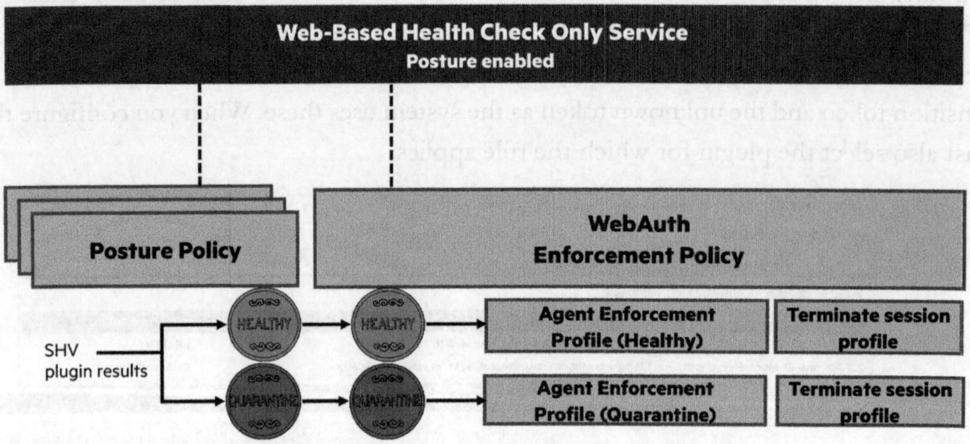

Figure 5-12: Example Configuration Web-Based Health Check Service

Note that the action specified in the WebAuth enforcement policy rule only applies when the client's posture changes to that posture. If a client had a Healthy posture and then receives a Healthy posture again, the rule action does not apply.

You pull together the posture policy and WebAuth enforcement policy in a Web-Based Health Check Only service, which is responsible for receiving incoming SHV reports. You can reference multiple posture policies in a service as long as the policies do not apply to the same client. You reference just one enforcement policy.

Example Configuration 802.1X Service

When you are using posture and 802.1X together, as in this example, it is important that you configure the 802.1X service's enforcement policy to take posture into account. Then, when a client is forced to reauthenticate by a change in its posture, the 802.1X service can use that posture to apply the appropriate profile.

HPE Aruba Networking generally recommends referencing the posture in the enforcement policy, rather than the role mapping policy. This ensures CPPM will take the latest posture token into account.

You can create an 802.1X service using similar guidelines to those discussed earlier in the study guide. You can also adjust an existing service. When configuring the enforcement settings, make sure to enable the option for caching roles and posture from previous sessions. Otherwise, CPPM will always consider clients to have unknown posture, and they will never be allowed to connect fully.

You can then create or adjust a RADIUS enforcement policy to apply the desired profiles to clients based on their posture and other context. In the example shown in Figure 5-13, an enforcement policy had a rule that assigned clients with the "employees-cppm" role to the "employee-full-access" enforcement profile. Now you add another condition to that rule. The client must have a healthy posture.

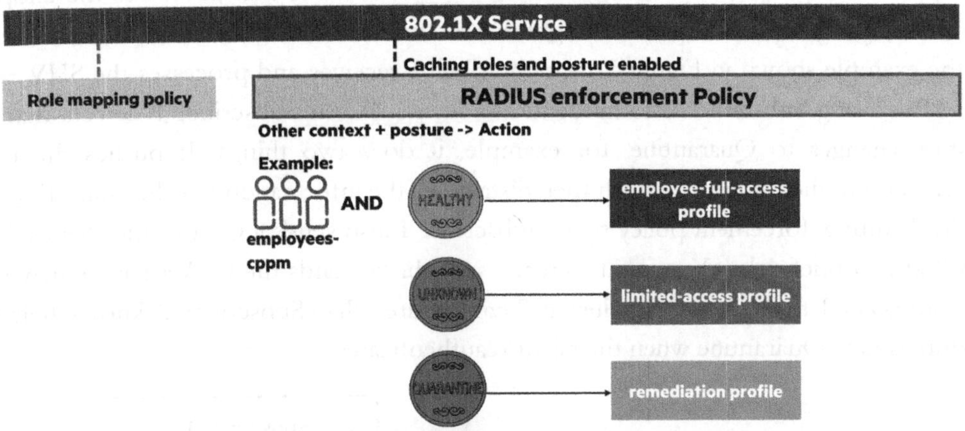

Figure 5-13: Example Configuration 802.1X Service

Because clients begin with an unknown posture, you need a rule for that posture. In this example, that posture applies a "limited-access" profile, which sends an Aruba-User-Role VSA. The NADs should be configured to allow very limited access to the corresponding role. For example, the role might only permit the traffic necessary for the OnGuard agent to communicate with ClearPass and obtain a posture (DHCP traffic, DNS traffic, HTTPS traffic to the ClearPass server, and TCP port 6558 to the ClearPass server). Some companies might also choose to permit Internet access for these clients.

You must also create a rule that specifies the quarantine posture and an appropriate profile for those clients. Again, you would typically have this profile send an ex VSA set to a name like "remediation," and the HPE Aruba Networking infrastructure devices apply appropriate firewall policies.

You could even create more complex rules, in which, for example, quarantined employees of different types and quarantined contractors receive different profiles, as required by the security team.

ClearPass Clusters and CoAs

This scenario features a common setup, in which the WebAuth enforcement policy applies a CoA profile. This profile triggers CPPM to send a CoA to the client's Network Access Server (NAS) so that the client can reauthenticate and be assigned a new enforcement profile, based on its new posture.

If you are using a ClearPass cluster, you need to consider the time that it takes for the endpoint's new posture to propagate across the cluster. The Subscriber that processes the endpoint's SHV report and assigns the new posture might not be the same Subscriber that authenticates the client. In the example shown in Figure 5-14, Subscriber 1 receives and processes the SHV report, while the AP is using Subscriber 2 to authenticate clients. When Subscriber 1 detects that the client's posture changes to Quarantine, for example, it does two things. It pushes the new posture information to the Publisher, which then distributes the information to other subscribers. Based on the WebAuth enforcement policy rules, Subscriber 1 also sends a CoA to the client's AP (or other NAS) after a brief delay. You want to make sure that it sends the CoA after the new information has propagated throughout the cluster. That ensures that Subscriber 2 knows that the client's posture is now Quarantine when the client reauthenticates.

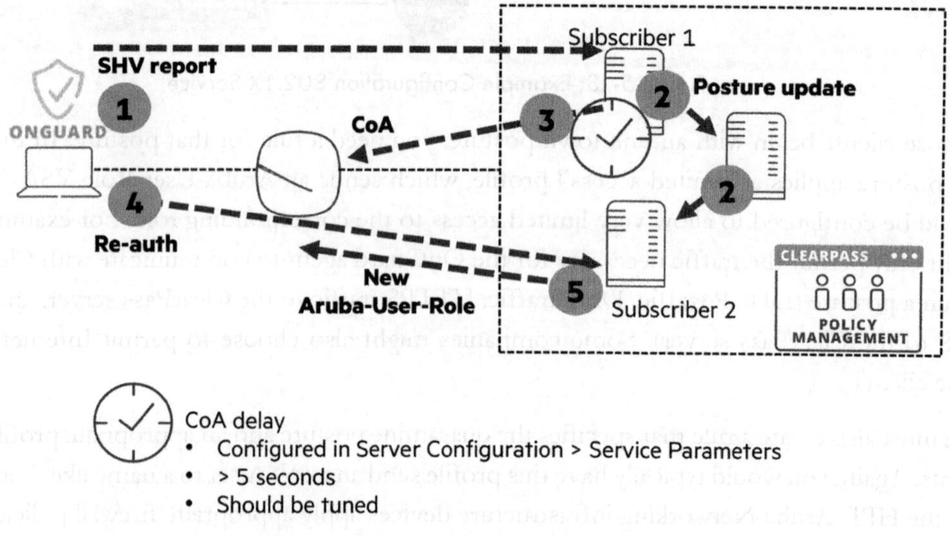

Figure 5-14: ClearPass Clusters and CoAs

By default, CPPM uses a 2-second delay between when it decides to issue a CoA and when it sends the CoA. But you can adjust the delay in each server's configuration under Service Parameters. Most solutions require at least a 5 second delay. However, the time required to propagate information depends on many factors such as cluster size and network latency. You might start with a 5 second delay, run some tests, and then increase the delay if required.

You should also note that the Subscriber that processes the SHV report sends the CoA, even if that is not the Subscriber with which the NAS is configured to use as its authentication server. So, make sure to configure your NASes to accept CoAs from all servers in the cluster.

Microsoft Endpoint Manager Integration

This section explains how CPPM can integrate with Microsoft Endpoint Manager.

Third-Party Integration for Endpoint Information

CPPM can also receive information about endpoints by integrating with third-party solutions in the 360 Security Exchange program. CPPM integrates with a variety of mobile device management (MDM) and mobile application management (MAM) solutions, including Microsoft Endpoint Manager, formerly called Intune, which is an Azure cloud-based MDM. It also integrates with AirWatch, Citrix XenApp, and Ivanti LANDesk. These solutions can send information about endpoints' settings, as well as applications running on them. CPPM can also integrate with McAfee ePolicy Orchestrator to receive information about endpoints' compliance with various policies. For a complete list of Exchange partners, visit https://www.arubanetworks.com/partners/technology-partners/partner-finder and filter for "360 Security Exchange" under programs.

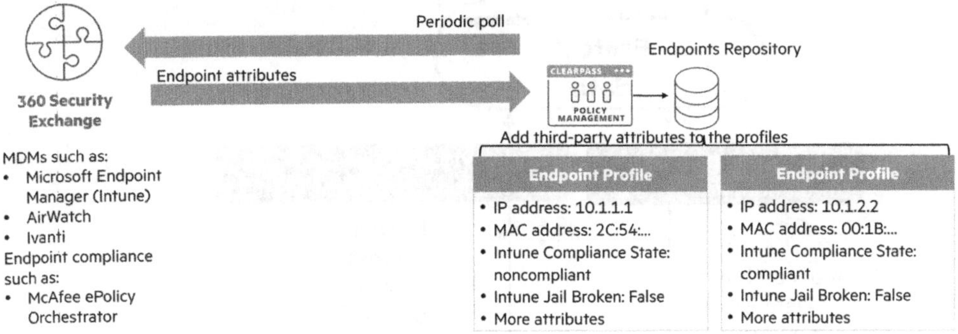

Figure 5-15: Third-Party Integration for Endpoint Information

The integrated MDM and endpoint compliance solutions do not specifically set CPPM posture tokens. Instead, CPPM periodically polls them for endpoint attributes, as shown in Figure 5-15. CPPM adds these attributes to the client's endpoint profile within the CPPM Endpoints Repository, which is a database that CPPM maintains with a profile for each client that has authenticated to it (or about which CPPM has learned in another way). These endpoint attributes provide rich information about the client's status and posture. ClearPass admins can then use these attributes in role mapping or enforcement policy rules. In this way, they can assign clients to CPPM roles and enforcement profiles based on information relevant to the company's specific security policies.

Microsoft Endpoint Manager

Microsoft Endpoint Manager is an MDM and MAM system, which runs in the Microsoft Azure cloud. Microsoft manages a broad array of operating systems, including Apple iOS, Apple iPadOS, Apple MAC, Google Android, and Windows. For a complete list, refer to Microsoft documentation.

Endpoint Manager integrates with Azure Active Directory to regulate users' access to the corporation's Microsoft Office 365 and other SaaS applications. Admins can create application protection policies to control which applications users in various groups can access and what users can do within those apps. The company can further create different policies for devices that are enrolled in Endpoint Manager and for unmanaged devices. Endpoint Manager also provides profiles and policies to configure mobile devices.

Of most interest to you for this study guide, though, is Endpoint Manager's ability to assess mobile device's compliance status. Endpoint Manager integrates with Mobile Threat Defense, which feeds it information about the devices' security compliance and posture. Endpoint Manager can then provide that information to a Network Access Control partner such as CPPM, as shown in Figure 5-16. (CPPM must run version 6.7.2 or above to integrate with Endpoint Manager.)

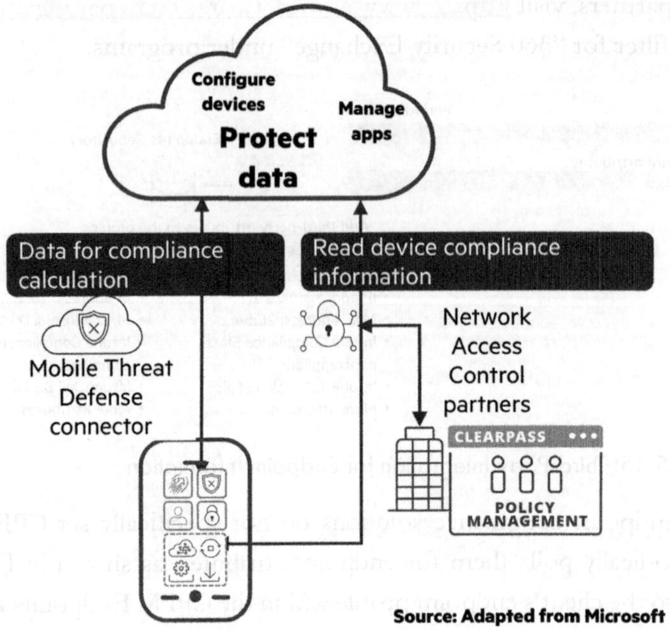

Figure 5-16: Microsoft Endpoint Manager

CPPM Integration with Endpoint Manager

When you integrate CPPM with Endpoint Manager, CPPM can use Endpoint Manager's information about the mobile device's posture to make its decisions. The extension for this integration is still called "Intune." The process, as it occurs when the Intune extension uses Periodic Sync mode (added in v5) is illustrated in Figure 5-17.

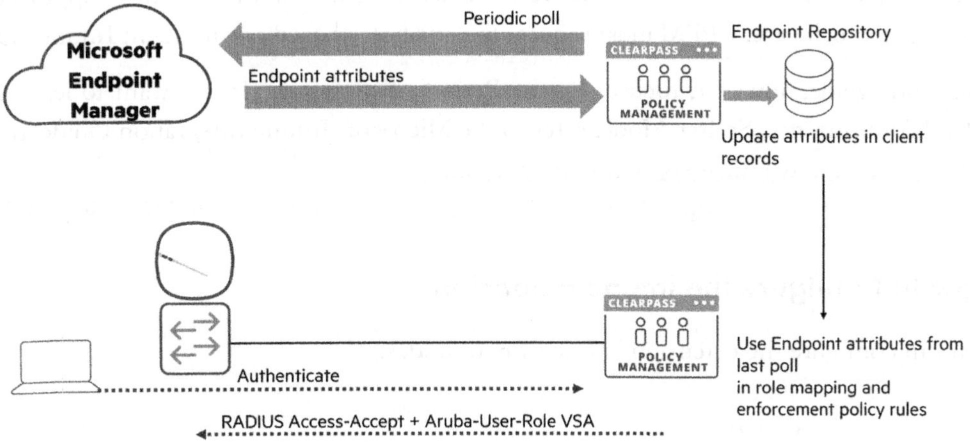

Figure 5-17: CPPM Integration with Endpoint Manager

In this mode, CPPM periodically polls Microsoft Endpoint Manager for endpoint attributes. The default polling interval is 30 minutes, but you can change this interval when you configure the Intune extension. At the polling interval, CPPM pulls down the attributes from Microsoft Endpoint Manager in batches. It then adds the attributes with their latest values to the corresponding endpoint record in its Endpoint Repository.

When a client authenticates, CPPM uses its role mapping and enforcement profile policies to assign enforcement profiles to the client. These rules can use Endpoint attributes received from Microsoft Endpoint Manager in their conditions. For example, you might have a role mapping rule that assigns clients with an Endpoint attribute of "Intune Compliance State EQUALS compliant" to a "compliant-cppm" role. Then you can edit enforcement profile rules to require clients to have the "compliant-cppm" role status to receive full access.

Note that CPPM uses the endpoint attributes that it received in the last polling period. If an endpoint's attributes change in a later poll, the endpoint remains authenticated with the current profile. But when the client reauthenticates, CPPM will use the new information.

HTTP Authorization Source Mode

The Intune extension v5 and later also supports HTTP Authorization Source Mode. When the extension uses this mode, you add Intune as an authorization source and specify that source in the services that authenticate clients. CPPM then dynamically calls Microsoft Endpoint Manager to receive the latest attributes during the authorization process. It can request the attributes for the client in question based on MAC address. However, MAC address is not supported for wired devices, in which case, CPPM must request the attributes based on the client Intune ID.

This study guide, however, focuses on the Periodic Sync mode. If you want to learn more about HTTP Authorization Source Mode, refer to the Microsoft Intune Integration Guide available here: https://community.arubanetworks.com/community-home/digestviewer/viewthread?MID=27076#bm4318f5bb-0e1b-45de-8f6e-c3ef8648bd0d

How to Configure the Intune Extension

You can configure the extension in one of two modes.

Periodic Sync Mode

To configure CPPM to use the Intune Extension in Periodic Sync mode, you follow this basic process (see Figure 5-18):

- You register CPPM in Microsoft Azure. During this process, you will also collect information, which you will later use to configure the Intune extension.
- You install the extension and configure settings for it.
- You can then use Intune attributes in rules within your enforcement policies.

Figure 5-18: How to Configure the Intune Extension

HTTP Authorization Source Mode

If you want to use HTTP Authorization Source Mode, you must also set up Intune as an authorization source. This option is covered in the expert level training. You can also refer to the Microsoft Intune Integration Guide available here:

https://community.arubanetworks.com/community-home/digestviewer/viewthread?MID=27076#bm4318f5bb-0e1b-45de-8f6e-c3ef8648bd0d

Register CPPM in Microsoft Azure

To integrate with Microsoft Endpoint Manager, you must register CPPM as an application within Microsoft Azure. These instructions assume that an Intune (Endpoint Manager) Tenant Admin account has already been set up in the company's Azure solution.

Access Azure at https://portal.azure.com. Log in with the Intune Tenant Admin account.

Open Azure Active Directory and select App Registrations. Click New registration. Specify an intuitive name like "CPPM-CompanyABC." Accept default settings and click Register. You will then see the registered application as well as information about it. Record the values for "application (tenant) ID" and "client (directory) ID."

Next you must assign the registered application several permissions. Click Add API permissions. Add at least these minimum permissions:

- Intune (or Endpoint Manager)
    - get_device_compliance
- Microsoft Graph
    - DeviceManagementApps.Read.All
    - DeviceManagementManagedDevices.Read.All
    - DeviceManagementManagedDevices.ReadWrite.All
    - User.Read

In Azure, access the client secrets for the applications and add a new secret. Take care to copy out the secret as soon as you create it because it will not be visible again. You can set an expiration period of your choice.

For more details on these steps, refer to the Microsoft Intune Integration Guide.

Install the Extension

You are now ready to install the Intune Extension. You complete this step using the ClearPass Guest pages. Go to **Administration > Extensions** and choose to install an extension, as shown in Figure 5-19.

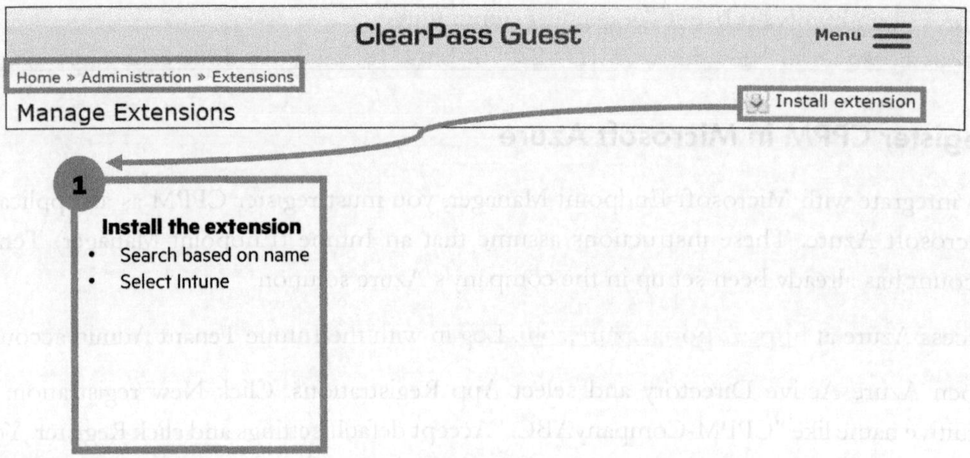

Figure 5-19: Install the Extension

You can then search for the extension by name. You will see the Intune extension, which you can then choose to install. You will also be asked to give Intune an extension IP address. When you use the v5 extension, you can let ClearPass set the extension IP address automatically.

The extension will then install.

Configure the Extension

After the installation, you should choose to edit the extension configuration. You will see XML settings like the ones shown in Figure 5-20.

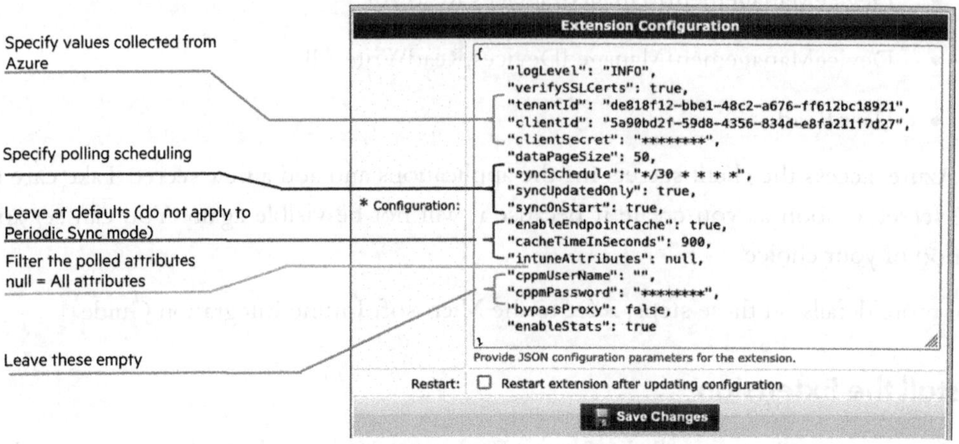

Figure 5-20: Configure the Extension

Specify the information that you previously collected from Azure in these attributes:

- "tenantID"—You collected the tenant ID when you registered CPPM as an application in Microsoft Azure Active Directory.
- "clientID"—You collected the client ID when you registered CPPM as an application in Microsoft Azure Active Directory.
- "clientSecret"—You set this value when you created the client secret in Microsoft Azure Active Directory.

If you want, you can edit the polling scheduling. The value for "syncSchedule" uses cron syntax:

- First value = Minute (0-59)
- Second value = Hour (0-23)
- Third value = Day of the month (1-31)
- Fourth value = Month (1-12)
- Fifth value = Day of the week (0-6 in which 0 is Sunday)

An asterisk permits all values. The default setting establishes polling at a 30-minute interval: "*/30 * * * *"

When "syncUpdatedOnly" is set to true, CPPM only pulls down attributes that have changed since the last poll. This setting speeds up polling, particularly when CPPM polls frequently. The "syncOnStart" attribute determines whether or not CPPM pulls down the current attributes when the extension first starts.

You can leave the "enableEndpointCache" and "cacheTimeInSeconds" attributes at their default values. These attributes only apply when you are using HTTP Authorization Source mode.

In version 6.9 and later, you should leave the "cppmUserName" and "cppmPassword" attributes undefined.

Use "intuneAttributes" if you want to specify the list of attributes that CPPM collects from Endpoint Manager. For example, you could set the value to the following: ["complianceState", "deviceEnrollmentType", "jailBroken"]. If you leave the default value, null, CPPM collects all attributes.

You can generally leave other attributes at their default values.

When you are done, select the check box for restarting the extension and save your changes. The extension will restart with the correct configuration.

Use Intune Attributes in CPPM Policies

CPPM will now receive endpoint attributes from Microsoft Endpoint Manager. It will store the attributes for each client in the Endpoint Repository record for that client. You can set up role mapping and enforcement policies that use those attributes. To use the attributes in a rule, set the condition Type to **Endpoint**, as shown in Figure 5-21. In the Name field, type **Intune** to see the list of attributes delivered by Endpoint Manager. For example, you could create a condition that checks the endpoint's compliance state: **Endpoint Intune Compliance State EQUALS compliant**. This is just one example of what you can do.

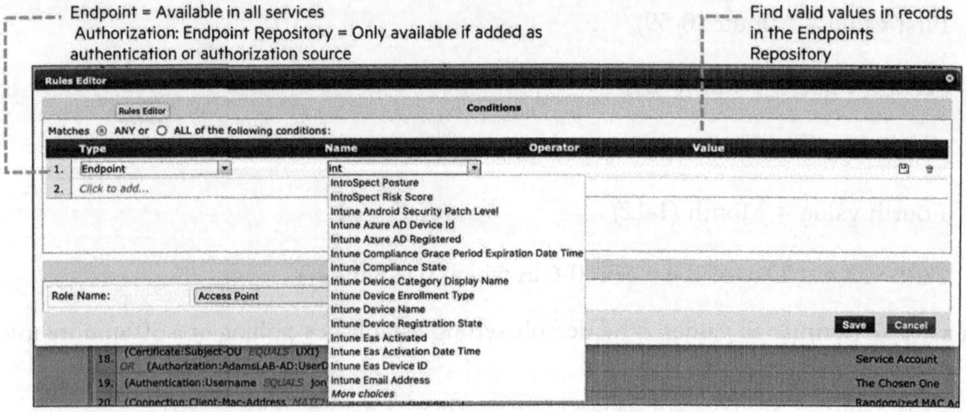

Figure 5-21: Use Intune Attributes in CPPM Policies

You can look in records within the Endpoint Repository to find examples of the Intune attributes and their values. Work with the IT security team to formulate policies that meet that particular organization's needs.

CPPM always has access to Endpoint attributes even if you have not enabled authorization in the service. However, if you want to specify attributes through the Endpoint Repository, you must enable authorization in the service to which you want to apply the endpoint enforcement policy. Then add the Endpoint Repository as an authorization source in the **Authentication** tab.

Summary

Congratulations! You can now set up ClearPass OnGuard agent-based health checks. You can also use the postures derived from these health checks in CPPM enforcement policies. You also learned how to integrate CPPM with Microsoft Endpoint Manager (Intune) to obtain rich information about the posture of clients managed by that third-party solution. You can now better control clients' access based on their compliance with key security policies.

Learning checks

1. What is one difference between the OnGuard persistent agent and the dissolvable agent?
    a. The persistent agent only works for Windows systems while the dissolvable agent supports Windows and Linux.
    b. The persistent agent supports ongoing monitoring while the dissolvable agent supports a one-time check.
    c. The persistent agent integrates with Microsoft Intune while the dissolvable agent offers no integration.
    d. The persistent agent posture overrides the posture delivered by any other source while the dissolvable agent posture is overridden by other sources.
2. Match each component with its role.

    Component

    a. Posture policy
    b. WebAuth enforcement policy
    c. Web-based Health Check Only service

    Role

    __ Provides rules that assign enforcement profiles based on system posture

    __ Provides SHV plugin configuration and rules for assigning posture based on SHV checks

    __ Responds to requests from OnGuard agents
3. You have added an Intune extension, and now you need to configure it. How do you determine what to specify for the "tenantId," "clientId," and "clientSecret?
    a. You should always leave the default values for these fields.
    b. You should consult with HPE Aruba Networking support to assess the right values for your environment.
    c. You should register ClearPass as an App in Azure AD and collect the values there.
    d. You should use XML files available in GitHub to look up the right values.

Note

For answers to all learning checks, refer to the Appendix.

---

6 Implement a Trusted Network Infrastructure

EXAM OBJECTIVES

- ✓ Set up secure authentication and authorization of network infrastructure managers, including,
- ✓ Advanced TACACS+ authorization
- ✓ Multi-factor authentication
- ✓ Secure L2 and L3 protocols, as well as other protocols such as SFTP

Assumed Knowledge

- Basics of managing AOS-CX switches
- Authentication, Authorization, and Accounting (AAA) framework and basics of protocols such as RADIUS and TACACS+
- Fundamental HPE Aruba Networking ClearPass Policy Manager (CPPM) components, such as services, roles, enforcement profiles, and enforcement policies

Overview

The foundation of trusted network communications lies on a trusted network infrastructure.

In the first section, you will learn how to implement secure management of AOS-CX devices. In this way you will ensure that only authorized managers can change switches' configuration, as well as control their rights.

It is also critical that you secure the fundamental protocols on which network communications rely. The next section covers AOS-CX features for protecting many Layer 2 and Layer 3 protocols.

Implement Secure Management of AOS-CX Switches

In this section, you will learn how to implement secure management of AOS-CX switches.

Review Management Access on AOS-CX Switches

AOS-CX switches offer two main options for accessing the management plane, as shown in Figure 6-1.

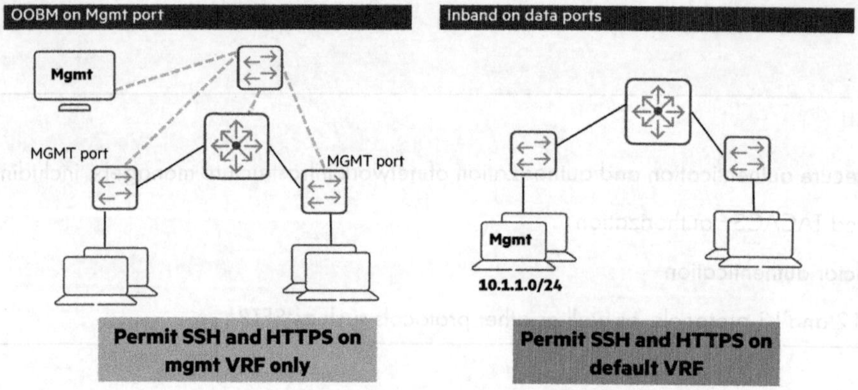

Figure 6-1: Review Management Access on AOS-CX Switches

The switches support out-of-band management through an Ethernet port called the management (mgmt) port. The management port operates at Layer 3 and has an IP address. Admins can access the CLI or Web UI at the mgmt IP address.

AOS-CX switches also support in-band management. With this method, admins reach the management plane through the same ports to which users and other endpoints connect. They contact the switch at an IP address configured on one of the switch Layer 3 VLAN interfaces or ports.

You control which access methods are available out-of-band and in-band by telling the switch on which VRFs particular protocols are enabled. The mgmt port is part the "mgmt" VRF while data ports are part of the "default" VRF, by default. If you want to confine the switch to out-of-band management, you should enable SSH and HTTPS on the mgmt VRF and disable them on the "default" VRF. This approach makes it easy to specify and determine which protocols the switch accepts from different network segments.

HPE Aruba Networking generally recommends out-of-band management as the more secure option. To achieve the best available security, you should ensure that the OOBM network and the production network are physically separated. In other words, you should connect the mgmt ports and the management stations to switches dedicate to the management network. Doing so does require extra effort and equipment. However, management switches can be inexpensive, since 1GbE will provide all the necessary bandwidth for management.

If the customer requires inband management, you should leave SSH and HTTPS enabled on the default VRF (or whatever VRFs the customer is using).

Note that enabling a protocol on a VRF or port applies to traffic using that protocol that is directed to the switch. It has no effect on traffic using that protocol but sent to another destination.

Adding a Control Plane ACL for Inband Management

When you use inband management, users can reach the switch management interfaces. You will implement authentication, as discussed in a moment, but it is best practice to implement an additional layer of security beyond manager authentication alone.

You can create an IP access control list (ACL) that specifies the types of traffic that you want to allow, as well as authorized sources. You can then apply that list to the control plane in a specific VRF. The ACL applies to all traffic from the VRF to which it is applied, destined to the switch's management interface or control plane.

The ACL uses the format shown in Figure 6-2.

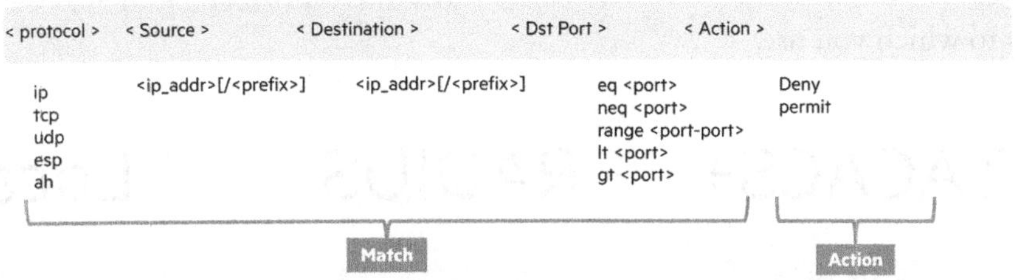

Figure 6-2: Adding a Control Plane ACL for Inband Management

Example Control Plane ACL

It is very important that you understand that the control plane ACL filters *all* IP traffic to the switch control plane as well as the management traffic. The ACL must permit *all* necessary traffic, including routing updates, RADIUS traffic, SNMP traffic, DNS traffic, and any other traffic that you want the switch control plane to accept. Carefully document all the traffic that the switch might need to receive before implementing the ACL. In Figure 6-3, you see an example control ACL that includes many necessary protocols.

```
access-list ip controlPlane
    10 permit tcp 10.1.10.0/24 any eq 22
    20 permit tcp 10.1.10.0/24 any eq 443
    30 permit udp 10.1.4.5 eq 161 any
    40 permit udp 10.1.19.10 eq 161 any
    50 permit udp 10.1.19.10 eq 1812 any
    60 permit udp 10.1.19.10 eq 1813 any
    70 permit udp 10.1.19.10 eq 3799 any
    80 permit tcp 10.1.19.10 eq 49 any
    80 permit ospf 10.1.1.1 10.1.1.3
    90 permit ospf 10.1.1.2 10.1.1.3
    100 permit udp 10.254.1.31 eq 53 any
    110 permit icmp any 10.1.1.3 icmp-type echo-reply
apply access-list ip controlPlane control-plane vrf default
```

Switch management traffic (SSH, HTTPS, SNMP)
Authentication traffic (RADIUS, TACACS+)
Routing traffic
DNS
Pings

Figure 6-3: Example Control Plane ACL

The control plane ACL filters traffic in the inbound direction. In other words, when you add a rule to the ACL, the source IP address is the device that you want to permit to communicate with the switch.

Review Management Authentication Options

You have several options for authenticating managers who reach the switch's management plane, as summarized in Figure 6-4. For the best scalability and manageability for authenticating SSH and HTTPS managers, you should generally choose an external authentication option: TACACS+ or RADIUS. Each of these protocols have their advantages, and often a company will have its own preference as to which you use.

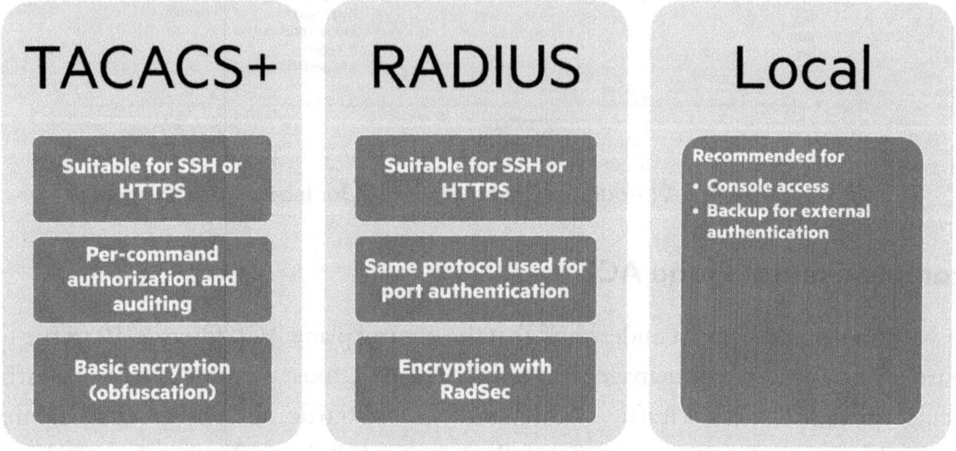

Figure 6-4: Review Management Authentication Options

TACACS+ provides the ability to control the commands that managers enter in a granular way. It also provides detailed auditing for each command that managers enter, which is a requirement for some companies. In addition, the TACACS+ packets exchanged between the switch and the authentication server are encrypted by a shared secret. While this encryption does not meet modern standards and can be cracked, it does provide a layer of obfuscation against casual eavesdropping.

RADIUS also has some features that companies might want. RADIUS is the standard authentication protocol used with port authentication like 802.1X. So if the company is also implementing port authentication, the company might choose to use the same RADIUS server to authenticate managers. RADIUS itself does not encrypt messages between the switch and the authentication server. However, if security is required, you can add RadSec.

Managers can also authenticate against a local list of users. However, HPE Aruba Networking primarily recommends local authentication as a backup method in case the TACACS+ or RADIUS authentication server is not available.

Review TACACS+ Server Setup

To use either of these external authentication methods, you must specify the authentication server on the AOS-CX switch.

First, you must identify the server by IP address or fully qualified domain name (FQDN), as shown in Figure 6-5. The latter assumes that you have configured the switch to communicate with a DNS server.

You must also specify a key, which sets a shared secret for communicating with the server. You can specify a global key used for all authentication servers configured on the switch, or you can specify an individual key when you define the server, which overrides the global key. Set the key as long as possible to enhance security. AOS-CX switches allow keys up to 31 characters in length.

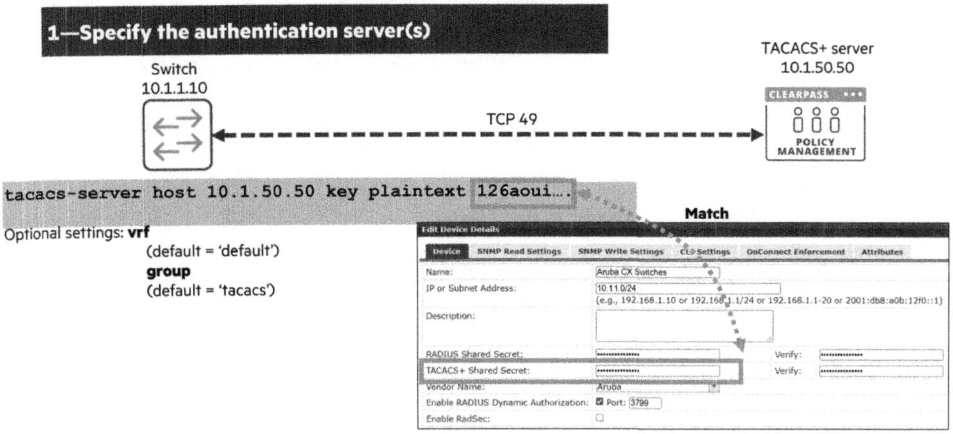

Figure 6-5: Review TACACS+ Server Setup

You must also be aware of the network topology and understand on which VRF the switch reaches the server. You can then specify this VRF when you define the TACACS+ server. If you do not define a VRF, the default VRF is used.

You can optionally place the server in a manually defined group. If you do not, the server is part of the default group for its type. For TACACS+ servers, this group is named "tacacs." For RADIUS servers, this group is named "radius."

You can also configure the authentication protocol used with RADIUS or TACACS+. The switches support CHAP or PAP with both TACACS+ and RADIUS; the default protocol is PAP. You do not need to set the protocol on CPPM. It will accept either CHAP or PAP for RADIUS messages, but only PAP for TACACS+ messages. Therefore, you should leave the switches' TACACS+ protocol as PAP.

You must verify that the authentication server is similarly set up to communicate with the switch. For the purposes of this study guide, this server is CPPM. CPPM must have the switch defined on it as a network device. The shared secret configured for this network device *must* match the key configured for the server on the switch. Otherwise, the server will not respond to the switch's messages.

Review Using TACACS+ for Manager Authentication

After you have specified the RADIUS or TACACS+ servers on the switch, in the default or other group, you can enable external authentication for specific access types. After specifying **aaa authentication**, you specify the access type (such as **ssh** or **https-server**), the primary authentication method (such as "group tacacs," "group radius," or "local"), and an optional secondary authentication method.

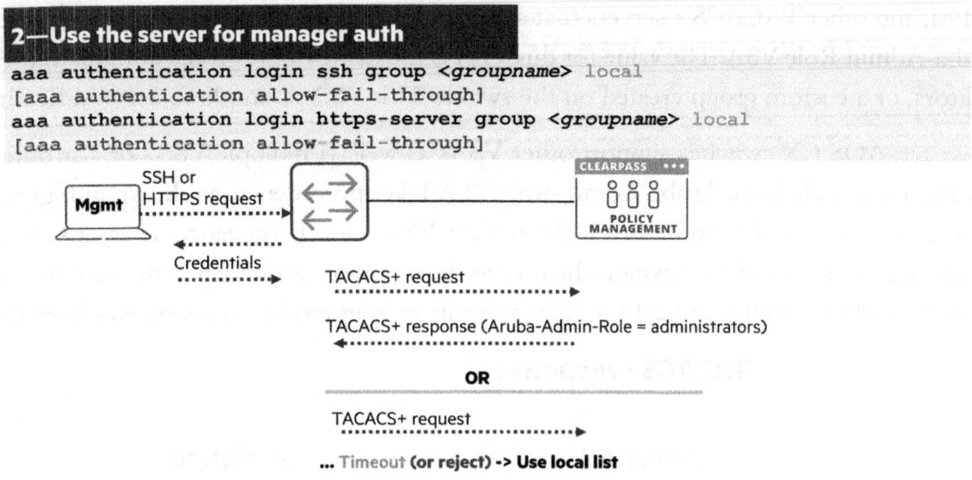

Figure 6-6: Review Using TACACS+ for Manager Authentication

When a manager attempts an SSH or HTTPS connection with the switch, the switch queries the manager for a username and password. The switch then submits them to a server in the primary authentication group specified in the **aaa authentication login** command. In the example shown in Figure 6-6, that server is CPPM. CPPM authenticates the client and returns a TACACS+ response, which rejects or accepts the login request. If CPPM accepts the request, it generally also includes an Aruba-Admin-Role VSA in the message to define the manager's role on the switch.

The switch uses the secondary authentication method if the switch does not receive a reply from the server specified in the primary method. Typically, you should specify **local** as the final option and create at least one local user account so that the switch remains manageable in case of network outages.

Setting the secondary authentication option only causes the switch to use that option when the first option times out. However, sometimes you want to use the secondary option when the first authentication server actually rejects the client. Then users can authenticate with either a remote or local account during normal operation, which some companies prefer. Also managers avoid issues in which they are not able to log into a switch because the authentication takes too long to time out. To enable the switch to use the secondary method when the primary one rejects the user, enter **aaa authentication allow-fail-through**.

Methods for Assigning Users to Manager Roles

The AOS-CX switch determines which group to assign a particular manager based on information in the TACACS+ server's response.

CPPM, and other TACACS+ servers that support vendor specific attributes (VSAs), can include an Aruba-Admin-Role VSA. The value for this VSA is the name of the group: administrators, operators, auditors, or a custom group created on the switch. This VSA is simple to use and flexible.

However, AOS-CX switches support other VSAs as well. If multiple VSAs or attribute value pairs (AVPs) are included, the Aruba-Admin-Role VSA takes precedence, as shown in Figure 6-7. If that VSA is not present, the Aruba-Priv-Admin-User VSA takes precedence. It assigns roles based on numbers from 1-15 and 19. 1 assigns the user to the operators group, 15 to the administrators group, 19 to the auditors group, and 2-14 to any manually configured local groups that have those IDs.

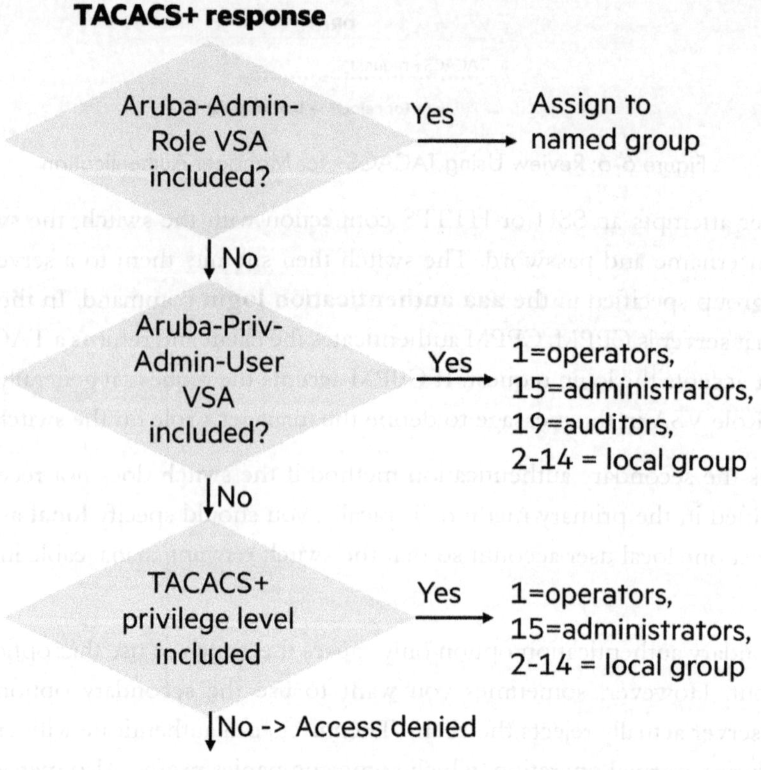

Figure 6-7: Methods for Assigning Users to Manager Roles

AOS-CX switches can also accept the standard TACACS+ privilege levels. The level takes effect for TACACS+ if no Aruba VSAs are sent. Privilege level 1 corresponds to the operators group. Privilege level 15 corresponds with the administrators group. Note that the standard TACACS+ privilege levels do not support assigning managers to the auditors group. You can also set up custom local groups to correspond with a privilege level.

Group Privileges

You can review the privileges for each group in the following sections.

Administrators

Administrators have full privileges, including full read-write access to the CLI and Web UI. They can configure any settings, including user accounts.

Operators

Operators have no configuration privileges. They can only view nonsensitive switch configuration information. In the CLI, operators can only issue show commands, and they can only view settings in the Web UI.

Auditors

Finally, auditors have access to CLI commands only in the auditor context (auditor>). Auditors can view event log contents and local accounting log contents. They can also copy command output to a remote server or a local USB drive.

Using CPPM as the TACACS+ Authentication Server

You will now look at how to configure CPPM to act as the TACACS+ server for an AOS-CX switch or AOS device.

As with all functions on CPPM, you must define a service to receive the incoming requests. You can choose to use a wizard for TACACS+ Enforcement, or you can create a service manually. In either case, the correct service type is TACACS+ Enforcement.

CPPM submits the manager's credentials to the authentication source, which is AD in the example shown in Figure 6-8. It also collects authorization attributes for authenticated clients.

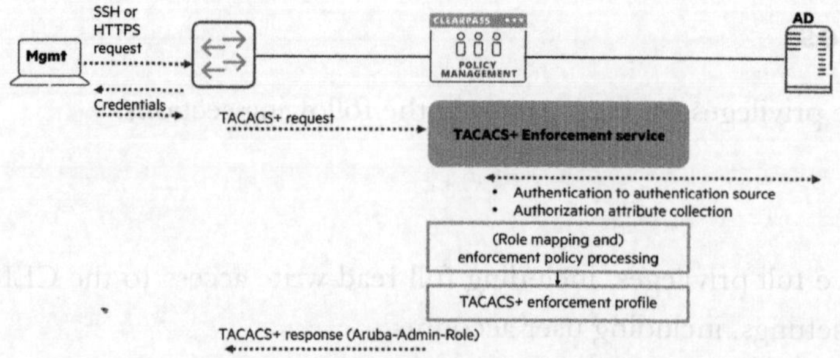

Figure 6-8: Using CPPM as the TACACS+ Authentication Server

CPPM then processes the service's role mapping and enforcement policies. In the end, it applies a TACACS+ enforcement profile. It is this profile which defines the specifications for the TACACS+ response, including the Aruba-Admin-Role VSA.

Defining TACACS+ Enforcement Profiles

You set up TACACS+ Enforcement services much like you do other CPPM services. The main new concept is the TACACS+ enforcement profile.

You will need one TACACS+ enforcement profile per admin role that you want to assign to managers on your infrastructure devices. For example, if you want to authenticate switch managers and assign them to the "administrators" and "operators" roles, you will need two profiles.

Much like the RADIUS enforcement profiles that you have already examined, a TACACS+ enforcement profile is defined by a name and type (TACACS+). The profile further specifies whether to accept or reject the request.

As shown in Figure 6-9, TACACS+ accept enforcement profiles have a **Services** tab, rather than an **Attributes** tab like RADIUS enforcement profiles. In this tab, you define the TACACS+ privilege level from 0 to 15. This privilege level determines the client's level of management access on the network infrastructure device. However, when CPPM also sends the Aruba-Admin-Role VSA, the role specified in that VSA overrides the privilege level.

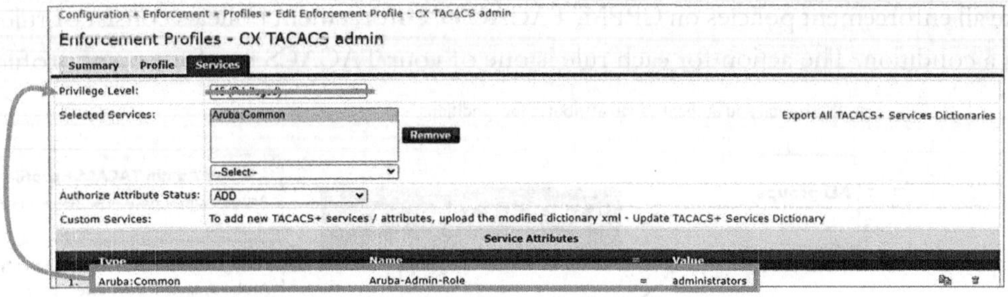

Figure 6-9: Defining TACACS+ Enforcement Profiles

In the **Services** tab, you can also add one or more services and then specify attributes or other settings for those services.

When you are using TACACS+ to authenticate managers of AOS-CX switches or AOS devices, you should add the "Aruba: Common" service. You then add the "Aruba:Common Aruba-Admin-Role" VSA and set the value to the name of the group to which you want to assign managers with this profile.

CPPM has two default TACACS+ enforcement profiles related to HPE Aruba Networking network infrastructure devices:

- **[AOS Wireless – TACACS+ Root Access]** adds the Aruba:Common service, sets the Aruba-Admin-Role VSA to "root"
- **[AOS Wireless – TACACS+ Read-Only Access]** adds the Aruba:Common service, sets the Aruba-Admin-Role VSA to "read-only"

If you want to assign managers to different roles on HPE Aruba Networking infrastructure devices, you will need to add similar enforcement profiles with those roles' names as the value for the Aruba-Admin-Role VSA.

Defining TACACS+ Enforcement Policies

Once you have defined the TACACS+ enforcement profiles, you can reference them in an enforcement policy.

You generally want one enforcement policy for each different type of network infrastructure devices that are using CPPM as a TACACS+ server. In this example, you have a policy for wired devices, your AOS-CX switches. And you have a policy for your wireless devices. This approach often works well because switches and mobility devices required different TACACS+ profiles for their managers.

Like all enforcement policies on CPPM, TACACS+ enforcement policies consist of rules. Each rule has a condition. The action for each rule is one of your TACACS+ enforcement profiles.

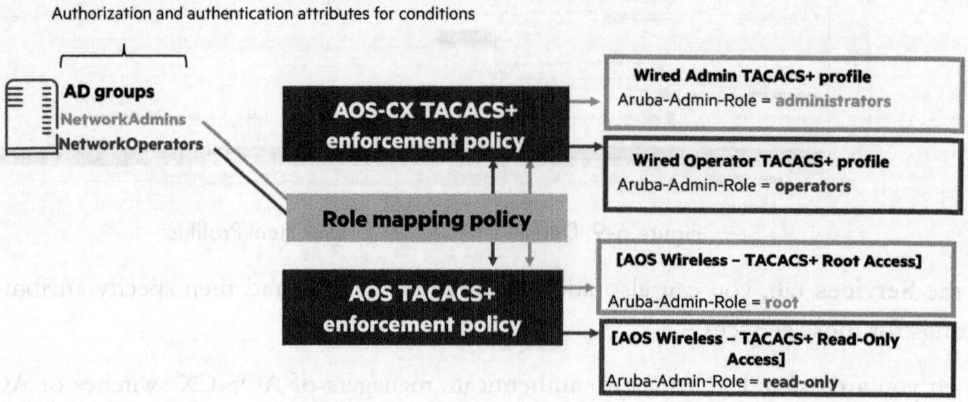

Figure 6-10: Defining TACACS+ Enforcement Policies

In the example shown in Figure 6-10, you are using the Authorization:*AD instance*:Groups attributes for conditions. In other words, managers are assigned to a profile and admin role based on AD security group. For example, the Wired TACACS+ enforcement policy assigns members of the NetworkAdmins group to the Wired Admin TACACS+ enforcement profile. And this profile assigns the managers to the "administrators" group on AOS-CX switches based on the Aruba-Admin-Role VSA.

However, you can use different attributes in conditions to fulfill different use cases.

You could also use role mapping policies and then reference ClearPass roles in the enforcement policies. However, in this example, you are not using role mapping policies; because you cannot reuse the same role mapping policy across multiple services, it is simpler to just use enforcement policies.

Example TACACS+ Enforcement Services

When you have created the TACACS+ enforcement profiles and policies, you are ready to combine those components in services. You will now look at one example way to set up these services. You can adjust the approach shown in Figure 6-11 to your particular requirements.

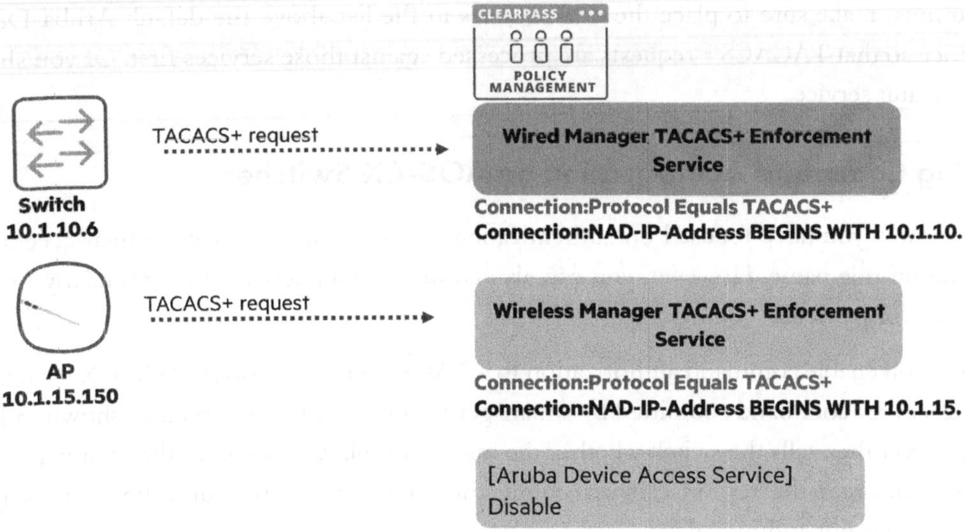

Figure 6-11: Example TACACS+ Enforcement Services

CPPM comes with a default service for authenticating TACACS+ managers on AOS devices. This service authenticates managers against CPPM's local repository. By default, this service matches all incoming TACACS+ requests. When you look at the service, you will not see any service rules. However, services with the TACACS+ Enforcement type always include an implicit rule of "Connection:Protocol EQUALS TACACS+."

If you want to authenticate managers on multiple types of devices, using accounts in a domain or directory, you will need to create new TACACS+ enforcement services. You often need different services and enforcement policies for wireless and wired managers because these devices use different names for their roles.

You can create the service manually or using a wizard. The considerations are much the same.

When you select the TACACS+ Enforcement type, the service will automatically select TACACS+ requests. You will need to add another rule that selects requests from specific infrastructure devices. One way to do this is by selecting ranges of NAD-IP-Addresses. For example, a company might give all switches at a site IP addresses in subnet: 10.1.10.0/24. You would create this rule to select TACACS+ requests from those devices: "Connection:NAD-IP-Address BEGINS with 10.1.10." You could create a similar rule in a TACACS+ Enforcement service for wireless managers that selects IP addresses on AOS devices.

For each service, you would then proceed to specify the correct authentication source and reference your enforcement policy.

You must make sure to place the new services in the list above the default Aruba Device Access Service so that TACACS+ requests are processed against those services first. Or you should disable the default service.

Using Command Authorization on AOS-CX Switches

Up to now you have focused on authenticating managers and controlling their access level based on admin role name. However, you can also control admin access more granularly by authorizing admins for particular commands.

When you enable command authorization to a TACACS+ server on an AOS-CX switch, the switch checks every command submitted by a manager to the TACACS+ server, as shown in Figure 6-12. The server then tells the switch whether the manager is allowed to enter that command or not. The switch enforces the response by executing the command or by outputting a message that the command is not authorized.

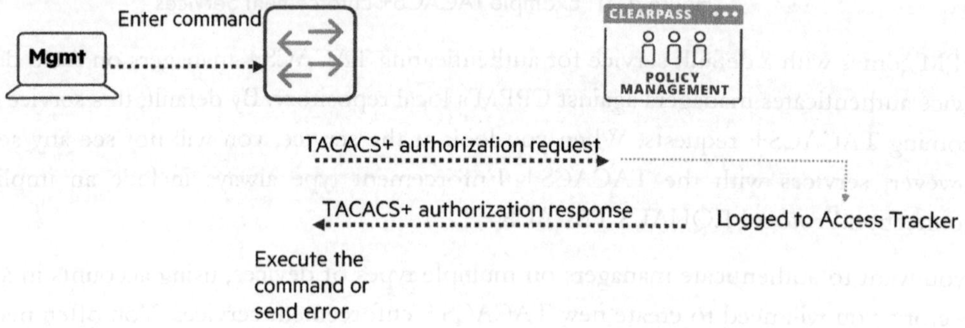

Figure 6-12: Using Command Authorization on AOS-CX Switches

In addition to providing granularity of access control over managers, command authorization also creates an audit trail on CPPM, which records in Access Tracker every command entered by each manager. You can even set up alerts in ClearPass Insight if admins enter denied commands.

Adding Command Authorization to a TACACS+ Enforcement Profile

Before you enable command authorization on the switches, you should adjust the TACACS+ enforcement profiles to implement the authorization.

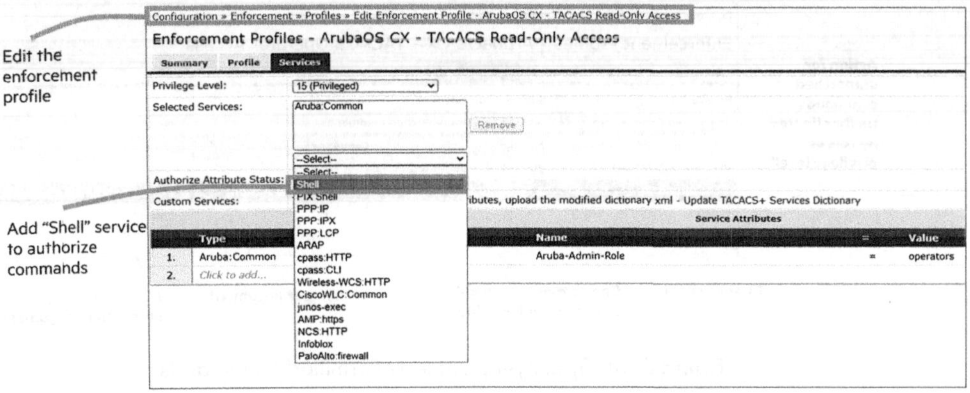

Figure 6-13: Adding Command Authorization to a TACACS+ Enforcement Profile

Edit the enforcement profile associated with the admin role to which you want to apply per-command authorization.

Then add the "Shell" service in addition to the "Aruba:Common" service. The Shell service makes the **Commands** tab appear, as shown in Figure 6-13.

Specifying Denied/Permitted Commands

The **Commands** tab includes a list of commands, which are either permitted or denied. You can choose to deny all unmatched commands (the default option) and add specific permitted commands. Or you can choose to permit all unmatched commands and add specific denied commands, as shown in this example. If you choose to permit unmatched commands, the manager is still bound by the specific limits of the assigned role. For example, this profile uses the "Aruba:Common" service to assign the operator role with a VSA and the "Shell" service to permit or deny specific commands. Permitting all unmatched commands will permit all operator commands that are not specifically denied; however, the operator will still be prohibited from entering manager commands even though those commands are not explicitly denied.

When you add a command, you specify only the first keyword in the command for the command value, as shown in Figure 6-14. For example, you could enter "show." This command entry will now act a bit like a permit or deny list itself. You can specify a list of attributes for the command and choose whether to permit or deny each of those attributes. You also select whether unmatched attributes are permitted or denied. You can use regular expressions to make the entries more flexible. For example, "show users information" will *not* match the "show" command "users" attribute because it is not an exact match to that attribute. However, "show users information" *will* match the "users .*" attribute with .* indicating a wildcard for multiple characters.

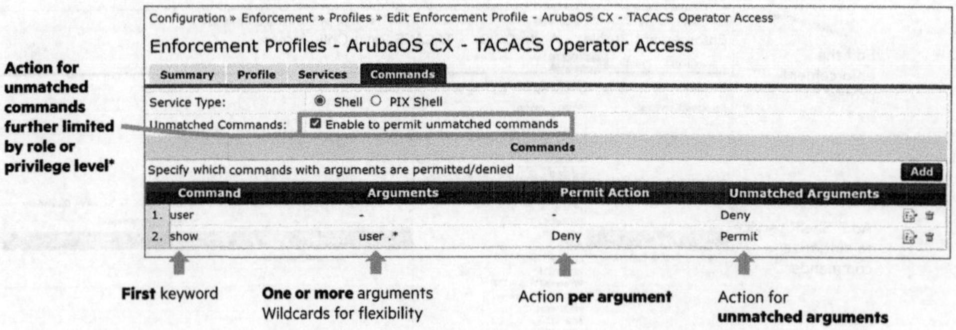

Figure 6-14: Specifying Denied/Permitted Commands

If you want to deny all arguments for a command, it is easiest to simply add the command, specify no arguments, and set the unmatched arguments setting to "deny" as shown in the first rule in this example. Similarly, if you are using a deny unmatched command approach, you can easily permit all arguments for a command by adding the command, leaving the arguments list empty, and setting the unmatched arguments to "permit."

You can add multiple commands to the list to create as granular a policy as the company requires.

This example is a profile for operators on AOS-CX switches. As a result of this configuration, operators cannot enter any command that begins "user" or "show user." They can enter all other *operator* commands. However, they cannot enter administrator or auditor commands because the role limitations continue to apply.

However, if you want you can take a different approach. You can assign all users to TACACS+ privilege level 15 and set no Aruba-Admin-Role. You can then use the commands rule list alone to control the users' privileges. This approach gives you more flexibility, but it does require you to be careful in defining precisely which commands are permitted and denied.

Enforcing Command Authorization

Once you have set up the enforcement policies with the desired command lists, you can enter this command on the AOS-CX switches to enable command authorization to CPPM: **aaa authorization commands <default | ssh | console> group <*groupname*> local**. (See Figure 6-15.) The option after **commands** specifies for which forms of access to the CLI this type of command authorization applies (default applies if either connection type does not have a method explicitly configured for it). You then specify group and the TACACS+ group to which you assigned CPPM.

```
aaa authorization commands <default | ssh | console> group <groupname> <local
| none>
```

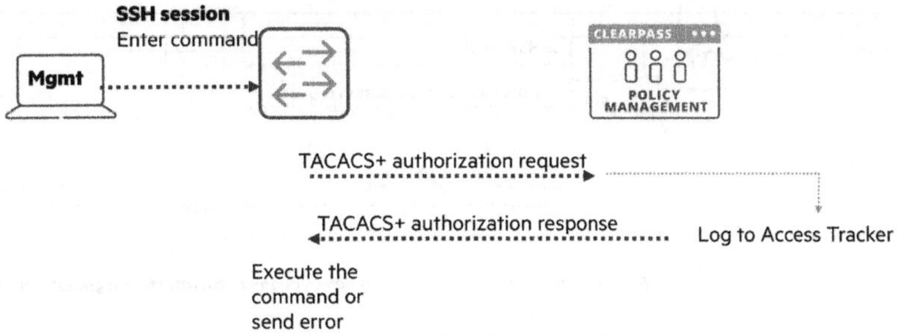

Figure 6-15: Enforcing Command Authorization

HPE Aruba Networking strongly recommends specifying **local** or **none** as the secondary authorization source. Otherwise, managers will not be permitted to enter any commands if the switch loses connectivity with CPPM. If you specify local, the switch checks the command list configured for the particular admin group to which the user is assigned. Therefore, if you choose that option, you *must* create local command lists and apply them to the group.

After you enter the **aaa authorization commands** command, the switch imposes a check each time a manager connected in the specified way (SSH or console) enters a command. Because CPPM's TACACS+ group is the source in the example in Figure 6-15, the switch sends a TACACS+ authorization request to CPPM. The request falls under the authentication request for login, which the user already passed. CPPM checks the user's enforcement profile and determines whether the command is allowed. It then sends an authorization response with its decision (permit or deny the command).

SSH Public Key Authentication

SSH also provides some additional authentication options.

You have been concentrating on password authentication. With this form of authentication, the switch checks the manager credentials—a username and password—against a local list, RADIUS server, or TACACS+ server based on the choice that you make with the **aaa authentication login ssh** command.

However, AOS-CX switches also support public key authentication for SSH. Public key authentication takes a similar approach to certificate authentication, but without a concept of a PKI. The SSH client is configured with one or more public/private key pairs, sometimes called SSH host keys. When the client is prompted to log in with public key authentication, the client submits the manager's username. It further proves its identity by also submitting that manager's public key and by signing data with the corresponding private key, as shown in Figure 6-16.

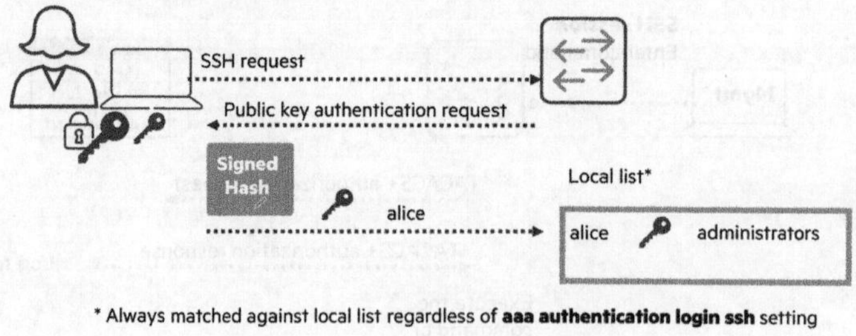

Figure 6-16: SSH Public Key Authentication

The key stands on its own without having a trusted issuer in the way a digital certificate does. Therefore, the SSH server on the AOS-CX switch must be set up with a list of authorized usernames and their public keys in advance.

Public key authentication on AOS-CX switches is always implemented against the local list of users, regardless of the setting in the **aaa authentication login ssh** command.

Setting up SSH Public Key Authentication

Public key authentication is enabled by default on AOS-CX switches. If it has been disabled, you can enable it with this command: **ssh public-key-authentication.** (See Figure 6-17.)

1—Enable public key auth (if disabled)

```
ssh public-key-authentication
```

2—Specify the user's public key

```
user alice group administrators
User alice is being created without a
password.
Do you want to continue (y/n)? y
user alice authorized-key ssh-rsa
AAAAB3NzaC1yc2EAAAADAQABAAABAQCn+/KP08mlx
Vsly02IBcKz/7+2BzUFzSPiQQDMHh3K1oP1sFM/C5
```

Single line no breaks

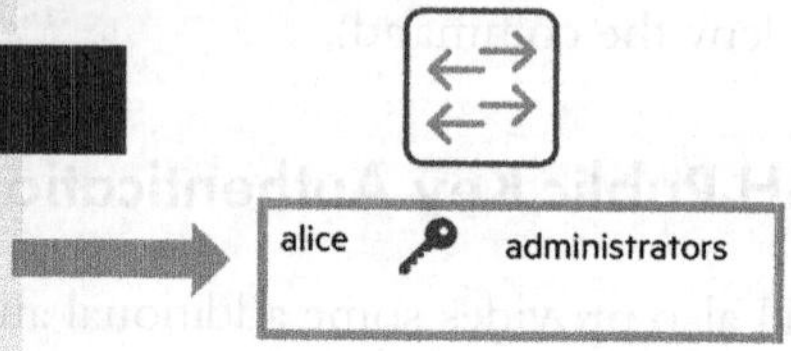

3—Disable password authentication (optional)

```
no ssh password-authentication
```

Figure 6-17: Setting up SSH Public Key Authentication

To use public key authentication, you must also create one or more user accounts and assign them to a managers group: administrators, operators, or auditors. In this example, you are adding the user without a password because you want the user to login with the public key.

You must then specify the public key that the user has installed on their SSH client. The correct syntax consists of the key algorithm, a space, and then the full key with no line breaks. After you create the account, the manager with the corresponding public and private key can connect to the switch. The manager receives the rights associated with that account on the switch without submitting a password.

If you want to use only public key authentication, you must disable password authentication with this command: **no ssh password-authentication**.

If you leave both password and public key authentication enabled, you can choose to enable two-factor authentication. The switch then requires a manager to pass *both* public key and password authentication. The switch first checks the SSH client's key against its local admin accounts. It then submits the manager's username and password to the authentication server specified for SSH.

Secure L2 and L3 Protocols

In this section, you will review how to secure L2 and L3 protocols.

Overview

Many of the protocols used to regulate Layer 2 and Layer 3 communications were developed in a time in which internal users and clients were considered inherently trustworthy. By default, these protocols accept all valid protocol packets without distinguishing between whether those packets arrive from legitimate network infrastructure devices or not.

But you need to deal with the reality that internal users might be untrustworthy or internal clients might be compromised. Hackers can exploit Layer 2 and Layer 3 protocols to manipulate where traffic is directed, eavesdrop on traffic, create DoS attacks, and launch man-in-the-middle attacks.

A Zero Trust Security network implements protection for the protocols. In this section, you will review Layer 2 and Layer 3 protections available on AOS-CX switches. As shown in Figure 6-18, these protections include BPDU protection, DHCP snooping, ARP inspection, and OSPF authentication.

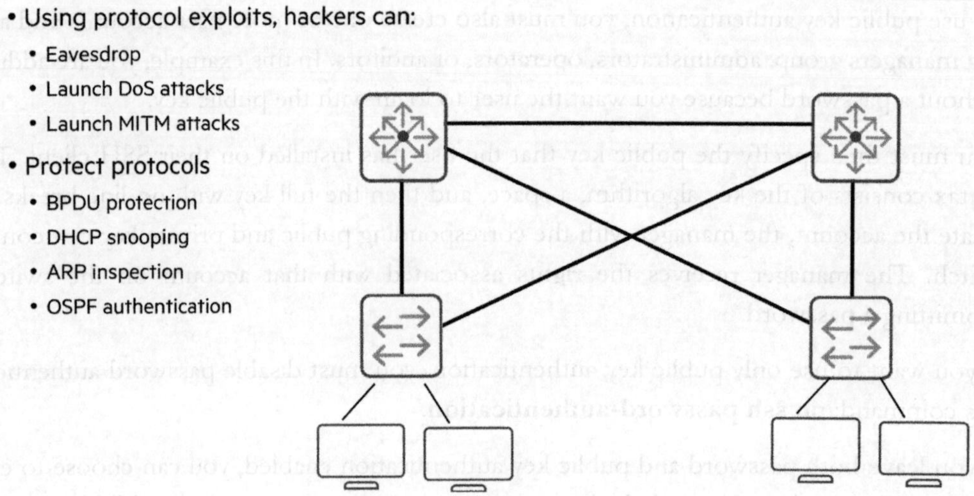

Figure 6-18: Overview

Review BPDU Protection

Without protections on spanning tree, rogue switches can connect to the network and, worse, become the spanning tree root. The new root alters paths, which can cause DoS attacks, as well as let the user on the rogue device eavesdrop on traffic or even launch an MITM attack.

BPDU protection (also called BPDU guard) prevents this situation from occurring. This feature operates at the port level and is best implemented on edge ports, which should not be receiving BPDUs.

BPDU protection monitors a port for incoming BPDUs. As shown in Figure 6-19, if the port receives a BPDU, the switch shuts down the port, protecting the network from the rogue device. You can configure the amount of time the port remains disabled using a number of seconds that ranges from 0 to 65535. Configuring a value of zero (0) specifies an infinite time out period. In other words, the port remains disabled until manually re-enabled.

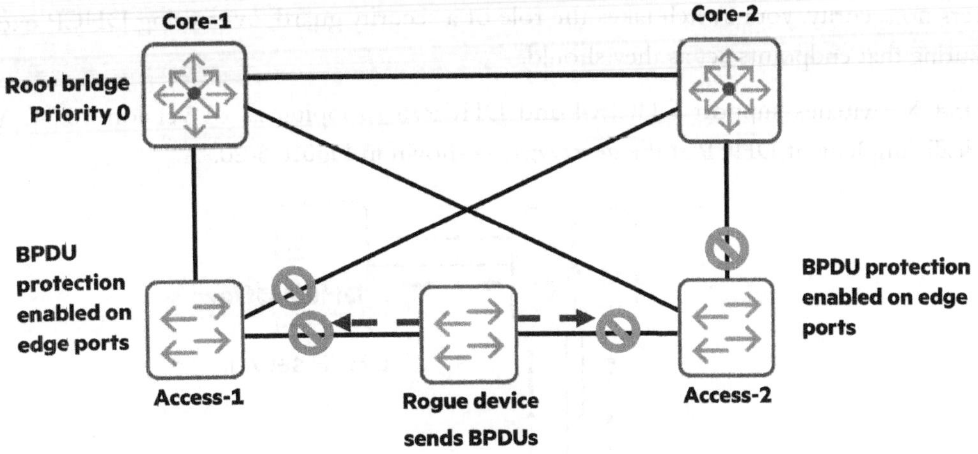

Figure 6-19: Review BPDU Protection

The switch also logs an event when BPDU protection blocks the port.

BPDU Protection Versus BPDU Filtering

Do not confuse BPDU protection with BPDU filtering.

BPDU filtering effectively disables spanning tree on a port. It drops all incoming BPDUs and also prevents the transmission of BPDUs on a port. The port also always remains in the forwarding state. Therefore, BPDU filtering, implemented on an edge port, would prevent a rogue switch from taking over as root by dropping the rogue switch's BPDUs. However, because BPDU filtering forces the port into a forwarding state, it could allow loops to occur.

Edge ports that are vulnerable to rogue switches are often also vulnerable to situations such as these, which could create loops:

- Users connect an unauthorized switch that does not support spanning tree. The users loop ports together on that switch.
- Users loop wall jacks together.

Therefore, you should NOT implement BPDU filtering on edge ports. BPDU filtering is best suited to disable spanning tree on *trusted* switch-to-switch ports that are implementing routing and that you know will not create loops.

Review DHCP Snooping

When possible, you should enable DHCP snooping to protect your network against address spoofing and address exhaustion attacks. Because, on its own, the straightforward DHCP process offers no security, your switch takes the role of a security guard, overseeing DHCP exchanges and ensuring that endpoints act as they should.

AOS-CX switches support DHCPv4 and DHCPv6 snooping as of version 10.04. You should typically implement DHCP at the *access layer*, as shown in Figure 6-20.

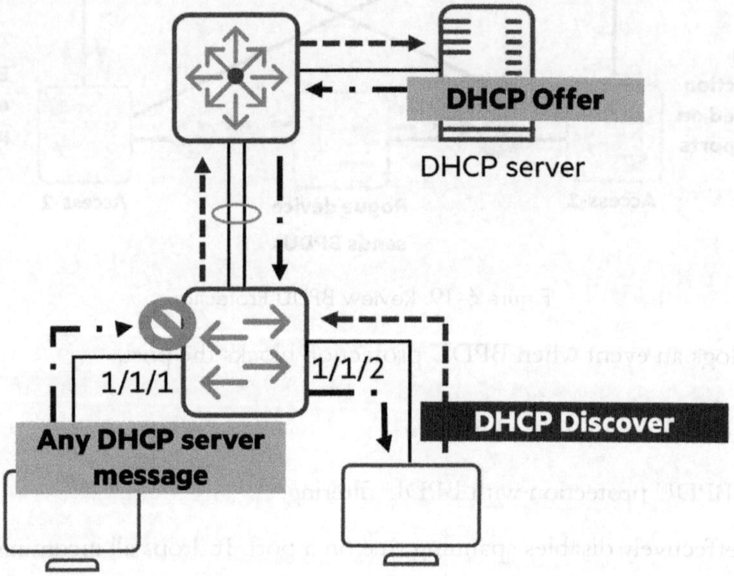

Figure 6-20: Access Switch Implementing DHCP Snooping

With DHCP snooping, the switch distinguishes between trusted and untrusted ports. When you enable DHCP snooping on a VLAN, all interfaces in that VLAN become untrusted by default. You must configure trusted ports manually.

Untrusted ports

As shown in Figure 6-21, the switch filters DHCP packets on untrusted ports and determines whether or not they are allowed.

Ports	Status	Port behavior
1/1/1 1/1/2	Untrusted (default)	Accepts **valid** DHCP client packets Drops DHCP server packets Does not flood DHCP client packets
lag1	Trusted	Accepts and forwards any DHCP packet

Figure 6-21: DHCP Snooping Port Behavior

For example, only DHCP servers should send Offer and Acknowledgement packets, so if the switch detects these types of packets on an untrusted port, it immediately discards them. On the untrusted ports, the switch only accepts DHCP client packets:

- Discover
- Request
- Decline
- Inform
- Release

The switch also verifies information in the DHCP clients' Discover and Request packets against the packet headers before allowing the packets onto the network. It drops packets in which the source MAC address does not match the DHCP client MAC address—a sign of spoofing. It also drops Release and Decline packets that arrive on a different port from one detected for the packets' source MAC. In other words, the packets have the signs of coming from a hacker trying to prevent a legitimate client from receiving an IP address.

In addition, although the switch accepts DHCP client packets on untrusted ports, it does not forward client packets on these ports. This behavior prevents other clients from hearing each other's DHCP traffic.

Trusted ports connect to the network's own trusted devices, such as your DHCP server or, if the server is centralized, an upstream switch. The switch allows DHCP packets to flow freely on these ports. It accepts any DHCP packet on these ports. The switch also forwards DHCP client broadcasts on these ports.

You can also set a max binding on a per-interface basis. This setting limits the number of DHCP addresses a client can obtain on a given port in order to prevent IP exhaustion attacks. In this example, you set the max binding to 4, but you could adjust that limit based on the needs of your network (for example, if the port connects to an AP).

Configuration

DHCP snooping operates on an individual VLAN level. Generally, you enable this feature on endpoint VLANs, especially untrusted ones such as guest VLANs, on access switches. To enable DHCP snooping, you must enable the feature both globally and on the desired VLANs. You then set the trusted port.

The example configuration in Figure 6-22 focuses on v4 for simplicity. But remember to also configure v6 snooping if your network uses IPv6.

```
vlan 11
   dhcpv4-snooping
interface lag 1
   dhcpv4-snooping trust
dhcpv4-snooping
interface 1/1/1
  dhcpv4-snooping max-bindings 4
interface 1/1/2
  dhcpv4-snooping max-bindings 4
```

Figure 6-22: Access Switch Config for DHCP Snooping

Supplemental Information

In the example shown in Figure 6-20, the access switch, which is implementing DHCP snooping, forwards the DHCP traffic to the core switch, which relays the traffic. However, in some networks, the access switch might be routing traffic and acting as the relay. In this case, the port (or link aggregation) that forwards DHCP Discover and Request packets and receives DHCP server packets does *not* support the VLAN on which DHCP snooping is enabled. However, you *must* still make that port a trusted port for DHCP snooping.

You can increase the level of security that DHCP snooping provides by configuring authorized servers. When you configure authorized server addresses, the switch must receive packets from DHCP servers on a trusted port *and* the packets must have a source address that is on the authorized server list. If the server address is not on the list, even if the port is trusted, the switch considers the packet invalid and drops it.

Configuring authorized servers is optional. If you do not configure them, your switch will consider all DHCP server packets that arrive on trusted ports to be valid. You can configure multiple authorized DHCP servers.

Review ARP Inspection

To review, ARP helps devices map each IP address on a subnet to the MAC address of the device that owns that IP address. Most IP devices update their tables every time they receive an ARP packet even if they did not request the information. This makes ARP vulnerable to several attacks.

Hackers can commit ARP spoofing by sending fraudulent ARP responses to other clients on the network. This can allow attackers to "poison" clients' ARP tables with the wrong MAC address for an IP address. In effect, the hacker is able to spoof an IP address. ARP poisoning and IP spoofing can lead to MITM attacks, in which hackers intercept traffic that does not belong to them. If a hacker masquerades as a network infrastructure device, the poisoning can create a DoS attack.

When you enable ARP inspection on a VLAN, all interfaces in that VLAN become untrusted by default. As shown in Figure 6-23, the switch checks ARP requests and responses on all untrusted ports. It:

- Examines all ARP requests and responses received on that port
- Verifies the packet's IP address and MAC address match an entry in the switch's IP-to-MAC binding table for that VLAN
    - If the binding is valid, the switch updates its local ARP cache or forwards the packet to the appropriate destination.
    - If the binding is invalid, the switch simply drops the packet, preventing other devices from receiving it and being tricked by the faulty information.

IP	MAC
10.1.10.10	18...aa
10.1.10.11	d8...bb
10.1.10.12	d4...cc

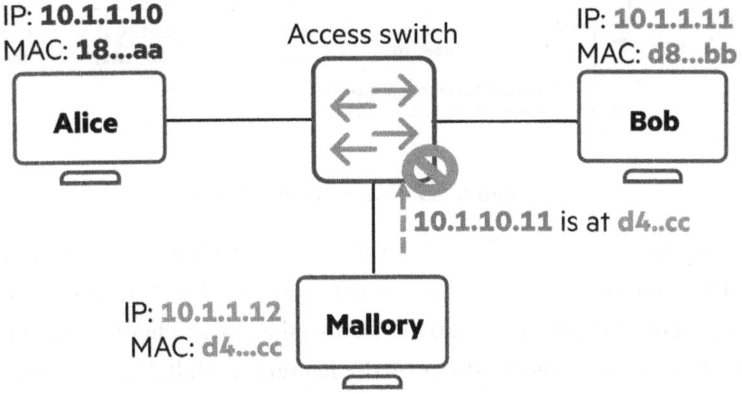

Figure 6-23: IP-to-MAC Binding Table Derived from DHCP Snooping

The IP-to-MAC binding table, shown in Figure 6-23, is typically built up by DHCP snooping. Therefore, you should enable DHCP snooping on your network at least one week before enabling

ARP inspection. You can also add static IP-to-MAC bindings if you have any devices with static IP addresses on ports that are untrusted for ARP inspection.

```
interface lag 1
  arp inspection trust
vlan 11
  arp inspection
```

Figure 6-24: Access Switch ARP Inspection Configuration

As shown in Figure 6-24, also make sure to configure any switch-to-switch ports, such as lag 1 in this example, as trusted ports. Otherwise, the switch will reject ARP responses on this port and interfere with communications across the network.

IP Source Lockdown

ARP inspection prevents ARP spoofing and ARP poisoning attacks. You can also add dynamic IP source lockdown to prevent hackers from spoofing their source IP address in any IP traffic. You enable this feature on a physical interface with this command: **ipv4 source-lockdown** or **ipv6 source-lockdown**. When you enable IP source lockdown on a port, the switch inspects all traffic that arrives on a port. It checks that the source MAC address and source IP address match the binding for that port and drops non-matching traffic, as shown in Figure 6-25.

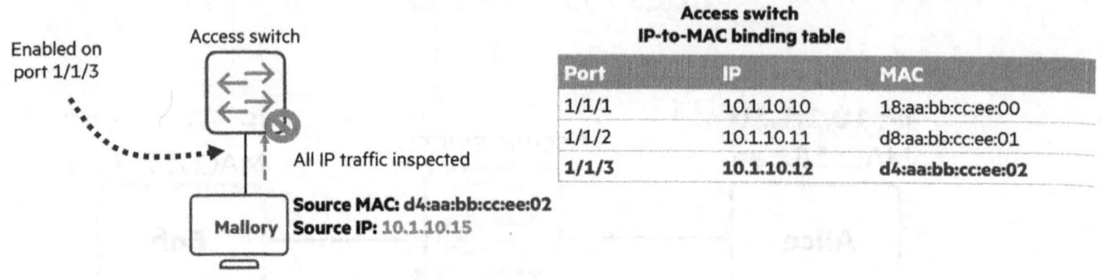

Figure 6-25: IP Source Lockdown

Like ARP inspection, IP source lockdown typically relies on DHCP snooping to build a table of valid IP-to-MAC bindings, so you should make sure to enable DHCP snooping well before enabling IP source lockdown. Then IP source lockdown ensures that wired clients send traffic only from the source IP address that was assigned to them by the legitimate DHCP server. You can alternatively create static binding entries for IP source lockdown, which manually associate a particular IP address with a particular MAC address on a particular VLAN and port.

Saving IP-to-MAC Bindings to External Storage

Because the IP-to-MAC binding table is so important for the ARP inspection and IP source lockdown features, you might want to preserve it in case the switch reboots. Otherwise, ARP inspection might interrupt traffic while the switch rebuilds the table.

To prevent this situation, you can configure the AOS-CX switch to back up the IP-to-MAC binding table to external storage, as shown in Figure 6-26.

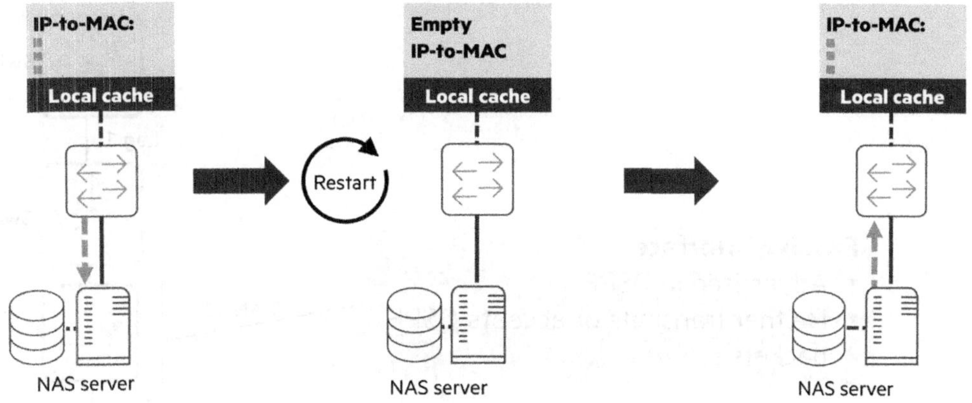

Figure 6-26: Saving IP-to-MAC Bindings to External Storage

First create an external-storage volume that references a Network Attached Storage (NAS) server. Then specify saving the IP-to-MAC binding table to a filename on that server. Use these commands, replacing the volume name, IP address, and filename with values suitable for your environment:

```
Switch1(config)# external-storage myNAS
Switch1(config-external-storage-myNAS)# address 10.254.1.13
Switch1(config-external-storage-myNAS)# enable
Switch1(config-external-storage-myNAS)# exit
Switch(config)# dhcpv4-snooping external-storage volume myNAS file
Switch1_iptomac
```

The switch then copies all the IP bindings to an external storage file so that they are retained after the switch restarts. When it restarts, the switch reads the IP bindings from the configured external storage file to populate its local cache.

Secure OSPF Communications on Client VLANs

Now you will briefly review how to secure routing protocol communications on client VLANs, focusing on OSPF. Often the routing topology is designed such that some Layer 3 interfaces are used to support endpoints, but not to connect to other routers (stub networks) while others are used to connect to other routers or routing switches (transit networks).

The stub networks are usually VLANs, and you can configure these VLANs as "passive interfaces," as shown in Figure 6-27. The subnet associated with a passive interface is advertised in the protocol, but the OSPF routing device does not accept OSPF packets on this interface and does not form neighbor relationships on it. This is a very easy way to prevent rogue devices from sending unauthorized OSPF packets.

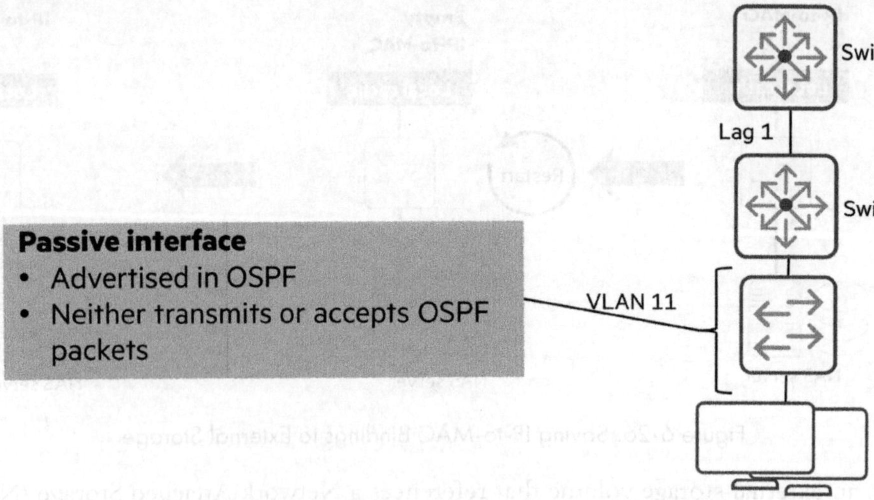

Figure 6-27: Secure OSPF Communications on Client VLANs

On AOS-CX switches, you should typically configure the OSPF instance to make all interfaces passive by default. You then explicitly disable the passive interface settings on the correct interface or interfaces—lag1, in the example configuration shown in Figure 6-28. The other interfaces, such as VLAN 11 in the example shown in Figure 6-27, remain passive interfaces.

```
router ospf 1
    passive-interface default
    area 0.0.0.1 stub
interface vlan11
    ip address 10.1.11.1/24
    ip ospf 1 area 0.0.0.1
interface lag1
    ip address 10.1.1.2/30
    no ip ospf passive
    ip ospf 1 area 0.0.0.1
```

Figure 6-28: Switch 2 OSPF Configuration

Secure OSPF Communications on Transit Interfaces

Occasionally a VLAN or other Layer 3 interface will support both endpoints and legitimate OSPF neighbors. Or you might sometimes have a routing-switch-to-routing-switch link (or routing-switch-to-router link) that you consider to be untrusted (perhaps it passes through a service provider network). In both of these cases, you cannot make the interface a passive interface. You can use OSPF authentication in HMAC-SHA mode to add security to the communications.

When operating in this mode, the OSPF interface hashes OSPF packet data and the key. It then places the hash at the end of the packet. The authentication field includes the key ID, a length for the hash, and a cryptographic sequence number; however, it does not include the key itself, protecting it from eavesdroppers. A router receiving the OSPF packet hashes the packet with its secret key, identified by the key ID. If the hash matches, the router knows that the transmitting router has the correct secret key for this key ID, so it accepts the packet.

To set up OSPF authentication on an AOS-CX switch, set the authentication algorithm, SHA key ID, and SHA key on the Layer 3 interface. Make sure to use exactly the same algorithm, key ID, and key on other routing switches or routers connected to this interface.

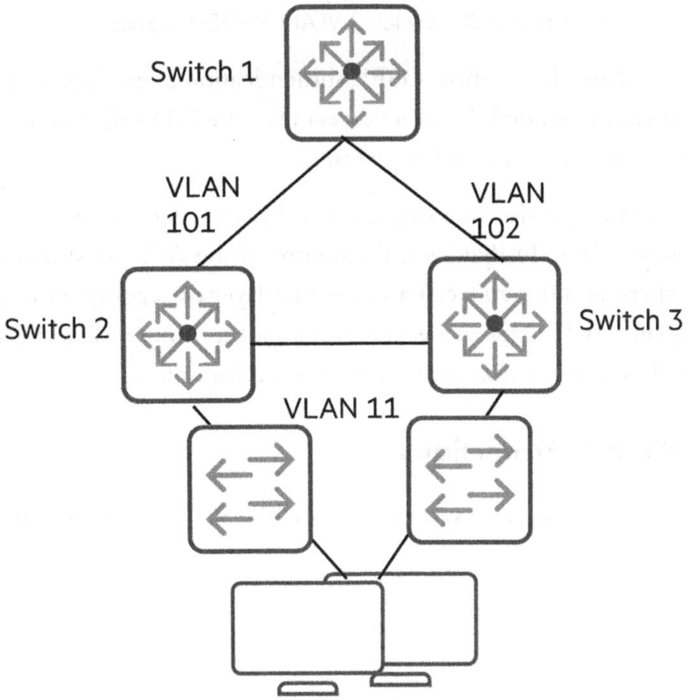

Figure 6-29: Secure OSPF Communications on Transit Interfaces

For example, you might need to secure the OSPF protocol between Switch 2 and Switch 3, shown in Figure 6-29. In Figure 6-30, you can see the example configuration for Switch 2.

```
interface vlan11
    ip address 10.1.11.1/24
    ip ospf 1 area 0.0.0.1
    ip ospf authentication hmac-sha-512
    ip ospf sha-key 1 sha plaintext &3bnr6E*
```

Figure 6-30: Switch 2 VLAN 11 OSPF Config

As you see in Figure 6-31, Switch 3 uses the same key ID and value.

```
interface vlan11
    ip address 10.1.11.2/24
    ip ospf 1 area 0.0.0.1
    ip ospf authentication hmac-sha-512
    ip ospf sha-key 1 sha plaintext &3bnr6E*
```

Figure 6-31: Switch 3 VLAN 11 OSPF Config

Note that AOS-CX switches also support OSPF authentication in message digest and simple mode. These modes are not recommended. Message digest uses the MD5 algorithm, which is deprecated, and simple mode transmits the password in plaintext.

As discussed earlier, you might also be using a control plane ACL to filter communications to the AOS-CX switch control plane. In that case, the control plane ACL must include a rule permitting legitimate OSPF packets, and this rule can add another layer of security. However, you should not rely on the control plane ACL alone, as hackers can spoof legitimate IP addresses. You should use the control plane ACL in conjunction with OSPF authentication.

Checklist of Additional Protections

You should ensure that you have followed a few more best practices on switches, as summarized in Figure 6-32.

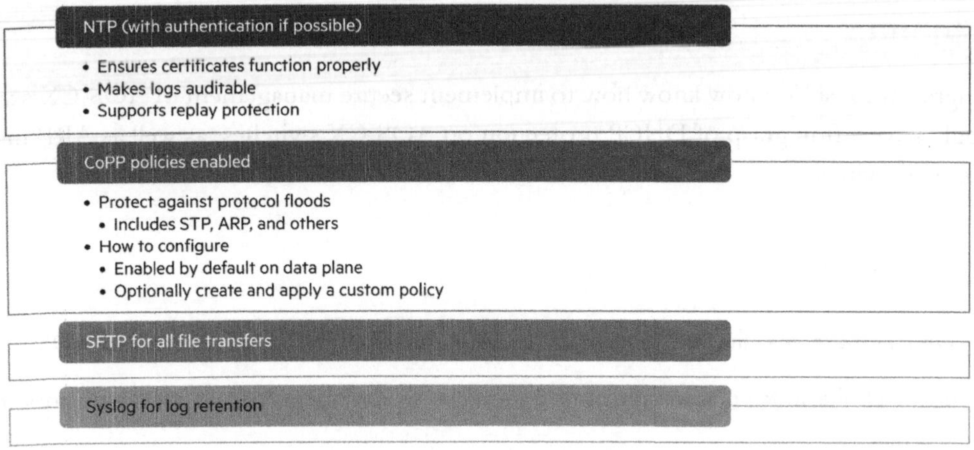

Figure 6-32: Checklist of Additional Protections

You should set the switch clock using Network Time Protocol (NTP), rather than manually. If the company has a private NTP server, you should use NTP authentication to further secure the communications. Ensuring that all network infrastructure devices have an accurate clock and take their time from a consistent source benefits security in many ways. It ensures that switches can properly validate certificates using the certificates' validity periods. It makes logs easier to read and correlate across multiple devices. The logs must also have a valid time in order to be used for auditing or investigating security incidents. NTP also helps switches to implement replay protection with various protocols; only with an accurate clock can the switch accurately check packets' timestamps and ensure that they are up-to-date.

AOS-CX switches also support control plane policing (CoPP). CoPP examines packets incoming to the switch control plane and rate limits them. This rate limiting protects against protocol floods, which could tie up the switch CPU and lock up the switch. The switch comes with a default policy, which is applied by default to the data plane. The default policy applies specific rate limits each traffic class, such as STP BPDUs, ARP messages, and many others. You should generally accept the default policy, but if necessary, you can create a custom policy and apply it instead of the default policy. Only one CoPP policy can be active.

AOS-CX switches use SFTP for transferring files. SFTP is a secure file transfer protocol, which uses authentication and encryption.

It is also best practice to enable sending logs to a central Syslog server for auditing and archival.

If you want more information about these features, you can refer to prerequisite training or HPE Aruba Networking manuals.

Summary

Congratulations! You now know how to implement secure management of AOS-CX switches. You should have a firm grasp of DHCP protection on AOS-CX switches, as well as ARP inspection on AOS-CX switches.

Learning checks

1. What purpose does the "aaa authentication allow-fail-through" command serve on AOS-CX switches?
   a. It lets the switch authenticate users against the secondary method only if the primary method times out.
   b. It lets the switch actively test whether a TACACS+ server is available.
   c. It lets the switch allow users operator level access if their authentication fails.
   d. It lets the switch authenticate users against the secondary method if the primary method rejects the user.
2. How does ARP inspection prevent MITM attacks?
   a. It discovers which IP and MAC addresses belong together
   b. It drops all ARP responses that arrive on untrusted ports
   c. It drops ARP responses on untrusted ports that do not match the IP-to-MAC table
   d. It stops clients from generating gratuitous ARP responses

---

Note

For answers to all learning checks, refer to the Appendix.

---

7 Implement 802.1X and Role-Based Access Control on AOS-CX Switches

EXAM OBJECTIVES

- ✓ Deploy AAA for wired devices using HPE Aruba Networking ClearPass Policy Manager (CPPM), including local and downloadable roles
- ✓ Explain Dynamic Segmentation, including its benefits and use cases
- ✓ Deploy Dynamic Segmentation using VLAN steering
- ✓ Configure 802.1X authentication for APs

Assumed Knowledge

- Authentication, Authorization, and Accounting (AAA) framework and basics of protocols such as RADIUS and TACACS+
- Forms of Layer 2 authentication such as 802.1X, including common EAP methods such as PEAP and EAP-TLS
- Basics of managing AOS-CX switches
- PKI and digital certificates
- VLANs
- Access control list (ACL) basics
- Fundamental ClearPass Policy Manager (CPPM) components, such as services, roles, enforcement profiles, and enforcement policies

Overview

Dynamic Segmentation allows you to optimize both internal security and employee performance. It lets you give each user the right level of access and to adjust that access as circumstances change.

In the first section of this chapter, you will learn about laying the groundwork for Dynamic Segmentation by implementing 802.1X for wired devices. Next, you will review AOS-CX roles and Dynamic Segmentation. After that, you will take a look at implementing local forwarding with VLAN steering and role-based access control lists (ACLs). You will then learn how to use 802.1X to authenticate APs.

802.1X for Wired Devices

In this section, you will review how to implementing 802.1X to authenticate wired devices.

Why 802.1X for Wired Clients

Dynamic Segmentation involves identifying each client as it connects to the network and then adjusting the point of access based on the client's identity as well as other context-based information. To implement Dynamic Segmentation on wired clients, you must set up port-based authentication, which allows the switch to authenticate clients against an external authentication server and then grant that user a particular level of network access, based on the authentication server's decisions.

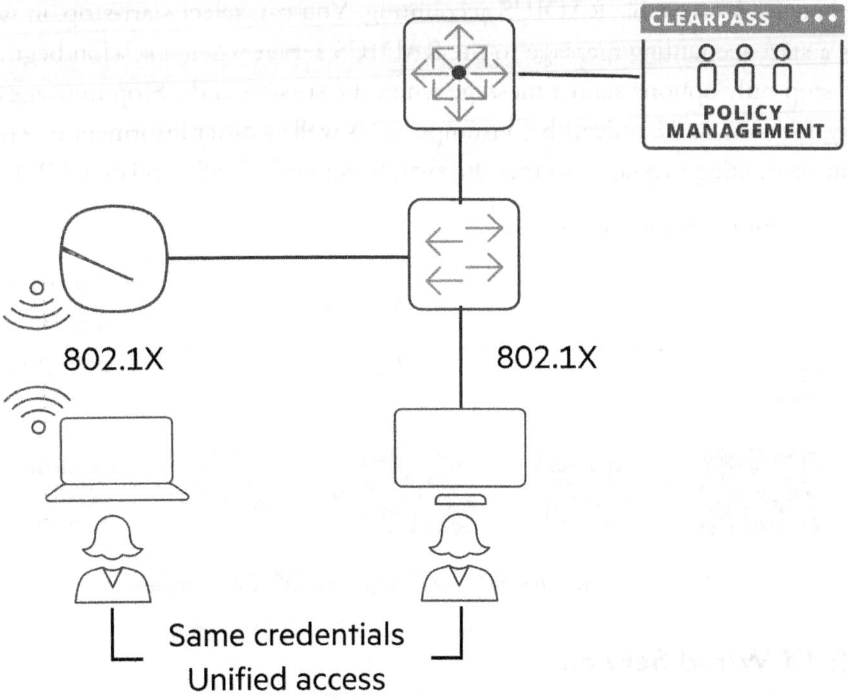

Figure 7-1: Why 802.1X for Wired Clients

Just as 802.1X provides the most secure authentication option for wireless networks, it also provides the most secure option for port-based authentication. As shown in Figure 7-1, you can configure AOS-CX switches to use 802.1X to authenticate wired clients to CPPM, much as APs do. Then users on wired devices can authenticate with the same credentials and a similar process that they use for connecting to the wireless network. In this way, users have a consistent authentication experience no matter how they connect, and they also obtain a consistent experience on the network.

Review 802.1X Setup on AOS-CX Switches

To implement 802.1X on an AOS-CX switch, you must specify the RADIUS server. If you are using CPPM, HPE Aruba Networking recommends that you specify the server by the hostname in CPPM's HTTPS certificate. This practice will help the switch successfully download user-roles from CPPM if you decide to use that feature.

If you placed the RADIUS server in a non-default group, you must specify that RADIUS group as the one that you want to use for 802.1X.

Then, you must enable 802.1X on a per-interface basis. You must also enable 802.1X globally. For 802.1X to operate on a port, it must be enabled both on the port and globally.

Finally, best practice is to enable RADIUS accounting. You can select start-stop, in which case the switch sends a start accounting message to the RADIUS server when a session begins. Both start-stop and the stop-only options send a message when the session ends. Stop messages describe the user's identity, location, and bandwidth consumption, as well as other information. You should also set up interim accounting messages so that the switch can periodically update CPPM.

These steps are summarized in Figure 7-2.

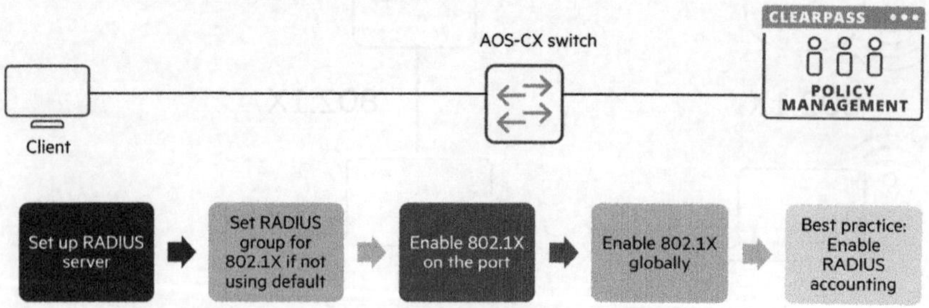

Figure 7-2: Review 802.1X Setup on AOS-CX Switches

CPPM 802.1X Wired Service

When a switch enforces 802.1X, it relies on a RADIUS server to make decisions about whether clients are allowed network access and what level of access they receive. For our purposes, this RADIUS server is CPPM.

As always, CPPM requires a service for processing the incoming RADIUS requests. A wired 802.1X service resembles a wireless 802.1X service in many ways.

Service

To create the service, you can create a service and select the 802.1X wired service template. Or you can use the 802.1X wired wizard, as shown in Figure 7-3. When you use either the template or the wizard, CPPM automatically creates service rules, which match 802.1X-related RADIUS requests from switches. Note that the value for the RADIUS NAS-Port-Type AVP is "Ethernet." This value helps CPPM distinguish the wired 802.1X service from a wireless 802.1X service.

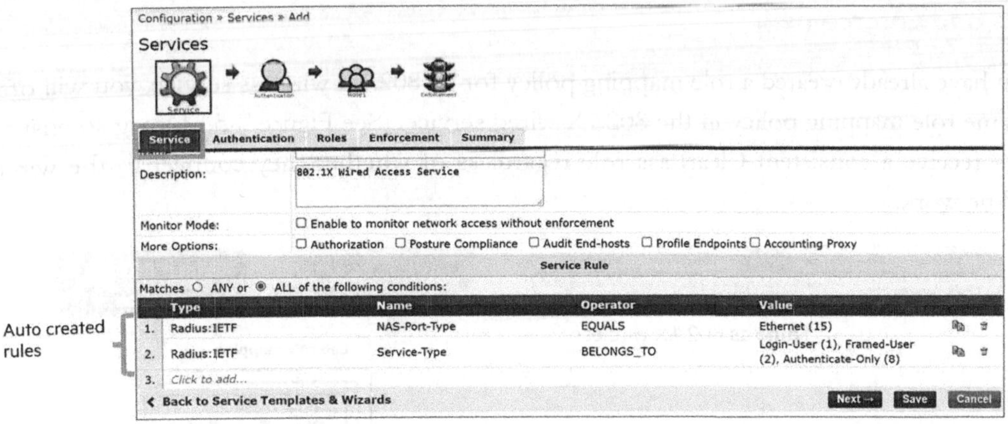

Figure 7-3: Service

Authentication

Choose authentication methods and sources following the same guidelines that you use for wireless 802.1X services. In the example shown in Figure 7-4, you are using EAP-TEAP and EAP-TLS. The order of the authentication methods indicate the order in which CPPM suggests them; in this example, EAP-TEAP is preferred.

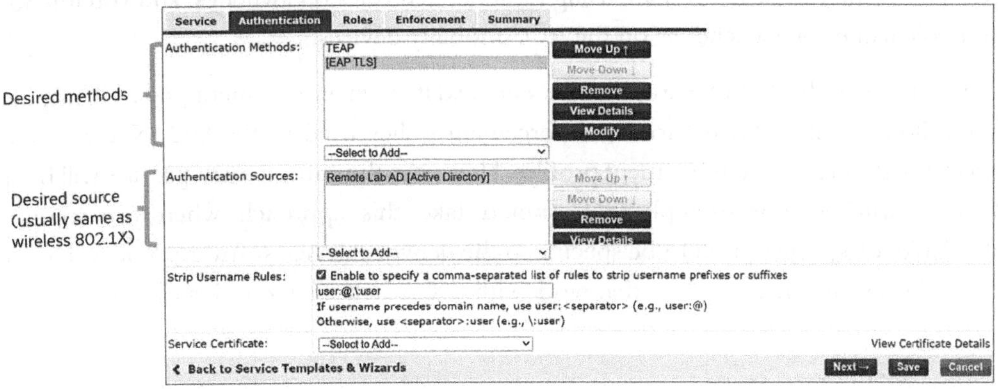

Figure 7-4: Authentication

You are also using the same AD instance as in the wireless service. When you select this AD instance as an authentication source, it is automatically set as an authorization source.

Remember to set up the rules for stripping usernames if users might enter their names with domain names or if certificates include domain names in the subject name or UPN.

Roles and Enforcement

If you have already created a role mapping policy for an 802.1X wireless service, you will often use the same role mapping policy in the 802.1X wired service. (See Figure 7-5.) Doing so ensures that clients receive a consistent ClearPass role regardless of whether they connect to the wireless or wired network.

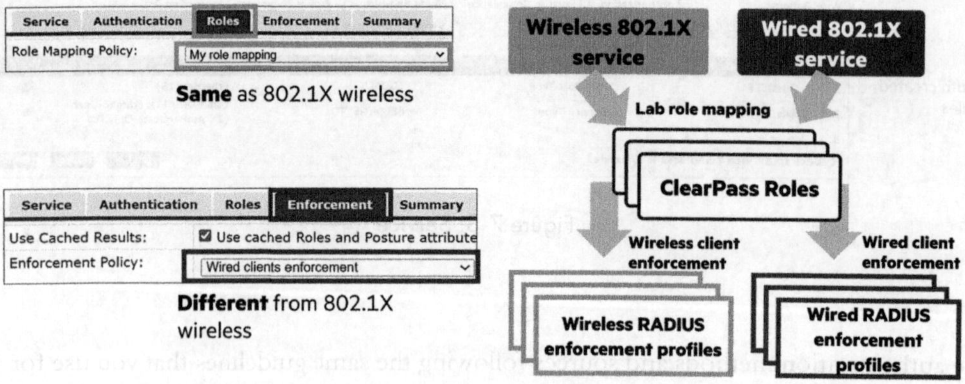

Figure 7-5: Roles and Enforcement

Sometimes you can even use the same enforcement policy for the 802.1X wired and 802.1X wireless services. This is the case when you are using local defined roles on switches, and you intend to use the same role names on switches as on the AOS mobility devices.

Sometimes, though, the 802.1X wired service will need its own enforcement policy. This policy will typically follow a similar format to the enforcement policy used in the 802.1X wireless service, mapping ClearPass roles to enforcement profiles. However, the enforcement profiles will be specific to the CX switches. For example, you should take this approach when you want to use downloadable roles, which need to be specific to the network device software. You will learn more about defining enforcement profiles that work with CX switches later in this chapter.

Review AOS-CX Roles and Dynamic Segmentation

In this section, you will review how AOS-CX switches assign clients to roles and enforce Dynamic Segmentation.

What Are AOS-CX User-Roles?

AOS-CX switches are designed to apply user-roles as part of the authentication process. An AOS-CX switch user-role, much like an AOS role, dictates how the network infrastructure device controls traffic from a client.

User-roles enable the switch to dynamically adapt how a port behaves based on the identity and context for the client that connects to the port. Such a port is sometimes called a "colorless" port because it does not require admins to pre-define, or color, its settings. Similarly, such a port is sometimes called a "multi-use" port because admins can define the same settings on all edge ports and then connect any client type to any port. Admins and users can also move clients around. As shown in Figure 7-6, the port dynamically adapts to the client that is currently connected to it.

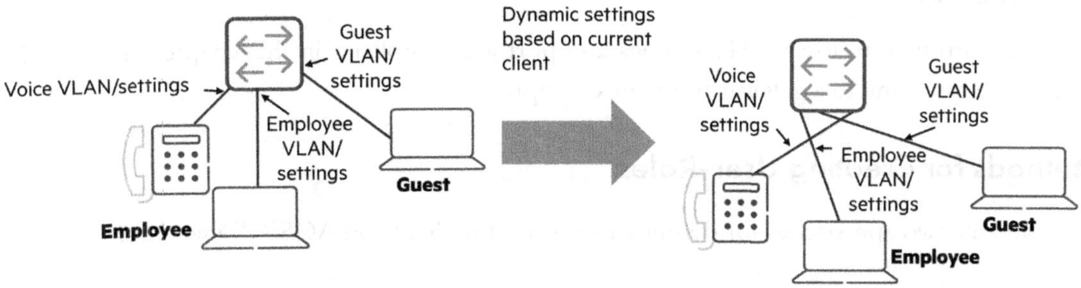

Figure 7-6: What Are AOS-CX User-Roles?

AOS-CX User-Role Benefits

The benefits of assigning "roles" to users and/or devices are well known within the HPE Aruba Networking wireless world. HPE Aruba Networking has simply adapted this approach for use with the wired switch port. Just as with AOS solutions, companies can now simplify the burden of configuration. They can group policies into a "role," which can be used for many clients on many network infrastructure devices. Additionally, when you use CPPM to manage the roles, you can add context (time of day, type of machine, device profiling) when deciding if the network will allow a user/device as well as what access rights it will grant or deny.

When you use downloadable user-roles (DURs), admins can centrally define the role settings, rather than have to define settings manually on all access switches across the network.

While HPE Aruba Networking recommends ClearPass for the highest levels of security, companies can use third-party RADIUS servers if they choose. In this case, only local user-roles (LURs) are supported.

Common Settings Defined in AOS-CX User-Roles

The user-role controls a client's access based on several settings.

The user-role dictates settings such as the VLAN for the client's traffic (tagged or untagged). It also defines whether traffic is locally switched or tunneled back to an HPE Aruba Networking gateway or mobility controller (MC).

You can also apply ACLs or quality of service (QoS) settings to clients through a port access policy. You can also configure reauthentication timers, which dictate how often the client must reauthenticate.

You can also set up a captive portal redirect for a role that you want to use for guests or for onboarding clients.

You will learn how to use VLAN settings and port access policies in this chapter. You will learn how to use tunneling to a gateway in the next chapter.

Methods for Creating User-Roles

You can take two approaches for defining user-roles for clients on AOS-CX switches.

Local User Roles (LURs)

LURs is local configuration. Admins must configure LURs on the switch in advance of clients connecting and being assigned to those user-roles.

As shown in Figure 7-7, when a user authenticates on the port, CPPM sends an Access-Accept with the client's assigned user-role in an Aruba-User-Role VSA. The switch matches that VSA's value to the LUR with the same name and applies the settings within.

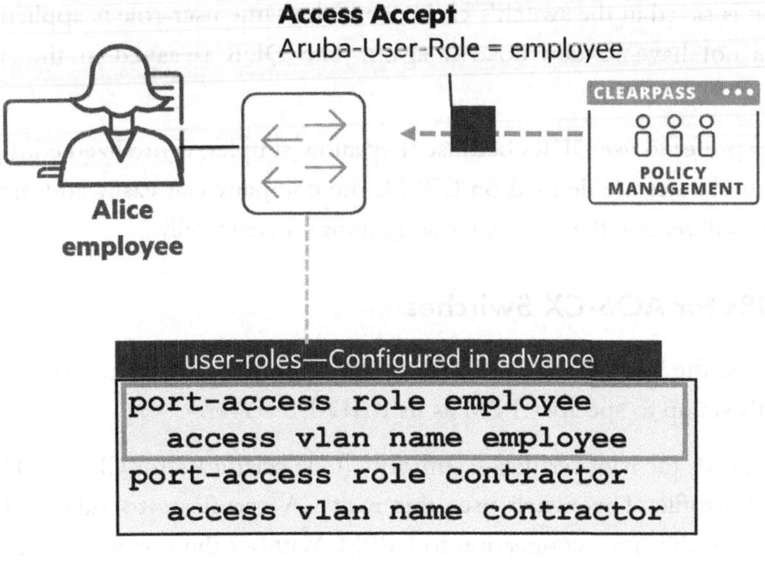

Figure 7-7: Local User Roles (LURs)

Downloadable User Roles (DURs)

With DURs, the switch does not need to have any user-roles configured on it in advance. Instead the user-role is defined centrally on CPPM, and the switch downloads the user-role from CPPM on demand.

When a user authenticates on the port, CPPM sends an Access-Accept with the name of the assigned DUR in the Aruba-CPPM-Role VSA, as shown in Figure 7-8.

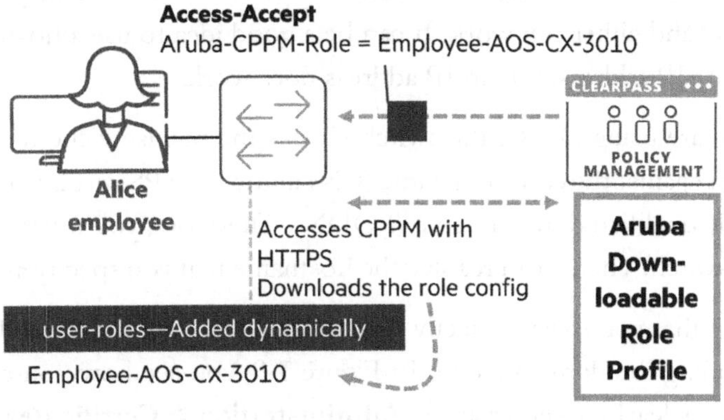

Figure 7-8: Downloadable User Roles (DURs)

If the switch has not downloaded that DUR yet, it accesses CPPM with HTTPS and downloads the user-role from the server. It then applies the settings in that user-role to the client. Once a user-role is downloaded, it is saved in the switch's cache, so if the same user-role is applied to another user, the switch does not have to download it again. The DUR is saved in the memory, not the configuration.

Some companies prefer to use DURs because they allow simpler, centralized configuration of user-roles. Once the user-roles are defined on CPPM, the company can easily add more switches, and the new switches will receive the user-role configuration dynamically.

Enabling DURs for AOS-CX Switches

To support DURs, the AOS-CX switch must meet a few key requirements. Make sure that the switch is correctly set up to specify CPPM as its RADIUS server.

This switch also needs the root certificate authority (CA) certificate for CPPM's HTTPS certificate installed in a TA profile. The switch uses this root CA certificate to validate CPPM's HTTPS certificate when it establishes a connection to CPPM. Without the certificate, the switch will abort the HTTPS connection and fail to download any user-roles.

To validate that the certificate has not expired, the switch also needs to have a correct clock. You learned about enabling Network Time Protocol (NTP) for this purpose earlier in this study guide.

The switch also needs the username and password for an account on CPPM.

You will now go over each of these requirements in more detail.

Correct RADIUS Server

Typically, when you define the RADIUS server on an AOS-CX switch, you can specify an IP address or hostname, and either can work. It can be a good idea to use a hostname to future proof against changes to the IP address, but an IP address does work.

However, when you are using DURs, the switch checks the precise value that you specify for the RADIUS server host against the common name (CN) in the HTTPS certificate's subject name. This CN should be a fully qualified domain name (FQDN). Therefore, you must make sure to define a DNS server on the switch. Then it can resolve the hostname that you specify as the RADIUS server.

You must make sure that you specify exactly that same FQDN for the RADIUS server host name on the switch, including the domain name. In Figure 7-9, you see how you can view the HTTPS server certificate on a ClearPass server at the **Administration > Certificates > Certificate Store** page. Make sure that you select the HTTPS usage. Then find the CN. You can also match to a DNS SAN within the HTTPS server certificate.

```
ip dns server-address X.X.X.X vrf <vrf>
radius-server host clearpassxx.training.arubanetworks.com key XXX [vrf <VRF>]
```

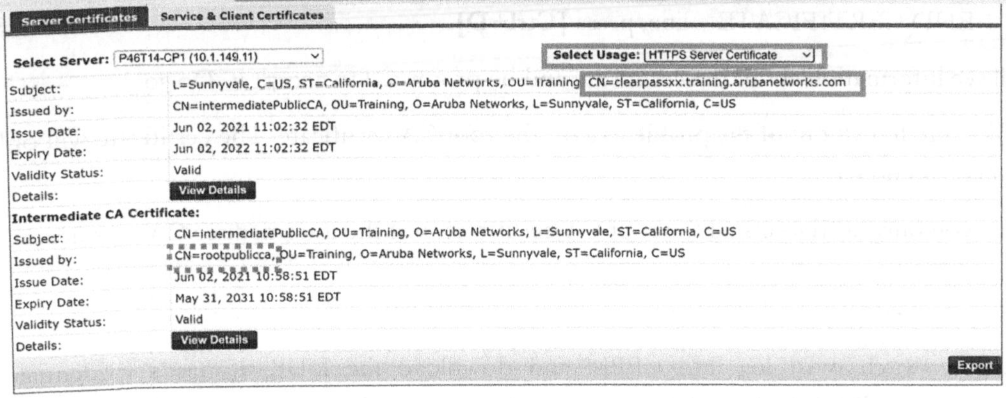

Figure 7-9: Correct RADIUS Server

Also take note of the root CA certificate for this certificate. In this example, the certificate was signed by an intermediate CA. That CA's certificate was signed by a root CA certificate. It is the root CA certificate that you will want to upload on the AOS-CX switch.

CA Certificate

To validate the CPPM HTTPS certificate, the switch also checks the certificate signature. This means that you must install the root CA certificate for this certificate on the switch. Obtain this certificate in Base64 format. Then create a TA profile on the switch and import the certificate into that profile. As you learned earlier, you can import from the terminal by copying the Base64 formatted text into the CLI. Or you can import with SFTP.

```
AOS-CX(config)# crypto pki ta-profile rootpublicca
AOS-CX(config-taprofile-rootpublicca)# ta-certificate import terminal
Paste the certificate in PEM format below, then hit enter and ctrl-D:
AOS-CX(config-ta-cert)# -----BEGIN CERTIFICATE-----
AOS-CX(config-ta-cert)# MIID3DCCAsSgAwIBAgIQep2sjAYy5rBA48oiGL7HCjANBgkqhk
AOS-CX(config-ta-cert)# YNYXJ1YmFuZXR3MRMwEQYKCZImiZPyLGQBGRYDY29tMR0wGwYK
<-output omitted->
AOS-CX(config-ta-cert)## -----END CERTIFICATE-----
[Ctrl+D]
The certificate you are importing has the following attributes:
Subject: <Root CA cert subject>
Issuer:  <Root CA cert subject>
Serial Number: <Root CA serial number>
TA certificate import is allowed only once for a TA profile
Do you want to accept this certificate (y/n)? y
AOS-CX(config-taprofile-rootpublicca)# exit
```

Figure 7-10: CA Certificate

The example in Figure 7-10 shows importing the certificate using the terminal. You copy the Base64 text directly into the terminal. As you see, each line will copy in separately. This is okay. After you see the ---END CERTIFICATE---line, press **[Ctrl+D]**.

You will see information about the certificate and be prompted to accept it. Do so.

Also make sure to exit out of the profile to save the root CA certificate. The certificate will only be saved when you do so.

Note that you only need to install the root CA certificate, not the intermediate CA certificate.

CPPM User Account

Because the switch must log into CPPM and download the DUR, it needs credentials for authenticating to CPPM. In Figure 7-11, you see the command for configuring the username and password for a ClearPass server on an AOS-CX switch.

```
AOS-CX(config)# radius-server host clearpassxx.training.arubanetworks.com
clearpass-username cxswitch clearpass-password plaintext @ruba123!
```

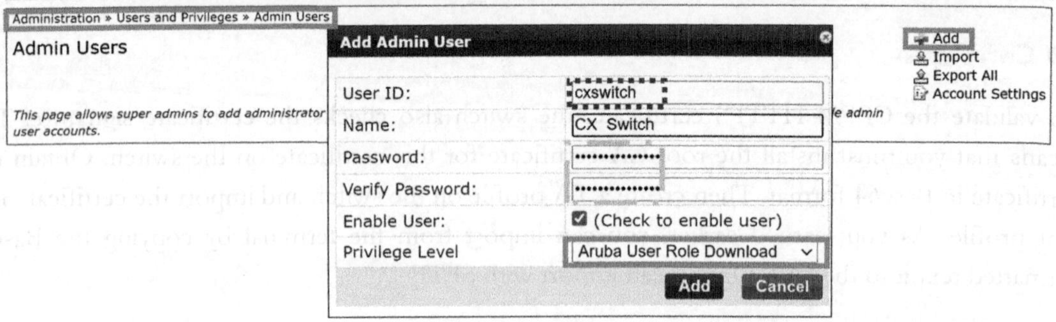

Figure 7-11: CPPM User Account

You must also set up a matching account on CPPM.

Log into CPPM and navigate to **Administrator > Users and Privileges > Admin Users**. Click the plus icon to create a new account.

Give the user any User ID and password that you desire, making sure, of course, to match the credentials configured on the AOS-CX switches. Note that the password shown in Figure 7-11 is for simplicity while in the real-world you should follow best practices for using long, random, complex passwords that are not based on dictionary words.

Give the account the **Privilege Level** of **Aruba User Role Download**.

Click **Save**.

Creating Downloadable Enforcement Profiles on CPPM

With the prerequisites for switches to download the user-roles in place, you can create the DURs themselves on CPPM. You do so by creating Downloadable Enforcement Profiles.

In CPPM, go to **Configuration > Enforcement > Profiles** and then click **Add Enforcement Profile**. Create a RADIUS profile.

Under the **Profile** tab, set the **Template** to **Aruba Downloadable Role Enforcement** and the **Action** to **Accept**. (See Figure 7-12).

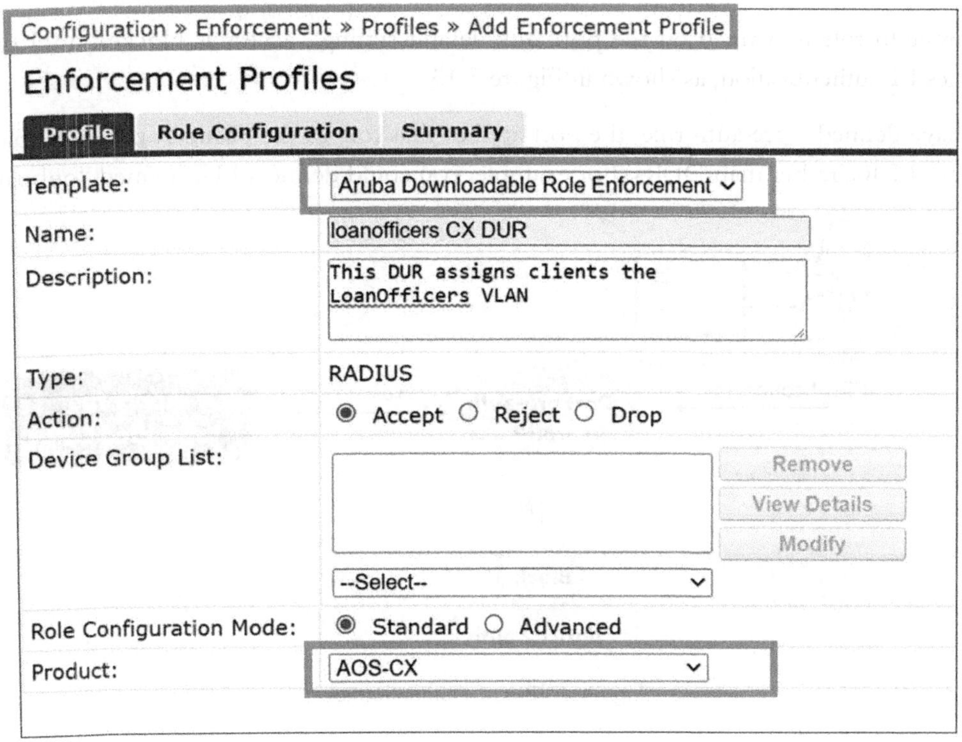

Figure 7-12: Creating Downloadable Enforcement Profiles on CPPM

Under Product, choose **AOS-CX Switch**. You will then see options for specific AOS-CX user-role settings such as VLAN assignment, UBT settings, and reauth interval. These settings are the practical elements that define users' access level. Later in this chapter, you will learn about which settings to choose for various use cases when constructing these user-roles.

More Details on User-Role Assignment

Now that you know how to ensure that user-roles are in place, you will look at how clients receive the user-role assignment.

Pre-Authentication

When you enable 802.1X or Mac-Auth on a port, the switch attempts to authenticate all clients that connect on the port. When the client first connects, before authentication occurs, one of two situations occurs.

If no pre-auth role is defined on the port (the default setting), access is blocked until the client completes L2 authentication, as shown in Figure 7-13.

If you have defined a pre-auth role, the port applies that role to the client. A pre-auth role always references LURs, rather than DURs. For example, you could define a LUR named "onboard."

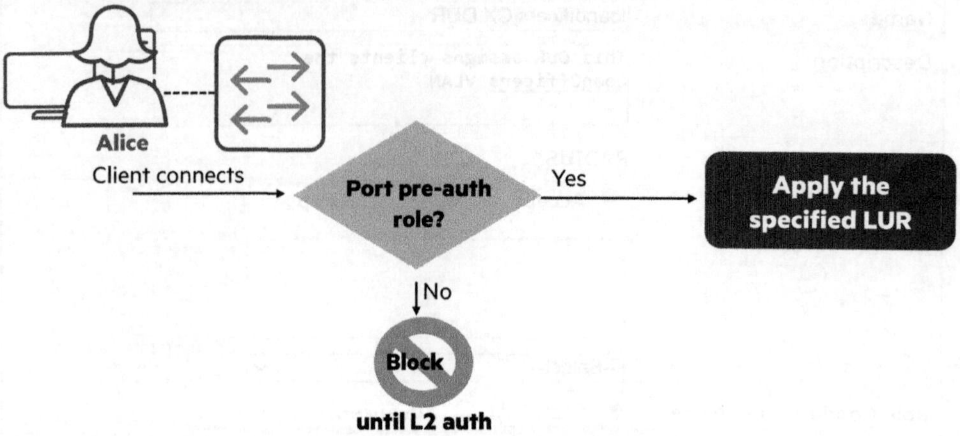

Figure 7-13: Pre-Authentication

Authentication Timeout

After about two minutes (in default settings), if the client does not attempt to authenticate, authentication times out. Note that this is different from the client attempting to authenticate, the switch sending an Access-Request to the RADIUS server, and the RADIUS server never responding. In the case that you are examining, the client never attempts to authenticate.

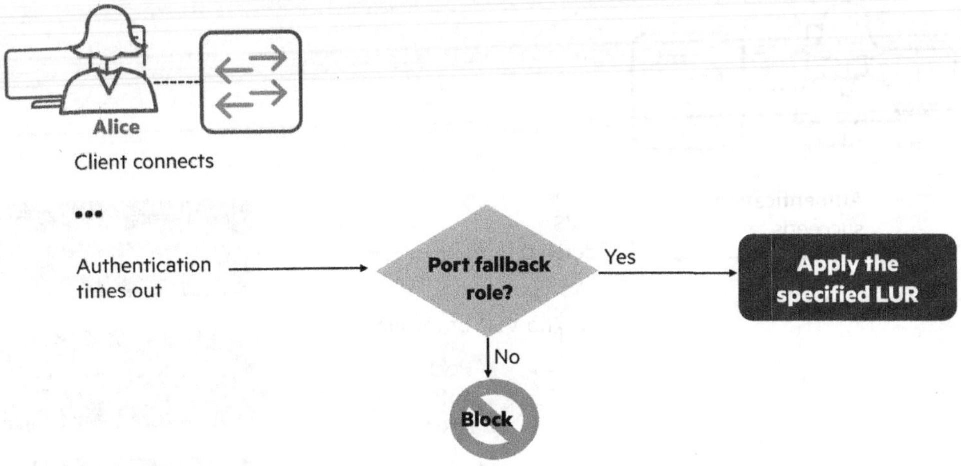

Figure 7-14: Authentication Timeout

When client authentication times out, the AOS-CX switch checks whether the port has a port fallback role assigned to it, as shown in Figure 7-14. This role is always a LUR. If the fallback role is specified, the switch applies the indicated LUR to the client. You might apply this role when you want to accommodate devices that need to be onboarded.

If no fallback role is specified for the port, the switch simply continues to block access after authentication times out.

Note that the fallback role was added in AOS-CX 10.07. Prior to that version if authentication times out, the pre-auth role continues to apply. Or if no pre-auth role exists, the client is blocked from access.

Post-Authentication

Now look at what happens when a client connects and succeeds in authenticating. (See Figure 7-15.)

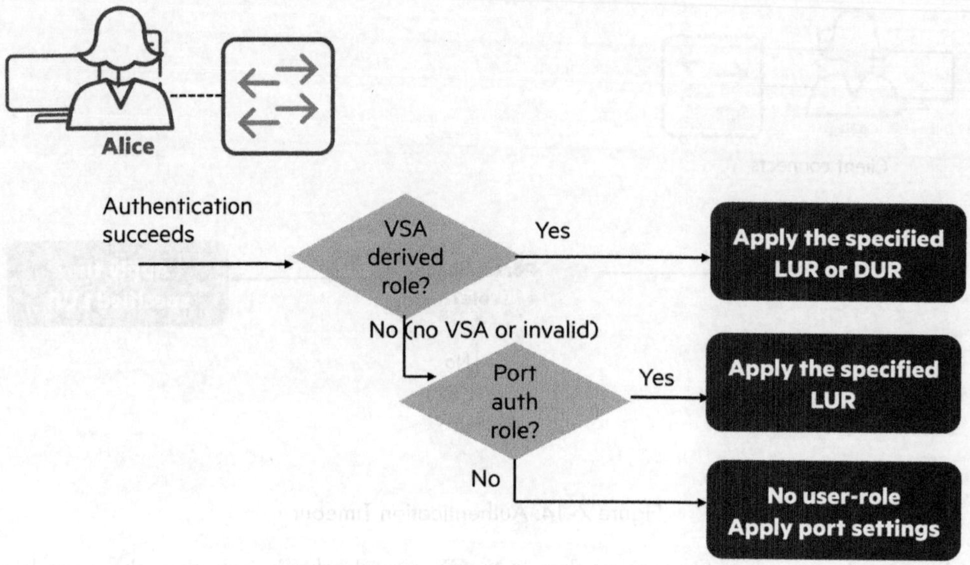

Figure 7-15: Post-Authentication

First, the switch attempts to apply a user-role communicated by the RADIUS server in an Access-Accept VSA. This VSA could be an Aruba-User-Role VSA, which must have a value that matches a LUR already configured on the switch. Or it could be a CPPM-User-Role VSA, which triggers the switch to attempt to download a DUR.

If the VSA is not present, the switch moves on to trying the next type of user-role. Similarly, if the Aruba-User-Role value does not match with a LUR, or the switch cannot download the DUR, the switch moves on. Next the switch attempts to apply the auth role defined on the port. This user-role always matches a LUR defined on the switch.

If no valid VSA is sent to define the user-role, and the port does not have an auth role defined on it (the default setting), the client is allowed to connect without a user-role. The switch simply applies the settings defined on the port to the client connection.

Best practice is to have the RADIUS server, such as CPPM, define the user-role. As a failsafe, define a user-role that denies access or permits minimal access as the port's auth role.

Authentication Failure

Now consider a situation in which the client attempts to authenticate and fails. (See Figure 7-16.)

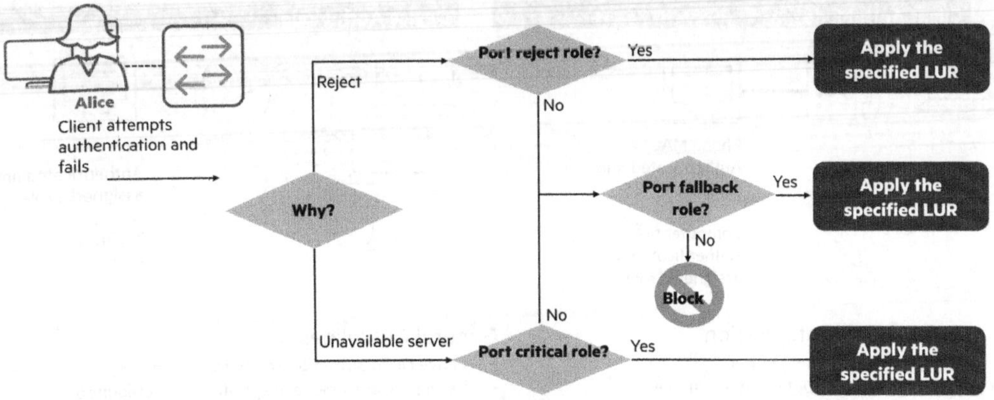

Figure 7-16: Authentication Failure

The authentication could fail because the server sends an Access-Reject. In that case, the switch checks whether a reject role is defined on port. If it is, the switch applies that role, which always corresponds with a LUR. If this type of role is not defined, the switch checks whether a port fallback role is specified. If it is, the switch applies that role. Otherwise, the switch blocks access.

The authentication could also fail because the switch cannot reach the RADIUS server. The switch would then check whether a critical role is defined on the port. If so, it applies that role. This role is intended to provide access in case of an outage on the RADIUS server or a disruption in connectivity. It should typically provide minimal access only to services that can never be down. As with all roles defined on the port, the role references a LUR. If no critical role is defined on the port, the switch again checks for a port fallback role. If that role is defined, the switch applies the specified LUR. Otherwise, the switch blocks the client's access.

Client vs Device Auth-Mode

By default, AOS-CX switches implement port-access in client auth-mode. As shown on the left of Figure 7-17, in client auth-mode, the switch tracks all MAC addresses on the port and enforces 802.1X for each separately. The switch can apply a different role to each client. You can also configure a limit for the number of clients permitted on a port.

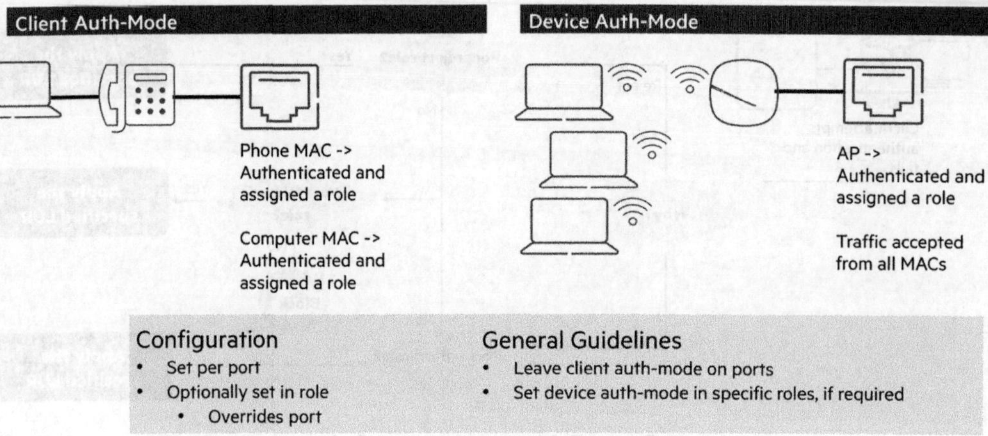

Figure 7-17: Client vs Device Auth-Mode

Alternatively, a port can operate in device auth-mode, as shown on the right of Figure 7-17. In this mode, the switch implements authentication on the first client that connects to the port. As soon as that client authenticates, the port is open for all clients connected behind the port, and the port uses the role applied to the first client.

The auth-mode is configured on the port. However, you can also configure an auth-mode in a role, and that auth-mode overrides the auth-mode configured on the port.

Generally, you should leave all ports operating in client auth-mode. This mode ensures that only authorized clients connect to the port. Or if the port connects to multiple clients, such as a phone and computer, each client receives its own role and access level. However, if you are implementing 802.1X on ports that connect to APs, you should use device auth-mode so that the switch does not try to authenticate wireless clients again. You can specify device auth-mode in the AP role to dynamically apply that mode when necessary. You will learn more about implementing 802.1X to authenticate APs later in this chapter.

Consideration questions

You will now review how users are assigned a role in a scenario.

Consider an example customer: Bank ABC, which has a corporate headquarters and multiple branch offices. The bank's network is built on HPE Aruba Networking APs and switches, and the bank uses CPPM to authenticate users on the wireless network. Now the bank wants to add 802.1X to wired ports to bolster the security of that infrastructure, while still ensuring that users on wired clients have the appropriate access level.

You have been asked to set up the port role settings on AOS-CX switches. The bank security team has given you these instructions.

Authenticated clients should receive their role assignment from CPPM, and the switch should also download the role configuration from CPPM. But if CPPM fails to send a valid role, the authenticated client should be given the "guest" role. The clients should also receive the "guest" role if the RADIUS server is unavailable. Clients that do not try to authenticate should receive a role called "unauthenticated." Clients that fail authentication should be denied access entirely.

How will CPPM assign roles to users that successfully authenticate? How should you configure switch ports to apply the "guest" and "unauthenticated" roles? How should they be configured to ensure that clients who fail authentication receive no access? Finally, which roles need to be configured locally on the switches?

Refer to the appendix for answers to these questions.

Methods for Handling Wired Client Traffic

Once the switch has assigned a client to a role, the switch has three broad ways to handle that client's traffic, based on the role settings.

- Local forwarding
- User-based tunneling (UBT)
- Virtual network-based tunneling (VNT)

Local Forwarding

Once the switch has assigned a client to a role, the switch has three broad ways to handle that client's traffic, based on the role settings.

The switch can use local forwarding, as shown in Figure 7-18. In this case, the role assigns the user to a VLAN (this is sometimes called VLAN steering), and switches direct the traffic toward its destination locally. The switch can also apply other settings, like a port access class, which filters traffic based on IP header.

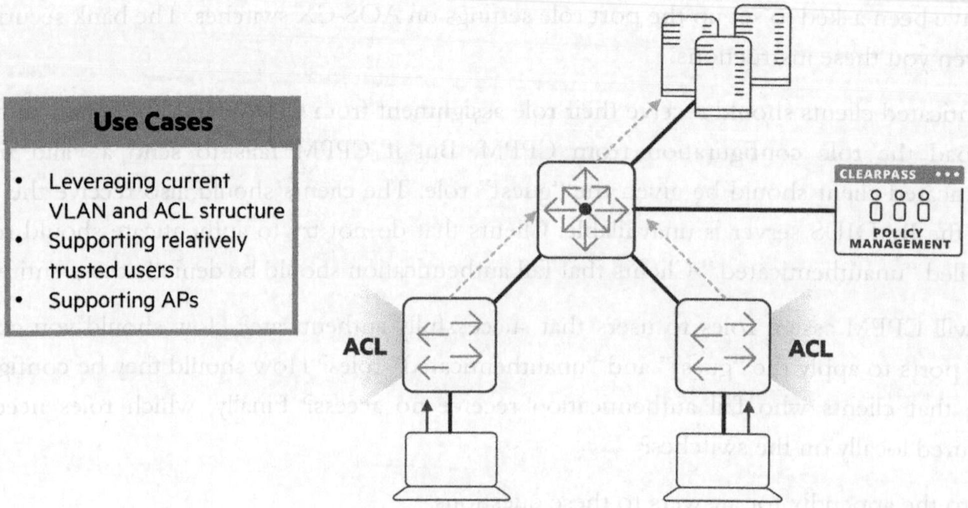

Figure 7-18: Local Forwarding

Local forwarding often works well for companies who want to start implementing authentication, but retain their current VLAN and ACL structure. The company might choose to use local forwarding for typical employees, whose traffic does not necessitate additional security measures. IP-based filtering policies are sufficient.

It is also typical to forward AP traffic locally. The AP itself might bridge wireless client traffic or tunnel that traffic to a gateway, but once the traffic reaches the switch, the switch should forward the traffic locally. (It does not make sense, for example, for the switch to tunnel already tunneled traffic.)

User-Based Tunneling (UBT)

Alternatively, the role can specify user-based tunneling (UBT), as illustrated in Figure 7-19. In this case, the role indicates a gateway profile, and the switch uses GRE to tunnel the traffic to an HPE Aruba Networking gateway (AOS-10) or HPE Aruba Networking Mobility Controller (AOS-8). The gateway or MC then applies security policies and forwards the traffic. The switch can also tunnel traffic to a gateway cluster or MC cluster for high availability.

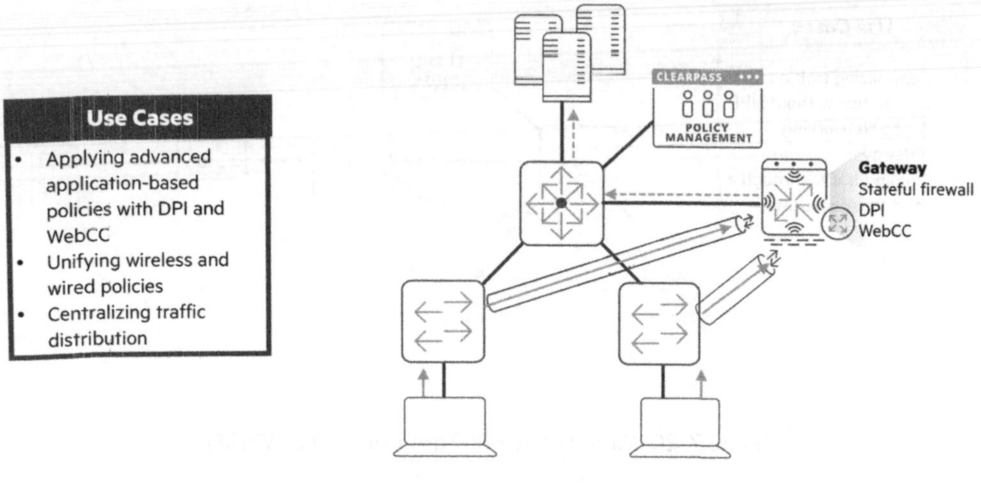

Figure 7-19: User-Based Tunneling (UBT)

You might want to use this method when you need to apply extra security with application-based policies that draw on deep packet inspection (DPI) and Web Content Control (WebCC) policies. (Note that AOS-CX version 10.11 added support for application recognition to switches on their own; however, this study guide focuses on version 10.08 and earlier.)

Regardless of the switches' own security features, UBT makes it easier to apply consistent security policies to wireless users and wired users. For example, a gateway might be handling traffic for employees that connect on both wireless and wired connections. You can define a wireless_employees role and a wired_employees role and assign the same firewall policies to both roles.

Some companies also like the fact that the UBT approach centralizes traffic distribution. Then admins do not need to worry about where they extend particular VLANs. Instead, they know that client traffic is always tunneled to a central location, where gateways reside, and the required VLANs only need to exist there.

Virtual Network-Based Tunneling (VNBT)

Finally, you can use virtual network based tunneling (VNBT), as shown in Figure 7-20. With VNBT the switch uses Virtual Extensible LAN (VXLAN) to tunnel the client's traffic to another VXLAN endpoint—typically an AOS-CX switch in the data center. In this case, the role settings are similar to those used for local forwarding—they include a VLAN assignment and can include other settings like a port access class. However, the VLAN to which the user is assigned has VXLAN enabled on it.

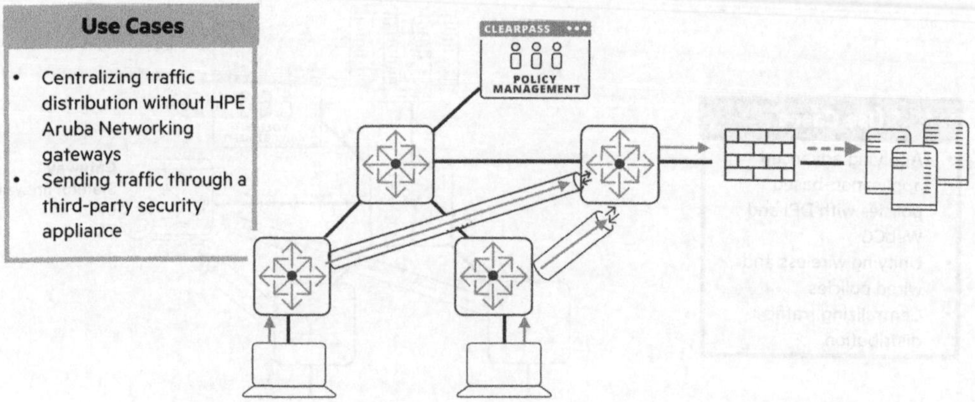

Figure 7-20: Virtual Network-Based Tunneling (VNBT)

VNBT is useful when you want to centralize traffic distribution without using HPE Aruba Networking gateways. Your main goal might be to assign clients to VLANs without regard to the underlying physical infrastructure. However, you do not care about receiving access to additional security features provided by HPE Aruba Networking gateways for this particular client group. In this case, it is more cost effective to tunnel the traffic to another AOS-CX switch.

You might also use VNBT if you want to send clients' traffic through a 3rd-party security solution in the data center.

Note that the role settings have no relationship to how the role is applied, as a LUR or a DUR. In other words, a DUR could apply local forwarding, UBT, or VNBT, just as a LUR could specify any of these forwarding types.

Implementing Local Forwarding with VLAN Steering

In this section, you will learn how to implement local forwarding with VLAN steering.

Overview

In previous chapters, you learned how Bank ABC can use ClearPass and HPE Aruba Networking mobility devices to authenticate employees who connect to the wireless network and control their access using role-based policies. Now the bank wants to similarly authenticate and apply role-based access controls to employees on wired connections.

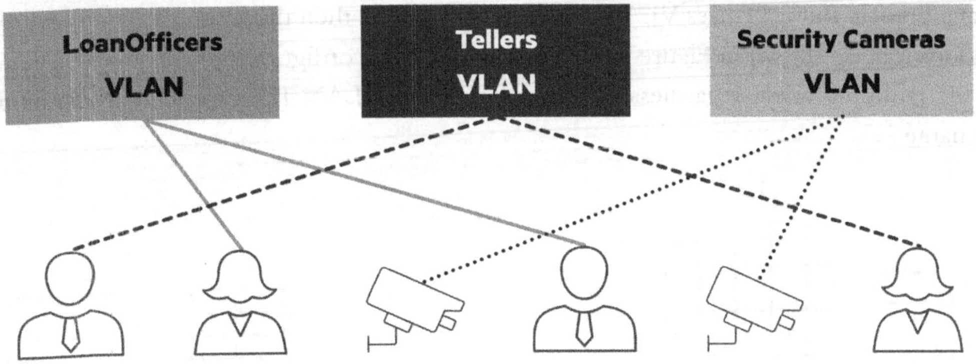

Figure 7-21: Overview

Currently the company uses VLANs to divide wired clients into different logical segments, as shown in Figure 7-21. A core routing switch then applies ACLs to the VLANs. IT has to manually specify the correct VLAN when setting up users' workstations or the connections for other wired devices like cameras. However, this approach is a management hassle. In addition, the bank is now letting some employees divide their time between the bank and their home offices, which means that some employees are now sharing a workspace. The company needs a way to ensure that each user and device is placed in the correct VLAN without IT intervention.

You will set up roles that use VLAN steering for this purpose. A teller could connect to a port one day, and the teller's computer is placed in the Tellers VLAN. The next day a loan officer could connect to the same port, and that user's computer is placed in the LoanOfficers VLAN. Employees can move security cameras where required, and they always receive the correct VLAN assignment too.

Best Practices for Role-based VLANs

You will now explore some best practices for implementing role-based VLANs.

ID versus Name

In order to apply a role that specifies a VLAN, the switch must already have a matching VLAN configured on it. If a role specifies a VLAN that does not exist on the switch, the application of the role will fail, and the client will not be able to connect.

The match can be by ID or by name.

Using the VLAN name in the role can provide easier management and greater scalability. Different areas of the network do not always support the same VLAN IDs. In this example, the VLAN ID for the LoanOfficers group at headquarters does not match the VLAN ID for the LoanOfficers group at a branch. But if IT uses VLAN names in their roles, then the same VLAN name will work in each location. As shown in Figure 7-22, a single user-role configuration will have the same effect across the entire network, regardless of which particular VLAN ID a switch associates with the VLAN name.

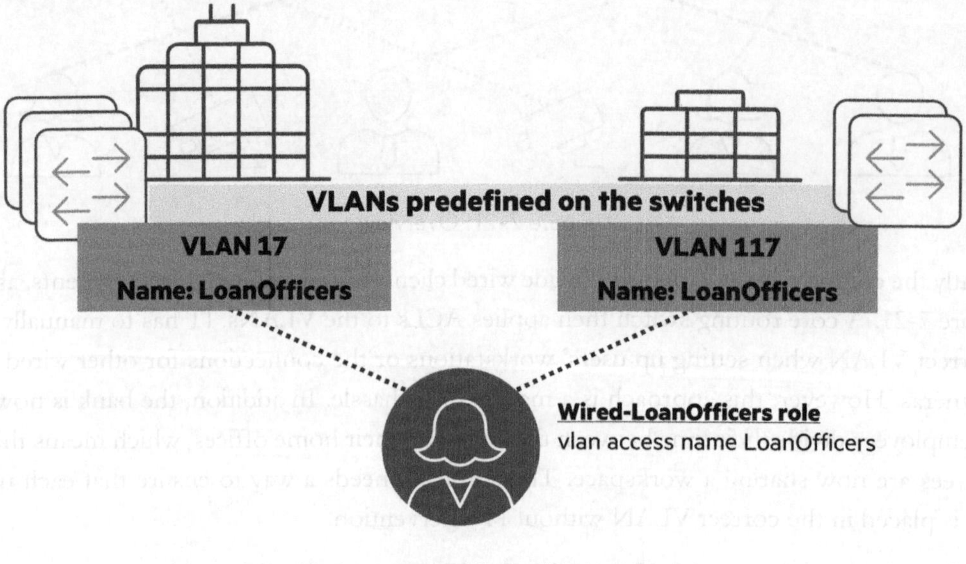

Figure 7-22: Named VLANs

Keep in mind that the VLAN name is derived from the VLAN ID, by default. If you want to use this approach in roles, switch admins and ClearPass admins must coordinate and assign the correct name to each VLAN.

Note that support for applying a VLAN by name in a user-role was added in 10.06. If your switches have earlier software, you must use the ID method.

Access and Trunk VLANs

When setting up the VLAN for a role, you also need to understand whether clients will send untagged or tagged traffic.

VLAN Assignments for Untagged Traffic

Most endpoints send untagged traffic, including most PCs, printers, POS devices, security cameras, and wired IoT devices. When setting up a role for these clients, specify the desired VLAN with the "VLAN Access" setting.

When the switch port operates in client mode, and a client is assigned to a role, the switch does not actually change the port access VLAN. Instead, the switch creates a MAC-based VLAN (MBV) for the port, as shown in Figure 7-23. An MBV specifies a VLAN that applies to untagged traffic from specific MAC addresses—in this case, the MAC addresses of clients assigned to the MBV by role. This approach lets multiple clients connect to the same port but be assigned to different role-based VLANs. (However, in most cases, only one client connects to the port.)

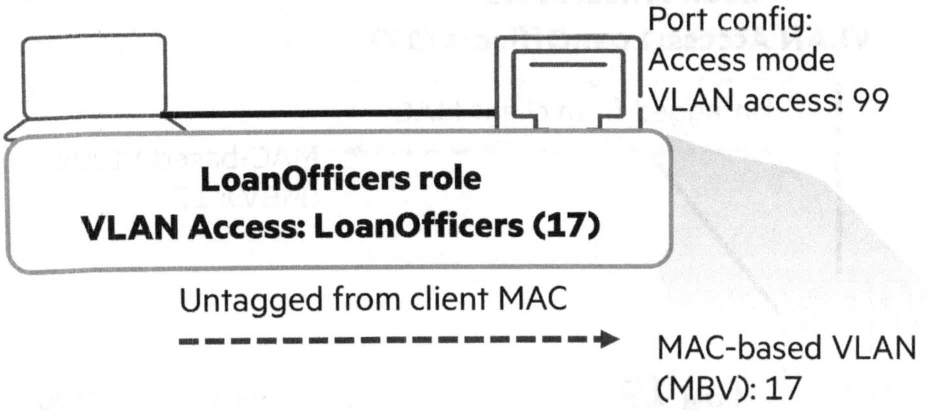

Figure 7-23: VLAN Assignments for Untagged Traffic

The port access VLAN applies to untagged traffic from any clients assigned to roles that do not specify a "VLAN Access" or "VLAN Trunk Native" setting. Typically, you should ensure that all roles specify a VLAN setting. (Roles that implement tunneling can be an exception, as you will learn later.) As a failsafe, you can place the port itself in a deadend VLAN, which is not carried on other ports.

VLAN Assignments for Tagged Traffic

Occasionally, clients will need to send tagged traffic. This scenario might occur when a computer and VoIP phone are daisy chained to a single port. The computer sends untagged traffic, and the VoIP phone sends tagged traffic.

In this case, you should continue to set up roles for users with "VLAN Access" settings. The switch port will create MBVs for clients in those roles, as you just learned.

For the VoIP role, specify a "VLAN Trunk Allowed" setting. When the switch applies a role with a "VLAN Trunk Allowed" setting, it begins accepting traffic tagged with the indicated VLAN ID, or with the ID associated with the indicated VLAN name, from clients with that role.

The switch can apply the "VLAN Trunk Allowed" setting even though the port is statically configured as an access port. This behavior simplifies configuration; admins do not need to know in advance which edge ports will need to accommodate clients that send tagged traffic.

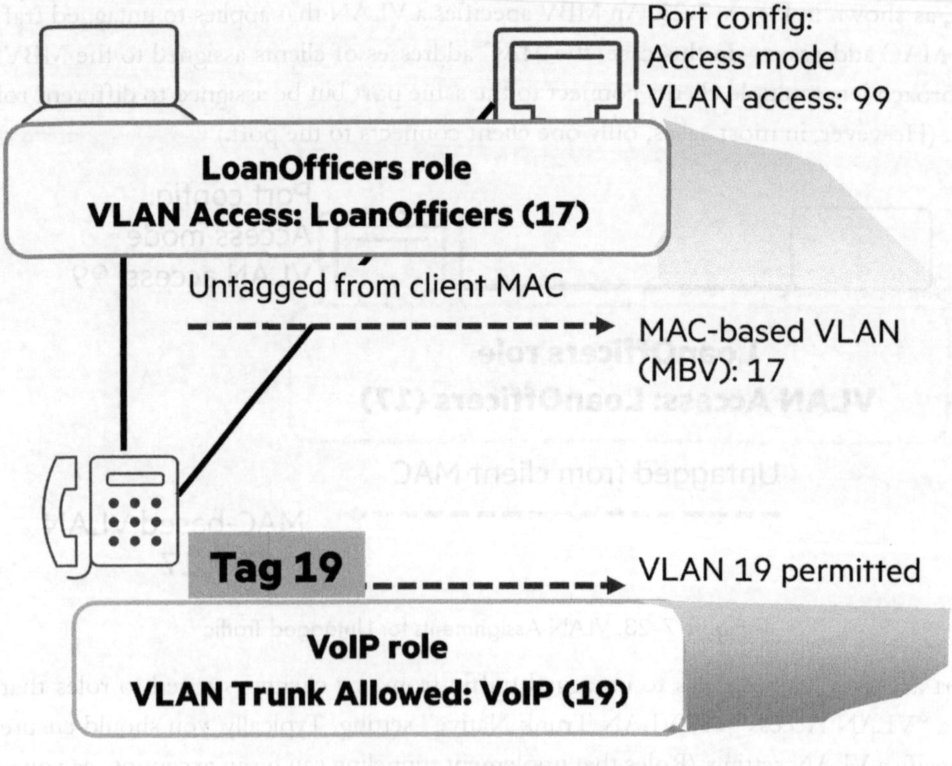

Figure 7-24: VLAN Assignments for Tagged Traffic

With the settings shown in Figure 7-24, if the VoIP phone happened to send any untagged traffic, that traffic would be assigned to the port access VLAN. If you do not want that behavior, you can add an additional VLAN setting to the VoIP role: "VLAN Access" or "VLAN Trunk Native."

VLAN Steering Roles

Now that you understand the VLAN settings, you can specify them within an AOS-CX switch local role or within a downloadable enforcement profile.

LUR

The first step for establishing VLAN steering is always to create the VLANs to which you want to assign clients. For the Bank ABC example, you might create two VLANs, one for "LoanOfficers" and one for "Tellers."

Rather than assign ports to the VLANs as you would do traditionally, you create two roles for your two types of clients. You then specify the **vlan access name** *<name>* within the role, as shown in

Figure 7-25. Now when a client is assigned to this role, the switch will assign the client's traffic to the role dynamically.

```
AOS-CX(config)# port-access role loanofficers-fullaccess
AOS-CX(config-pa-role)#   vlan access name LoanOfficers

AOS-CX(config-pa-role)# port-access role tellers-fullaccess
AOS-CX(config-pa-role)#    vlan access name Tellers
```

Figure 7-25: VLAN Assignments in LURs

Aruba Downloadable Role (Standard Config)

When you are using DURs, the define the user-role, including its VLAN settings, within an enforcement profile on CPPM, as shown in Figure 7-26.

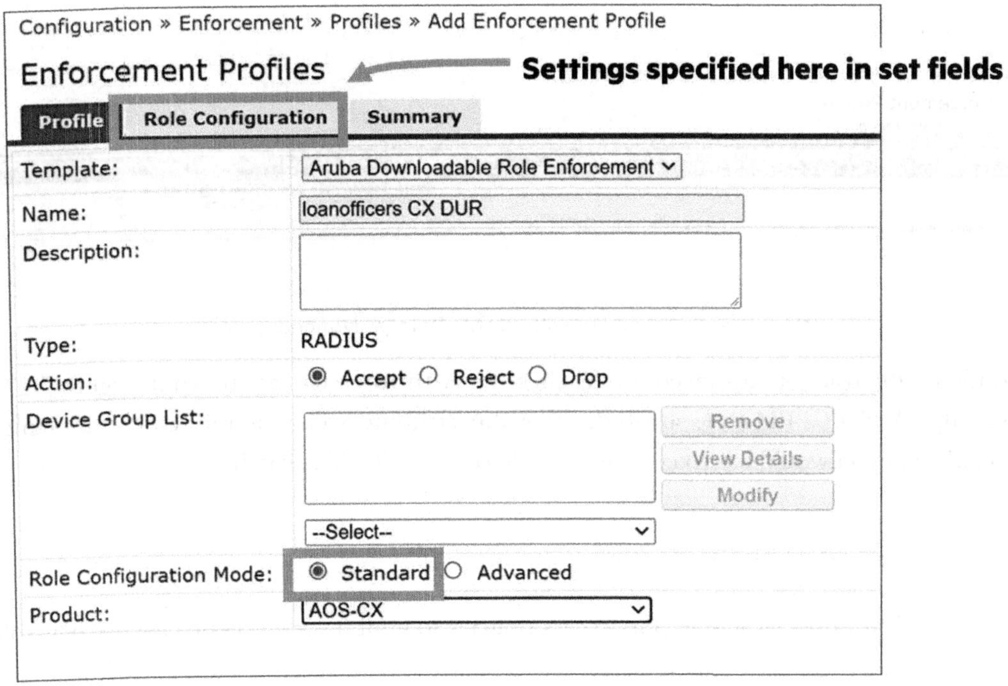

Figure 7-26: Aruba Downloadable Role (Standard Config)

As you learned earlier, you set the profile template to **Aruba Downloadable Role Enforcement**. You then have two choices for the Role Configuration Mode, **Standard** and **Advanced**. Choose **Standard** to select options from set fields. In CPPM v6.9, you can use **Standard** to define the VLAN by ID, but not by name. In that version, standard configuration also allows you to specify only one VLAN ID for the VLAN Trunk Allowed setting. (However, in v6.10 and above, CPPM provides fields for specifying VLAN names and also permits specifying more than one VLAN for the VLAN Trunk Allowed setting.)

After choosing **Standard**, click the Role Configuration tab. You can then fill out the VLAN settings, as well as other settings, using fields presented by the interface.

Aruba Downloadable Role (Advanced Config)

Choose **Advanced** for the Role Configuration Mode if you want complete flexibility in configuring any settings that are supported on your local switch. In CPPM v6.9, use **Advanced** mode to specify VLANs by name or to allow multiple trunk VLANs.

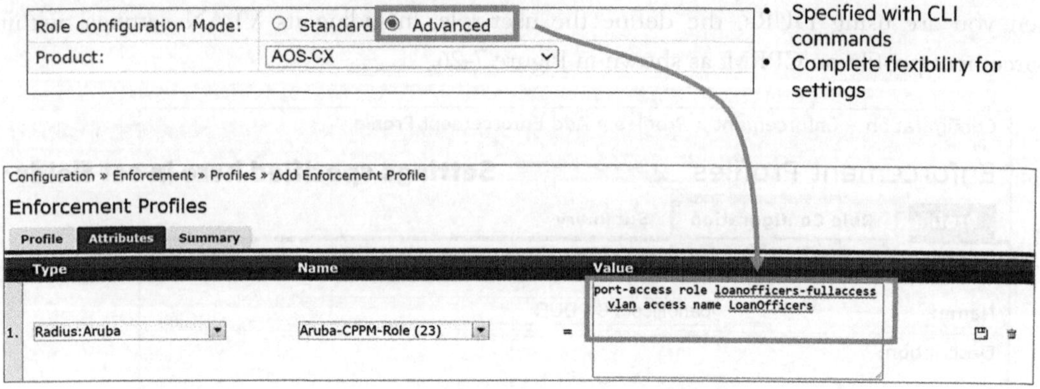

Figure 7-27: Aruba Downloadable Role (Advanced Config)

To configure the role for Advanced mode, click the **Attributes** tab, as shown in Figure 7-27. You will see the Aruba-CPPM-Role attribute. For the attribute value, simply enter the same CLI commands that you would use to configure a role on an AOS-CX switch.

Adding Port-Access Policies

Now you will learn how to adding port-access policies to limit users' access based on their roles.

Approaches for Applying ACLs to Wired Devices

Dividing wired clients into different VLANs provides a first step towards segmenting their traffic. However, if you want to truly assign different rights to different users, you must add ACLs to enforce those rights. AOS-CX switches support MAC and IP ACLs, which consist of rules that filter traffic based on its MAC or IP header.

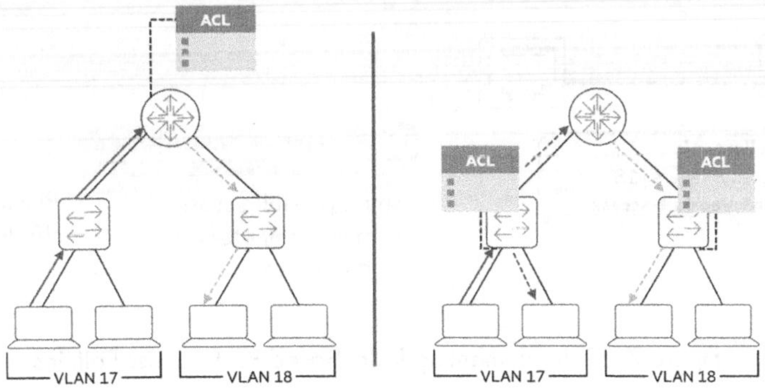

Figure 7-28: Approaches for Applying ACLs to Wired Devices

You can take two general approaches for using switch-based ACLs to boost security, as shown in Figure 7-28.

You can assign users to different VLANs based on role, and then configure a core routing switch to apply ACLs as the traffic is routed from one VLAN to another VLAN.

Alternatively, you can continue to assign users to different VLANs based on role, but configure each access layer switch to apply ACLs to all traffic incoming on edge ports. You can use roles to apply the ACLs so that the port enforces a different set of rules for each client, based on that client's role. In this case, the ACLs are IP ACLs.

While both approaches add another layer of security beyond VLAN steering, the second approach has some additional advantages. By using ACLs on access layer switches, traffic is filtered immediately, rather than passing through the network before reaching a core switch. Similarly, traffic destined to another device within the VLAN, as well as traffic that will be routed to another VLAN, is filtered.

Many companies prefer a centralized approach to configuring ACLs to simplify setup and ensure consistency. However, with DURs you can apply ACLs at the edge port while also centralizing configuration.

Implementing Role-Based Port Access Policies

To implement a role-based ACL on a port, you reference the name of a port access policy as a policy within the role. (See Figure 7-29.) When a client is assigned to that role, the switch will use the rules defined within the port access policy to filter all ingress traffic from the client.

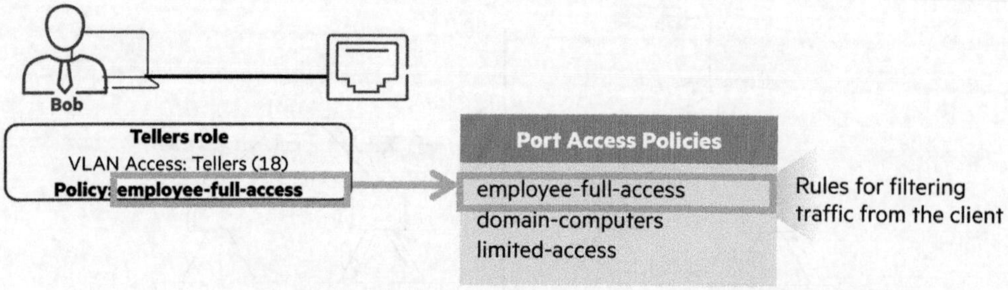

Figure 7-29: Implementing Role-Based Port Access Policies

Sources for Port Access Policies

An AOS-CX switch can have a port access policy defined on it in three ways, illustrated in Figure 7-30.

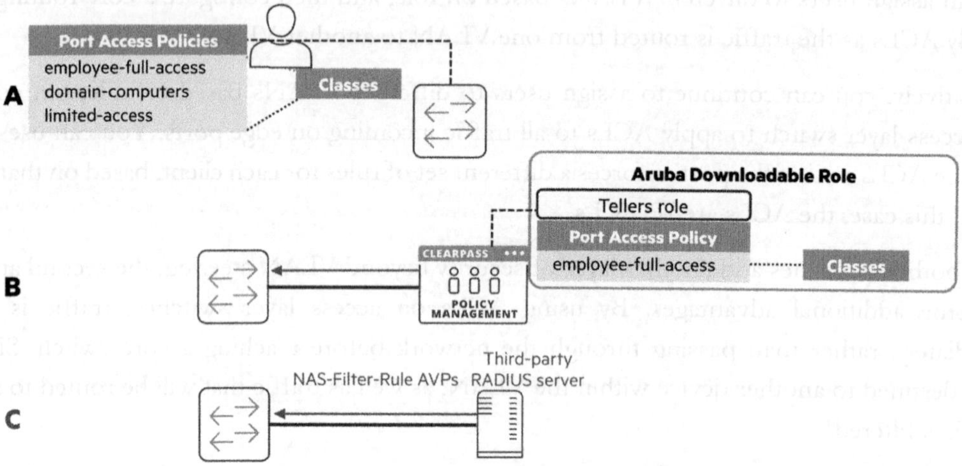

Figure 7-30: Sources for Port Access Policies

You can configure the port access policies locally on the switch. As you will see in more detail later, port access policies reference classes. When you use local configuration, you must configure these classes manually on the switch as well.

Alternatively, if CPPM is applying DURs, the Aruba Downloadable Role enforcement profile can also include the port access policy and class configuration. The switch will dynamically download the classes and port-access policy as well as the role configuration.

The switch can also receive rules sent in RADIUS attribute value pairs (AVPs). AOS-CX switches support the standard RADIUS NAS-Filter-Rule AVP or the Aruba-NAS-Filter-Rule VSA. You might use this option when you want to centralize the policy configuration, but you have a third-party RADIUS server. However, this last option is not the focus of this study guide.

Port Access Policy Components

To implement a port access policy, you must first create IPv4 or IPv6 classes. Each class selects one or more types of traffic. You then reference those classes within the port-access policy and specify how you want to treat the traffic.

For example, you might want to enforce a simple policy that permits most traffic but prohibits a few applications and destinations. As shown in Figure 7-31, you could create an IP class named "prohibited" for selecting the traffic to drop. In the example, prohibited traffic includes SSH traffic and traffic to a sensitive subnet in the data center. However, you can define whatever traffic you want to prohibit based on your organization's policies.

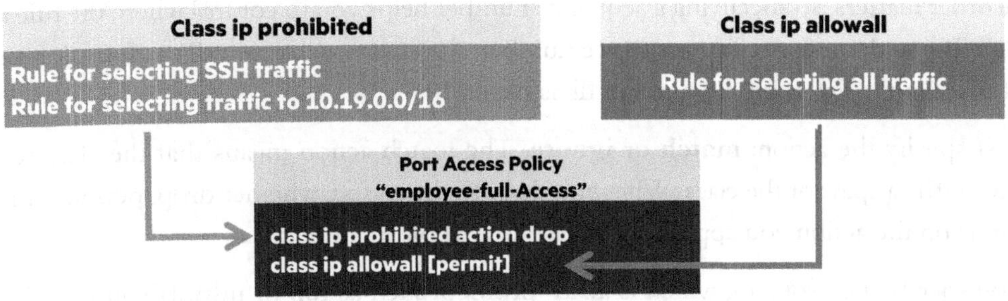

Figure 7-31: Port Access Policy Components

Then you could create an IP class name "allowall," which selects all traffic.

Then you could specify in a port-access policy that you want to drop IP class "prohibited" but permit IP class "allowall."

Configuring Classes Locally

Whether you create a policy locally or in a CPPM downloadable role, the syntax for the setup is the same. You first use commands to create IPv4 or IPv6 classes, which select traffic based on criteria within the IP header. You can see the syntax for IPv4 in Figure 7-32, and the IPv6 syntax is very similar.

```
AOS-CX(config)# class {ip | ipv6} <CLASS-NAME>

AOS-CX(config-class-ip[v6])# [sequence number] {match|ignore}
<protocol> <source> <dest> [TCP/UDP dest port] [options]
```

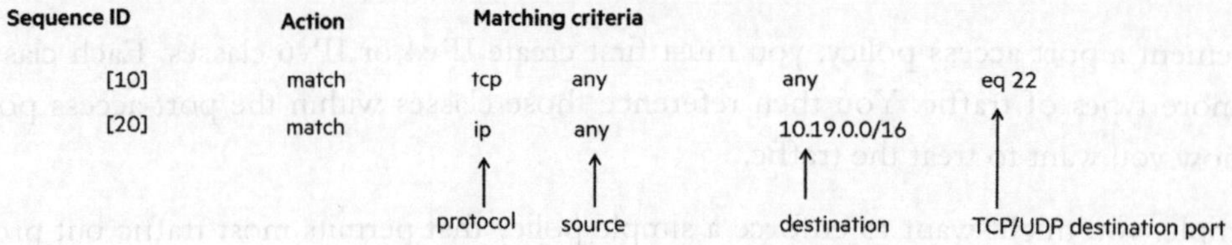

Figure 7-32: Configuring Classes Locally

First specify this command: **class ip <*class-name*>**. You can then enter a sequence of rules.

A sequence number at the beginning of the rule is optional. As with access lists on AOS devices, the rule order matters, so specifying a sequence number helps you to control where the rule is placed in the list. If you do not specify a sequence number, the switch automatically assigns the rule to the highest place in the list, rounding up to a divisible of 10.

You next specify the action: **match** or **ignore**. The match action means that the class selects the specified traffic as part of the class. What you do with the traffic, whether drop, permit, or rate limit it, depends on the action you apply to the class as a whole in a policy.

Next you specify the protocol, which is an IP protocol such as **tcp** or **udp**. If you specify **ip**, then the rule applies to all IP traffic that meets the rest of the rule's criteria.

You then specify the source, which can be **any** or an IP address. You can specify a range of IP addresses using a prefix length or subnet mask.

Next, you specify the destination, which follows the same format as the source.

If you specified **tcp** or **udp** for the protocol, you can specify a destination port, which helps to specify the particular application such as SSH or HTTPS. You begin with a keyword such as **eq** to specify an exact port or **range** to specify a range of ports.

You can then specify a number of extra options such as **vlan** to specify the VLAN to which the traffic must belong. You can refer to the switch documentation for more details.

In Figure 7-32, you see example rules for the scenario introduced earlier. The **10 match tcp any any eq 22** rule matches any TCP traffic to port 22 (for SSH). The **20 match ip any 10.19.0.0/16** rule matches all IP traffic destined to subnet 10.19.0.0/16.

As a second example, you might create another class and add this rule to it:

10 ignore ip any 10.101.10.0/24

This IP class now ignores IP traffic coming from any source destined for 10.101.10.0/24.

If, in the same context, you enter **20 match tcp any eq 443**, then the class will match TCP traffic from any source going to any destination on port 443 (for HTTPS). Note, though, that any HTTPS traffic to 10.101.10.0/24 will not be matched by this class because the previous rule already ignored that traffic.

If you enter **30 match tcp any 10.101.20.0/24 range 50000 50100**, then the switch will further use this class to identify TCP traffic from any source, destined for 10.101.20.0/24 on any port between 50000 and 50100.

Configuring Port Access Policies Locally

Once you have created classes for the traffic you want to control, you can create a policy, which assigns actions to each class. In the **config** context, you enter **port-access policy** followed by the policy name.

```
AOS-CX(config)# port-access policy <POLICY-NAME>

AOS-CX(config-pa-policy)# [sequence number] class
{ip|ipv6} <CLASS-NAME> [action {drop | redirect |
<police-actions> | <remark-actions>}]
```

```
AOS-CX(config)# port-access policy employee_access

AOS-CX(config-pa-policy)# 10 class ip prohibited action drop
AOS-CX(config-pa-policy)# 20 class ip allowall
```

Implicit deny at end of all policies

Figure 7-33: Configuring Port Access Policies Locally

Then you create a series of rules for the classes. Begin with an optional sequence number, followed by **class {ip | ipv6}** and the class name.

If you want to permit the class with no rate limit or change in priority, you simply end the command there, as shown in rule 20 in Figure 7-33.

If you want to deny the class, enter **action drop**, as shown in rule 10 in Figure 7-33.

The **redirect** action sends all traffic of a class to a captive portal server. Or you can apply a police action, which rate limits the traffic. After **action**, enter **cir kbps** followed by the rate in kilobits per second you want to enforce.

It is important that you understand that the switch processes packets against each rule in order. If the packet matches the class in the rule, the switch applies the specified action. Otherwise, it processes the next rule. If the switch reaches the end of the policy without finding a match for the packet, the switch drops the packet based on an implicit deny.

You can also apply actions that are more related to QoS than to security, applying a remark action to change the traffic's priority.

Defining Classes in an Aruba Downloadable Role on CPPM

If you are using DURs, you define the classes directly within an Aruba Downloadable Role enforcement profile. Add a new enforcement profile of this type and configure the other settings as you learned how to do previously. Or edit an existing profile.

Then click **Manage Classes**.

You will see the Class Configuration window (see Figure 7-32). Choose a name for the class and set its type to IPv4 or IPv6.

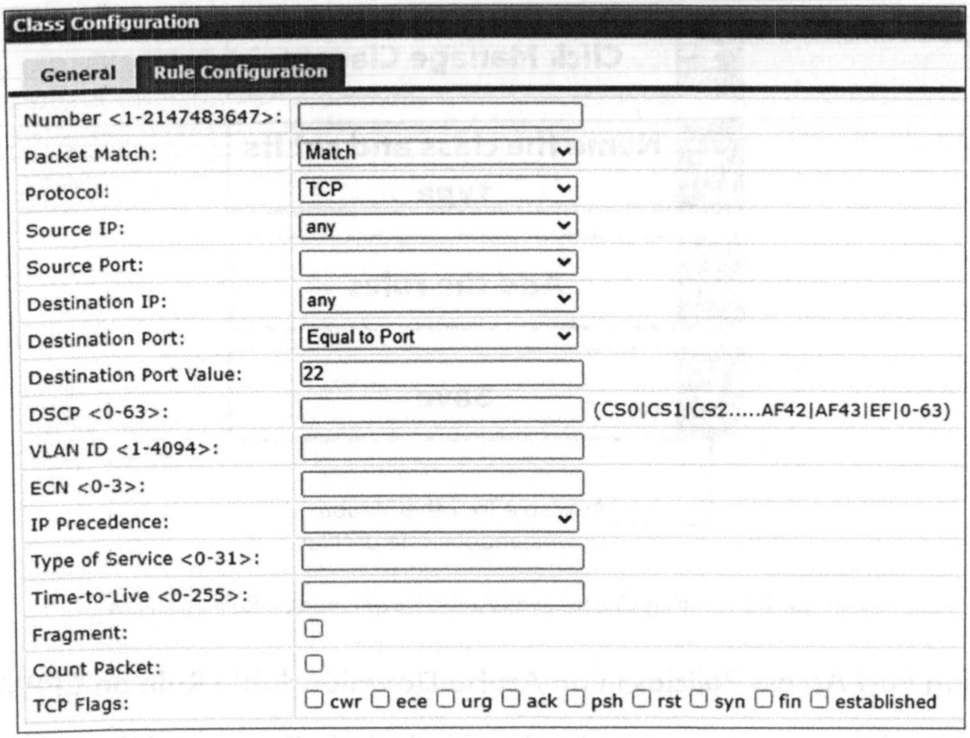

Figure 7-34: CPPM's Class Configuration Window

You can then add one or more rules. You define the rules in the **Rule Configuration** tab. As you see, the fields available for the rule resemble the options that you can configure in a locally defined class. If you do not want to select traffic based on a particular type of criteria, leave that field blank or set to "any." The example shown in Figure 7-34 selects all SSH traffic.

After you have added the desired rules, you can save the class. The class is now available for use in a port access policy in this enforcement profile or in other Aruba Downloadable Role enforcement profiles.

These steps are summarized in Figure 7-35.

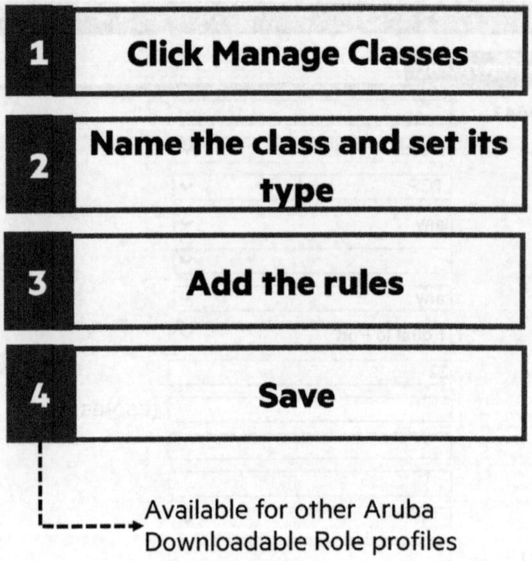

Figure 7-35: Defining Classes in an Aruba Downloadable Role on CPPM

Defining Port Access Policies in an Aruba Downloadable Role on CPPM

You can also define port access policies directly within the Aruba Downloadable Role enforcement profile.

Click **Add Policy**. Then use the windows presented by CPPM to configure the policy. You will name the policy. You can then add a rule by either selecting **Add Rule** or click the **Rule Configuration** tab. For each rule, choose a class and define an action for it. (See Figure 7-36.)

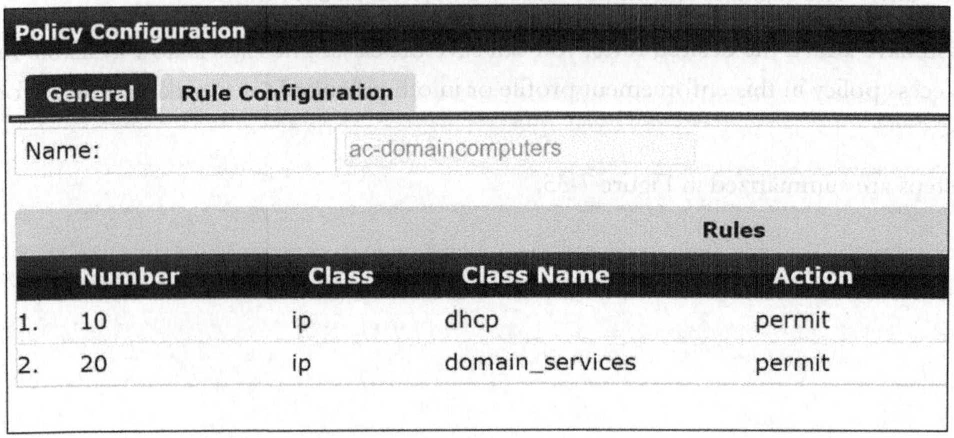

Figure 7-36: CPPM's Policy Configuration Window

When you have finished configuring the policy, click **Save Policy**.

You can then choose the policy in the Policy drop-down menu for this enforcement profile. The policy will also be available for selection in other Aruba Downloadable Role enforcement profiles.

These steps are summarized in Figure 7-37.

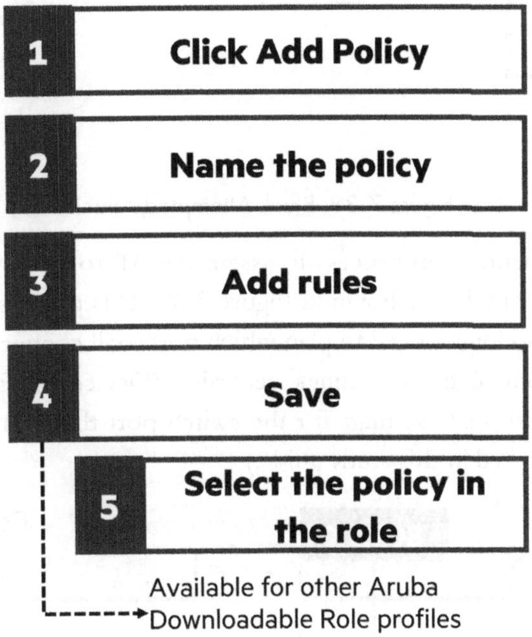

Figure 7-37: Defining Port Access Policies in an Aruba Downloadable Role on CPPM

802.1X Authentication of APs

In this section, you will learn how to use 802.1X to authenticate APs to switch ports.

Why Authenticate APs with 802.1X

It is best practice to implement 802.1X on all switch edge ports, including ones that connect to APs, particularly when the APs are connected in locations accessible to users. Enforcing 802.1X prevents hackers from disconnecting an AP and connecting their own device to the switch port intended for the AP. This port could provide hackers with a backdoor into the network, which they might be able to leverage into access to other systems. By implementing 802.1X, you stop the hacker immediately at the port, as shown in Figure 7-38.

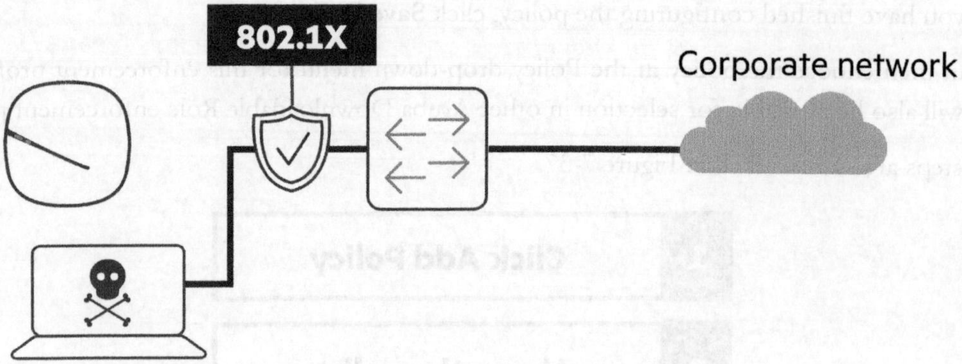

Figure 7-38: Block Attempted Attack

You can also use the authentication process to assign the AP to a role and apply the appropriate settings for the port dynamically, as shown in Figure 7-39. This process can simplify setup on the access switch, as admins no longer need to plan which ports will connect to APs in advance. Using DURs, they can also control those settings centrally. (Device profiles offer another way to dynamically apply the appropriate settings for the switch port that connects to an AP. However, device profiles are not covered in this study guide.)

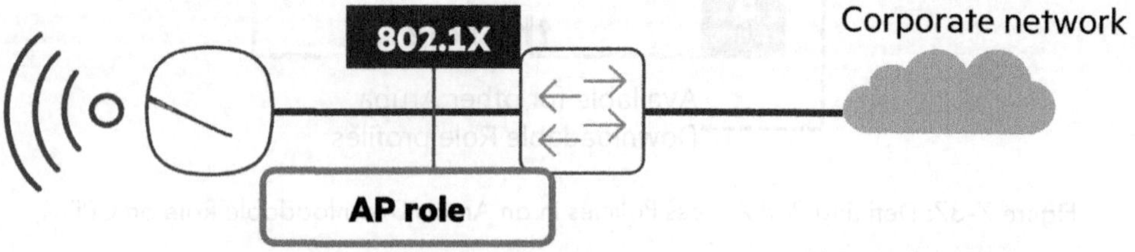

Figure 7-39: Ensure Proper Switch Port Configuration

Configuring 802.1X on HPE Aruba Networking AP Uplinks

With the certificates in place, you can enable 802.1X on the APs' uplink.

If you are managing your APs in HPE Aruba Networking Central, you generally configure the settings in the group-level device configuration for Access Points. (In version 2.5.5, you would then navigate to **Interfaces > Uplinks** and find the **AP1X** settings.)

You have two choices for the AP1X Type: PEAP or TLS (EAP-TLS). You must match the method configured on CPPM. EAP-TLS is recommended. When you select TLS, you must further select the certificate that the AP uses to authenticate. You can choose between the User certificate, which is a CA-signed certificate that you must load on the APs, or the factory-installed TPM certificates. You learned about the advantages and disadvantages of the certificate types in Chapter 2.

When you configure APs managed by HPE Aruba Networking Central and select a user certificate, you can select the Validate Server check box. The AP then validates the certificate submitted by the RADIUS server. You must make sure to configure the APs with CPPM's RADIUS/EAP certificate as a Trusted CA for 802.1X in the certificate usages.

When you have finished the configuration, make sure to save the settings.

You can also configure 802.1X on uplinks for APs managed by MCs. In that case, you use the provisioning settings for the APs' uplink (**Configuration** > **Access Points** > **Provision** >**Uplink**). You then configure settings much like the ones just discussed. On APs managed by MCs, you can also specify a domain name, which the APs append to the hostname that they submit as their username.

Note that for APs managed by HPE Aruba Networking Central to support TLS, the APs must run AOS 10.3 or later and HPE Aruba Networking Central must be at version 2.5.4 or later.

Example AOS-CX Roles for APs

You need to set up roles to which APs will be assigned after the complete authentication.

```
port-access role tunnel-ap
  description "Tunneling_AP"
  vlan access 106
  poe-priority critical
  trust-mode dscp
  session-timeout 28800
```

```
port-access role bridge-ap
  description "Bridging_AP"
  auth-mode device-mode
  vlan trunk native 106
  vlan trunk allowed 107-109
  poe-priority critical
  trust-mode dscp
  session-timeout 28800
```

LURs or DURs

Figure 7-40: Example AOS-CX Roles for APs

The role should implement local forwarding. It should further assign the AP's traffic to the correct VLANs. If the AP is tunneling all client traffic to a gateway or MC, you only need to assign the role to the VLAN which the AP uses to communicate with the gateway or MC and to tunnel the traffic, as shown on the left of Figure 7-40. But if the AP is bridging client traffic, it often bridges traffic into one or more tagged VLANs, as shown on the right of Figure 7-40. You need to configure the bridging APs' native VLAN, on which it sends untagged traffic as well as the tagged VLANs in which it forwards user traffic. Remember that you configure the tagged VLANs with the **vlan trunk allowed command**. You must also configure the bridging APs' role to set the port in device-mode. This mode prevents the switch from attempting to authenticate the client MAC addresses that it sees the AP bridging on the port.

For both types of APs, you can assign settings that give the AP privileges suited to its role. You can set the port PoE priority to critical, which gives the port priority for delivering PoE power if the switch's power is oversubscribed. You can also configure the switch to trust DSCP, which are quality of service (QoS) marks within IP headers. APs can set the DSCP for IP traffic to indicate its priority.

And you can specify other supporting settings such as a session timeout.

As always you can configure the roles locally on switches or centrally on CPPM using DURs. If you use DURs, you can set the configuration mode to advanced and use exactly the same syntax that you would for a LUR.

Example CPPM Service for Authenticating APs

You will now look at an example service which CPPM can use to authenticate the APs. In this example, the APs are authenticating with their TPM certificates, but you could take a similar approach to configure CPPM to authenticate APs using CA-signed user certificates. Keep in mind this service is only one example; you can adjust some of these settings based on companies' particular requirements.

Generally, you want to create a dedicated service for authenticating the APs. Then you can configure settings that are appropriate for authenticating APs but might not be appropriate for corporate users.

Service Rules for an Example CPPM Service

To distinguish between RADIUS requests related to authenticating APs and requests related to authenticating clients, you can add another rule to the standard Wired 802.1X rules, as shown in Figure 7-41. Remember that you can configure APs to append a domain name to their username. This rule looks for a username that ends with the domain name you configured. In this example, that domain name is "aps.example.com," which is designed to be different from the domain name used by normal clients.

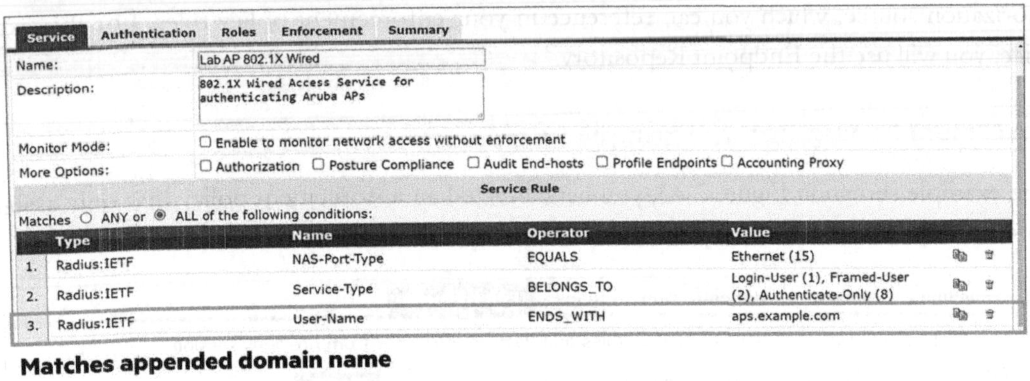

Figure 7-41: Service Rules for an Example CPPM Service

Authentication Settings for an Example CPPM Service

For the authentication settings in this example, you add a custom EAP-TLS method, as shown in Figure 7-42. In this method you disable authorization because your typical authentication sources, such as an AD instance, do not have accounts for the APs. You also disable OCSP because the TPM certificates do not provide OCSP.

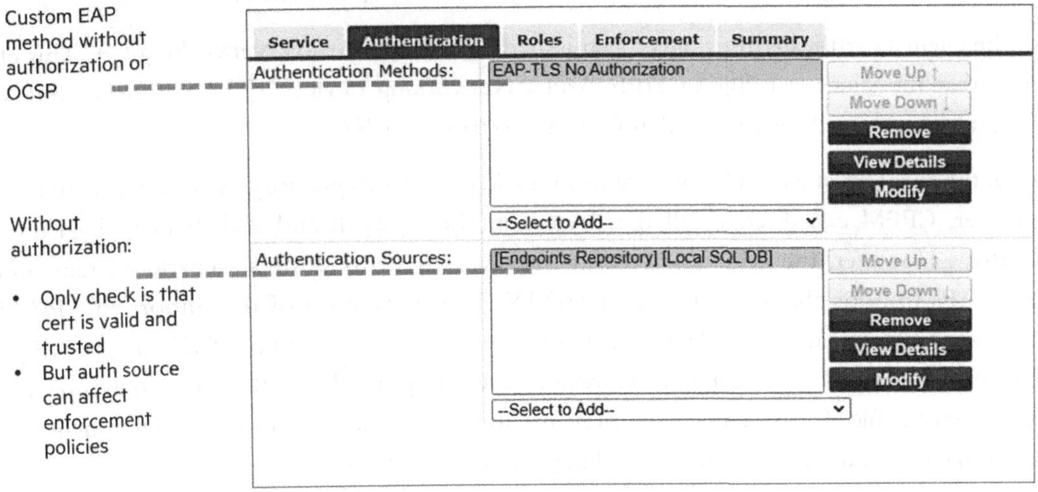

Figure 7-42: Authentication Settings for an Example CPPM Service

Because you disabled authorization, CPPM does not check the APs' usernames against an authentication source. It simply verifies that the certificates are signed by a trusted CA, which has the EAP usage. However, you must specify an authentication source. You can specify any source as a placeholder. However, keep in mind that the authentication source will also become an authorization source, which you can reference in your enforcement policy rules. For this example service, you will use the Endpoint Repository.

Enforcement Settings for an Example CPPM Service

In the example shown in Figure 7-43, you have created an enforcement policy that only assigns an enforcement profile if the client passes two checks.

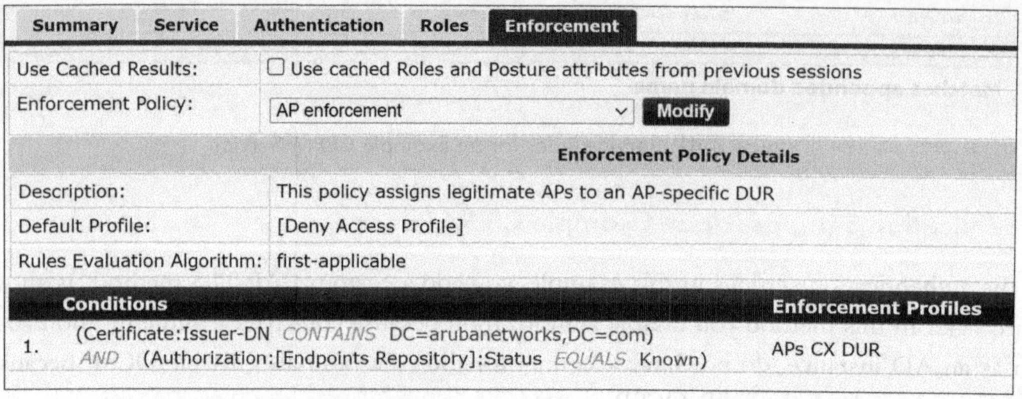

Figure 7-43: Enforcement Settings for an Example CPPM Service

First, the client's certificate must have been issued by an HPE Aruba Networking CA. This check ensures that the client is using an HPE Aruba Networking device (TPM) certificate and not a certificate signed by a different CA that CPPM also trusts for EAP.

Second, the client must have a known status in the Endpoints Repository. As you will learn in more detail later, CPPM can discover all the devices in the network and add them to its Endpoint Repository. However, the devices start with an unknown status. In this scenario, admins initially connect APs to ports that do not enforce 802.1X. They have a list of the authorized APs' MAC addresses, and they mark those devices as known. After that they set up 802.1X on the AP ports. CPPM detects whether the client is known in the Endpoint Repository and only applies the enforcement profile if it is. This check prevents users or hackers from connecting unauthorized APs that have a valid HPE Aruba Networking Device Certificate.

Clients that do not match both of these conditions are denied.

CPPM's CA Trust List

CPPM checks the APs' certificate during authentication. Remember to install the corresponding root CA certificate and set its usage to EAP. When the APs are using their TPM certificates, you must enable the Aruba Networks Trusted Computing Root CA, which installed on CPPM by default.

You can quickly identify this certificate in CPPM's Certificate Trust List because it has the default usage of Aruba Infrastructure. (See Figure 7-44.) You must enable the CA and add the EAP usage.

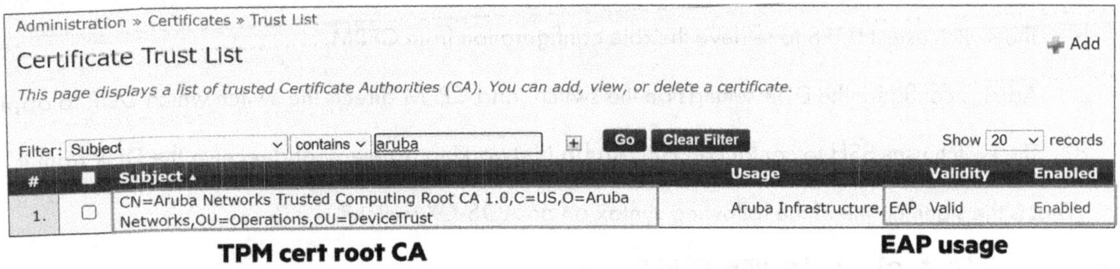

Figure 7-44: CA Trust List

If a different CA signs your APs' certificates, you should simply install that root CA certificate and similarly set its usage.

Summary

Congratulations! You should now understand how to implement 802.1X for wired clients. You should have a renewed grasp of AOS-CX roles and Dynamic Segmentation. You should be knowledgeable about implementing local forwarding with VLAN steering and role-based ACLs, as well as enforcing 802.1X on ports that connect to HPE Aruba Networking APs.

Learning checks

1. CPPM already acts as the authentication server for wireless 802.1X. Now the company wants to add wired 802.1X. Which CPPM components does HPE Aruba Networking generally recommend should be the same for wired and wireless 802.1X? (Select two.)

   a. Service

   b. Authentication source

   c. Role mapping policy

   d. Enforcement policy

2. How does a switch receive the configuration for a DUR?
   a. CPPM uses SSH to log into the switch and configure the DUR on it.
   b. The switch uses HTTPS to retrieve the role configuration from CPPM.
   c. Admins configure the DUR settings on the switch, and CPPM directs the switch which DUR to apply.
   d. The switch uses SSH to contact an HPE Aruba Networking gateway and receive the DUR from it.
3. What is the significance of the following syntax on an AOS-CX switch?

```
AOS-CX(config)# class ip NEW-CLASS
AOS-CX(config-class-ip)# 10 match tcp any 10.101.12.0/24 eq ssh
```

   a. The switch will drop all traffic from any IP address destined for an address in 10.101.12.0/24.
   b. The switch now has a class that matches all SSH traffic from devices within 10.101.12.0/24.
   c. The switch now will permit all IP traffic destined for SSH ports on 10.101.12/24.
   d. The switch now has a class that matches all SSH traffic destined to addresses in 10.101.12.0/24.

Note

For answers to all learning checks, refer to the Appendix.

8 Implement Dynamic Segmentation on AOS-CX Switches

EXAM OBJECTIVES

- ✓ Explain Dynamic Segmentation, including its benefits and use cases
- ✓ Deploy Dynamic Segmentation, including:
- ✓ User-based tunneling (UBT)
- ✓ Virtual network-based tunneling (VNBT)

Assumed Knowledge

- Authentication, Authorization, and Accounting (AAA) framework and basics of protocols such as RADIUS and TACACS+
- Forms of Layer 2 authentication such as 802.1X, including common EAP methods such as PEAP and EAP-TLS
- Basics of managing AOS-CX)switches
- AOS-CX user-roles
- PKI and digital certificates
- VLANs
- IP address and routing fundamentals

- Fundamental HPE Aruba Networking ClearPass Policy Manager (CPPM) components, such as services, roles, enforcement profiles, and enforcement policies

Overview

When you implement Dynamic Segmentation with tunneling, you have even more power to segment and control clients. You can direct the traffic to the correct destination, which then applies powerful access control policies.

In this chapter, you will review user-based tunneling (UBT) and how to configure it using local user-roles (LURs) and downloadable user-roles (DURs).

You will then learn about use cases for virtual network based tunneling (VNBT). And you will receive a basic overview of VNBT configuration.

Implementing UBT on AOS-CX Switches

In this section, you will learn more about implementing UBT on AOS-CX switches.

UBT

You will now focus in on user-based tunneling (UBT).

As you learned earlier, UBT uses the following basic process.

As shown in Figure 8-1, a client connects to a switch port that implements L2 authentication. Based on the authentication, the client is assigned to a user-role. The user-role has a gateway setting defined within it. This setting tells the switch to tunnel client traffic to an HPE Aruba Networking gateway or Mobility Controller (MC). The switch also informs the gateway or MC of the client's gateway role. The gateway or MC then assigns the client to that role, applies the policies and settings configured for that role, and forwards the traffic.

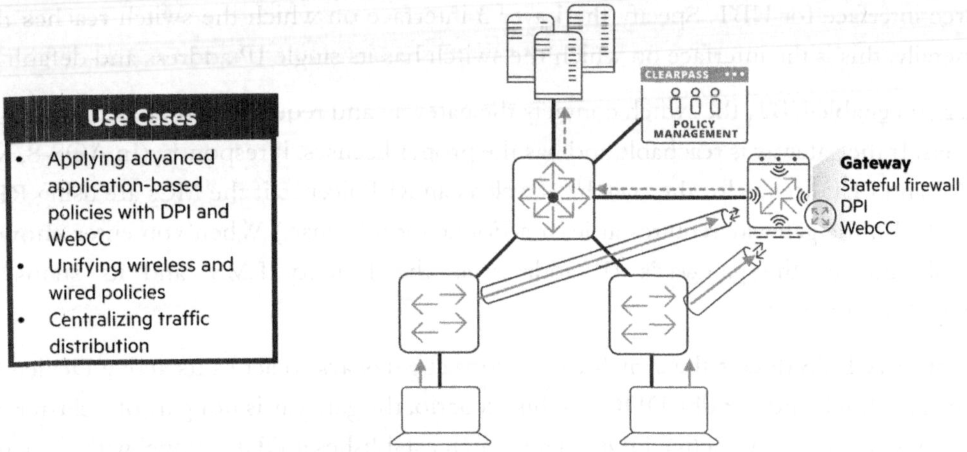

Figure 8-1: UBT

Because AOS gateways and MCs support more advanced, application-based policies than the AOS-CX switches, UBT is useful for companies who want to apply deep packet inspection (DPI) and Web Content Classification (WebCC) to wired clients. UBT also enables companies to unify wireless and wired policies, as well as centralize traffic distribution.

Review UBT Tunnel Configuration

For UBT to function, you must create a UBT zone on each AOS-CX access switch. This zone tells the switch where to access the gateway (AOS-10) or MC (AOS-8).

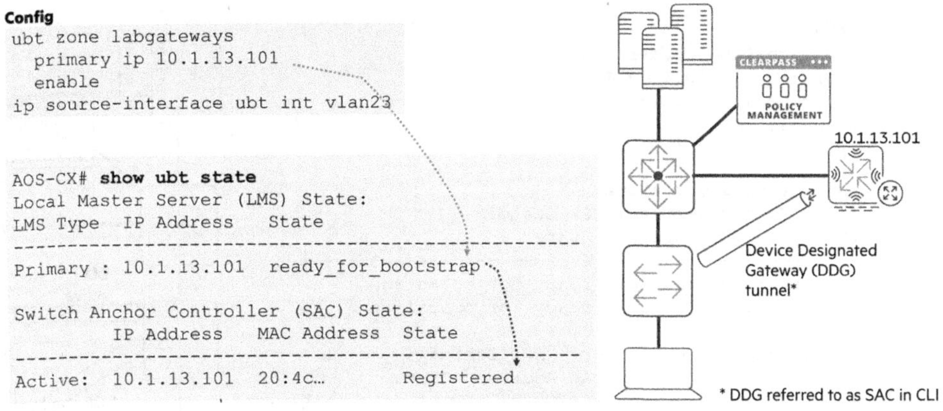

Figure 8-2: Review UBT Tunnel Configuration

In this first simple example, illustrated in Figure 8-2, the company has a single gateway. The switch UBT zone configuration references that gateway's IP address as in the **primary-controller ip** command. For the switch to contact the gateway, you must also enable the UBT zone and set the source interface for UBT. Specify the Layer 3 interface on which the switch reaches the gateway. Generally, this is the interface on which the switch has its single IP address and default route.

After you enable UBT, the switch contacts the gateway and requests that the gateway bootstraps its tunnels. If the gateway is reachable and has the proper licenses, it responds. (In AOS-8, MCs require an AP and PEF license for the switch, as well as an RFP license if the MCs are using RFP for APs. In AOS-10, the gateway requires at least a Foundation license.) When you enter **show ubt state**, you should see the gateway's IP address as the Primary LMS, and its status should be "ready_for_bootstrap."

The primary LMS directs the switch to the correct gateways to act as its active Device Designated Gateway (DDG) and standby DDG. In this scenario, the gateway is not part of a cluster, so it directs the switch to use it as its active DDG. The switch establishes a DDG tunnel with the gateway. This tunnel will carry management traffic such as heartbeats. (Note that the switch CLI might refer to the DDG as a Switch Anchor Controller [SAC], but the concept is the same.)

Review UBT Tunnel Configuration to a Gateway Cluster

In the scenario illustrated in Figure 8-3, the company has a gateway cluster. When you configure the ubt zone, you can choose any of the gateway's IP addresses to add as the primary controller. As before, the switch contacts the primary gateway and is bootstrapped. As you see, the primary gateway can assign the switch to any gateways in the cluster as the active and standby DDGs. The switch establishes tunnels with them. When you enter the **show ubt state** command, you should see a "Registered" state for both of those gateways.

Config

```
ubt zone labgateways
  primary ip 10.1.13.101
  backup ip 10.1.13.102
  enable
ip source-interface ubt vlan 23
```

```
AOS-CX# show ubt state
Local Master Server (LMS) State:
LMS Type  IP Address     State
-----------------------------------------------
Primary : 10.1.13.101  ready_for_bootstrap
Secondary:10.1.13.102  ready_for_bootstrap

Switch Anchor Controller (SAC) State:
      IP Address    MAC Address   State
-----------------------------------------------
Active:  10.1.13.102  20:4c...      Registered
Standby: 10.1.13.103  20:4c...      Registered
```

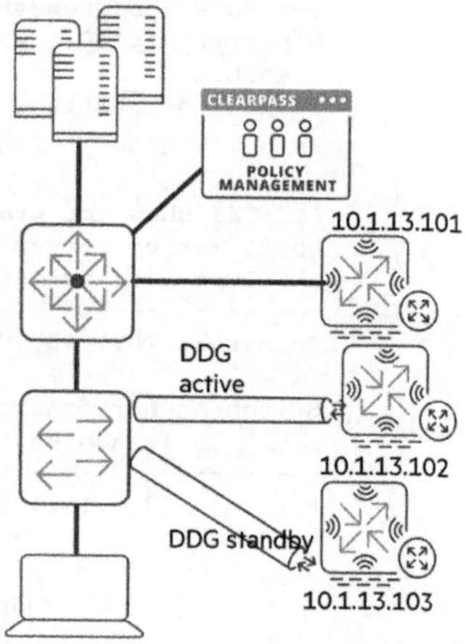

Figure 8-3: Review UBT Tunnel Configuration to a Gateway Cluster

You can specify just the **primary ip** command, and the switch still creates active and standby DDG tunnels. However, it is best practice to also specify another gateway IP address in the **backup-controller ip** command in the gateway zone. Then, even if the primary gateway is down, the switch can still become bootstrapped.

Triggering UBT for a Client's Traffic

After the switch is registered with its DDG or DDGs, it can tunnel client traffic to gateways in the zone. You trigger tunneling for a client's traffic using port-based authentication and port-access roles. Enable 802.1X or MAC-Auth on the port as you learned how to do in the previous chapter.

As shown in Figure 8-4, the client authenticates to HPE Aruba Networking ClearPass Policy Manager (CPPM), and CPPM applies an enforcement profile. This profile assigns the client to a role (which might be a LUR or a DUR). In any case, that role has a **gateway zone** setting within it. That setting tells the switch to tunnel the user's traffic to a gateway in the specified zone. The **gateway zone** setting further specifies a gateway role, which the switch tells the gateway to apply to the client.

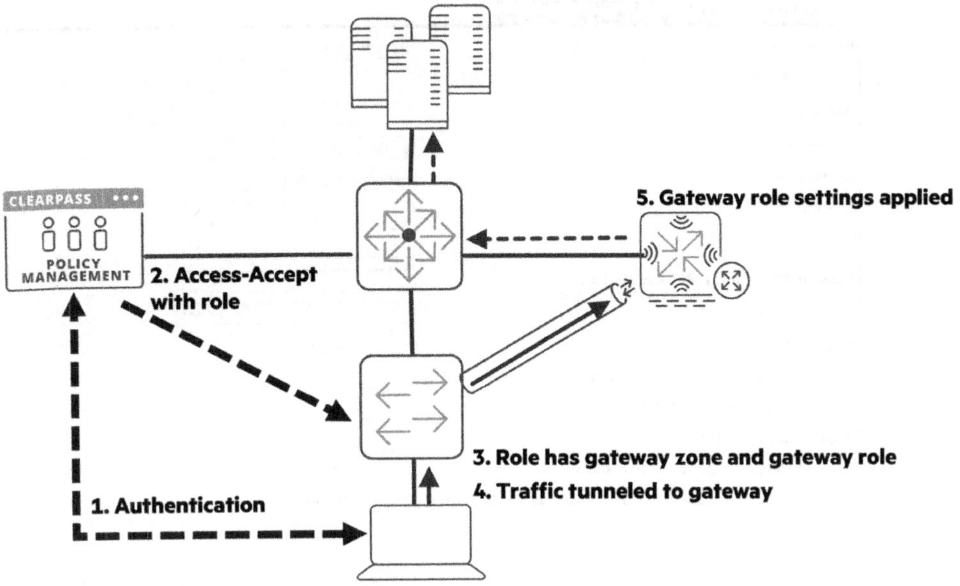

Figure 8-4: Triggering UBT for a Client's Traffic

The gateway receives the traffic in the tunnel. It applies the gateway role settings, which might include firewall policies and a VLAN assignment. It then forwards the traffic.

Role Configuration for UBT

The role that triggers UBT can be an LUR or a DUR. The minimum configuration for the LUR is shown in Figure 8-5. After you create the LUR, you can specify as little as one command: **gateway zone <*zone_name*> gateway-role <*gateway_role_name*>**.

```
port-access role ubt-loanofficers
    gateway zone labgateways gateway-role loanofficers-gateway
```

Figure 8-5: LUR Configuration for UBT

The corresponding settings for an Aruba Downloadable Enforcement Profile include the Role Name, Gateway Zone, and Gateway Role. (See Figure 8-6.) You can add other settings, such as a session timeout, as you desire. In some cases, VLAN configurations are also required. You will learn more about that in a moment.

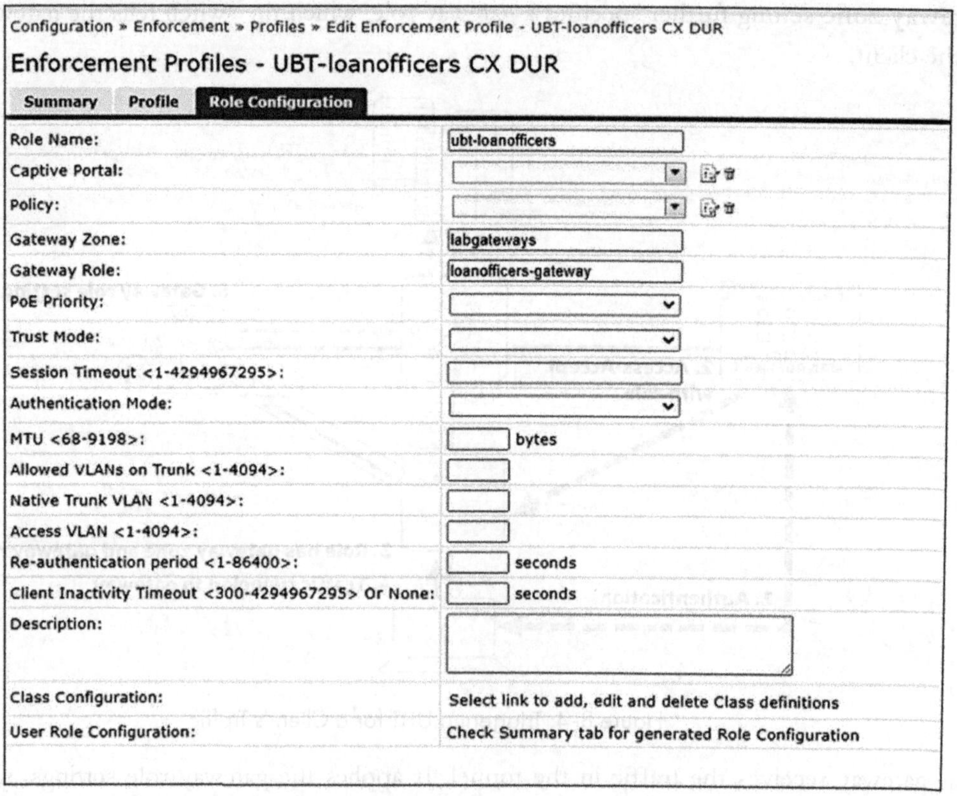

Figure 8-6: DUR Configuration for UBT

Client UDG Assignment in AOS-10

When a role activates tunneling on a client, the switch sends the client's traffic over a different tunnel from its DDG tunnels. It sends the traffic on a User Designated Gateway (UDG) tunnel to the client's active UDG, as shown in Figure 8-7.

Each client is assigned to an active UDG and standby UDG. In AOS-10, the cluster intelligently determines the UDG based on several factors. If Virtual Router Redundancy Protocol (VRRP) is enabled on the user's VLAN, the VRRP master for that VLAN is always assigned as the active UDG. This configuration ensures that traffic follows a sensible flow, arriving directly on the gateway that acts as the default router for the client's VLAN. If the client's VLAN does not use VRRP, it is assigned to an active and standby UDG more or less randomly. The primary LMS gateway sends the switch a "bucket list," when it bootstraps the switch. This bucket list tells the switch which MAC addresses to assign to which gateways as active and standby UDG; it load shares the assignments more or less evenly. (Note that the switch CLI might refer to the UDG as a User Anchor Controller [UAC] depending on the version.)

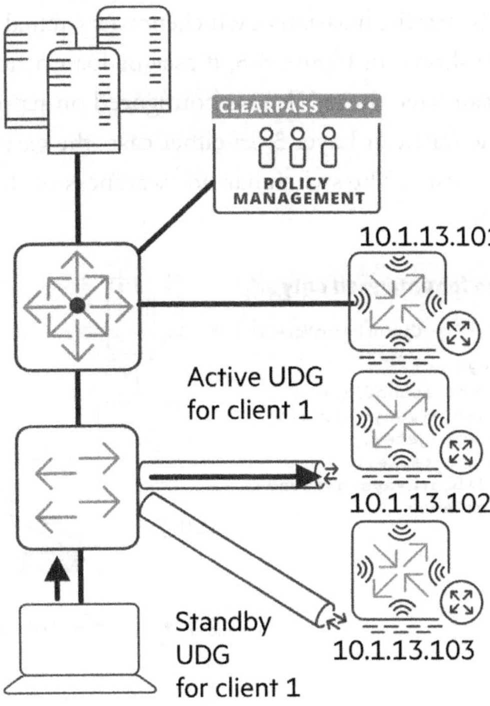

Figure 8-7: Client UDG Assignment in AOS-10

Note that AOS-10 branch gateway clusters can operate in default gateway mode. In this mode, the gateway assigned as the cluster's active gateway always becomes the VRRP master regardless of VRRP priorities set on the gateways.

Reserved (Local) VLAN Mode

For you to configure the port-access role and switch UBT settings correctly, you must also understand whether the switch is operating in reserved VLAN mode or extend VLAN mode.

In AOS-CX 10.04 and above, switches support reserved VLAN mode. In AOS-CX 10.04 and 10.05, this mode is the only option. In AOS-CX 10.06 and later, you can choose reserved VLAN as one option. Reserved VLAN mode is sometimes also called UBT 2.0 or local VLAN mode.

In this mode, you specify a UBT client VLAN on the switch. This VLAN must exist on the switch, but it must not be configured on any ports. The switch assigns this VLAN ID to traffic that it encapsulates and sends over a UDG tunnel. It uses the same reserved VLAN ID for all UBT clients. The VLAN simply acts as a placeholder.

When the gateway receives the traffic, it assigns each client to its actual VLAN based on the client's gateway role. In the example shown in Figure 8-8, it assigns loan officers to VLAN 17 and tellers to VLAN 18. The gateway can have these VLANs configured on its ports and forward the traffic at Layer 2. Or it can route the traffic at Layer 3. In either case, the gateway is solely responsible for setting the VLANs for UBT clients. The switch has no awareness of these VLANs, and you do not configure them on the switch.

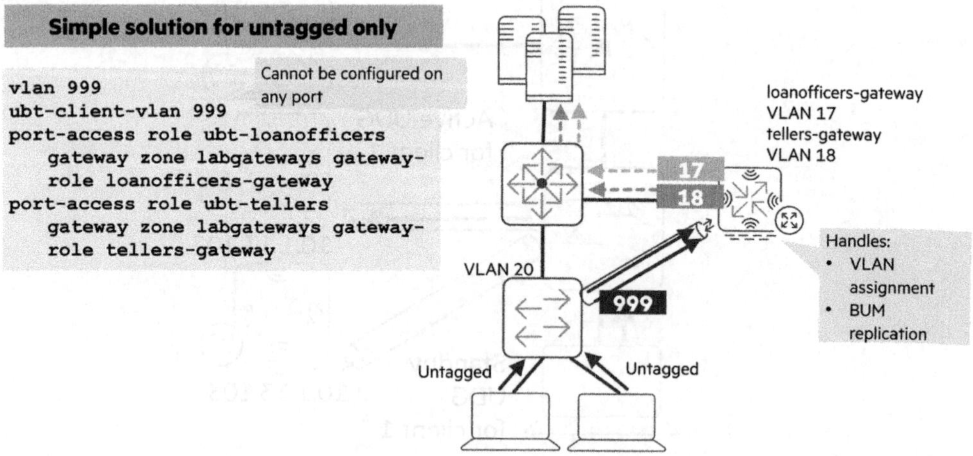

Figure 8-8: Reserved (Local) VLAN Mode

Reserved VLAN mode offers a simple set up. Admins can configure new VLANs on the centralized gateways without having to coordinate with switch admins. However, this mode also has some drawbacks for some situations. In reserved VLAN mode the switch can only carry untagged traffic with UBT. Companies sometimes want to tunnel tagged traffic, such as from VoIP phones. The switch cannot do so in reserved VLAN mode. In addition, the gateway handles flooded traffic—broadcasts, unicasts with unknown destinations, and multicasts (BUM). It replicates the BUM traffic as unicasts for each client to which the traffic should be flooded. It then sends the separate unicasts over the UDG tunnels. Sometimes this approach causes additional overhead on the gateways and tunnels.

VLAN Extend Mode

Alternatively, switches with AOS-CX version 10.06 and later can operate in VLAN extend mode. You enable this mode by entering this command: **ubt-mode vlan-extend**. Note that switches with AOS-CX versions earlier that 10.04 always operated in VLAN extend mode without the use of this command.

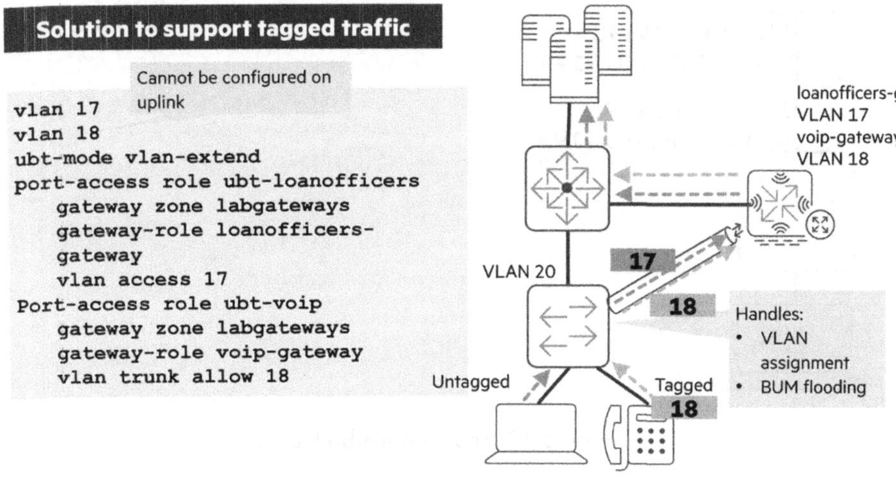

Figure 8-9: VLAN Extend Mode

In VLAN extend mode, UBT clients' roles must include a VLAN setting. This VLAN can be untagged (access) or tagged (vlan trunk allow). In this way, VLAN extend mode supports clients that send tagged traffic such as VoIP phones. Just like when you use VLAN steering with local forwarding, the switch must have the VLANs specified in the roles configured on it. Otherwise, the role application fails. However, you must not configure the VLANs in the UBT client roles on the switch uplink.

The gateway must have corresponding VLANs configured on each corresponding gateway role, as shown in Figure 8-9.

When a switch operates in VLAN extend mode, its gateways establishes separate multicast tunnels to it. The switch receives multicast traffic destined to its clients on the multicast tunnel. It then replicates the traffic for flooding.

Enabling Jumbo Frames

The tunnels encapsulate traffic with a GRE header. This extra header means that a packet can exceed the typical maximum transmit unit (MTU) for a packet. To prevent issues, you should raise the MTU on the path between the access switches and their gateways or, in other words, enable jumbo frames.

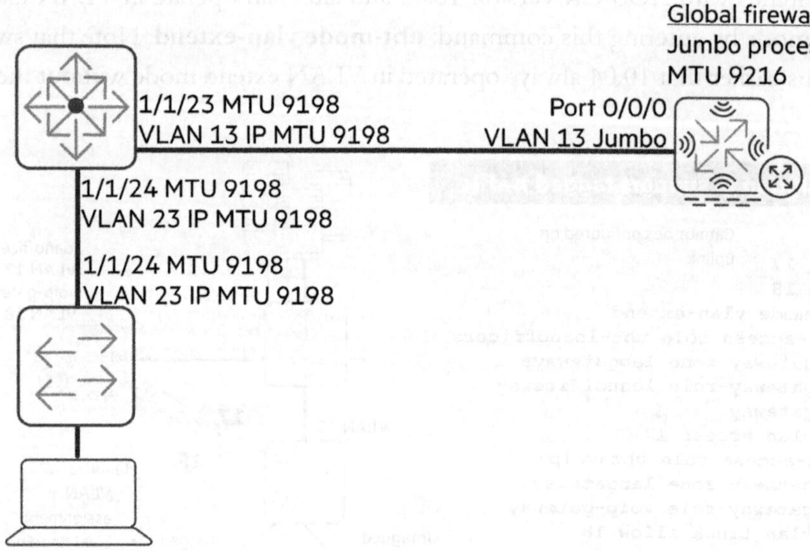

Figure 8-10: Enabling Jumbo Frames

On AOS-CX switches, you enable jumbo frames by increasing the IP MTU on the Layer 3 interface. You must also raise the MTU on the physical interfaces to match or exceed the Layer 3 MTU, as shown in Figure 8-10.

On gateways you must enable jumbo frames on the port on which the switch contacts the gateway. You must also enable jumbo processing in the global firewall settings and set the MTU. AOS devices count the frame header in the MTU, so 9216 is the same as 9198 on the AOS-CX switches. Also make sure to reboot the gateways after enabling jumbo frames.

You must take care to set the MTU to equal values on all devices on the path. In example shown in Figure 8-10, you are using the maximum value for simplicity.

Troubleshooting

You might run into issues in which a client's UBT state remains down, or other problems occur.

First validate that the client authenticated successfully. If not, you must troubleshoot the authentication settings. Make sure that 802.1X and MAC-Auth are correctly enabled, and the switch is accessing the correct CPPM IP address or fully-qualified domain name (FQDN). Also make sure that the switch meets the prerequisites for obtaining DURs, if you are using that configuration option.

Be aware of whether your switch is using reserved VLAN mode or VLAN extend mode. If the switch is using reserved VLAN mode, you must specify the UBT client VLAN, and that VLAN must not be used for any other purpose. If you are using VLAN extend mode, you must specify a VLAN in each role. Those VLANs must exist on the switch, but not be on uplinks.

If the user tunnel fails to activate, the gateway-role configured in the switch role might not match the name of a role configured on the gateway. Match up the settings carefully.

Also double-check that CPPM is applying the correct roles and DUR configuration, if DURs are in use.

Implementing VNBT on AOS-CX Switches

This section explains how to implementing VNBT on AOS-CX switches.

VNBT

Like UBT, VNBT enables AOS-CX switches to tunnel clients' traffic to a remote destination, rather than switch that traffic locally. Both of these technologies make it simple for companies to group clients together based on identity without regard to the underlying network architecture. For example, you could use VNBT to group all IoT devices into the same Layer 2 network even if those devices are connected to switches divided by a Layer 3 boundary.

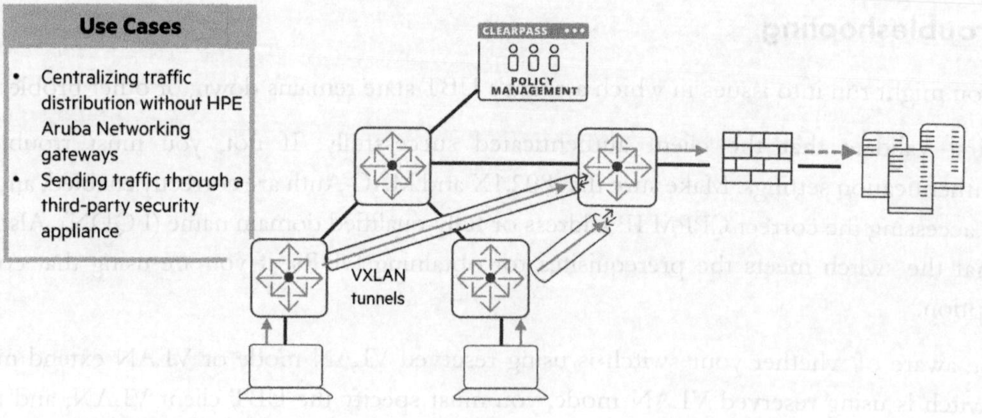

Figure 8-11: VNBT

However, UBT always tunnels client traffic to an HPE Aruba Networking MC or gateway, while VNBT tunnels traffic to another switch using Virtual Extensible LAN (VXLAN) tunnels. VNBT works well for companies that want to group clients together at Layer 2, based on role, and then send the clients' traffic directly to the data center. As shown in Figure 8-11, VNBT also works well for companies that want to send clients' traffic to an application-layer firewall, before allowing the traffic to continue towards its destination. You can deploy UBT and VNBT in parallel, applying UBT to some roles and VNBT to other roles, based on the company's particular security requirements for each role.

For example, the bank in this chapter's scenario might have IoT devices that manage lighting, temperature, and HVAC. Because those devices are critical to the business, and also at high risk for a security incident, the company wants to run all of their traffic though a security appliance in the data center. VNBT allows the switches to dynamically assign the IoT devices to a virtual network instance (VNI) and send their traffic to the DC security appliance before the traffic continues toward its destination.

Finally, note that VXLAN, like GRE, does not apply authentication or encryption. It is intended to tunnel traffic between switches that connect over relatively trusted links.

VXLAN Overview (Underlay and Overlay)

VXLAN is called an overlay technology because it uses tunnels to establish virtual networks "over" an "underlay" network (usually an IP network). The underlay network is simply the traditional network, responsible for routing traffic from one point to another. The example underlay network, illustrated in Figure 8-12, consists of routing switches in several buildings and a data center. It features multiple routing hops from end to end. (However, in Figure 8-11, you only see a couple access switches while a real network would have many more access switches.)

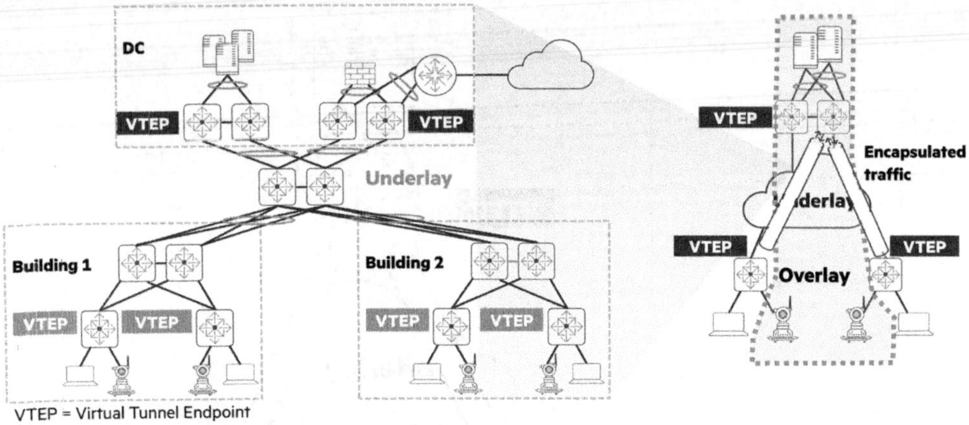

Figure 8-12: VXLAN Overview (Underlay and Overlay)

Each virtual "overlay" network extends over the underlay network without awareness of the IP addressing and architecture in the underlay network. Two clients in a virtual network can communicate at L2 even though they are actually divided by several routing hops in the underlay network.

Similarly the underlay network infrastructure devices have no awareness of the IP addressing used within overlay networks. They only see traffic exchanged between Virtual Tunnel Endpoints (VTEPs).

VTEPs maintain the distinction between underlay and overlay by encapsulating traffic that flows over the overlay. The VTEP receive packets from clients connected to overlay networks. They add a VXLAN header, a UDP header (because VXLAN is a UDP protocol), and a delivery IP header. The delivery IP header includes the source VTEP IP address and the destination VTEP IP address. The underlay network delivers the packet to the destination VTEP, which decapsulates the packet and forwards it to the destination client. In this way, the underlay only needs to know how to forward traffic between VTEPs and needs no visibility into the overlay networking, which is handled by the VTEPs.

VXLAN Virtual Network ID (VNI)

Each virtual network in VXLAN is identified by a Virtual Network Identifier (VNI). Often the virtual network is called a VNI.

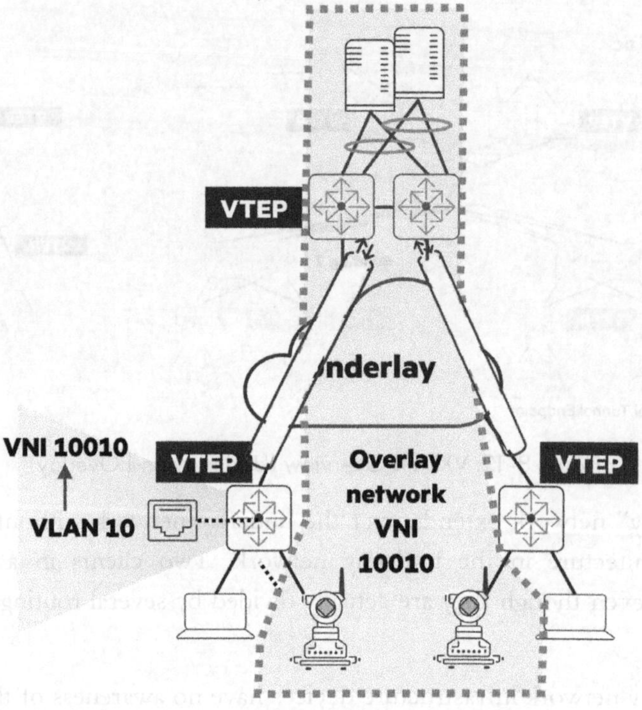

Figure 8-13: VXLAN Virtual Network ID (VNI)

A VNI is similar to a VLAN in that it defines a broadcast domain. All endpoints assigned to the same VNI have IP addresses in the same subnet, and their broadcasts and multicasts reach each other.

When you set up a VNI on an AOS-CX switch, you map a VLAN to the VNI, as shown in Figure 8-13. The switch then places any traffic that is assigned to that VLAN within that VNI. For example, client 1 is assigned to VLAN 10, which is mapped to VNI 10010. Therefore, the switch forwards client 1's traffic within VNI 10010.

Different VTEPs within a VXLAN domain can map the same VLAN IDs to the same VNI or different VLAN IDs to the same VNI. Only the VNI is visible between the VTEPs.

Centralized vs Distributed L3 Gateways

VXLAN can operate with centralized or distributed L3 gateways.

With the centralized approach, a central router, such as the border VTEP or even a firewall connected to that VTEP, acts as the default gateway for VNIs. (See Figure 8-14.)

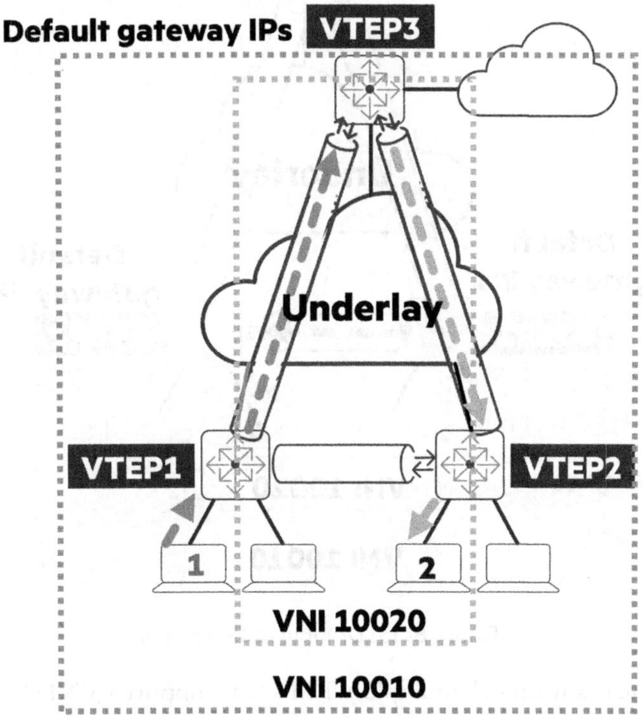

Figure 8-14: Centralized Gateways

As shown in Figure 8-15, with distributed L3 gateways, each VTEP routes traffic for any VNIs that it supports.

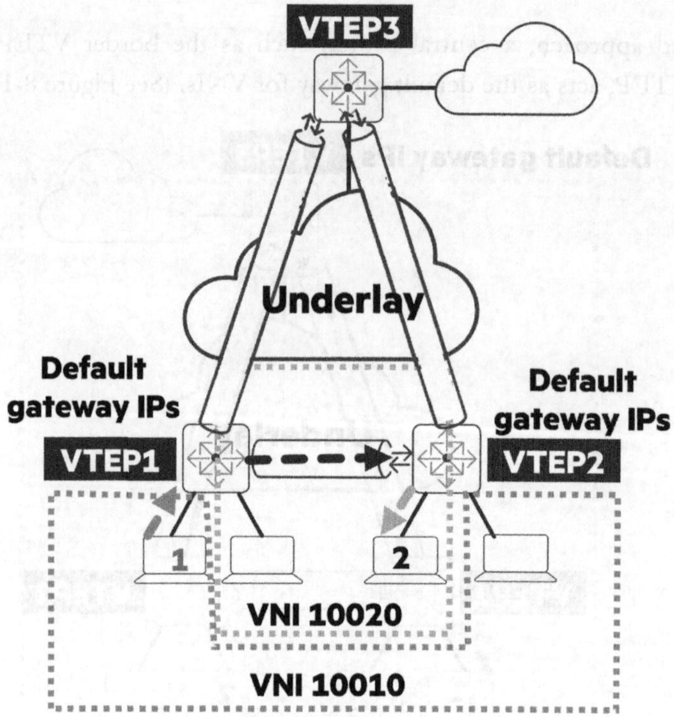

Figure 8-15: Distributed Gateways

In the distributed L3 gateway model, multiple VTEPs that support a VNI share the default gateway address for that VNI. Each VTEP responds to ARP requests for that IP address from its locally connected clients. It does not forward those ARP messages to other VTEPs, so each VTEP can act as if it owns the IP address for its own clients. AOS-CX switches use a technology called symmetric integrated routing and bridging (IRB) to route VXLAN traffic as distributed gateways. (For more details about this technology, refer to other HPE Aruba Networking training).

The distributed approach offers a few advantages. Each VTEP only needs to maintain ARP table entries for the clients connected behind it, while a centralized gateway would need a very large ARP table to support all of the clients in the VXLAN solution. In addition, the failure domain is reduced. If one VTEP fails, routing continues undisturbed for clients connected to other VTEPs, whereas if a centralized gateway fails, routing is disrupted for the entire solution. Finally, the distributed approach can reduce latency for client-to-client traffic flows. If two clients in different VNIs on the same switch, or on switches at the same site, need to communicate, distributed routing lets them do so without sending their traffic through a central location.

However, some companies prefer the centralized approach because they want to send all traffic through a centralized router. For example, if the company wants to send traffic through a third-party firewall in the DC, the company might prefer a centralized approach.

VNBT Configuration

To set up VNBT, you must first establish the VXLAN tunnels between the VTEPs. As you have seen, VTEPs might include access layer switches that implement VNBT as well as any centralized VTEPs.

Next, you create user-roles that trigger VNBT for specific users and devices.

Establish VXLAN tunnels

At a high-level you follow the process shown in Figure 8-16 to establish VXLAN tunnels.

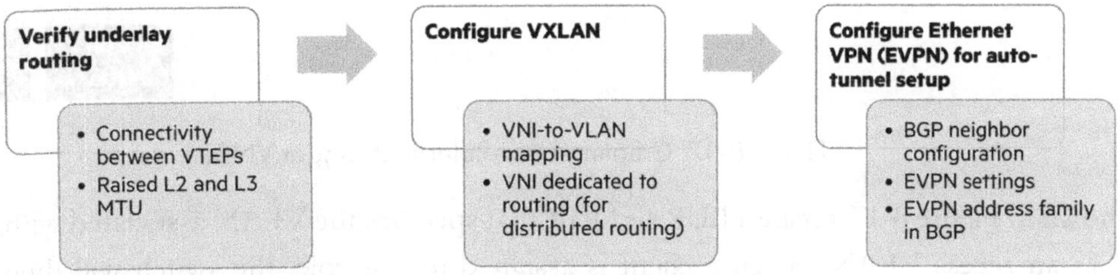

Figure 8-16: Establish VXLAN tunnels

1. Verify underlay routing. All VTEPs require connectivity between each other. Much as with UBT, you should also raise the L2 and L3 MTUs across the underlay fabric.

2. Configure VXLAN, which includes:

   a. Mapping VLANs to VNIs

   b. If using distributed routing, creating a VNI dedicated to routing

3. Configure Ethernet VPN (EVPN) to enable switches to set up tunnels automatically. EVPN uses BGP advertisements to allow VTEPs to discover each other and advertise information such as their VNIs. The EVPN setup includes:

   c. Configuring BGP neighbors

   d. Configuring EVPN settings for each VLAN mapped to a VNI

   e. Creating an EVPN address family in BGP and adding your BGP neighbors to that family

Creating User-Roles that Trigger VNBT

To apply VNBT to clients' traffic, you simply need to place the clients within a VLAN that is mapped to a VXLAN VNI. You could assign this VLAN to edge ports statically. However, it is most secure and flexible to apply the VLAN dynamically, as you learned how to do in the previous chapter.

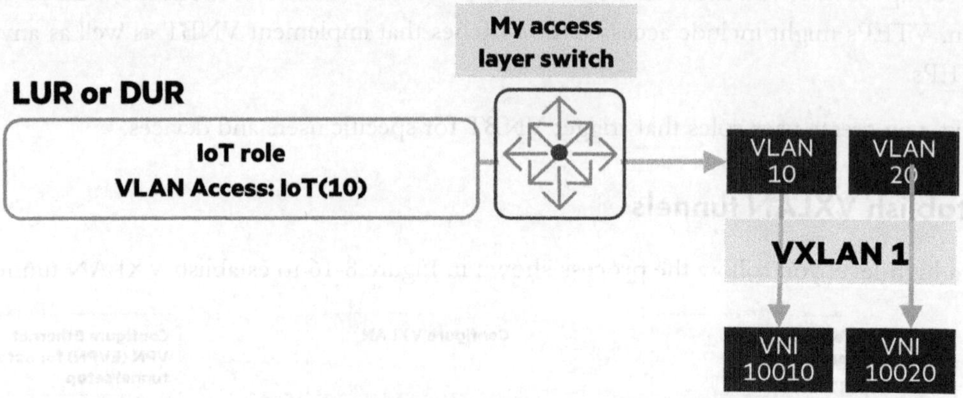

Figure 8-17: Creating User-Roles that Trigger VNBT

As shown in Figure 8-17, create a LUR or DUR that specifies the VLAN associated with the desired VNI as an access VLAN. When a client is assigned to this role, the switch will then tunnel the client's traffic in that VNI.

Summary

Congratulations! You should now know all about implementing UBT on AOS-CX switches. You should also have a basic knowledge about implementing VNBT on AOS-CX switches.

Learning checks

1. How do you apply UBT to a client's traffic?

   a. Use the gateway zone setting in the role to which the client is assigned

   b. Enable the UBT (gateway) zone globally, which applies it to all clients' traffic

   c. Apply the UBT zone to the client's port

   d. Apply the UBT zone to the client's VLAN

2. How can VNBT help a company secure clients' traffic?

   a. By tunneling the clients' traffic to an HPE Aruba Networking gateway, which performs advanced security functions

   b. By redirecting the clients' traffic to a captive portal server, which authenticates the clients

   c. By redirecting the clients' traffic to a cloud security solution, which performs advanced security functions

   d. By tunneling the clients' traffic to a switch in the data center, which passes the traffic to an application-level firewall

Note

For answers to all learning checks, refer to the Appendix.

9 Monitor with Network Analytics Engine (NAE)

EXAM OBJECTIVES

✓ Deploy and use Network Analytics Engine (NAE) agents for monitoring

Assumed Knowledge

- Basics of managing AOS-CX switches

Overview

The AOS-CX NAE helps admins take a new approach towards monitoring and troubleshooting the network. You will begin this chapter by learning about NAE and ways that you can use this feature in your network. You will then look at NAE agents in more detail, exploring how they monitor various resources on an AOS-CX switch and how they trigger actions.

You will then learn how to deploy NAE agents. You will see how all agents are based on a script, and you will explore ways to install scripts and then create agents from them. You will also learn how to use the Analytics dashboard, which displays information collected by agents.

Overview of AOS-CX NAE

This section presents an overview of AOS-CX NAE.

Introduction to Network Automation with AOS-CX

AOS-CX switches come with a powerful tool for monitoring the health of the switch and the network. This tool is the Network Analytics Engine (NAE).

As shown in Figure 9-1, NAE is integrated with the AOS-CX system configuration and time series databases (TSDB). It uses Python-based agents, developed by HPE Aruba Networking and the HPE Aruba Networking community, to help companies better monitor and troubleshoot their network environment through an automated process. Agents monitor the state, settings, and statistics around specific system components within the current state database. For example, an agent can monitor the number of bytes transmitted on an interface or the switch's operating temperature. The agent then writes the ongoing results to the time series database for ongoing analysis.

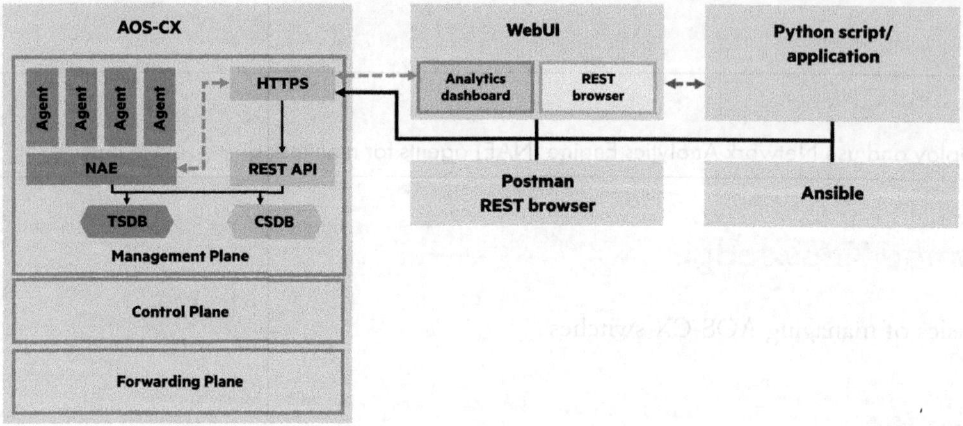

Figure 9-1: Introduction to Network Automation with AOS-CX

Admins can interact with the NAE through the Web UI's analytics dashboard. They can also automate the NAE through the REST API using tools such as Postman, Ansible, or another choice. They can also create or obtain Python scripts for new agents and load the agents on the switch using the Web UI or the REST API.

HPE Aruba Networking NAE Areas of Focus

As summarized in Figure 9-2, NAE agents can address many use cases. These use cases include monitoring system health, analyzing the network and alerting you to issues or changes, providing insight into application performance, and automatically optimizing the network based on current conditions.

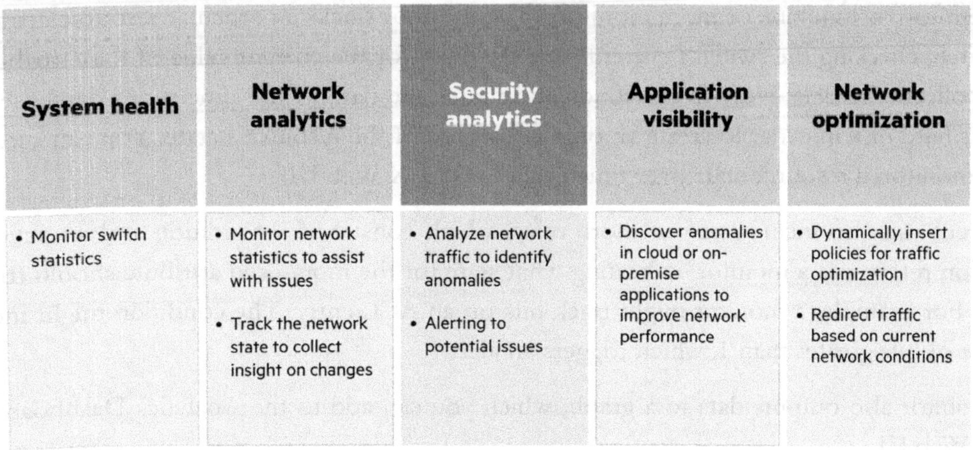

Figure 9-2: HPE Aruba Networking NAE Areas of Focus

The focus for this study guide, though, is security. You can develop NAE agents to analyze network traffic passing through the core and learn about anomalies. For example, normally most traffic might flow north-south from clients to the core. A new burst of east-west traffic flows could signal an issue such as a camera attempting to access a different network segment, indicating that it has potentially been compromised. Or a flood of invalid ARP packets could indicate a DoS attack.

NAE Agent Components

An NAE agent consists of several components, illustrated in Figure 9-3.

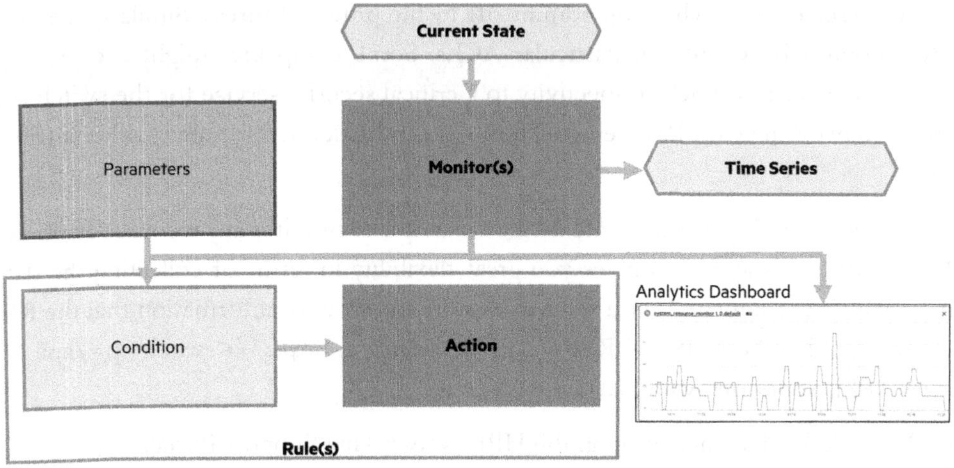

Figure 9-3: NAE Agent Components

An agent can include one or more monitors. The monitor tracks a particular resource attribute on the switch, checking the switch's current state database for the current value of that attribute. The agent collects the data every five seconds and writes the data to the time-series database on the switch's hard disk in order to create an ongoing record of the resource's state. You can see a graph of the monitored resource state over time in the AOS-CX Web UI.

The agent can also include one or more rules, which consist of a condition and an action. The condition references a monitor, indicating what state for the monitored attribute should trigger an action. For example, a monitor might track hits on an ACL entry. The condition might indicate a number of hits greater than 1, which triggers an alert.

The monitor also outputs data to a graph, which you can add to the Analytics Dashboard in the switch Web UI.

An agent might also have specific values of parameters, which can be referenced in monitors, conditions, and actions.

You will learn more about each of these components over the rest of this section.

What NAE Agents Monitor

An NAE agent can monitor any state or statistic defined within the AOS-CX REST API. For example, it could monitor the number of transmitted bytes on a particular interface or the CPU utilization on a management module. A monitor can track the packets passed and dropped by control plane policing (CoPP) policies. If the NAE agent detects a surge in the number of dropped packets, it can create alerts, which tip admins off to the potential threat. Similarly, you can create agents that monitor hit counts on particular ACLs. Some companies might use the application health monitoring agent to track connectivity to a critical security service for the switch such as an HPE Aruba Networking ClearPass server. Monitors can be defined for many other attributes on a CX switch.

In fact, companies' developers can create their own scripts to monitor any resource attribute defined in the REST API. This approach gives you great flexibility in terms of collecting the data that is relevant for your environment. If you want to explore the types of information that the REST API exposes, you can visit the AOS-CX REST API Reference at https://<switch_ip>/api. Note that only the v1 API is supported with NAE.

You can also look for documentation in the HPE Networking Support Portal.

How Actions are Triggered (No Clear Condition Defined)

You now understand a bit about how monitors collect data and make it available for graphs and rules. Next you will look at the rules in more detail. Rules let the agent take automatic action based on the information obtained by the monitor.

A rule must have a condition, which defines a particular state for the monitored value. At any given moment, the state is either true or false. Generally, the state being true indicates a potential issue that admins should know about. For example, an agent might monitor the number of bridge protocol data units (BPDUs) dropped by a CoPP policy on the switch. A condition in a rule might be that the monitored value is greater than 5 per second.

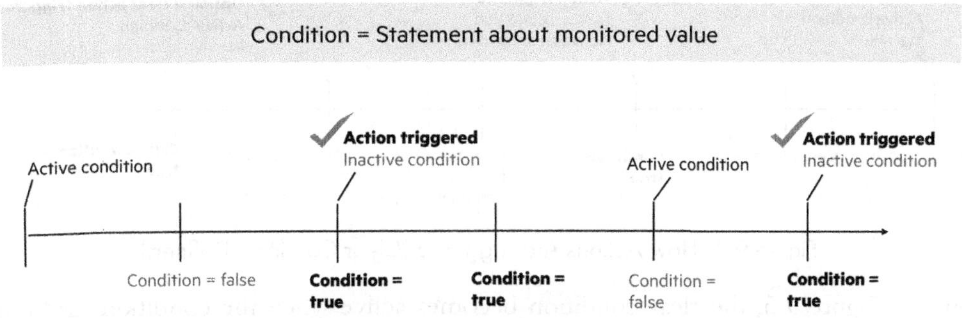

Figure 9-4: How Actions are Triggered (No Clear Condition Defined)

The agent checks the state against the condition every 5 seconds. As shown in Figure 9-4, the rule's action executes when the condition changes from false to true. At that point, the condition becomes inactive. The agent does not re-execute actions if the condition remains true. This prevents constant alerts for the same ongoing issue. If the condition becomes false again (for example, the number of dropped BPDU is lower than 5 per second), the condition becomes active again. The agent then will execute the action if the condition becomes true again.

This example shows just one rule, but agents can customize their behavior to nuanced conditions by using multiple rules, each with a different condition and action. For example, a condition in one rule might indicate a less severe problem and a condition in another rule might indicate a more severe problem.

How Actions Are Triggered (Clear Condition Defined)

In some cases, a condition might toggle frequently between true and false. For example, consider a condition configured to trigger when a CoPP policy drops more than 5 BPDUs per second. The policy might drop 5 BPDUs per second over one interval and change the condition from false to true. Then it drops 4 BPDUs per second over the next interval, and the condition transitions from true to false. Then the policy drops 6 BPDUs per second over the next interval, and the agent triggers the action again.

To avoid this problem, you can define a clear condition, which must be met to free the main condition. The "clear" condition is a low threshold that the monitored attribute must fall below before the condition triggers the action again.

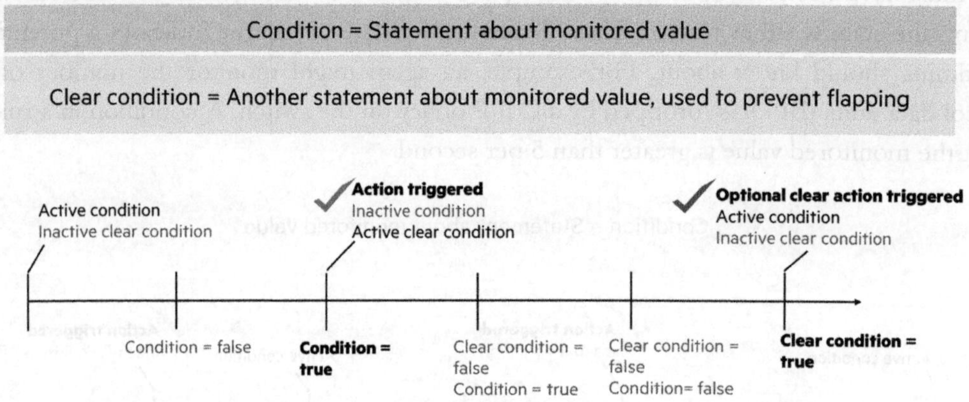

Figure 9-5: How Actions Are Triggered (Clear Condition Defined)

As shown in Figure 9-5, the clear condition becomes active when the condition first transitions from false to true, and the agent begins checking the state of the clear condition every five seconds. Even if the condition becomes false again, the clear condition remains active, and the condition inactive, as long as the clear condition is also false. Only when the clear condition becomes true does the condition transition to active. In the example discussed previously, the clear condition could be that the policy drops 0 BPDUs per second. Now the rate must fall to 0 before the original condition can become active again. You can optionally define a clear action, which also executes when the clear condition changes from false to true.

Agents That Use Baselines to Trigger Actions

Often it can be difficult to determine the best value to define in a condition. For example, you are trying to define a condition that indicates excessive traffic on an Ethernet interface. Is a rate of 500 million bytes per second excessive, or is a higher or lower value more appropriate? The answer often depends on the particular network, switch, and interface. The AOS-CX NAE can help to automate and remove the guesswork for defining conditions through the use of baselines. The agent can construct a baseline of normal values for the monitored resource attribute. A condition then triggers if the value deviates too much from the baseline. It is important that you understand, though, that agents do not create baselines by default; whoever creates the agent must specifically script the agent to create the baseline.

The high-level process of creating a baseline is illustrated in Figure 9-6.

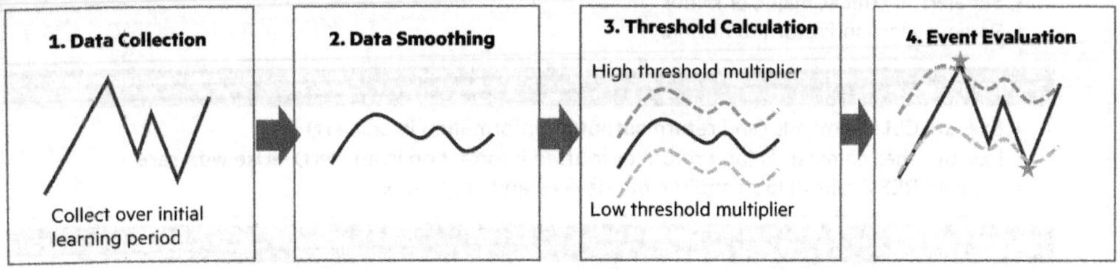

Figure 9-6: Agents That Use Baselines to Trigger Actions

When configured to do so, the agent constructs the baseline by collecting data over an initial learning period, which is configurable. The initial learning period should be long enough that all representative patterns are included. For example, it should be at least a day and maybe as much as a week. The agent then uses an algorithm to smooth the data and create the baseline. The agent next uses a configurable threshold multiplier to determine the high threshold, which indicates when a condition using the baseline changes to true and triggers an associated action. The larger the multiplier, the more deviation allowed and the smaller the multiplier, the less deviation allowed. The low multiplier lets the agent construct a low threshold, which indicates when the condition is cleared and a clear action triggered. You could create several conditions with different multipliers that trigger different actions; the condition with the larger multiplier should have a more significant action because it indicates a greater deviation from normal.

In addition to the initial learning period, the agent also uses a continuous learning period to continue to recalibrate the baseline. This lets the agent respond to gradual changes that create a new normal. However, data from a continuous learning period that includes an anomaly that triggered a condition is discarded. This lets the agent detect the anomalies, but not use them to change the baseline.

Action Options

NAE supports several action types, summarized in Figure 9-7.

The agent can send an alert with a critical, major, or minor level. These alerts are visible at a glance in the Analytics dashboard, helping admins find issues quickly. The agent can also remove the alert and return to the normal level.

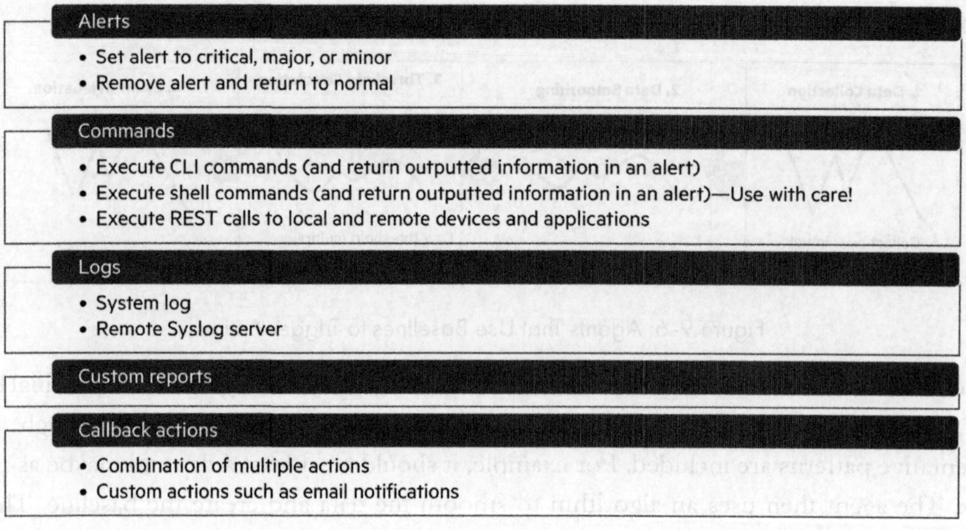

Figure 9-7: Action Options

An agent can also execute CLI commands or shell commands. The output for the command is collected in the alert for admins to view. You should be very careful with using shell commands in an agent because mistakes can damage the switch functionality and render the switch unusable. In addition, an agent can make REST API calls to itself or remote devices and applications. This allows for interaction with the HPE Aruba Networking portfolio of applications and devices, which have their own REST interfaces. Through REST calls, the NAE can also integrate with third party applications, including ticket generation systems like ServiceNow and TOPdesk.

An action can also send log messages to a system log, indicating information about the condition that has occurred. If the switch has log forwarding set up on it, the log is also sent to that server.

An action can also generate a custom report with multiple pieces of relevant information about the condition.

NAE agents can also implement more complex actions using "callback" actions. Callback actions permit a condition to trigger multiple actions such as setting an alert level, using CLI commands to collect information, and creating a log. They also permit the agent developers to create custom actions such as email notifications. And they let developers define more complex scripts with features such as error handling. The purpose of this study guide is not to teach you how to define such actions. However, you should be aware that NAE agents are capable of them.

Note that AOS-CX uses a safe sandbox environment to execute actions, which prevents agents from using excessive CPU or obtaining access to sensitive files such as certificates. The sandbox has access to the default VRF, not the management one. If actions need access to the Internet to execute properly, make sure that the default VRF has such access.

Deploying NAE Agents

In this section, you will learn how to deploy NAE agents.

Relationship Between NAE Scripts and Agents

Each NAE agent is defined by a Python script that is loaded on the AOS-CX switch. The script defines the agent, including which attributes it monitors and rules for any actions that the agent might take. The script might also define parameters, which are user configurable settings. For example, instead of specifying that an agent monitors transmitted bytes per second (Bps) on a specific interface, the script might specify that the agent monitors transmitted Bps on a variable interface parameter. Then users can choose the interface that they want when they create agents.

NAE script

- Defines the agent:
  - Monitors
  - Rules for actions
- Defines parameters (user-configurable settings)

NAE agents

- Instantiate the script with particular parameter values
- Perform the actual monitoring and actions

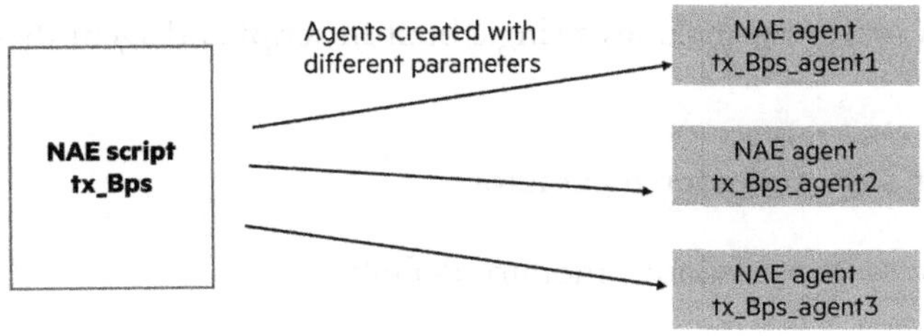

Figure 9-8: Relationship Between NAE Scripts and Agents

As shown in Figure 9-8, an NAE script has no effect until it has an agent associated with it. The NAE agent is a specifically-configured, executable instance of the script. Once the agent is created from the script, the agent performs the tasks defined by the script. They monitor the resource and take actions when conditions are triggered. Note that agents run with administrator rights.

When you create an agent, you define the values that you want for the parameters. You can create different agents from the same script using different parameters. For example, if one script parameter indicates the interface to monitor, you could create multiple agents, each monitoring a different interface.

Sources for Scripts (Standalone Switches)

There are three sources for an agent script:

- Built-in (system-created)
- HPE Aruba Networking Solutions Exchange (ASE)
- User-created

Built-in scripts

Built-in scripts and agents are installed on the switch before the switch is shipped from the factory. All scripts and agents have information about their origin associated with them. Built-in scripts and agents have an origin of "system" and are marked System Created in the Web UI.

Updating the switch software updates the system-defined agents as well.

Built-in scripts and agents cannot be deleted. But you can enable, disable, and change the configuration of built-in agents. Built-in scripts are not displayed in the switch config unless you have changed and saved one or more parameters.

If you create an additional agent from a built-in script, that agent is considered a user-created agent, which can be deleted.

The current software release includes a single built-in script and agent that monitors several system resources:

- Built-in script: system_resource_monitor
- Built-in agent: system_resource_monitor.default

Installed from HPE Aruba Networking Solution Exchange (ASE)

The primary source for additional NAE scripts is HPE Aruba Networking Solutions Exchange (ASE). You can access this portal at https://ase.arubanetworks.com/. No login is required to browse the exchange or look at information about solutions. However, to download solutions, view solution source code, and download NAE agents, you must log in. If you do not have an existing account, register for a free Airheads Social account at: http://community.arubanetworks.com. You can also access ASE directly through the AOS-CX switch UI, as long as your browser has Internet connectivity. In this case, you don't need to log in.

ASE provides a variety of HPE Aruba Networking-certified NAE agent scripts. These scripts are written by HPE Aruba Networking developers, and HPE Aruba Networking tests and validates that the scripts do what they advertise.

When you install scripts from ASE and create agents from them, you can later disable and remove these agents if you want. Before you complete a software upgrade, you may need to remove an old version of the agent and script. Then, after the upgrade, you can upload new ones.

User-created

You can also obtain agent scripts from community sources and load them on your switch using the WebUI. You should use these scripts at your own risk. You should always evaluate a script before loading it on the switch to ensure that the actions executed are what you expect.

Or you or developers in your organization can create agents on your own with Python script.

In either case, HPE Aruba Networking does not provide technical support for user-created scripts.

These agents are managed like the ones that you install from ASE. You can disable, remove, and update them separately from the software.

The Airheads community provides a place for members or participants to search for information, read and post about topics of interest, and learn from each other. Guests (unregistered visitors) can browse or search the community for information. Members (registered users) can post messages or comments, track discussions, and get email notifications on posting activity and other community actions. The Airheads community is the glue that holds together all the components of the NAE solutions. Airhead community members can share their expertise and tips and point each other toward relevant agents. There is now a Developer Community forum under the technology category in the Airheads with discussion forums dedicated to automation and NAE questions and support. HPE Aruba Networking is committed to expanding the NAE value with extensive research and development (R&D) investment in building NAE scripts and supporting the community. Within the Airheads community, there is a Developer Community group that is specific to APIs, programming, and automation. The Developer Community is the recommended place to post questions about NAE. See the Developer community at: https://community.arubanetworks.com/t5/Developer-Community/bd-p/DeveloperCommunity. See the Airheads community at https://community.arubanetworks.com/.

For more information about using ASE, see https://ase.arubanetworks.com/docs.

Analytics Dashboard

It is typical to manage NAE agents and other functions in the Web UI, rather than the CLI. This approach lets you view graphs and alerts produced by agents, as well as install agents from ASE. You will monitor and manage NAE from the Analytics Dashboard (see Figure 9-9). You can access this dashboard in two ways: click the Analytics section in the main dashboard or select Analytics in the sidebar.

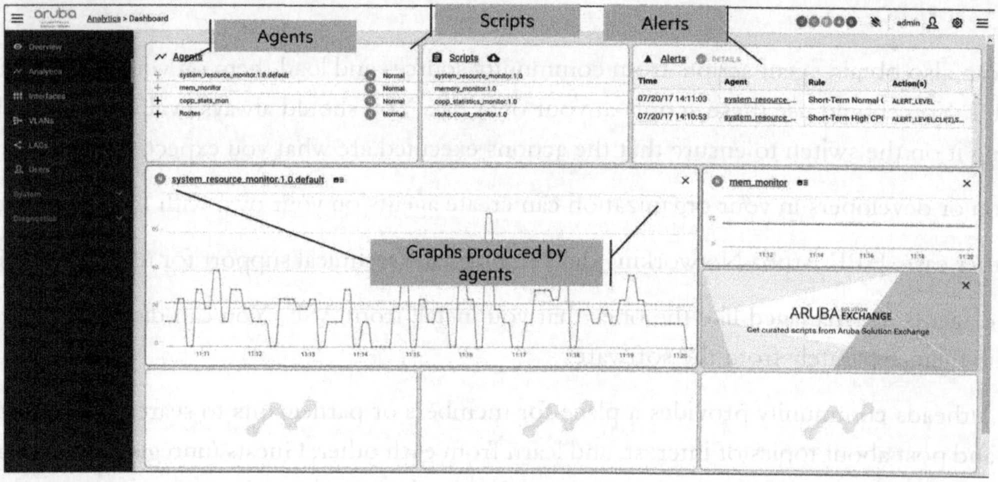

Figure 9-9: Analytics Dashboard

On the Analytics dashboard, the Agents section lists all of the agents on NAE and indicates whether they are operating normally. The Scripts section shows the Python scripts associated with those agents. You can click the Scripts link to go to the Scripts Management page. Look to the Alerts section to see more details on any minor, major, or critical alerts that the NAE agents have generated.

You can select the name of these sections to move to a page with details. For example, click the Agents link to go to the Agent Management and create new NAE agents. Or you can select an individual agent to move to that agent's detail page. Or click the Scripts link to go to the Scripts Management page and view and manage your NAE scripts.

The dashboard also shows graphs created by NAE agents. You will learn more about adding and using these graphs in a moment.

You can also change the layout for the dashboard. Click this icon at the top of the bar and choose **Unlock Page Layout**. Each section in the dashboard is then outlined with dashed lines. You can click a section and move it around. When you are done editing, you can click the icon again to lock the page layout.

Accessing ASE from the Analytics Dashboard

You can access ASE from the AOS-CX switch Web UI. Browse to the switch's IP address or hostname and log in with valid credentials. (As you learned in an earlier chapter, you can set up authentication to a local list or to an external authentication server like CPPM.)

Click **Analytics** in the left menu to access the Analytics dashboard.

As shown in Figure 9-10, then click in the ASE section. Or you can click the download icon in the Scripts section.

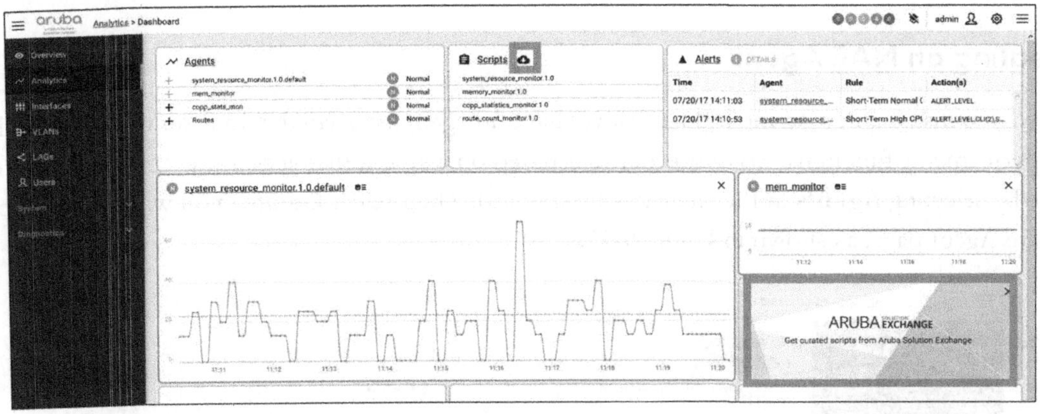

Figure 9-10: Accessing ASE from the Analytics Dashboard

Obtaining NAE Scripts Using ASE

You will be connected to ASE. You will not need to login because you connected through the switch UI.

You can browse through the exchange using tags to narrow your search. Examples of tags include the name of the resource that you want to monitor. Tags also indicate the switch model for which a script is intended. In ASE, HPE Aruba Networking-certified scripts have the following tag: nae-aruba-certified.

If you decide that you are interested in a script, you can select its check box and click **View Script** to see the script. Or you can click **Download** to view and possibly customize the script offline. You can also select **Install** to immediately install the script on your switch. (See Figure 9-11.)

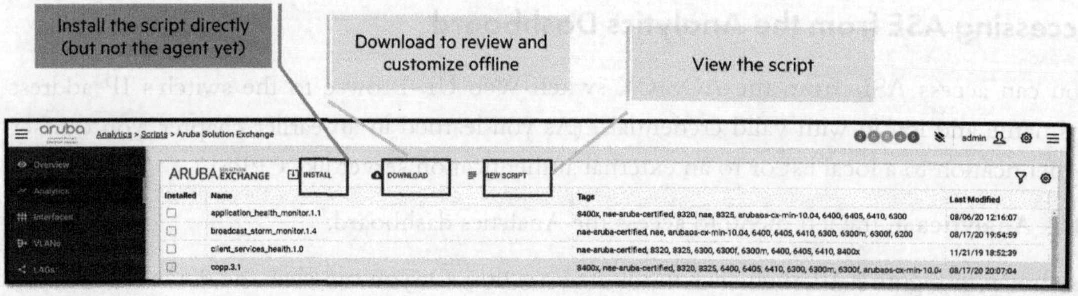

Figure 9-11: Obtaining NAE Scripts Using ASE

Creating an NAE Agent

When you install a new script, it does not yet have an agent associated with it, so the script will not yet perform any functions. To create an agent based on a script, first access the Agent Management page by selecting **Agents** in the Analytics Dashboard. Then select **Create**. You will be taken to the Create Agent page, as shown in Figure 9-12.

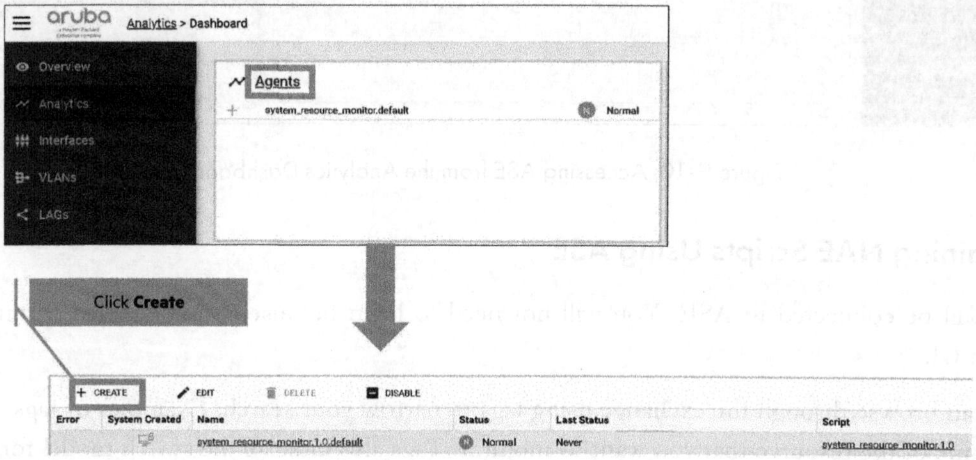

Figure 9-12: Creating an NAE Agent

Alternatively, you can go to the Scripts Management page by selecting **Scripts** in the Analytics Dashboard. Then you can click the plus icon (+) next to the script. This action also takes you to the Create Agent page.

Defining and Saving the New Agent

When you install a new script, it does not yet have an agent associated with it. To create an agent based on a script, first access the Agent Management page by selecting Agents in the Analytics Dashboard. Then select **Create** (you can also select **Create Agent** in the Script Management page).

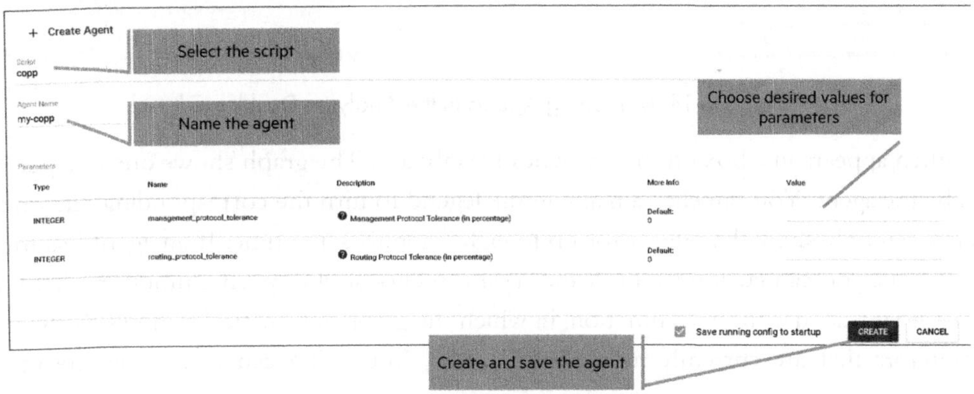

Figure 9-13: Defining and Saving the New Agent

As shown in Figure 9-13, a window appears where you can choose the script on which you want to base the agent. Then define the agent name. Finally, customize the parameters and save.

You can also select an agent in the Agent Management page and select Edit to adjust the parameters on an existing agent.

Managing Agents in the Analytics Dashboard

When you create a new agent, it is automatically enabled. Each enabled agent uses its monitors to collect data and uses rules to trigger actions. You can see a list of agents on the switch in the Analytics Dashboard.

If you want to see a graph of the data being collected by the agent, you can click the plus icon (+) next to the agent, as shown in Figure 9-14.

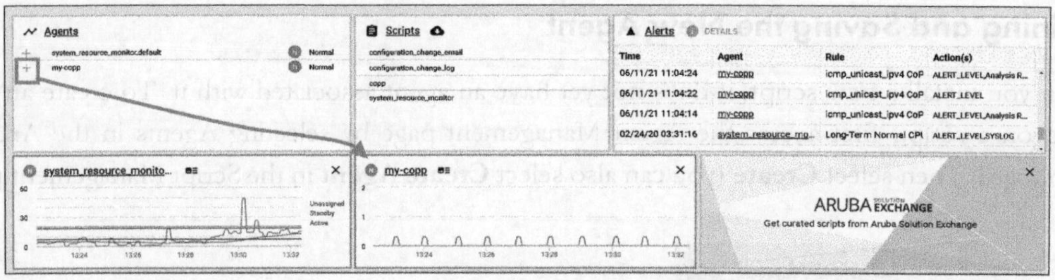

Figure 9-14: Managing Agents in the Analytics Dashboard

A graph then appears in a box on the Analytics Dashboard. The graph shows one line per monitor defined by the agent. You can click a name in the legend to turn the corresponding line on and off. The graph can only show the output for up to eight monitors at a time. If an agent has more than eight monitors, you can customize the view. You can choose the specific monitors that you want to show. Or you can use the auto function, in which the graph automatically shows the lines for the eight monitors that are currently most "interesting." Interesting can mean that the monitor is detecting changing data.

Even if you do not add the agent's graph to the Analytics Dashboard, you can see the graph by clicking the agent name and viewing the agent details.

If an agent rule triggers an action that sets an alert level, you will see an alert in the Alerts pane. You can see more details about the alert by clicking it.

NAE Agent Details

The Agent Details page provides more detailed information about the agent, including its version and status (see Figure 9-15). If the agent has multiple monitors, this page shows all of the graphs associated with those monitors. The parameters section shows agent settings that you can configure. You can also click icons to edit the agent or look at its script.

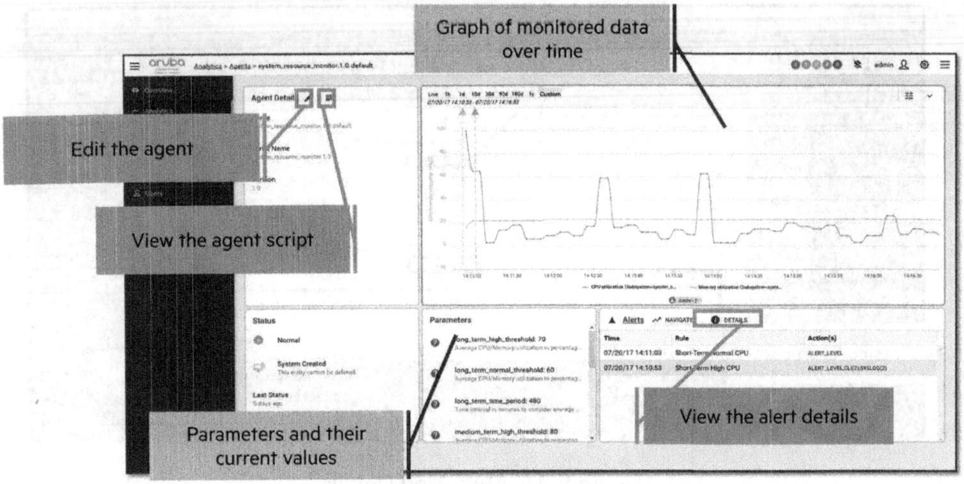

Figure 9-15: NAE Agent Details

Check the alerts section to see information about the alerts this agent generates so that you can take the proper steps to resolve them. You can click a particular alert and then select Details to see a window with Alert details.

NAE Alert Details

The Alert Details window shows details about when and why the alert occurred, as well as the alert level (see Figure 9-16). You can view a history of the actions taken in association with the condition that triggered this alert. In this example, these actions include syslog messages and CLI commands. You can also collect any information that those actions generated. Select Output to see the output from the CLI commands. If the alert generated a custom report, you can click and view that report.

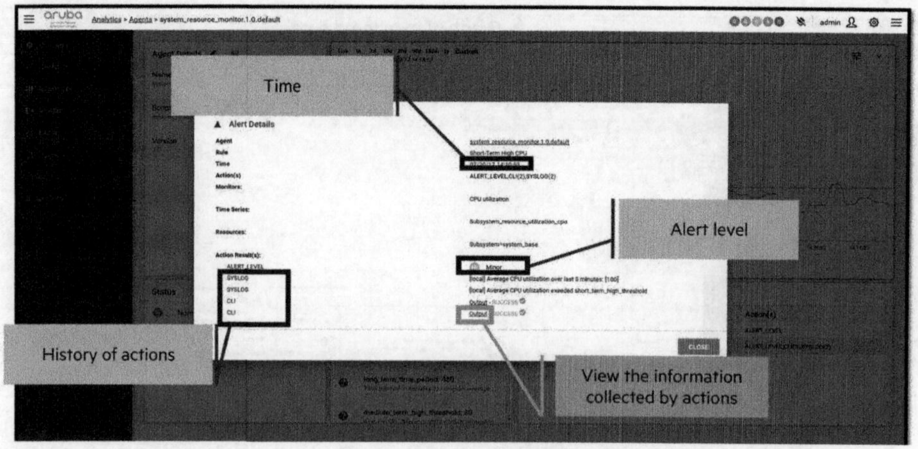

Figure 9-16: NAE Alert Details

You can also use the REST API to obtain information about NAE and NAE agents. Refer to the documentation and reference interface for more details.

Considering NAE Maximums

As you plan which NAE scripts to install and agents to deploy, you need to keep in mind that NAE naturally consumes resources on the switch. Switches support a maximum number of scripts, agents, and monitors. The maximums apply to agents, and the monitors within those agents, even if they are not enabled. If you try to install more scripts than allowed, or create more agents than allowed, you will receive an error. Different agents include different numbers of monitors, depending on the definitions in the script. One agent might use just one monitor, while another uses dozens. If creating an agent will make the switch exceed the monitor maximum, an error will occur and the agent will not be enabled, even if the switch has not reached its maximum number of agents.

Different switch models have different limitations. You can view the limitations on your switch, as well as the current consumption, with the command shown in Figure 9-17.

```
AOS-CX-6300# show capacities-status nae

System Capacities Status: Filter NAE
Capacities Status Name                                            Value   Maximum
---------------------------------------------------------------------------------
Number of configured NAE agents currently active in the system     2       10
Number of configured NAE monitors currently active in the system   99      130
Number of configured NAE scripts currently active in the system    3       10
```

Figure 9-17: Considering NAE Maximums

Use the maximums to plan which agents you want to run at all times and which you might deploy to investigate particular circumstances.

Managing NAE Agents with HPE Aruba Networking Central

Up to this point you have learned how to manage NAE on a single switch. HPE Aruba Networking also offers tools for managing your AOS-CX switches and their NAE functions at scale. If you are managing the switches within HPE Aruba Networking Central, you can use Express Config to deploy NAE agents to multiple switches. The process is summarized in Figure 9-18.

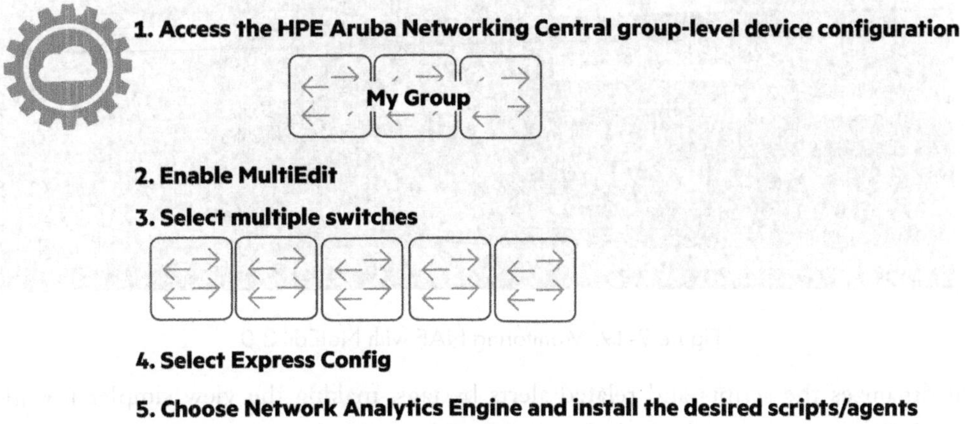

Figure 9-18: Managing NAE Agents with HPE Aruba Networking Central

Go to the group level for the switches that you want to configure. Access the switches' configuration (navigate to **Devices > Switches** and click the **Config** icon). Then enable MultiEdit mode.

When that mode is enabled, you will see a list of switches in the group. You can search and filter for particular switches. Select all of the switches on which you want to deploy agents. Choose to complete an Express Config.

Then choose Network Analytics Engine settings. You can then install scripts from the choices of scripts loaded in HPE Aruba Networking Central. When you install the script, it installs on all switches that you selected for Express Config. You can then create agents from the scripts, and again, the configuration applies to all of the selected switches.

Monitoring NAE with NetEdit 2.0

If you are managing your AOS-CX switches with NetEdit, you should be aware that NetEdit 2.0 provides integration with NAE. You can view the scripts installed on the AOS-CX switches from the Topology or Network view (see Figure 9-19). NetEdit also shows alerts that NAE has triggered, which can help admins discover the alerts more quickly.

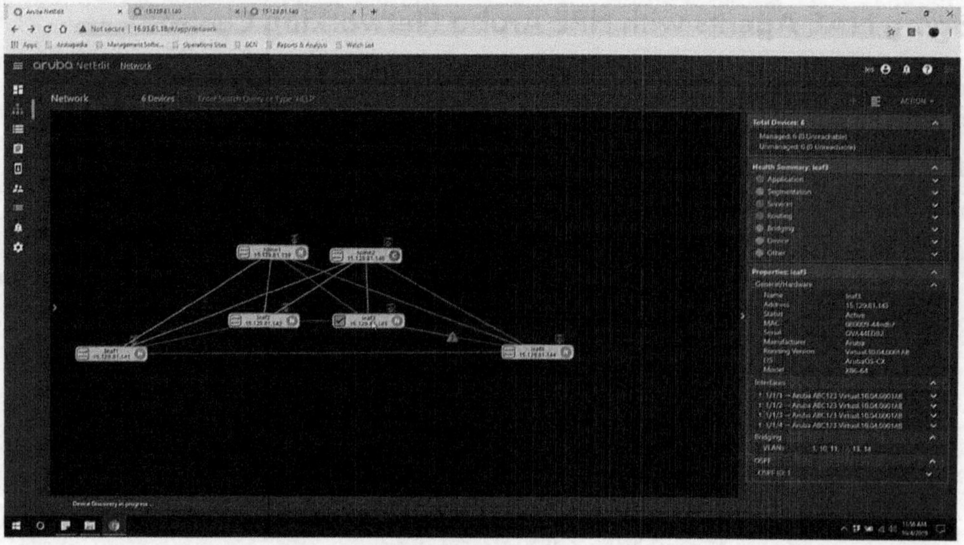

Figure 9-19: Monitoring NAE with NetEdit 2.0

NetEdit organizes the scripts and related alerts by tags, making the view simpler for admins to understand. These tags derive from the script tags. (Note that the tags are informational only.)

If you see an alert in NetEdit and want to follow up on it, you simply click a link from the network view. If you have logged in to the associated switch's Web UI recently, the browser is directed straight to the switch's Analytics Dashboard or to the alert that is showing on NetEdit. If you have not logged in recently, you will be taken to the switch's login page. After logging in, you can proceed to the Analytics Dashboard.

Summary

Congratulations! You now understand the benefits of AOS-CX NAE and, at a high level, how NAE agents work. You also learned how to obtain NAE scripts, install them, and create agents from them. You also practiced using agents and analyzing the rich set of data and alerts that they produce.

Learning checks

1. What functions does an NAE agent perform? (Multiple answers can be correct.)
   a. Collects data from the current state database and records it in a time-series database
   b. Performs authorization control of management users
   c. Allows you to virtualize legacy switch code within an AOS-CX switch
   d. Triggers actions based on whether a monitored attribute meets a particular condition
2. Which correctly describes HPE Aruba Networking NAE scripts?
   a. All HPE Aruba Networking-certified scripts are loaded on the AOS-CX switch at factory defaults.
   b. You can find HPE Aruba Networking-certified scripts in ASE and Github.
   c. HPE Aruba Networking provides support for all NAE scripts, as long as the scripts use Python code.
   d. You are only allowed to load HPE Aruba Networking-certified scripts on AOS-CX switches.

Note

For answers to all learning checks, refer to the Appendix.

---

10 Implement WIDS/WIPS

EXAM OBJECTIVES

- ✓ Explain the AOS WIPS and WIDS technology
- ✓ Configure AP rogue detection and mitigation

Assumed Knowledge

- Basics of managing AOS mobility devices
- WIDS/WIPS definition

Overview

Inherently open, the wireless medium can be vulnerable to a broad array of attacks from malicious attackers, including ones who manage to connect to the wired network as well as those who are simply in the vicinity of the network you are protecting. In this chapter, you will review how AOS WIDS/WIPS detects all APs and clients in the area and classifies them into various categories.

You will also learn about configuring WIDS/WIPS policies, focusing on practical guidelines for which options to enable or not.

Review AOS WIDS/WIPS

In this section, you will review AOS WIDS/WIPS.

Why Wireless Intrusion Detection/Protection?

Uncontrolled wireless devices, such as rogue access points (APs), can pose a large security threat to the network. Rogue APs are unauthorized APs that connect to the wired network infrastructure and can potentially provide backdoor access into the wired LAN. Laptops acting as ad-hoc networks, or in bridge mode, are also potential security risks to the wired infrastructure.

Denial of Service (DOS), Man-In-the-Middle (MITM) attacks, and forged traffic can make the WLAN slow or unusable for legitimate users.

These threats are summarized in Figure 10-1.

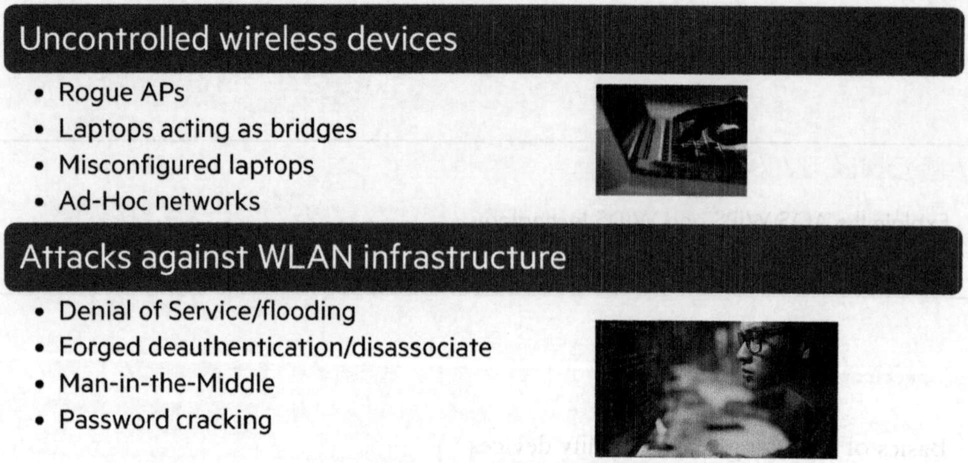

Figure 10-1: Why Wireless Intrusion Detection/Protection?

A Wireless Intrusion Detection System (WIDS) provides the network with a level of preventive security against any suspicious activity. The IDS achieves this objective through early warnings aimed at systems administrators. Sometimes these alerts are the best tool for the security team, helping the team to take the next steps to find and mitigate the threat. In a few cases, which you will examine in more detail later, you can enable Wireless Intrusion Prevention System (WIPS) features to mitigate threats automatically.

Wireless Threat Protection Framework

HPE Aruba Networking provides Wireless IDS/IPS features in wireless devices managed with HPE Aruba Networking Central that have Foundation or Advanced licenses. If the company is using an AOS-8 architecture, some basic rogue detection and containment features are provided with just AP licenses; however, RFProtect (RFP) licenses are required for the full feature set, including spectrum analysis.

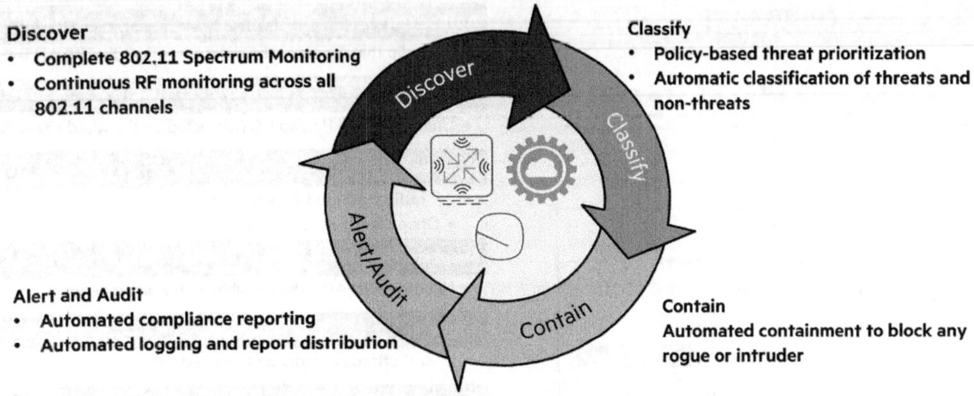

Figure 10-2: Wireless Threat Protection Framework

HPE Aruba Networking's Wireless Threat Protection Framework consists of four main parts: Discover, Classify, Contain, and Alert/Audit (see Figure 10-2). Each process feeds into another for a seamless framework.

Discover provides complete 802.11 spectrum monitoring and provides continuous monitoring of wireless devices, activity, and configuration across all 802.11 channels.

This carries over into **Classify**, which applies the policy-based threat prioritization and automatic classification of threats and non-threats to RF security.

The **Contain** process automates threat mitigation and containment to block any rogue APs or intruders.

Lastly, **Alert and Audit** complete the compliance reporting by logging and distributing reports to ensure you are in compliance with wireless security policies and regulations.

Review AP Classifications

WIDS/WIPS classifies all AP radios detected in the RF environment, as shown in Figure 10-3. It makes its decisions based on several factors, such as if the HPE Aruba Networking AP is part of the enterprise solution, if the AP is visible in the air, and if the AP is visible on the wire. WIDS/WIPS automatically classifies devices, but an administrator can override classifications when necessary.

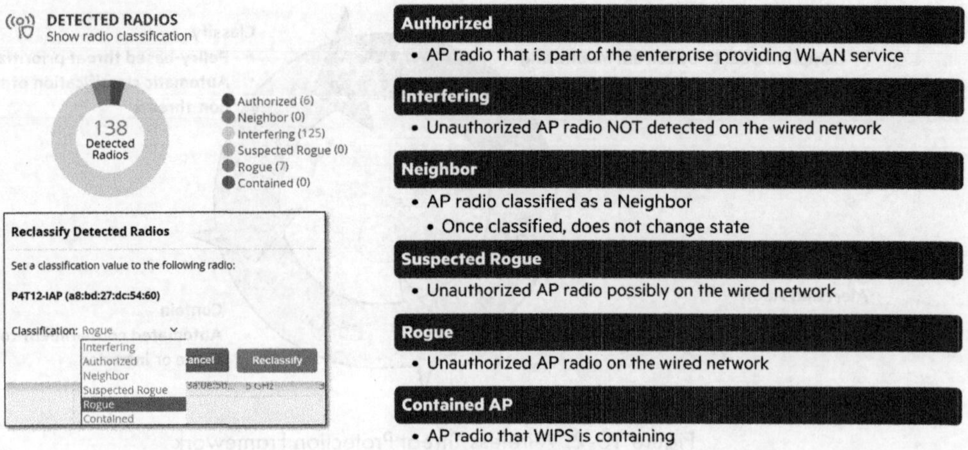

Figure 10-3: Review AP Classifications

An authorized AP radio belongs to an HPE Aruba Networking AP that is part of the enterprise solution. In an AOS-8 architecture, the AP is controlled by a HPE Aruba Networking Mobility Controller (MC). In an AOS-10 architecture, it is managed by HPE Aruba Networking Central. Administrators can also manually classify AP radios as authorized.

An Interfering AP radio is detected in the RF environment, but has not been detected on the wired network. All non-authorized APs begin with this classification. You can classify the BSSIDs of Interfering APs as Neighbors. After you classify a radio in this way, the state does not change unless an admin changes it manually.

A suspected rogue AP radio is detected in the RF environment. There are also indications that point towards the AP being attached to the wired network. However, WIDS is not able to determine for certain whether the AP is connected to the wired network. A rogue AP is an AP that has been detected wirelessly *and* has been determined with confidence to also be connected on the wired network.

A contained AP is an AP radio that WIPS is containing. An admin might have manually contained the AP, or a WIPS policy might mark the AP for containment.

Review Client Classifications

WIDS/WIPS discovers clients during scanning of the wireless medium, and it automatically classifies clients into various groups, as shown in Figure 10-4. (Again, administrators can override the automatic classifications when necessary.)

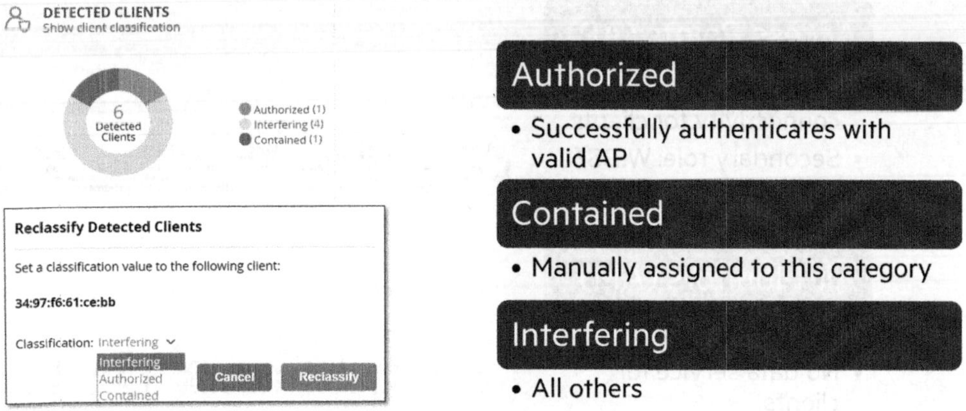

Figure 10-4: Review Client Classifications

WIDS/WIPS classifies any client that successfully authenticates with a valid AP as authorized. However, clients that connect to an open WLAN are not classified as authorized, so typical guest clients do not receive this status.

Admins can manually assign clients to the Contained category if they want the system to contain them (using the configured wired or wireless methods).

All other clients are classified as interfering.

Access Point (AP) & Air Monitor (AM) Considerations

APs primarily serve clients but can also provide WIPS/WIDS information to its management system, which might be MCs or HPE Aruba Networking Central. As you see in Figure 10-5, an AP can only briefly scan each channel because it must usually stay on its own channel to serve clients. Therefore, it can take an AP a while to detect a threat such as a rogue AP operating on a different channel. APs also prioritize client traffic, especially voice calls and video traffic, over scanning or containment.

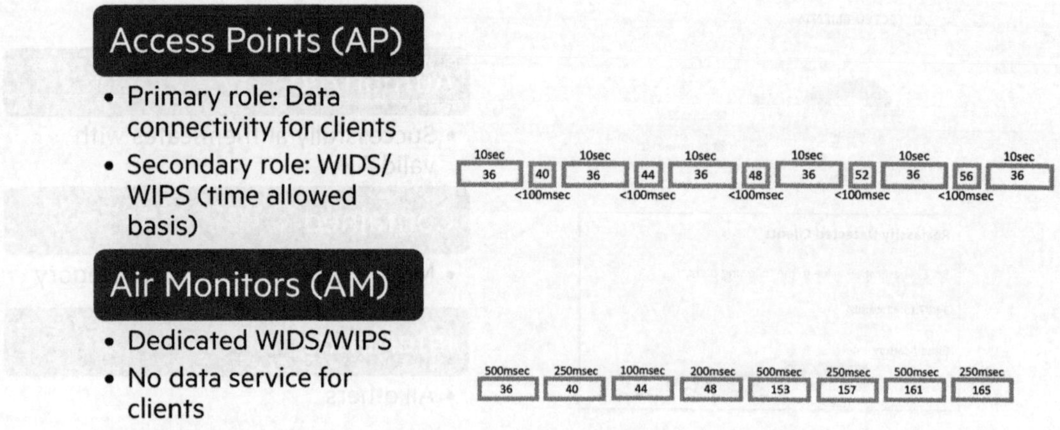

Figure 10-5: Access Point (AP) & Air Monitor (AM) Considerations

As shown in Figure 10-5, Air Monitors (AM) act as sensors and continuously scan the air, sending all captured information back to the managing system. The benefit of an AM is that it can scan nonstandard channels and provide continuous scanning.

You should deploy AMs in extremely high-security environments or environments that require wireless containment. AMs will detect attacks and rogues faster since they are dedicated to scanning channels. AMs scan the 2.51.2-2.484 and 4.9- 5.895 GHz spectrum in 5 MHz increments. They scan the active channels more frequently than regulatory channels, and both those channels more frequently than rare channels. The exact time to cycle through all the channels varies, but generally scanning through all regulatory channels takes approximately 5 minutes.

If the company plans to use wireless containment, they should typically use AMs because APs cannot spend enough time on the contained device's channel to operate as effectively.

HPE Aruba Networking recommends a 4:1 AP to AM ratio. However, the precise ratio depends on the density of APs and the type of building. For example, if you deploy the APs densely for a voice deployment, a 5:1 ratio will work. Note that when you deploy AMs, WIDS/WIPS can draw on data from both AMs and APs.

To create an Air Monitor, you will need to complete the following steps:

- First, create a new AP Group dedicated to Air Monitors.
- Then, provision the new group as air monitors in the RF Management Profile.
- Then you will need to assign appropriate access points to the new AM Group.
- Lastly, confirm AM mode with **show ap database** and **show ap active**.

Review AM Deployment Recommendations for Rogue AP Detection

As mentioned earlier, you should generally deploy about one AM per four or five APs to obtain good WIDS/WIPS coverage (see Figure 10-6).

In addition to considering RF coverage for the WIDS/WIPS solution, consider how the solution will detect whether unauthorized APs are connected on the wired network. Exactly how WIDS/WIPS detects rogue APs might differ slightly based on the AOS architecture. However, in general, WIDS/WIPS builds a list of "wired MACs" detected in the air. These are the source MAC addresses that APs transmit within 802.11 frames, which belong to clients in the wired LAN behind the AP. WIDS/WIPS compares the air-detected wired MACs against MAC addresses that it knows exist in the corporate LAN. If it finds matches or close matches, it knows that the AP is or might be connected to the corporate LAN. WIDS/WIPS can learn the corporate LAN MAC addresses in several ways, including APs/AMs reporting their default gateway MAC addresses, APs/AMs snooping ARP traffic on their VLANs, and even from MCs in an AOS-8 architecture.

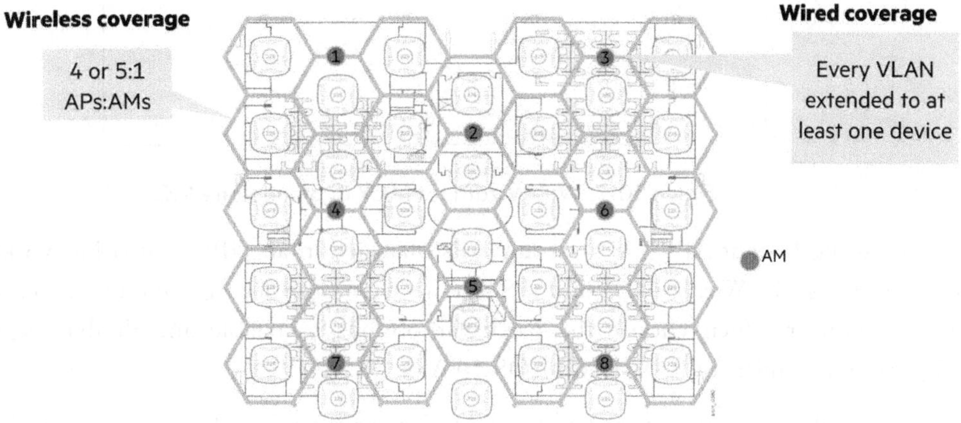

Figure 10-6: Review AM Deployment Recommendations for Rogue AP Detection

To ensure rogue AP detection works correctly, make sure that the HPE Aruba Networking solution has at least one device in each VLAN to which rogue APs might connect. One common way to achieve this goal is to extend all the VLANs in each area to AMs in that area.

Configure WIDS/WIPS

This section provides guidelines for configuring WIDS/WIPS.

How to Configure IDS/IPS

HPE Aruba Networking recommends the WIP Wizard for configuring IDS profiles and policies on a Mobility Conductor (MCR) or standalone MC (see Figure 10-7). Additionally, advanced users can use the CLI to configure IDS profiles.

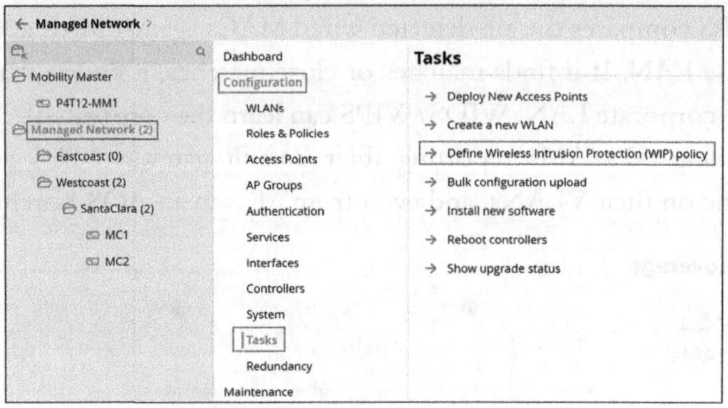

Figure 10-7: WIP Wizard for MCR or Standalone MC

If you are using Instant APs (IAPs) or any APs managed in an HPE Aruba Networking Central group, you access the Wireless IDS and IPS settings within the Security settings. You can select the custom option and then choose the particular options to enable and disable. Note that the configuration is similar for AOS-8 and AOS-10 APs.

This chapter will suggest which settings to enable for security minded customers and which to leave disabled.

The Risks of False Positives

It might seem that the more IDS settings that you enable—the more possible threats that the WIDS system can detect—the more secure the system. However, the reality is different. Some settings are more prone to creating false positives than others in some environments. False positives occur when normal behavior creates a similar effect as a potential threat. False positives create risks of their own because they distract admins and train them to not believe or investigate alerts. In some cases, particular settings search for threats that no longer pose a great threat in modern networks. It is best to disable settings that are unlikely to detect real threats of concern to most customers in most environments.

When determining which IDS and IPS settings to enable, consult with the company's security team, and use their recommendations. If the team is uncertain which settings to enable and asks you for guidance, you can use the HPE Aruba Networking recommendations as a starting point. The rest of this section covers which settings HPE Aruba Networking recommends that you should enable. Disable all other IDS settings to prevent distracting false positives.

Recommended Infrastructure IDS Settings

First, you will look at the recommended Infrastructure IDS settings to enable.

Detect AP Spoofing

An HPE Aruba Networking AP detects the AP spoofing attack when it detects another AP using its BSSID, as shown in Figure 10-8. An AP should never hear traffic from its own radio since it cannot send and receive at the same time. If it does, then an intruder is spoofing the AP. WIDS could classify the spoofing AP as an Interfering AP, if it is not connected to the LAN, or as a Rogue AP, if it is connected to the LAN.

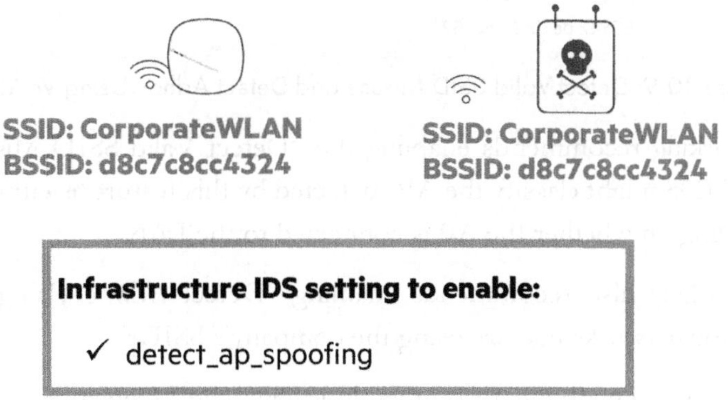

Figure 10-8: Detect AP Spoofing

This attack indicates a hacker who is attempting to deceive devices into connecting to it. Spoofing a BSSID is trivial with common software, and posing as a legitimate AP forms the foundation for many attacks. The spoofing AP can lure clients to connect to it in an attempt to obtain the users' credentials or to launch an MITM attack. Note, however, that if clients are properly configured to use 802.1X with certificate-based authentication, as well as to validate the RADIUS server, they should be protected from some of these attacks. However, AP spoofing does indicate a hacker who is attempting to cause harm, and you should enable the Detect AP Spoofing setting, as shown in Figure 10-8.

You can optionally enable Detect AP Impersonation if the customer is highly security minded. This feature detects much the same attack as the Detect AP Spoofing feature; however, it can be prone to false positives.

Detect Valid SSID Misuse and Detect Adhoc Using Valid SSID

As shown in Figure 10-9, the WIDS solution can detect any stray APs masquerading as valid APs and advertising one of a company's SSIDs, although not necessarily with the same BSSID. Again, such a circumstance often indicates a hacker who is attempting to steal users' credentials or launch other attacks.

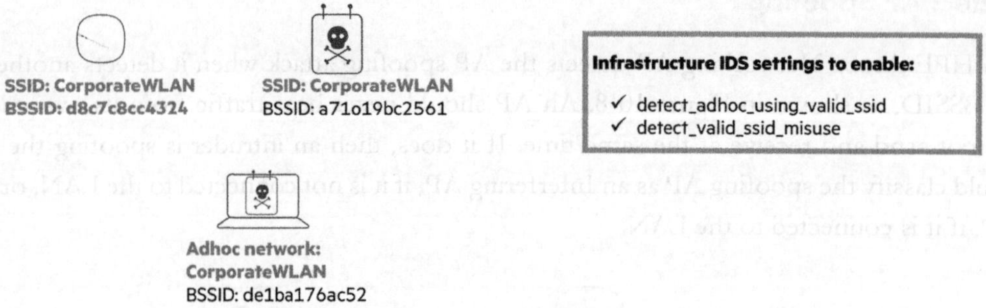

Figure 10-9: Detect Valid SSID Misuse and Detect Adhoc Using Valid SSID

HPE Aruba Networking recommends enabling the "Detect Valid SSID Misuse" IDS setting to detect these APs. WIDS might classify the APs detected by this feature as either Interfering APs or Rogue APs, depending on whether the AP is connected to the LAN.

HPE Aruba Networking also recommends enabling "Detect Adhoc Using Valid SSID." This feature detects ad hoc networks that are using the company's SSIDs.

However, HPE Aruba Networking recommends leaving the "Detect Adhoc Network" setting disabled. This setting detects all ad hoc networks regardless of the SSID. Because many devices can create these networks by default, this setting can produce many false positives.

Log output for this setting

This section provides some example log outputs associated with the "Detect Valid SSID Misuse" setting. The first log is for an interfering AP that is misusing the SSID, but is not connected to the wired infrastructure. The second is for an unauthorized AP that *is* connected to the wired infrastructure, making it a rogue AP.

1. The network does not detect AP connected to wired infrastructure:

```
(wips) #show log security all
```

```
May 17 06:15:17 :103062: <INFO> |ike| Starting cryptoPOST

May 17 06:15:54 :118004: <ERRS> |certmgr| Received unknown message

May 17 06:16:28 :125022: <WARN> |aaa| Authentication failed for User admin, Logged in from 192.168.50.248 port 49338, Connecting to 192.168.50.3 port 22 connection type SSH

May 17 06:17:58 :126005: <WARN> |wms| |ids| Interfering AP: The system classified an access point (BSSID 00:1c:10:a5:5e:30 and SSID wips-test on CHANNEL 1) as interfering. Additional Info: Detector-AP-Name:WIPS-AP-7d:3c; Detector-AP-MAC:d8:c7:c8:87:d3:c0; Detector-AP-Radio:2.

May 17 06:17:58 :126007: <WARN> |wms| |ids| AP(d8:c7:c8:87:d3:c0@WIPS-AP-7d:3c): Multi-tenancy SSID Violation: An AP detected an access point (BSSID 00:1c:10:a5:5e:30 and SSID wips-test on CHANNEL 1) is violating Valid SSID configuration by using a protected SSID.
```

2. AP connects wired infrastructure:

```
(wips) # show log security all

May 17 06:38:41 :103062: <INFO> |ike| Starting cryptoPOST

May 17 06:39:18 :118004: <ERRS> |certmgr| Received unknown message

May 17 06:43:48 :126005: <WARN> |wms| |ids| Interfering AP: The system classified an access point (BSSID 00:1c:10:a5:5e:30 and SSID personal-AP on CHANNEL 1) as interfering. Additional Info: Detector-AP-Name:WIPS-AP-7d:3c; Detector-AP-MAC:d8:c7:c8:87:d3:c0; Detector-AP-Radio:2.

May 17 06:43:49 :106000: <NOTI> |AP WIPS-AP-7d:3c@192.168.50.250 sapd| |ids-ap| AM d8:c7:c8:87:d3:c0: Potentially rogue AP detected BSSID 00:1c:10:a5:5e:30 SSID personal-AP MATCH MAC 00:1c:10:a5:5e:30

May 17 06:43:49 :126002: <ERRS> |wms| |ids| Rogue AP: The system classified an access point(BSSID 00:1c:10:a5:5e:30 and SSID personal-AP on CHANNEL 1) as rogue. Additional Info: Detector-AP-Name:WIPS-AP-7d:3c; Detector-AP-MAC:d8:c7:c8:87:d3:c0; Detector-AP-Radio:2.
```

Detect Broadcast Deauthentication and Disassociation

A de-authentication broadcast attempts to disconnect all stations in range, as shown in Figure 10-10. Rather than sending a spoofed deauth to a specific MAC address, this attack sends the frame to a broadcast address. In a similar attack, hackers send disassociation frames to the broadcast address. HPE Aruba Networking recommends enabling the settings for detecting both of these threats (see Figure 10-10).

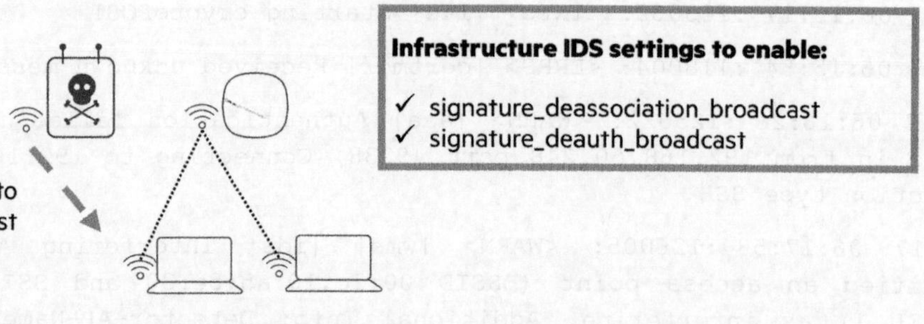

Figure 10-10: Detect Broadcast Deauthentication and Disassociation

Detect Windows Bridge

Many endpoints with both wired and wireless interfaces, including Windows clients, are capable of bridging between the interfaces. When a Windows client is connected to the corporate LAN and bridges between its wireless and Ethernet NIC, it can open a backdoor into the corporate LAN, as shown in Figure 10-11. A hacker who gains temporary access to the corporate LAN might set up such a client to provide remote access later. Or a user might implement this setting without realizing the risks. Hackers can then use the client's insecure wireless network to gain internal access to the corporate network, perhaps even from outside the building. In either case, the Windows bridge poses a risk to the network.

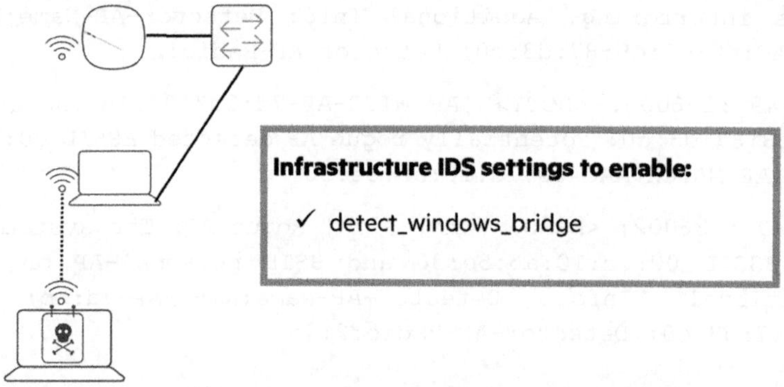

Figure 10-11: Detect Windows Bridge

The "Detect Windows Bridge" setting enables the HPE Aruba Networking solution to detect this situation. HPE Aruba Networking recommends enabling this setting, as it produces few false positives and can alert admins to issues.

Log output for this setting

```
(wips) #show log security all
```

```
May 17 11:28:44 :103062: <INFO> |ike| Starting cryptoPOST

May 17 11:29:15 :118004: <ERRS> |certmgr| Received unknown message

May 17 11:37:46 :126042: <WARN> |wms| |ids| AP(d8:c7:c8:87:d3:c0@WIPS-AP-7d:3c):
Windows Bridge: An AP detected a bridge on CHANNEL 6 between access point (BSSID
d8:c7:c8:87:d3:c0 and SSID and a node 24:77:03:2a:70:20.
```

Detect Wireless Bridge

A wireless bridge establishes a point-to-point link between two 802.11 radios, as shown in Figure 10-12. If a hacker manages to connect a rogue AP to the LAN, the hacker might establish a wireless bridge from that AP to a second AP outside the corporate office. In this way, the hacker can obtain access into the network from an external location. HPE Aruba Networking recommends enabling the "Detect Wireless Bridge" IDS setting to alert admins to potential security holes.

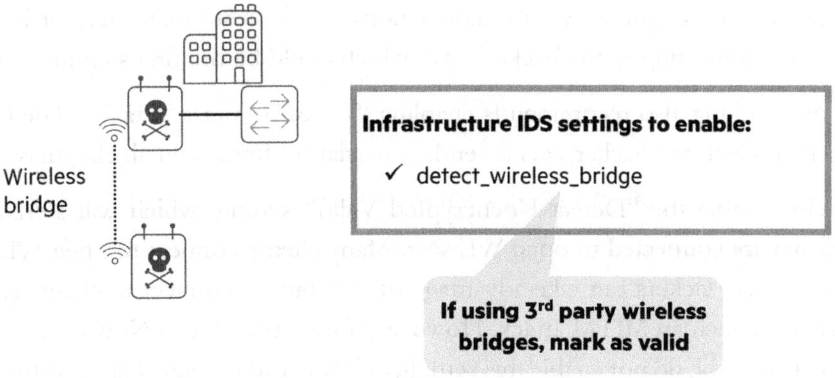

Figure 10-12: Detect Wireless Bridge

If the HPE Aruba Networking mobility solution includes wireless meshes, those meshes will not trigger the "Detect Wireless Bridge" feature. However, if the company is using authorized third-party wireless bridges, you must mark those bridges as valid manually.

Recommended Client IDS Settings

HPE Aruba Networking recommends enabling three client IDS settings, shown in Figure 10-13.

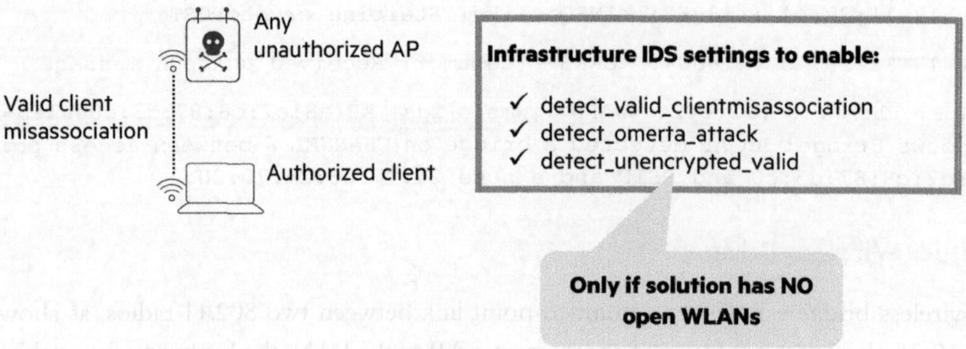

Figure 10-13: Recommended Client IDS Settings

With "Detect Valid Client Misassociation," WIDS detects whenever an authorized client connects to any unauthorized AP (see Figure 10-13). WIDS might classify that unauthorized AP as a Rogue AP, Interfering AP, or Neighbor AP. Misassociations can indicate that a hacker has successfully lured the client to connecting to the hacker's AP, which could be the first step in another attack.

HPE Aruba Networking also recommends enabling "Detect Omerta Attack." The Omerta attack is a DoS attack in which the hacker device sends dissociation frames on all channels.

You should also enable the "Detect Unencrypted Valid" setting, which will alert admins when authorized clients are connected to open WLANs. Many clients connect to open WLANs without the users' awareness. Hackers can take advantage of that fact to connect to clients and attempt to install malware or launch an MITM attack. However, if the HPE Aruba Networking solution itself includes an open WLAN, do not enable this setting, as it will lead to many false positives. In addition, you should be aware that Apple devices often create open WLANs by default, so, if the environment includes these devices, you will receive many false positives.

Possible IDS Settings for Highest Security Environments

Next, you will examine some IDS settings that you might possibly enable in limited scenarios.

Settings for Which Tuning Is Required

WIDS/WIPS can detect a variety of issues that might indicate DoS attacks, such as floods of various types of frames. However, these settings are also prone to triggering many false positives. For example, issues with client drivers or drivers on neighboring APs can cause the devices to send many RTS or CTS frames. Many clients also save power aggressively, which can look like a power save attack.

The same holds true for the following settings; they might indicate a DoS attack, but are also likely to trigger false positives:

- Infrastructure
    - Detect AP Flood
    - Detect RTS Rate Anomaly
    - Detect CTS Rate Anomaly
    - Detect Client Flood
- Clients
    - Detect Power Save DoS Attack
    - Detect Disconnect Station
    - Detect Block ACK Attack
    - Detect Rate Anomalies
    - Detect EAP Rate Anomalies

You should leave these settings disabled in many circumstances. However, some highly security conscious customers will want to enable these settings. In this case, you should enable the settings only if you also tune them by configuring their thresholds.

These settings all come with thresholds. For example, the RTS Rate Anomaly triggers when a wireless device sends more RTS frames within a certain time period than the configured threshold allows. You should not leave the default thresholds, as doing so is likely to trigger many false positives. Instead, you must analyze the customer's particular environment over a period of time to create baselines. How many frames of each type do wireless devices normally send over various time periods, such as one minute and five minutes? What is truly anomalous? You can then set the thresholds based on these baselines.

Possible IDS Settings for Highest Security Environments

The "Detect Overflow EAPOL key," "Detect Overflow IE," "Detect Malformed Frame Association Request," "Detect Malformed Frame Auth," and "Detect Malformed Frame-HT IE" settings detect various malformed frames. These frames might indicate DoS attacks but can also occur due to driver issues such as on third-party APs. Again, HPE Aruba Networking recommends leaving these settings disabled unless a highly security minded company requires them. In that case, only enable them if WIDS/WIPS is monitoring an AOS mobility solution.

The "Detect Invalid Address Combination" triggers when wireless frames have invalid addresses such as a broadcast or multicast address for the source. While an invalid address combination can indicate a DoS attack, it much more often occurs innocently due to bad drivers. Only enable this setting if required by the company and if the company manages, tests, and tracks all client devices.

Optional Client IDS Settings for Particular Use Cases

The settings shown in Figure 10-14 are for specific use cases.

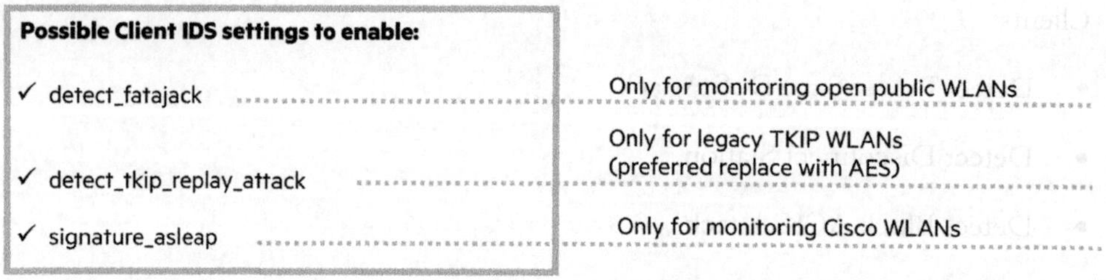

Figure 10-14: Possible IDS Settings for Highest Security Environments

"Detect FataJack" detects attacks from FATA-Jack, an 802.11 client DoS tool. This tool disconnects targeted stations using spoofed authentication frames that contain an invalid authentication algorithm number. HPE Aruba Networking recommends enabling this setting only if you are monitoring open WLANs in a public setting.

The TKIP replay attack only affects WLANs that are using TKIP, which is an outmoded encryption method. It is best to remove TKIP and use AES instead. However, if legacy devices require retaining support for TKIP, you can enable the "Detect TKIP Replay Attack" feature.

The "Signature ASLEAP" feature detects attacks on ASLEAP, which is a Cisco protocol. Only enable this setting if you are using WIDS/WIPS to monitor a Cisco WLAN.

If you are using an AOS-8 architecture, WIDS has a "Detect Misconfigured AP" setting. This feature checks whether authorized APs are using the settings indicated with a parameter list. Because the HPE Aruba Networking solution handles configuring APs correctly, you would only enable this setting if you are using WIDS/WIPS to monitor a third-party WLAN solution.

Review Intrusion Prevention Mechanisms

WIDS/WIPS supports methods for containing detected threats from APs or clients.

You can enable wired containment, wireless containment, or both.

Wired Containment

As shown in Figure 10-15, wired containment uses ARP cache poisoning for each IP address of the contained device (AP or station) at the rate of one per second. Specifically, the hybrid AP/AM sends an ARP response that claims to be from the contained device's default gateway but has an incorrect MAC address. As a result, the contained device cannot forward traffic out of the subnet. For wired containment to work, the hybrid AP/AM must be able to transmit traffic on the VLAN on which the contained device is connected. You should deploy hybrid APs/AMs on trunk ports that allow all VLANs to which devices might connect.

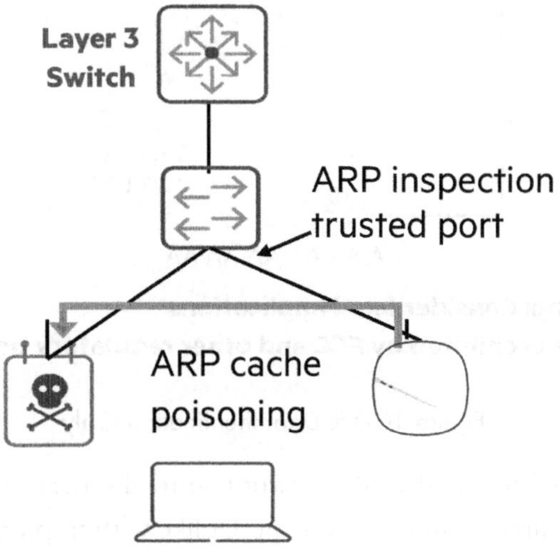

Figure 10-15: Wired Containment

Be aware of the settings on the switch to which the AP/AM implementing wired containment connects. If the switch implements ARP inspection on this port, the switch will drop the AP/AM's message. (It is an ARP poisoning attack.) Make sure the AP/AM's switch port is configured as a trusted port for ARP inspection.

Wireless Containment

If you choose to enable wireless containment, you can choose between deauthentication only or tarpit containment. Wireless containment of an AP attempts to disconnect all clients from that AP. Wireless containment of a client attempts to disconnect just that client.

You should be aware of the legal implications of both of wireless containment methods. Wireless containment in certain circumstances is prohibited by the FCC. You must always consult with a company's security and legal teams before enabling these mechanisms.

Deauthentication Only

With this method, illustrated in Figure 10-16, the hybrid AP/AM sends de-authorization messages to disconnect a client from the rogue AP. The AP impersonates the MAC address of the rogue APs when it sends the de-authorization messages to the client. The AP also impersonates the MAC address of the client when it sends de-authorization messages to the rogue AP.

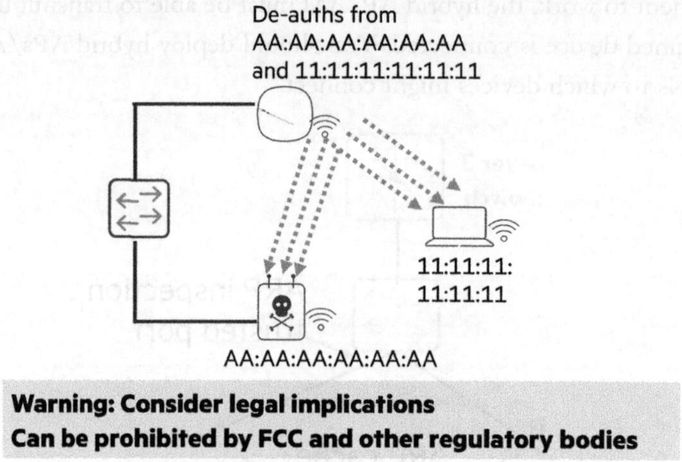

Figure 10-16: Deauthentication Only

Though this procedure is effective, it is also disruptive to the nearby stations. Each of these de-authorization messages requires airtime to transmit. Unlike with tarpitting, the client will detect that it disconnected from the rogue AP and probably attempt to reconnect automatically. Thereover, the AP/AM must continuously send wireless de-authorizations.

Tarpit

With tarpit shielding, the AP/AM deauthorizes clients from the contained AP. (Or if containing a client, it deauthorizes that one client.) But tarpit involves further steps to prevent clients from trying to continually reconnect to the contained device. The AP/AM copies the contained AP's probe response, but with a different BSSID, different channel, or both. This probe response gets clients to connect to the AP/AM instead of the contained AP, as shown in Figure 10-17. The AP/AM then ignores the traffic from that client. When a client is successfully tarpitted, most client drivers report that the client is "connected." Users can see that their device did not get an IP address and cannot pass data, so they may attempt to reconnect to the rogue network. However, without user intervention, the client remains in the tarpit.

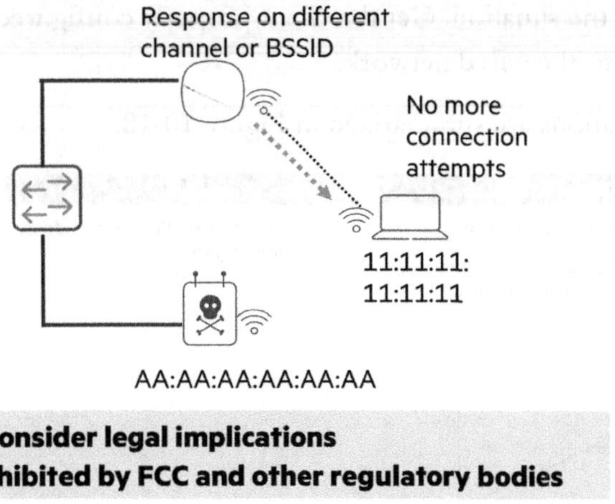

Figure 10-17: Tarpit

The client eventually stops trying to send data to the AP/AM. If the user tries to connect to the contained AP again, the client is lured into the tarpit again. This method is very efficient, because each AP, or AM, near the client can participate without spending much time on the channel. The network can also spread containment across multiple nearby APs or AMs, so that whichever AP/AM is available to handle the containment can do so.

For tarpit containment of an AP, you can choose to have APs/AMs tarpit all clients that attempt to connect to the contained AP or just non-authorized clients.

Recommended IPS Settings

You will now learn some recommendations for IPS settings.

Wired Containment Against Rogue APs

HPE Aruba Networking recommends that, among the Infrastructure protection options, you enable only "Rogue containment." Leave other options such as protecting your SSIDs, protecting against AP impersonation, and protecting against ad hoc networks disabled. HPE Aruba Networking also strongly recommends that you leave wireless containment disabled and enable wired containment instead. Wired containment poses fewer legal risks than wireless containment because the wired network belongs to the organization in question. However, you should still consult with the security and legal teams for authorization to use wired containment.

If you are using only wired containment, you might also enable protecting against Windows bridges within the Clients Protection settings. Enabling this feature would disconnect the bridging client from the wired network and leave it connected to the wireless network only. However, you would need to investigate the situation. Get the client properly configured so that it no longer bridges and can safely connect to the wired network.

These recommendations are summarized in Figure 10-18.

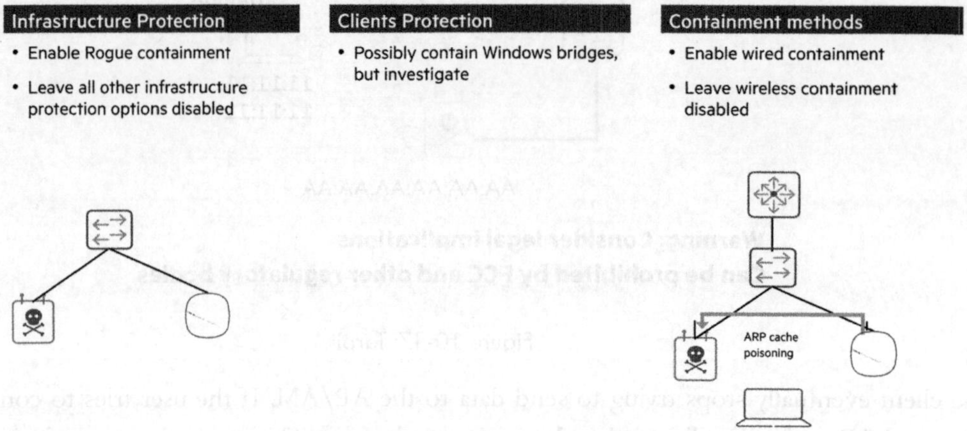

Figure 10-18: Recommended IPS Settings

AOS-8 architectures support several more Infrastructure Protection mechanisms. HPE Aruba Networking strongly recommends disabling these options for several reasons:

- Risks of disruption and large amounts of hybrid AP/AM time spent protecting against false positives:
  - Protecting 40MHz 802.11 High Throughput Devices
  - Protecting 802.11n High Throughput Devices
  - Protecting Against Ad hoc Networks
- Risks of regulatory fines:
  - Protecting SSIDs
  - Protection Against Misconfigured APs
  - Protection Against Wireless Hosted Networks
  - Suspected Rogue Containment at any confidence level

Considerations for Protect Valid Stations

For some highly security-minded customers, you can enable the "Protect Valid Stations" option for Clients Protection. As shown in Figure 10-19, this setting deauthenticates authorized clients from unauthorized APs, guarding the clients from malicious APs set up by hackers. For this setting to work, you must enable wireless containment, as the containment is applied to the wireless client. You should use the de-auth option if you want to avoid your clients getting "stuck" in a tarpit. Because this setting targets only authorized clients, which presumably belong to the company, it comes with fewer legal considerations than some of the other protection options.

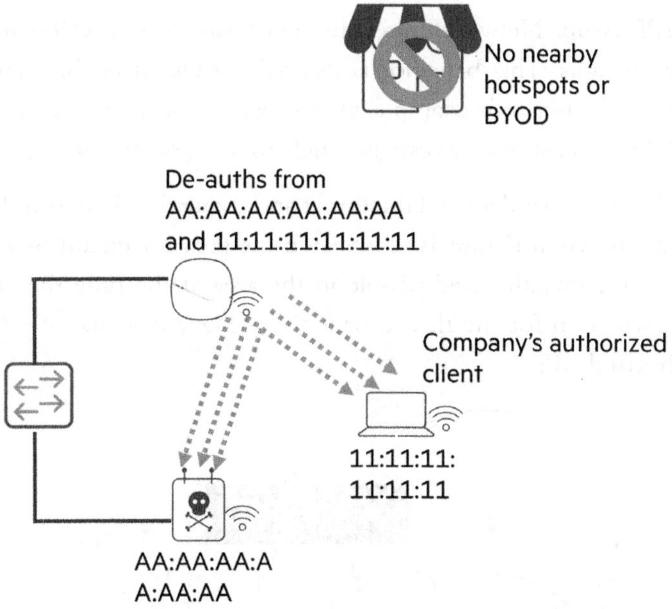

Figure 10-19: Considerations for Protect Valid Stations

However, you must keep several caveats in mind. First, carefully consider the customer environment. Is the corporate environment relatively isolated? If not, this feature could pose problems. Authorized clients might attempt to connect to hotspots at nearby restaurants, coffee shops, and similar locations. If legitimate hotspot APs are within range of the WIDS/WIPS solution, you should not enable this setting.

Also consider whether the company allows BYOD. If users are bringing their own devices into the corporate LAN, those devices belong to the users, but are marked as authorized clients. The company has less legal standing to control whether the BYOD devices connect to non-authorized APs.

Finally, consider the following situation. You have enabled rogue containment and wired containment, as recommended previously. You then enable "Protect Valid Stations" and a wireless containment option. Now HPE Aruba Networking WIPS will also target Rogue APs with wireless containment because the containment methods apply to all enabled protection options. However, you must consult with both the company's security team and legal team before you can enable wireless containment for any of the Infrastructure Protection options. So, you might not want to enable wireless containment at all.

Investigation Rather Than Automatic Containment

When you follow HPE Aruba Networking recommendations, WIDS will detect more threats than WIPS automatically contains. This behavior is desired because it enables you to investigate and mitigate threats with fewer risks. For example, rather than enable automatic protection against APs impersonating valid APs, you should investigate such APs as you are alerted to them.

Make a list of the APs/AMs that detected the threat and triangulate between them to find the likely area for the threat, as shown in Figure 10-20. Search the area for unauthorized devices. Question employees about possible unauthorized people in the area at the time the alert was triggered. Is there a legitimate explanation for the threat, or is this a likely instance of a hacker attempting to impersonate an authorized AP?

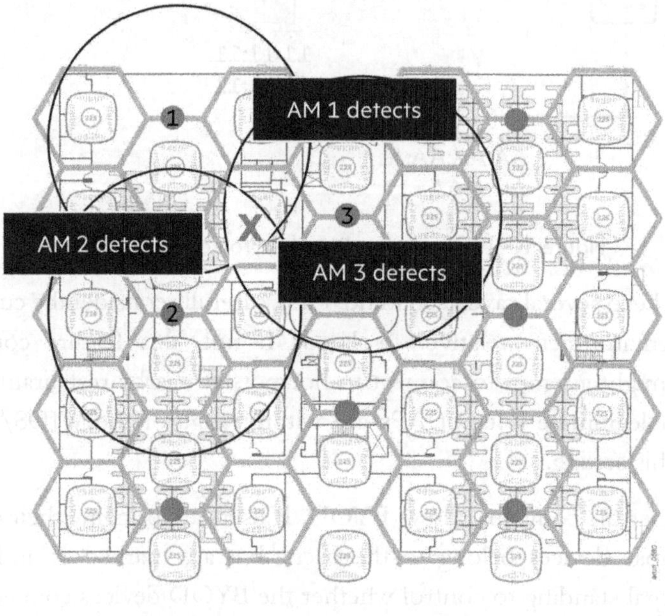

Figure 10-20: Locating the Rogue AP

Only after a thorough investigation and approval from the security and legal team, you might choose to manually mark the AP that triggered the alert as Contained. However, containing is not always the right approach. Consider an AP that is spoofing an authorized AP's BSSID. Attempting to contain that AP could cause a DoS against the legitimate network. Instead, you should remove the interfering AP manually. In addition, you generally want to remove rogue APs rather than contain them over the long term.

Remember that you can implement 802.1X on all Ethernet ports to reduce the risks of unauthorized devices connecting to the network in the future.

Summary

Congratulations! You now have a better understanding of how to formulate and implement WIDS/WIPS policies for particular environments. You have seen how AOS WIDS can alert you to threats, and you can take next steps to investigate those threats.

Learning checks

1. What distinguishes a Rogue AP from an Interfering AP?
   a. A Rogue AP has launched a DoS attack.
   b. A Rogue AP is connected to the corporate LAN.
   c. A Rogue AP uses unauthorized security settings.
   d. A Rogue AP is connected to Interfering clients.
2. What are some HPE Aruba Networking recommendations for IPS and IDS settings to enable? (Select two.)
   a. Enable all IDS settings.
   b. Enable all IPS settings.
   c. Enable IDS settings that are unlikely to produce false positives.
   d. If automatic containment is required, enable rogue AP containment with wireless deauthentication.
   e. If automatic containment is required, enable rogue AP containment with wired containment.

Note

For answers to all learning checks, refer to the Appendix.

---

11 Use CPPM and Third-Party Integration to Mitigate Threats

EXAM OBJECTIVES

- ✓ Describe log types and levels
- ✓ Use the HPE Aruba Networking ClearPass Policy Manager (CPPM) Ingress Event Engine to integrate with third-party logging solutions
- ✓ Set up integration between the HPE Aruba Networking infrastructure and CPPM, allowing CPPM to take action in response to events

Assumed Knowledge

- Fundamental CPPM components, such as services, roles, enforcement profiles, and enforcement policies
- Syslog basics

Overview

By integrating with third-party security solutions in HPE Aruba Networking 360 Security Exchange, CPPM enables companies to respond to threats more quickly and comprehensively.

This chapter explains how to use the CPPM Ingress Event Engine to interpret messages from event sources such as firewalls and other security appliances. You will examine Ingress Event Dictionaries and see how they enable CPPM to parse Syslog messages and extract meaningful attributes from them.

In the second section, you will learn about using ingress events to respond to threats. You will see how CPPM can "read" a Syslog message about a particular endpoint and use an event service to update the endpoint profile. You will further learn how you can configure your existing services to take the new information into account and block endpoints that pose a threat to your system.

CPPM Ingress Event Engine

This first section introduces you to the CPPM Ingress Event Engine.

Review of HPE Aruba Networking ClearPass 360 Security Exchange

Recall that HPE Aruba Networking solutions integrate with more than one hundred and fifty third-party solutions in the ClearPass 360 Security Exchange.

Through this exchange, HPE Aruba Networking solutions, including CPPM, can share information with the security appliances of companies' choice. The solutions operate in a mutually beneficial partnership, in which they provide information of interest to each other and enhance each other's abilities. As you will see, for example, an IDS can inform CPPM that an endpoint is generating an attack. CPPM can then order the network infrastructure to lock the endpoint out of the network.

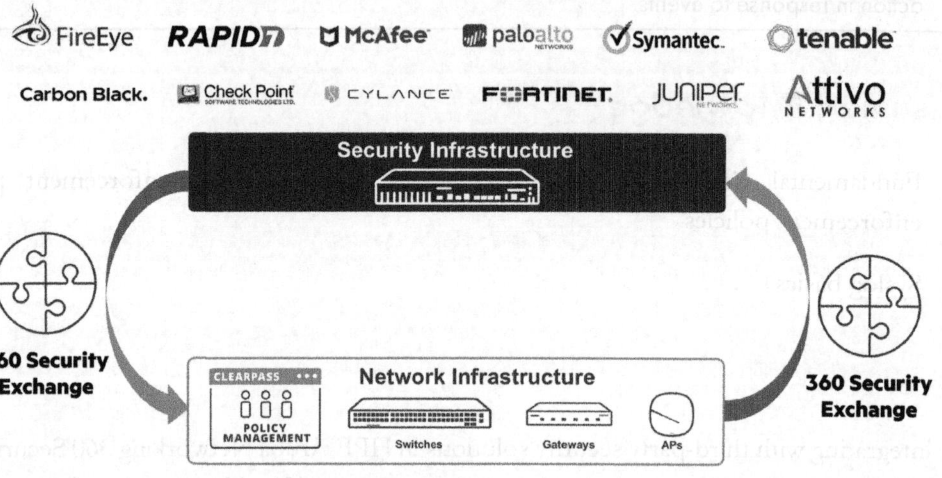

Figure 11-1: Review of HPE Aruba Networking ClearPass 360 Security Exchange

You can see some of the 360 Security Exchange partners listed in Figure 11-1. For a more comprehensive and up-to-date list, visit the HPE Aruba Networking Technology Partner Finder and search for partners in categories of interest to you. Use this link:

https://www.arubanetworks.com/partners/technology-partners/partner-finder

Categories of Third-Party Integration

CPPM can integrate with third-party security appliances in many ways—all based on the third-party solutions' capabilities and what the company wants to achieve. But you can start to think about the capabilities in two broad categories.

Context Servers

CPPM can integrate with "context servers," which include mobility device management (MDM) solutions, enterprise mobility management (EMM) solutions, device management solutions, and security solutions. As shown in Figure 11-2, these solutions can send CPPM information about endpoints, which CPPM can use as context in role mapping and enforcement policies. For example, you learned how CPPM can learn about endpoints' security posture from third-party devices in a previous chapter.

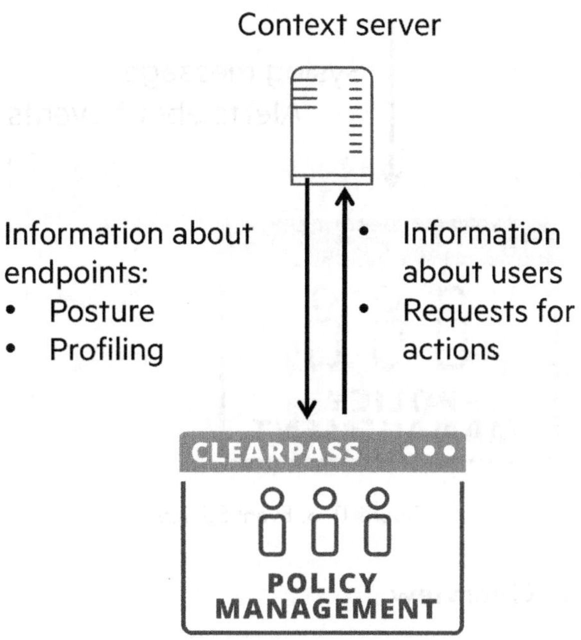

Figure 11-2: Context Servers

CPPM also collects a broad array of information about clients, using its authentication and authorization capabilities, as well as the device profiling abilities that you will learn about in the next chapter. CPPM can send this information to third-party solutions, which are called "context servers." Those solutions can then use that information to enhance their insight into clients and users and take more informed actions.

CPPM can also send requests to the context server to perform particular actions using the API of the particular context server in question.

Event Sources

CPPM can also receive Syslog messages from third-party solutions set up as event sources, as shown in Figure 11-3. It can use its Ingress Event Engine to parse those messages and take actions. You will focus on this second category of integration for the rest of the chapter.

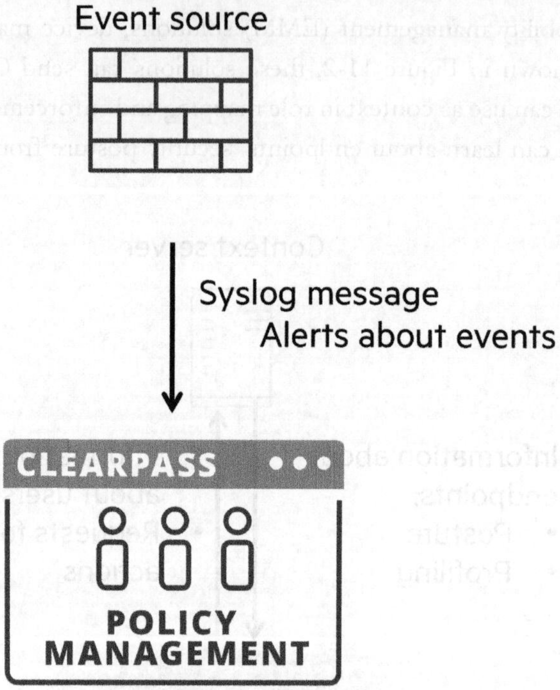

Figure 11-3: Event Sources

Ingress Event Engine Overview

Event sources, such as third-party firewalls, send Syslog messages to CPPM. In the example shown in Figure 11-4, a firewall has detected an attack that originates with client A. The firewall is configured to log attacks of this type to CPPM.

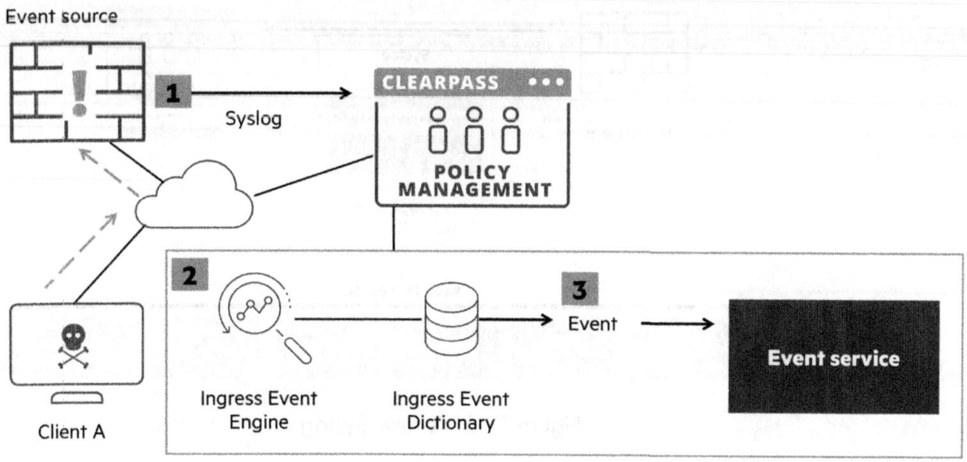

Figure 11-4: Ingress Event Engine Overview

CPPM's Ingress Event Engine receives the message and determines whether it comes from a defined source. If so, it processes it using an Ingress Event Dictionary. Assuming that a valid dictionary exists for the message, CPPM generates an event.

CPPM can then process the event with an event service. Like all services on CPPM, this enforcement service can take various actions based on the policies you define within it.

For the rest of this section, you will explore how to enable ingress event processing, define event sources, and use ingress event dictionaries. In the next section, you will learn more about creating and using event services.

Review Syslog

Syslog is an open standard used by many devices. Syslog uses a client-server framework. For the use case on which you are focusing, the Syslog client is the security appliance. It sends log messages to a Syslog server, which is CPPM in this use case.

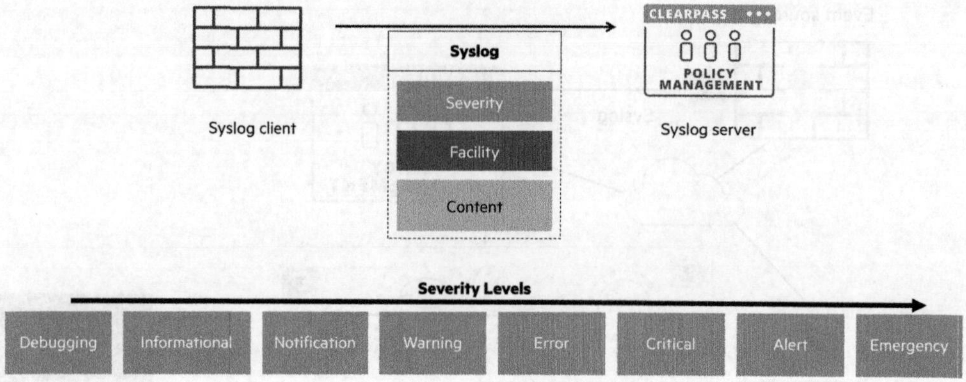

Figure 11-5: Review Syslog

As shown in Figure 11-5, Syslog messages include a severity rating. The categories range from debugging at the lowest level of severity through emergency. Individual vendors decide how they assign severity levels to logs for various events. Generally, the levels are understood to mean:

- **Debugging**—Very detailed logs about processes, intended only for troubleshooting or debugging a function
- **Information**—Logs about routine events and processes
- **Notification**—Logs about normal occurrences, but ones that might require notice (such as an interface coming up)
- **Warning**—Logs about events that might cause issues
- **Error**—Logs about events that are very likely to cause issues
- **Critical**—Logs about critical issues on a system
- **Alert**—Logs about critical issues that need immediate attention
- **Emergency**—Logs about issues that are causing widespread issues and need immediate attention

Syslog messages also include a facility, which indicates the general type of source that creates the log. For example, facility 13 is for security programs.

Considering the Effects of Ingress Event Processing

Using the Ingress Event Engine can be CPU-intensive. You need to consider how you use it to avoid overwhelming a ClearPass server and preventing it from responding to other requests.

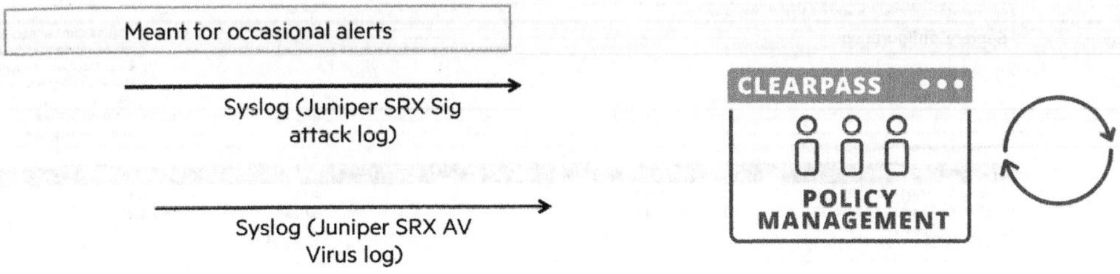

Figure 11-6: Recommended Ingress Event Engine Usage

As shown in Figure 11-6, you can safely enable ingress event engine processing on a CPPM server that will only receive a few Syslog messages per hour. This is the scenario you would expect if security appliances are just sending logs when threats occur.

CPPM is not intended for processing bulk Syslog messages, as shown in Figure 11-7. If you need a server to receive and process many Syslog messages per hour, you should add another server to the cluster and dedicate it to ingress event processing.

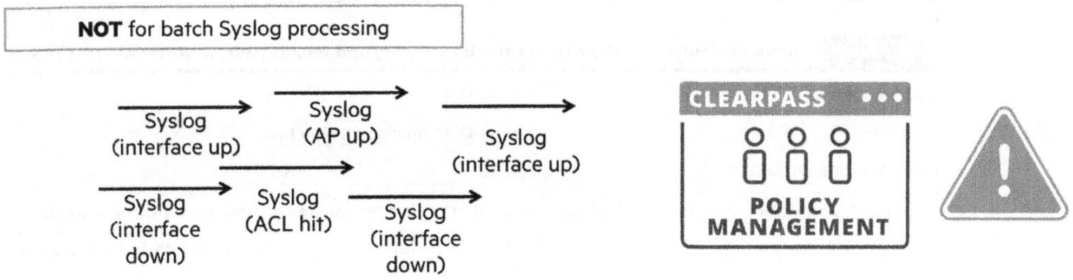

Figure 11-7: Non-Recommended Ingress Event Engine Usage

Enabling Ingress Event Processing and Insight

Once you have determined which servers you want to process ingress events, you can enable ingress event processing on them.

To enable ingress event processing, use the left menu in the CPPM UI to navigate to **Administration > Server Manager > Server Configuration**. Select the server that you want to use, as shown in Figure 11-8.

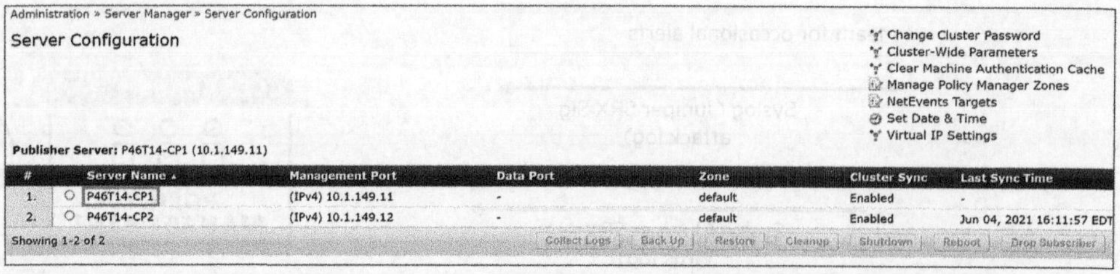

Figure 11-8: CPPM Administration > Server Manager > Server Configuration Page

In the page that appears, make sure you are in the **System** tab, as shown in Figure 11-9. Click the **Enable Ingress Events Processing** check box, and save the new configuration. You will see a warning related to the CPU intensity previously discussed. Click **Yes** to dismiss the warning.

You must also enable Insight, which enables CPPM to perform tasks like correlating IP addresses and MAC addresses. In that way, it helps CPPM figure out which endpoint is associated with an event.

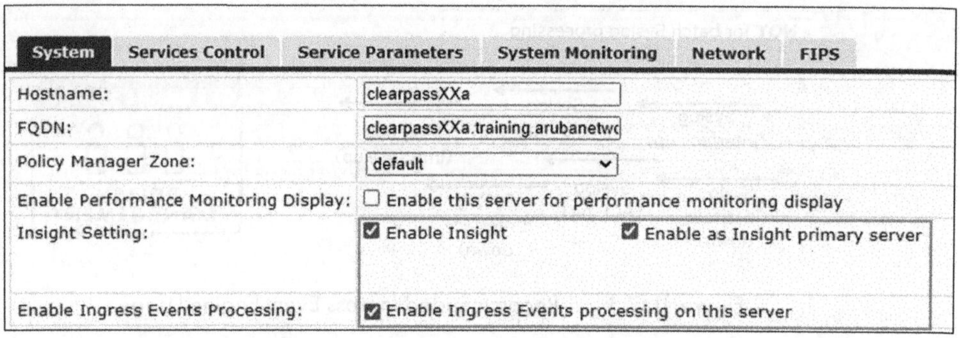

Figure 11-9: CPPM Server Configuration > System Settings

Repeat this step on any other ClearPass servers in the cluster that you want to process ingress events.

Examples of Event Sources

The event sources from which CPPM receives Syslog messages are third-party security appliances. CPPM v6.10 comes loaded with dictionaries for interpreting Syslog messages for several third-party devices, including

- Check Point firewalls
- Palo Alto Next Generation Firewalls (NGFWs)
- Juniper SRX firewalls

- Infoblox platform, which provides network device management, as well as detection of malware

You can also create and add your own custom dictionaries to enable CPPM to process Syslog messages from any system set up as a Syslog client. However, creating custom dictionaries is an advanced topic beyond the scope of this study guide.

Defining an Event Source

Even when a ClearPass server has ingress event processing enabled on it, CPPM only accepts and processes Syslog messages from a defined event source. If a Syslog message comes from an unknown source, CPPM will not act on the message.

To define an event source, use the left-hand menu in CPPM's interface to navigate to **Configuration > Network > Event Sources**. Click **Add**. Then fill in the information as shown in Figure 11-10. Click **Add** in the pop up window to confirm the new event source.

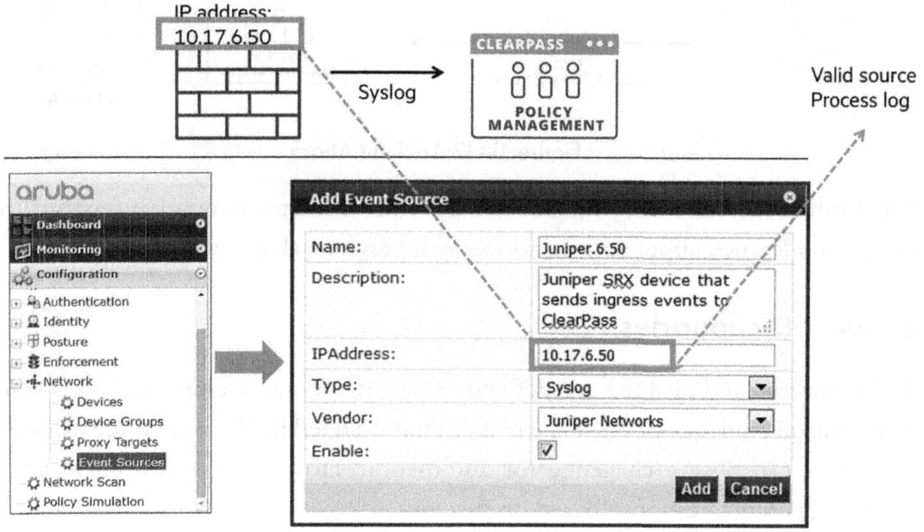

Figure 11-10: Defining an Event Source

Avoiding Spoofed Event Sources

Ironically, the Syslog messages that are meant to alert CPPM to threats can also become an attack vector. By spoofing their IP address, bad actors can pose as a legitimate event source.

As shown in Figure 11-11, a hacker could launch a DoS attack by sending a high volume of Syslog messages to CPPM. Processing the Syslog messages could compromise the ClearPass server's ability to respond to other service requests, which could cause authentication issues throughout the network.

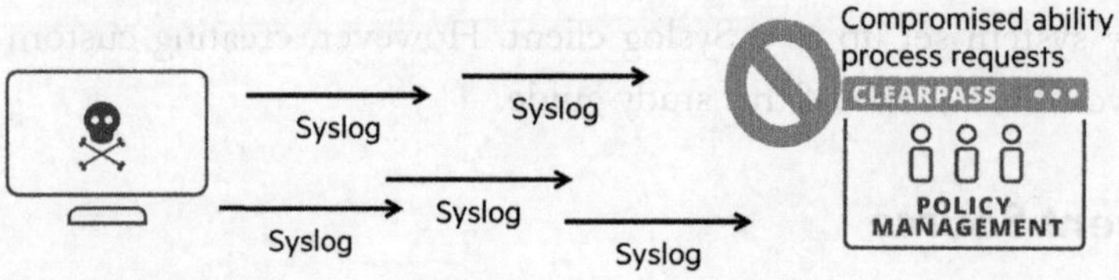

Figure 11-11: DoS Attack

Alternatively, as shown in Figure 11-12, a hacker could poison CPPM by sending a message that designates a legitimate client as a threat, thereby causing CPPM to deny or restrict access to that legitimate client.

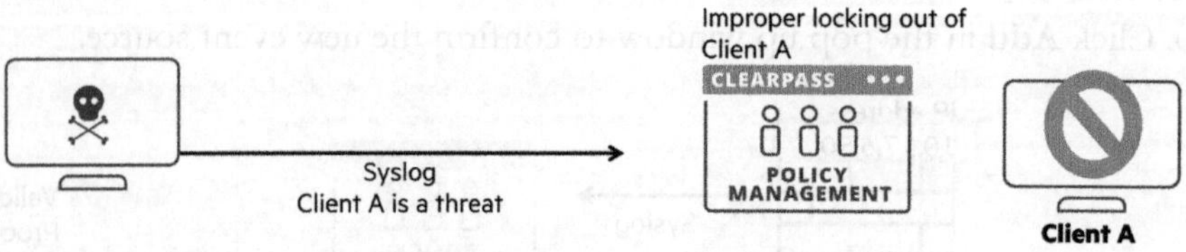

Figure 11-12: Lockout Attack

To prevent both of these attacks, you need to take steps within the network infrastructure to prevent IP spoofing. You learned about such measures in a previous chapter.

Ingress Event Dictionaries

While Syslog severity and facility have defined values, the message content itself is simply a string, meant to be human readable. Because the content has no fixed fields or attribute value pairs (AVPs), interpreting logs can pose a challenge for automation. How can you automate interpreting the message and triggering a response based on the content?

This is where CPPM's Event Ingress Dictionaries come in. These dictionaries enable CPPM to interpret Syslog messages from supported event sources and parse them into events with AVPs, as shown in Figure 11-13. You can then use the known attribute names to create profiles and policies to automate responses.

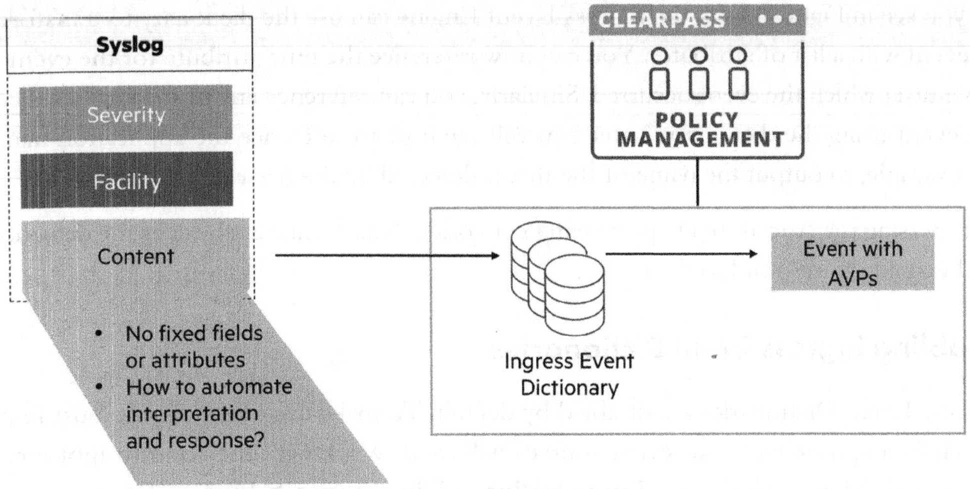

Figure 11-13: Ingress Event Dictionaries

How the Ingress Event Dictionary Works

When CPPM receives a Syslog message from an event source, its Ingress Event Engine finds an Ingress Event Dictionary for that source type that can parse the message. The dictionary breaks the message into pieces and defines an attribute with which each piece is associated.

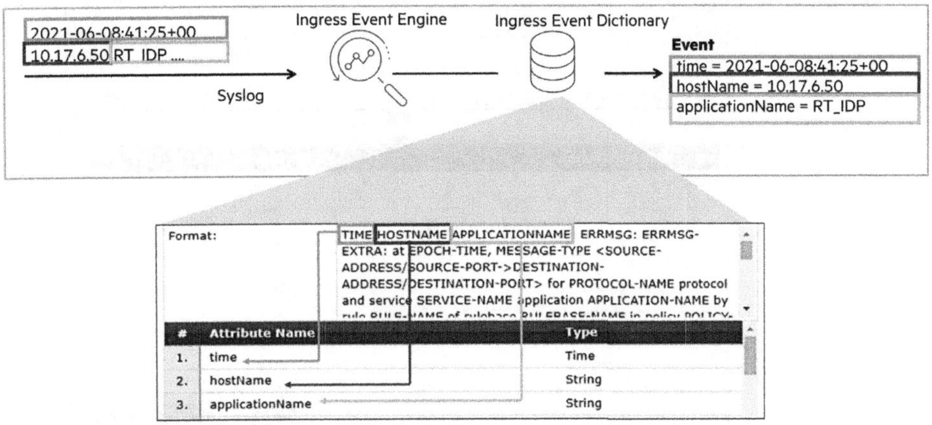

Figure 11-14: How the Ingress Event Dictionary Works

For example, the dictionary shown in Figure 11-14 indicates that the first element in a particular type of log is the time. The second element is the hostname of the firewall. And the third element is the application name.

As you see in Figure 11-14, the Ingress Event Engine can use the dictionary to transform a log into an event with a list of attributes. You can now reference the time attribute for the event and receive the time at which the event occurred. Similarly, you can reference any of the attributes compiled for the event using the dictionary. Later you will see how to reference the application-name attribute, for example, to output the name of the threat detected by the firewall.

This parsing is a crucial step in processing network events, because it bridges the detection of events to the ability to respond to them.

Enabling Ingress Event Dictionaries

Ingress Event Dictionaries are disabled by default. To make use of them, you must first determine which Syslog messages your event sources will send. Work with the security appliance admins to coordinate. Then navigate to **Administration > Dictionaries > Ingress Events**.

There you can find the dictionaries associated with your appliance vendor and enable them, as shown in Figure 11-15.

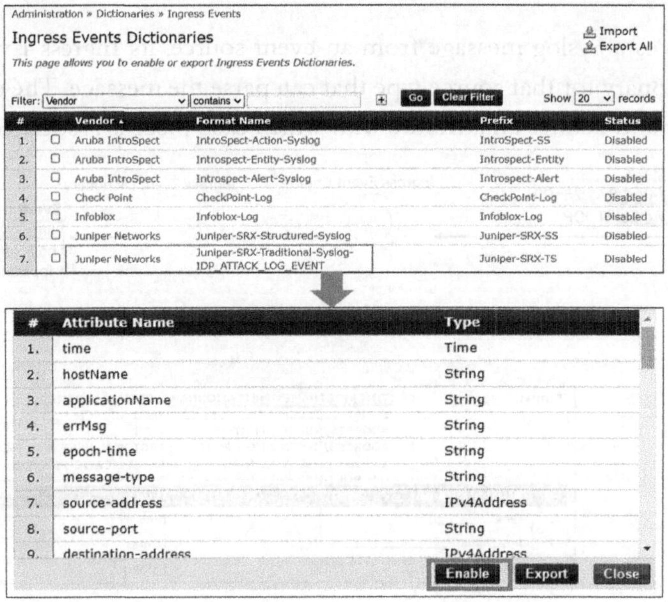

Figure 11-15: Enabling Ingress Event Dictionaries

Using Ingress Events to Respond to Threats

Next, you will learn how to use ingress events to respond to threats.

Overview of Using Events to Mitigate Threats

Once the Ingress Event Engine has generated an event based on the Syslog message, CPPM can use that event to perform actions. You will now look at an example of this process in more detail.

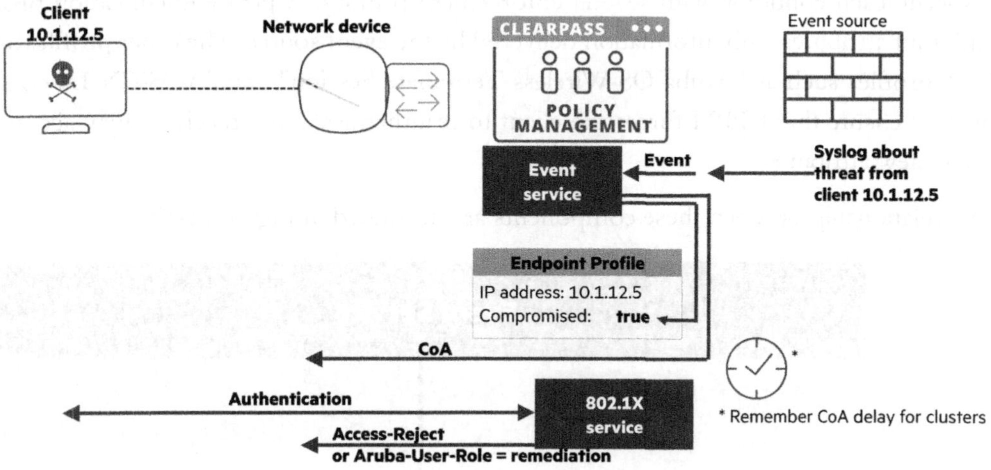

Figure 11-16: Overview of Using Events to Mitigate Threats

As shown in Figure 11-16, when CPPM receives an event from an event source, it processes the event with an event service. This service includes an event policy with a rule that matches an attribute in the event. This rule applies a post-authentication profile to the client in question. In this example, the post-authentication profile updates the client's endpoint profile on CPPM, marking the client as being compromised.

The policy rule also applies a second profile, which terminates the client session. CPPM sends a change of authorization (CoA) to the network device to which the client is connected. The network device then disconnects the client, and the client is forced to re-authenticate.

The client reauthenticates against an 802.1X service. This service has an enforcement policy with a rule that selects compromised clients. This rule denies the client access. Or it might apply a profile that sets the Aruba-User-Role to a remediation role set up in the network infrastructure. That remediation role would provide very limited access.

Remember the importance of tuning the CoA delay if you have a cluster deployment, as you learned when enforcing posture and health checks. One ClearPass Subscriber might process the event and issue the CoA while another Subscriber reauthenticates the client. Make sure that the first Subscriber delays the CoA long enough for the endpoint's attributes to be updated across the cluster.

Event Service Configuration Overview

To set up CPPM to function as just discussed, you must configure an event service, which references an event enforcement policy. The event enforcement policy matches attributes in an event with the enforcement profiles that CPPM should apply when that event occurs. Typically, you should associate each condition with several enforcement profiles. A post-authentication profile updates endpoint attributes with information delivered by the event source. The other profiles are RADIUS CoA profiles such as "Aruba OS-Wireless Terminate Session" and "AOS-CX Disconnect;" these profiles ensure that CPPM forces the client to reauthenticate and receive a new access level based on its new attributes.

The relationship between these components are illustrated in Figure 11-17.

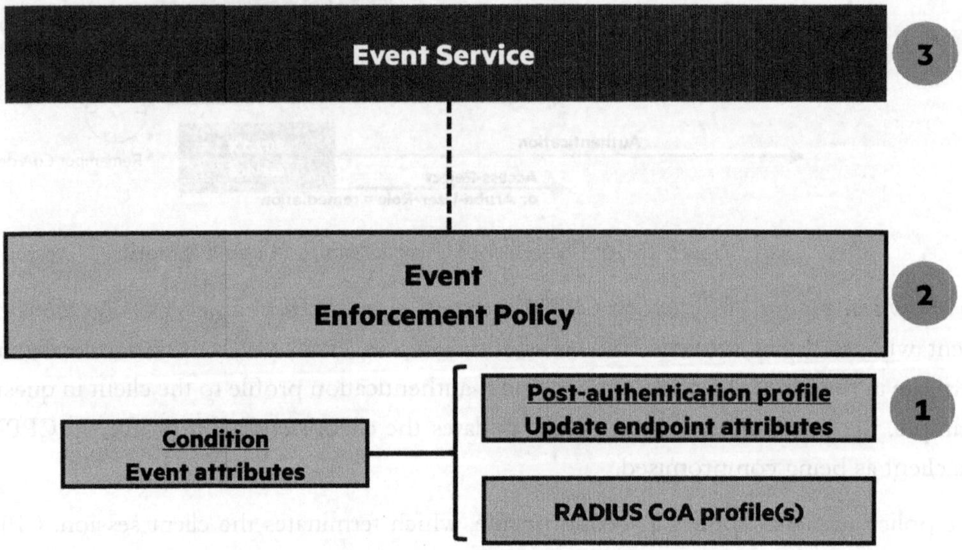

Figure 11-17: Event Service Configuration Overview

Begin by creating the post-authentication profiles. Then create the event enforcement policy, which maps events to the profiles to apply. Finally, create an event service and reference the event enforcement policy in that service.

Defining a Post-Authentication Enforcement Profile to Update Endpoint Attributes

Now you will look at the enforcement profile that CPPM will use to update the endpoint profile.

In this study guide so far, you have concentrated primarily on RADIUS and TACACS+ enforcement profiles, which CPPM applies while enforcing 802.1X or other authentication services. A post-authentication profile, on the other hand, applies to clients after they have authenticated.

CPPM supports several kinds of post-authentication profiles. For this scenario, you want to update a client's endpoint attributes based on AVPs within the event being processed by the event service. The correct template is "ClearPass Entity Update Enforcement," which updates the information and attributes in an endpoint's profile. In Figure 11-8, you see the proper settings.

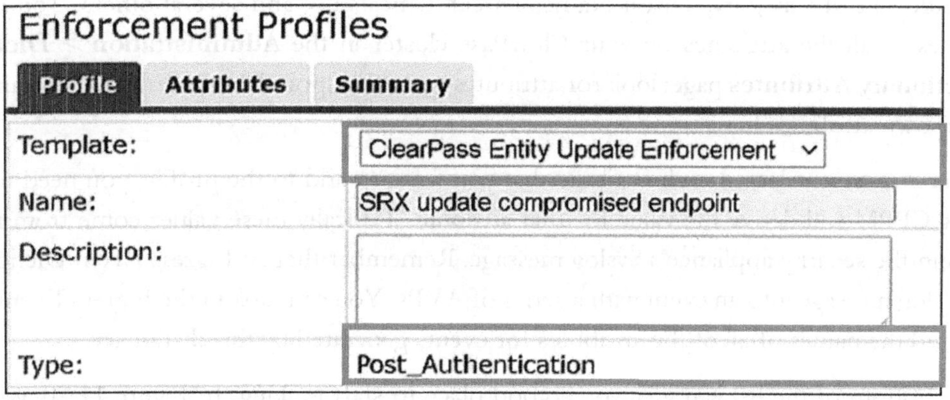

Figure 11-18: ClearPass Entity Update Enforcement Profile

Click the **Attributes** tab to configure the attributes that you want to add to the endpoint profile. For this scenario, you want to add the "Compromised" attribute and set it to true. As shown in Figure 11-19, Set the Type to **Endpoint** and the Name to **Compromised**. Then specify the value. For "true"/"false" values, selecting the checkbox sets the value to **true**. Click **Save** once you finish.

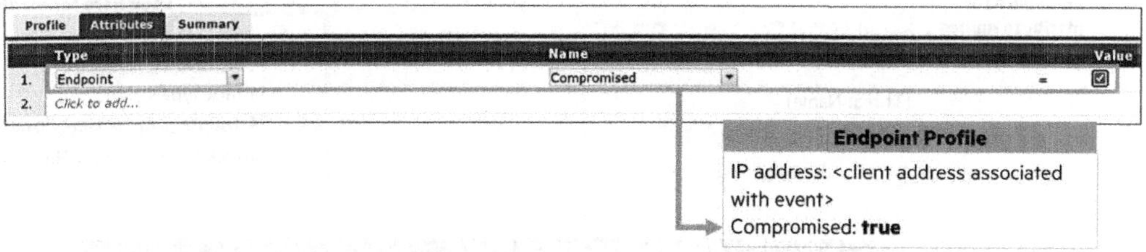

Figure 11-19: Attributes for a ClearPass Entity Update Enforcement Profile

CPPM supports several other types of post-authentication profiles as well, including:

- Session Notification Enforcement—This type of profile sends information to an external context server, such as a firewall.
- Session Restrictions Enforcement—This type of profile enables CPPM to enforce restrictions on the resources that clients consume.

Gathering Additional Attributes for the Post-Authentication Profile

Often you will also want to record additional information about the threat in the endpoint profile. This information is also recorded in endpoint attributes.

Valid attributes for endpoint profiles are defined in CPPM's Dictionary Attributes. Examples include Threat Category, Threat Name, Threat Timestamp, and several others. You can find the names of all the attributes on your ClearPass cluster in the **Administration > Dictionaries > Dictionary Attributes** page; look for attributes with "Endpoint" as the Entity. You can also define your own attributes here.

After you have selected each attribute that you want to add to the profile, you need to determine how CPPM will derive the value for that attribute. Typically these values come from information within the security appliance's Syslog message. Remember that an Ingress Event Dictionary parses a Syslog message into an event with a series of AVPs. You can look in the Ingress Event Dictionary to find the names of all of the attributes for events generated by this dictionary.

The bottom of the list can give you a good place to start looking. In Figure 11-20, you see that a few of the attributes are additionally mapped to "Generic Name" attributes, which are used by multiple event types. For example, a Juniper-SRX-TS Ingress Event Dictionary creates a "Juniper-SRX-TS:attack-type" AVP from the Syslog message and also creates a generic "Threat-Name" AVP with the same value.

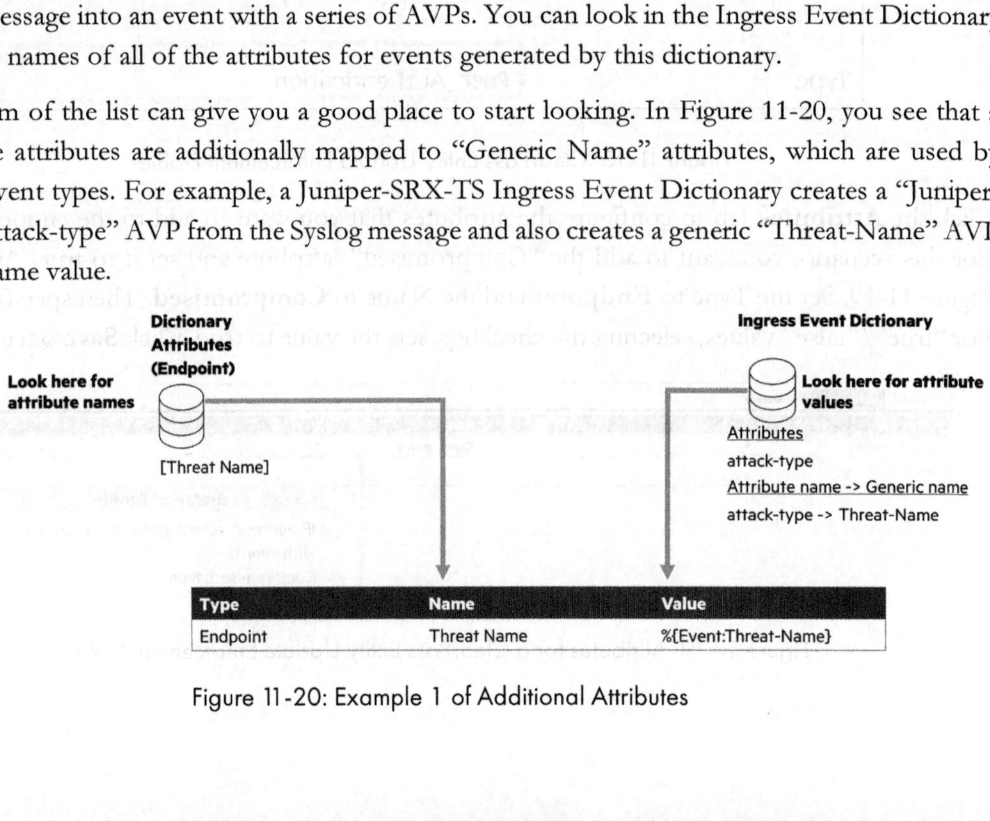

Figure 11-20: Example 1 of Additional Attributes

Now you can put this information together to define an attribute for the post-authentication profile. Say that you want to add the Threat Name attribute to an endpoint profile, as triggered by Juniper-SRX-TS events. Within the post-authentication enforcement profile, set the attribute type to **Endpoint** and the Name to **Threat Name**. For the value, specify a variable: **%{Event:Threat-Name}**

You can also reference an event's non-generic attributes in variables. In this case, you include the prefix in the variable. The format is: **%{Event:prefix:attribute}**.

Sometimes it might not be obvious to you which non-generic attributes for an event correspond to which type of information. You can meet with security appliance admins to get help. For example, you could tell the admins that you want to extract the threat category from a log from a Juniper SRX firewall. The admins could point out the piece of the log that provides this information. You can then look in the corresponding Ingress Event Dictionary's Format section to find the attribute that corresponds with the pinpointed piece of the log. In the example shown in Figure 11-21, that is the "message-type" attribute.

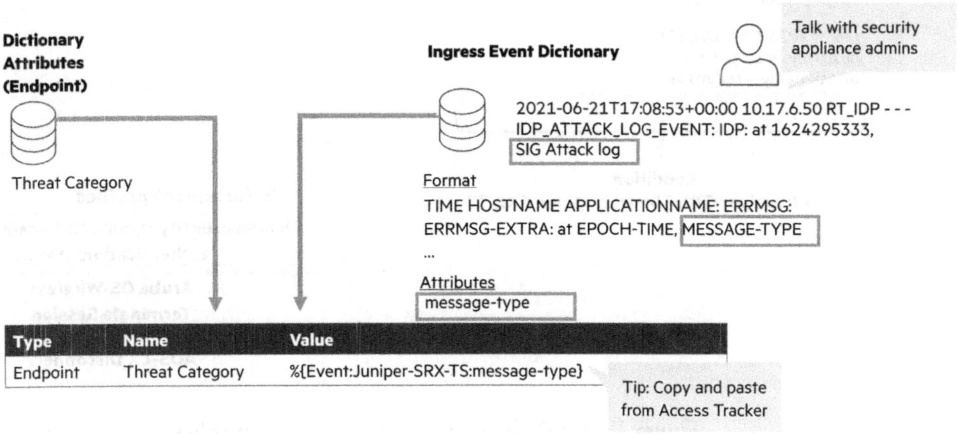

Figure 11-21: Example 2 of Additional Attributes

Therefore, when you define the Endpoint Threat Category attribute in the post-authentication profile, you use this value: **%{Event:Juniper-SRX-TS:message-type}**.

You can also work with security admins to perform test runs of the solution. Access Tracker will show the event processed by the event service. You can look in the Access Tracker record to find the event attributes. Then you can copy the attribute name from there to avoid making mistakes.

Defining an Event Enforcement Policy

You next need to create an event enforcement policy in which you define the rules to trigger the post-authentication profiles. Navigate to **Configuration > Enforcement > Policies** and add a new policy. Set the Enforcement Type to **Event**.

Then click the **Rules** tab. Here you will define the condition that the event must match in order for CPPM to apply the post-authentication profile.

For this scenario, you want your post-authentication profile to apply whenever CPPM receives a particular type of Syslog message. So, as illustrated in Figure 11-22, you create a condition that sets the Type to **Event**: and the prefix associated with the Ingress Event dictionary that you enabled. Because you want the event to trigger the rule for any Syslog message of this type, you simply choose an attribute that this type of message always produces and specify that this attribute must exist. In this example, you use the "attack-name" attribute.

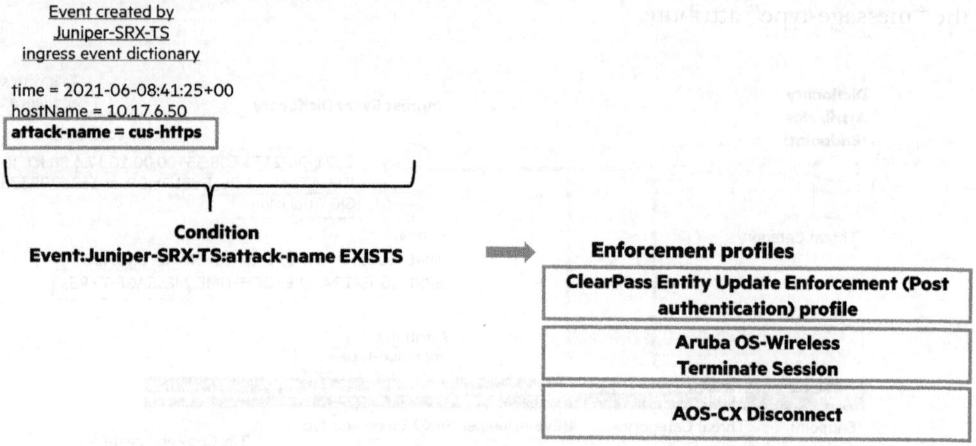

Figure 11-22: Defining an Event Enforcement Policy

Then select your post-authentication profile. Also select profiles for terminating wireless and wired sessions. Remember that you want CPPM to terminate the session so that it can take the endpoint's compromised status into account when the endpoint authenticates again.

If you wanted to create more complex policies, in which you apply different actions based on the threat type, you could define multiple rules. Each rule would have a condition that specifies a different value for "attack-name" and a different post-authentication profile. However, for this scenario, you want to apply the same actions for all events of this type, so one rule suffices.

Defining an Event Service

An event service processes events as they occur. You can create multiple event services with service rules for processing different types of events, or for processing events from different sources. For this example, though, you are creating one service, which processes all events.

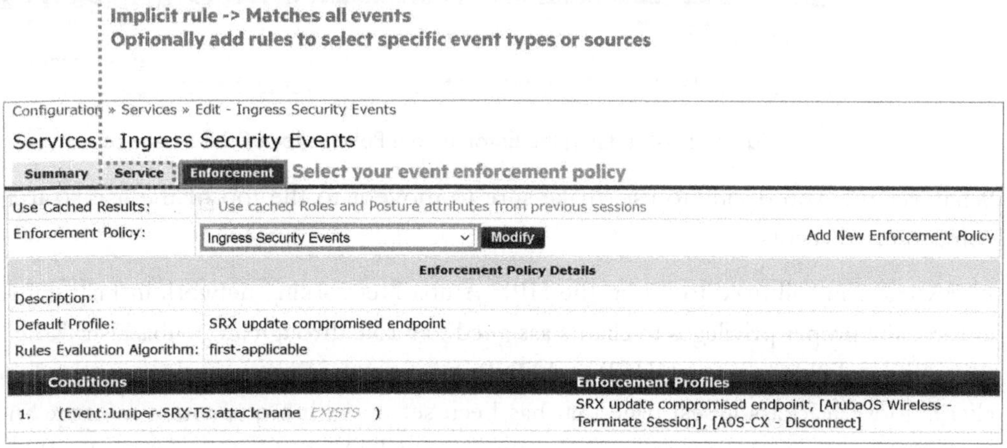

Figure 11-23: Defining an Event Service

You then click the **Enforcement** tab and select your event enforcement policy as shown in Figure 11-23.

Editing the Enforcement Policies for 802.1X Services

When CPPM terminates a client's session due to a security event, the client will try to re-authenticate. To ensure that CPPM responds correctly based on the client's compromised status, you need to edit the enforcement policy that you are using for authenticating clients. For example, you could create a rule that selects endpoints with a Compromised attribute set to "true" and applies a profile that sets Aruba-User-Role to a value like "remediation." In the example shown in Figure 11-24, that rule also specifies a second enforcement profile, which tells the Network Access Server (NAS) the client's username if the client uses TEAP to authenticate.

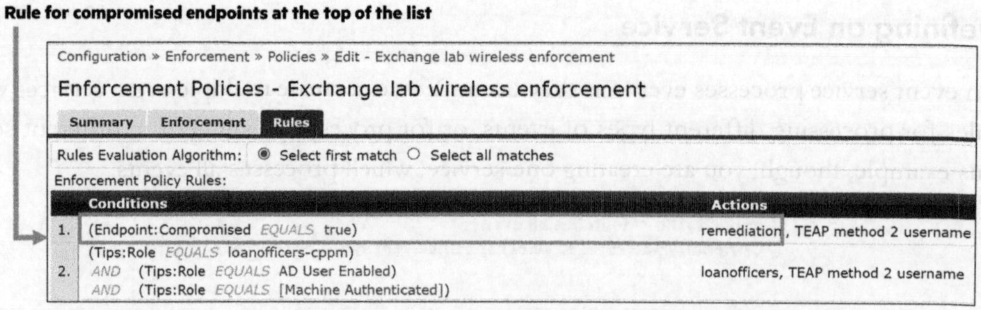

Figure 11-24: Editing the Enforcement Policies for 802.1X Services

Whatever rule you decide to use, make sure to move it to the top of the list, so it applies to all compromised clients.

As always, you will need to set up the HPE Aruba Networking network infrastructure devices to enforce the proper privileges to clients assigned to a role through an Aruba-User-Role VSA. Often the rights include access to DHCP and DNS. You can then use a captive portal rule to direct the clients to a ClearPass Guest page that has been set up with a quarantine message and links. For example, the page might link users to a site at which the users can update the client's antivirus application. The "remediation" role would also need to permit access to those links.

Summary

Congratulations! You can now integrate CPPM with third-party solutions. You began this chapter by learning about the CPPM Ingress Event Engine. You saw how the Ingress Event Engine enables CPPM to process Syslog messages from security appliances and generate events from them.

Then you learned about how to use the events to respond to threats. You took a high-level look at the solution before learning in greater detail about each of its components and how to configure those components.

Learning checks

1. What is an event source on CPPM?

   a. An MDM from which CPPM receives information about endpoint posture

   b. A security appliance to which CPPM provides user information

   c. A device from which CPPM will accept and process Syslog messages

   d. An endpoint on which malware or another threat has been detected

2. What purpose does the Ingress Event Dictionary serve?

   a. It lists the valid event sources from which CPPM is allowed to process events.

   b. It transforms Syslog content into an event with attributes that can be referenced.

   c. It specifies the actions that CPPM should take in response to an event.

   d. It defines the Syslog message that CPPM should send in response to specific requests.

3. Match the definition to the component.

   Definitions

   a. Defines which attributes to update in the endpoint profile

   b. Defines the condition that triggers applying the post-authentication profile

   c. Processes events

   Components

   __ Event service

   __ Event enforcement policy

   __ ClearPass Entity Update Enforcement profile

Note

For answers to all learning checks, refer to the Appendix.

---

12 Implement Device Profiling with CPPM

EXAM OBJECTIVES

- ✓ Explain benefits and methods of endpoint classification on HPE Aruba Networking ClearPass Policy Manager (CPPM), including active and passive methods
- ✓ Deploy and apply endpoint classification to devices
- ✓ Analyze endpoint classification data on CPPM to identify risks

Assumed Knowledge

- Fundamental CPPM components, such as services, roles, enforcement profiles, and enforcement policies
- Definition of device profiling and basic understanding of profiling methods
- Authentication, Authorization, and Accounting (AAA) framework and basics of protocols such as RADIUS and TACACS+
- Forms of Layer 2 authentication such as 802.1X

Overview

CPPM can collect information about endpoints and use that information to profile the endpoint type. CPPM can then use that information as part of the rich context that it uses to make access decisions.

In this chapter, you will review what you learned about endpoint classification in prerequisite training. You will then learn more about how to set up device profiling on CPPM using several types of collectors and both passive and active methods.

Next, you will look at ways to analyze and use the device information.

You will then learn how to use device profiling information within CPPM services so that you can assign different access levels to users on different types of devices. You will also learn how you can use device profiling and Conflict attributes to mitigate threats.

Review Endpoint Classification

In the first section, you will review endpoint classification.

Why Profile Endpoints?

Since the modern network hosts so many types of devices, you need a way to classify devices based on their type and place different device types into different security roles.

In a typical network a single user may bring a corporate issued laptop, smart phone, and tablet or lightweight device (such as a chrome book) onto the network. It is not uncommon for a single user to have between three and five devices on the business network. More than five devices is common in settings like universities. These devices often do not serve the same functions on the network, and therefore don't need the same security access. A Zero Trust Security (ZTS) strategy dictates that you grant access to each device based on the needs of users on that device. You should also consider the vulnerability of the device type. For example, smartphones might not be as secure as corporate-managed laptops. Providing users on smartphones the same level of access on a smartphone as on a laptop could potentially expose the network to threats.

Device profiling is also useful for detecting specialized devices such as IoT devices. CPPM can then take the device type into account when determining the device's access level. Because IoT devices often support lower levels of security, you want to limit their access and the extent of damage that a compromised device can do.

Collecting information about device types can also be helpful from a management standpoint. IT admins then have greater insight into the types of devices they need to support. This can be helpful when a particular device proves to be problematic on the network or requires a patch or upgrade. The administrator can look and quickly begin to assess the impact of a change to that type of device.

Using Device Profiling to Control Devices

Throughout this study guide, you have learned about using 802.1X authentication and a variety of contextual information to implement granular access control policies for users on various devices.

Now consider the broad range of devices available today and how companies might want to apply access control policies to them.

Many companies have many of these devices, including building access systems, badge scanners, multi-function printers, cameras and security systems, VoIP phones, and even vending machines. (See Figure 12-1 for just a few of the many categories of devices.) The network needs to support and control IoT devices based on their type. Most of these devices are not actually user inhabited, but require access to the network to do their job. You need to consider how best to implement different access controls for each type.

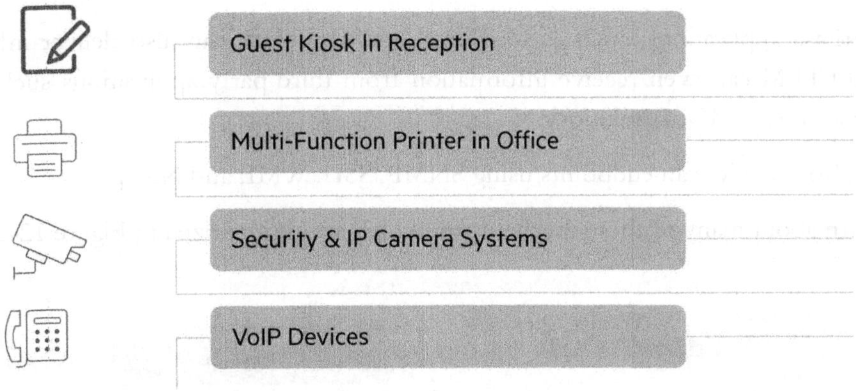

Figure 12-1: Using Device Profiling to Control Devices

One traditional method could be to create a different network for each device type and configure the device to connect to its own network. However, this approach is incredibly inefficient. Imagine how many wireless networks you would need to support just the short list in this example.

The better solution is to configure a single wireless network. CPPM and, as you will learn later, HPE Aruba Networking ClearPass Device Insight (CPDI) implement profiling to dynamically discover the type for each device. CPPM can then use role mapping and enforcement policies to apply the correct profile to each client. As you have learned, these profiles often place devices in an AOS or AOS-CX role, and the HPE Aruba Networking infrastructure devices will then enforce the correct role-based policies for the client.

Consideration questions

Consider the following questions, which you should answer based on your associate-level knowledge of security. are provided for your reference in the appendix.

1. What is the difference between passive and active profiling?
2. How can DHCP packets help CPPM determine the device type?

3. What useful information can CPPM find in clients' HTTP/HTTPS traffic?

CPPM Collectors

CPPM collects information about endpoints from a variety of sources, or collectors. It can listen in on DHCP and HTTP/HTTPS traffic passively, and it can also collect clients' MAC OUIs from this traffic.

CPPM can also receive traffic on a SPAN port, although this feature is less commonly used, or by receiving NetFlow or sFlow traffic samples from network infrastructure devices.

Other ClearPass applications, such as OnGuard and Onboard, can also deliver information to CPPM, and CPPM can even receive information from third-party applications such as Mobility Device Management (MDM) solutions.

CPPM can also actively scan endpoints using SNMP, SSH, WMI, and Nmap.

You will learn about many of these mechanisms, which are summarized in Figure 12-2, in the next section.

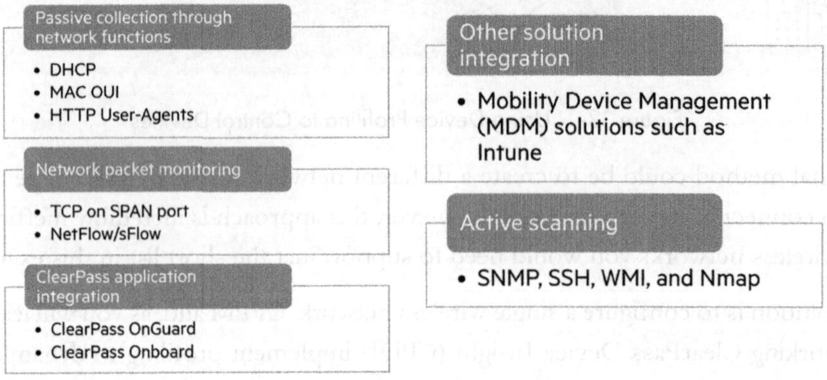

Figure 12-2: CPPM Collectors

Set Up Device Profiling on CPPM

In this section, you will learn how to set up device profiling on CPPM.

Review the Endpoints Repository

As shown in Figure 12-3, whenever CPPM learns about a client device on the network, it will create an entry for it in the Endpoints Repository and attempt to profile the device.

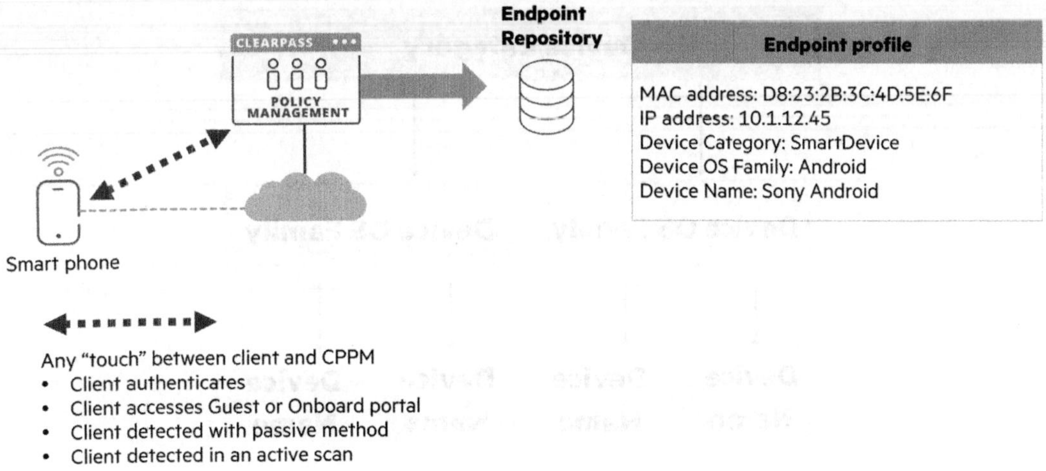

Figure 12-3: Review the Endpoints Repository

CPPM can learn about a client in several ways. The client might authenticate. Or it might access the ClearPass Guest or Onboard portal. If you have set up CPPM as an IP helper for DHCP relay, then CPPM will learn about all devices that obtain DHCP addresses. You can also set up active scans, about which you will learn more later. By using a combination of active and passive profiling methods, CPPM can build a complete list of every device on your network.

To view the endpoints in CPPM, navigate to **Menu > Configuration > Identities > Endpoints**.

Endpoint Profile

In more detail, as shown in Figure 12-4, an endpoint profile includes a hierarchy of information about the device:

- Device Category, which identifies the device type such as Computer, Smart Device, or Printer
- Device Family, which further classifies the device by OS or vendor; for example, a Computer might be in the Windows, Linux, or Mac OS family
- Device Name, which further segments the family; for example, an Apple Mac computer might have Mac OS X for its Device Name

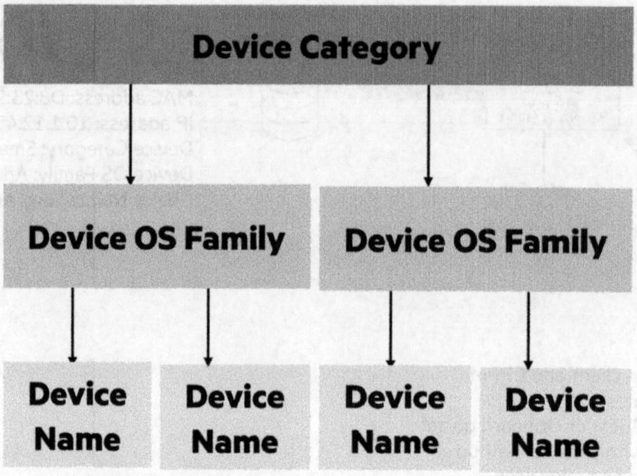

Figure 12-4: Device Category Hierarchy

Other information in the profile includes:

- IP address
- Hostname
- MAC Vendor
- Timestamp when the device was first discovered
- Timestamp when the device was last seen

An example endpoint profile is shown in Figure 12-5.

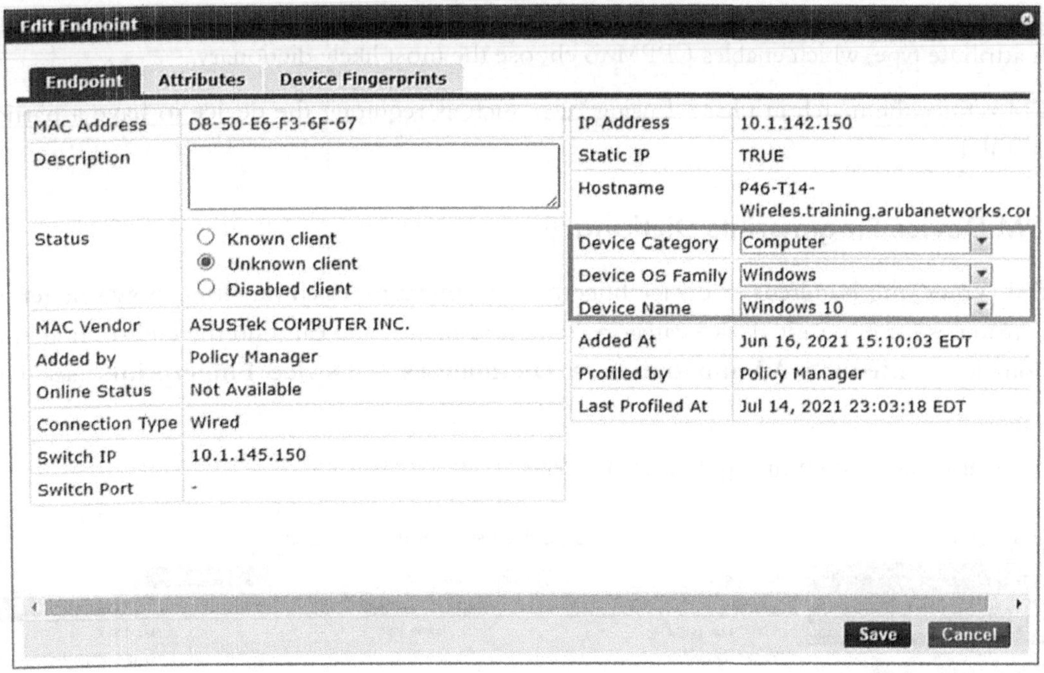

Figure 12-5: Endpoint Profile

Overview of How CPPM Profiles Devices

To categorize endpoints, CPPM collects information as attributes that it knows about the device. It then profiles the device in two phases, as shown in Figure 12-6.

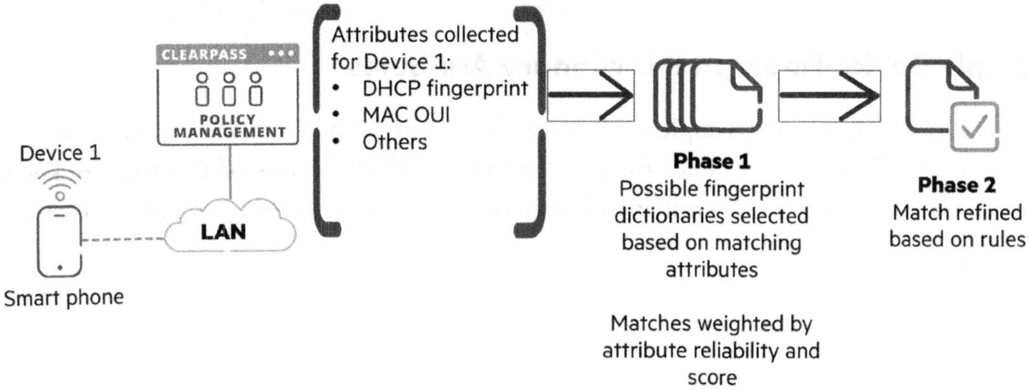

Figure 12-6: Overview of How CPPM Profiles Devices

In the first phase, CPPM matches attributes against device fingerprint dictionaries. These dictionaries list multiple attributes characteristic of devices of a particular type. Sometimes the device attributes might point in different directions. CPPM also assigns a reliability and score to each attribute type, which enables CPPM to choose the most likely dictionary.

CPPM refines the match in phase 2 using rules such as requiring the device to have a particular MAC OUI.

CPPM Device Fingerprints Dictionary

CPPM comes with hundreds of device fingerprint dictionaries. Each one lists known fingerprints and attributes for a specific device category, family, or name. You can find the built-in fingerprints dictionaries in **Menu > Administration > Dictionaries > Device Fingerprints** (see Figure 12-7).

You can also add custom fingerprints at this page.

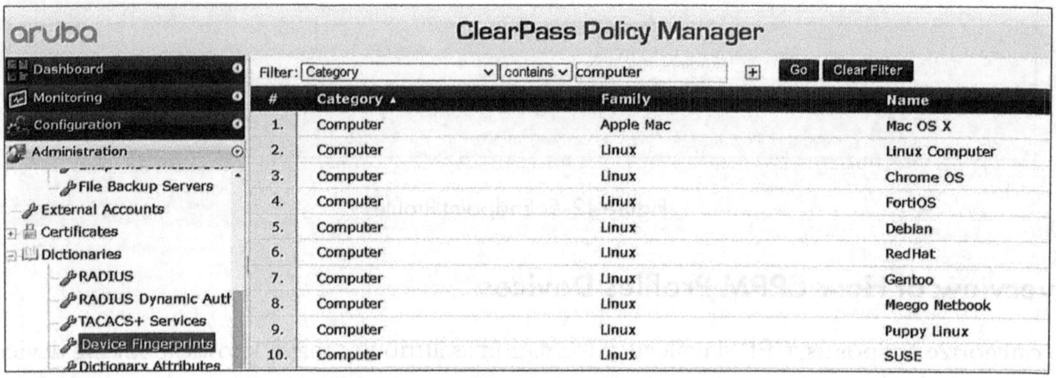

Figure 12-7: CPPM Device Fingerprints Dictionary

Example Device Fingerprint Dictionary Attributes

In Figure 12-8, you see an example of a device fingerprint dictionary. As you see, it lists several attributes, as well as a list of values for those attributes, which devices of this type are known to have. These attributes act as identifying characteristics by which CPPM can match a client to the dictionary.

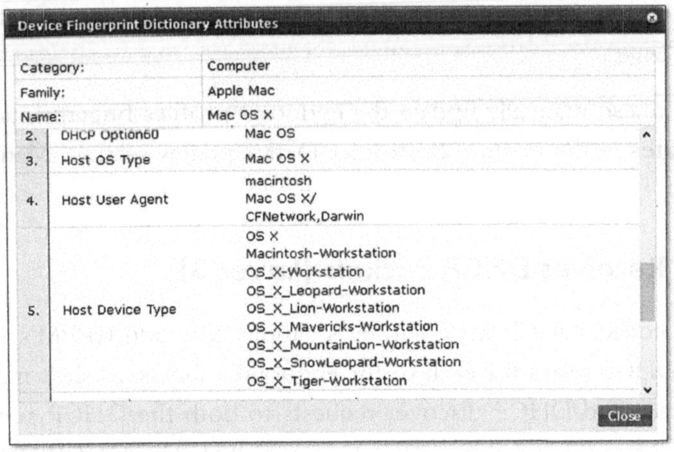

Figure 12-8: Example Device Fingerprint Dictionary Attributes

As indicated previously, the dictionary might have rules, as well, which enable CPPM to refine a match.

Some devices, like this Mac OS X computer, have a large number of attributes and rules while other devices have a smaller list.

Updating Fingerprints Separate from the Software

When you update the ClearPass software, CPPM receives the latest set of device fingerprints. However, HPE Aruba Networking also makes new device fingerprints available separate from a complete software update. Using these updates, you can enable CPPM to profile new types of devices more accurately without waiting for a software upgrade.

If you want your servers to receive these updates automatically, you can enable the **Automatically download Endpoint Profiler Fingerprints** setting in your standalone server's or cluster's cluster-wide parameters, as shown in Figure 12-9. Click **Cluster-wide parameters** in the **Administration > Server Manager > Server Configuration** page. Then find the setting in the **General** tab.

Figure 12-9: Updating Fingerprints Separate from the Software

You can check the status of updates in the **Administration > Agents and Software Updates > Software Updates** page.

In that page, you can also manually update the endpoint profiles fingerprints by uploading a file. Click **Import Updates** in the Posture & Profiler Data Updates section. Manual installs work for offline servers.

Ensuring CPPM Receives DHCP Packets (Layer 3)

Enabling CPPM to collect DHCP information is simple. You add CPPM's IP address to the IP helper list on your DHCP relays for each of the endpoint subnets, as shown in Figure 12-10. The DHCP relay will then send DHCP discover requests to both the DHCP server and CPPM, and CPPM can use those requests to obtain DHCP fingerprints. The DHCP traffic will also contain MAC addresses for the client, so CPPM can learn the MAC OUI.

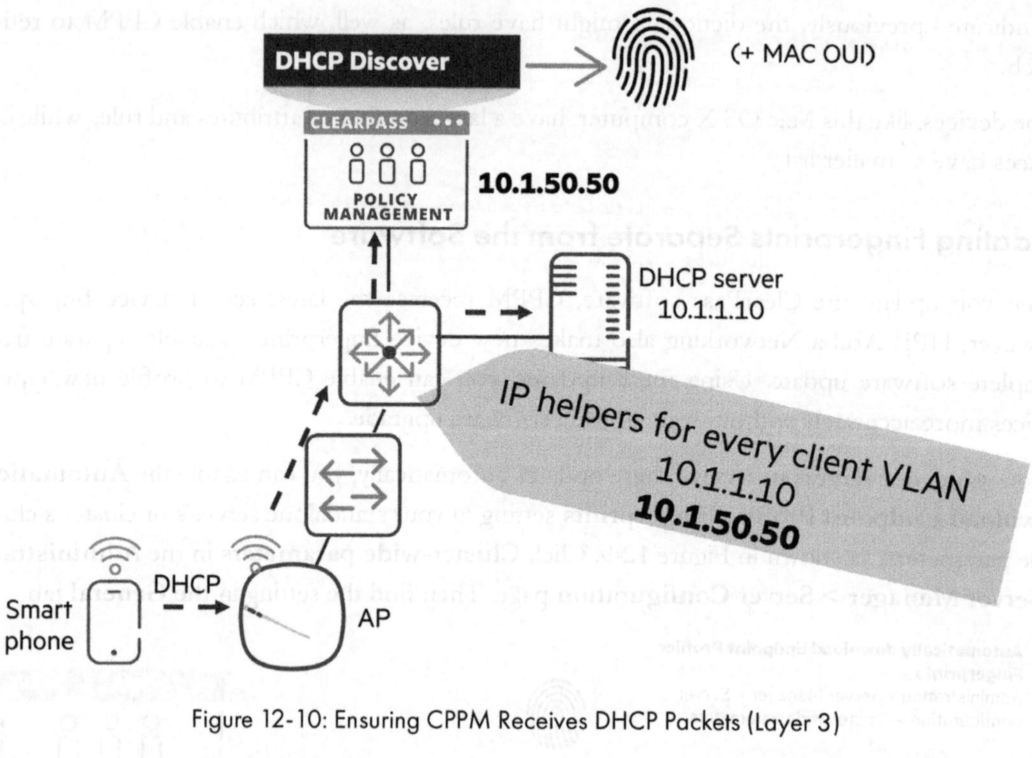

Figure 12-10: Ensuring CPPM Receives DHCP Packets (Layer 3)

CPPM can also learn the MAC OUI from the authentication process if clients authenticate to it.

Note that CPPM profiles devices with DHCP Discovers. In typical networks, clients come up and send out Discovers to obtain their IP addresses. However, if you are demonstrating CPPM's profiling feature, you might be templated to simply renew DHCP addresses, which causes the client to send out a DHCP Request instead of a Discover. For the best results, when you want to demonstrate re-profiling a device, you should fully shut down the device's connection and then bring the connection back up.

Ensuring CPPM Receives DHCP Packets (Layer 2)

Occasionally a company might have some endpoints that receive DHCP addresses from a server on the same subnet without a DHCP relay being involved, as shown in Figure 12-11. You can still enable CPPM to receive the DHCP traffic by adding IP addresses to the VLAN in question on the Layer 2 switch or switches that support that VLAN. in that subnet. Then specify the DHCP server and CPPM as helper addresses in that VLAN.

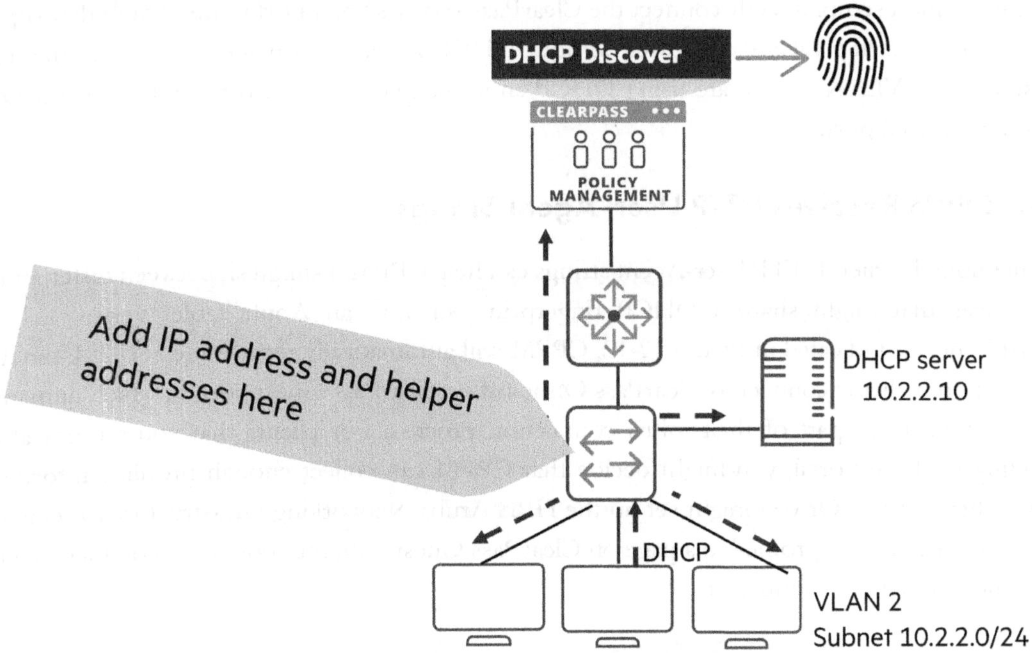

Figure 12-11: Ensuring CPPM Receives DHCP Packets (Layer 2)

You can alternatively connect the switch in question to CPPM's Span port. The sections below describe that alternative configuration.

Span Port Configuration on ClearPass Server

First you need to select the port to act as the Span Port.

If you are using only the Management Port on your ClearPass server, you can use the Data Port as the Span Port. Otherwise, you must use one of the unused ports on the appliance. If you are using a VM-based appliance, you must add additional virtual NICs to the VM.

To enable the Span Port, you must go to CPPM's **Administration > Server Manager > Server Configuration** page. Select the ClearPass server in question. In the **System** tab, select the port that you choose as the Span Port. Make sure to leave the **Enable TCP/ARP Fingerprinting** check box unselected. HPE Aruba Networking generally recommends against using this feature because it is processor intensive.

Switch Configuration

For the simplest setup, directly connect the ClearPass server's Span Port to the switch that supports the devices receiving DHCP addresses without DHCP relay. Then make that port a member of those devices' VLAN. If you are using DHCP snooping, you must also make this port a DHCP snooping trusted port.

How CPPM Receives HTTP User-Agent Strings

As mentioned earlier, HTTP User-Agent strings can help CPPM distinguish between different types of devices that might share a DHCP fingerprint, such as an Apple tablet versus an Apple smartphone. As indicated in Figure 12-12, CPPM will automatically receive the HTTP User-Agent information if clients connect to ClearPass Onboard or ClearPass Guest portals, which unmanaged clients often do as part of their initial connection process. For clients that you are not already directing to those portal, you might decide that CPPM can collect enough profiling information from other sources. Or you might set up the HPE Aruba Networking infrastructure to redirect all clients that need to be profiled to a page on ClearPass Guest until the profiling is complete; you will learn more about that setup later.

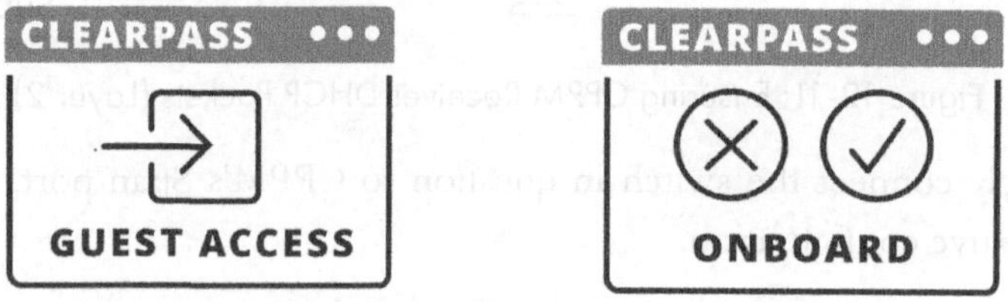

Figure 12-12: How CPPM Receives HTTP User-Agent Strings

Use Cases for Network/Subnet Scanning

You will now learn about setting up active scanning on CPPM using network or subnet scans.

Active scanning is also useful for detecting and profiling clients that have static IP addresses, so CPPM receives no DHCP packets from them.

When CPPM uses WMI or SSH as part of the scan, it can also collect more information about endpoints with both dynamic and static IP addresses. It can learn which ports and services are enabled on the client and other such information.

You can specifically use network scans, sometimes called discovery scans, to discover network devices and then add them as RADIUS/TACACS+ clients on CPPM. Some companies like this approach for quickly discovering network devices when they first deploy CPPM. Other companies prefer methods such as adding the devices manually or importing XML files.

Scan Methods

CPPM supports three methods for scanning devices, as summarized in Figure 12-13.

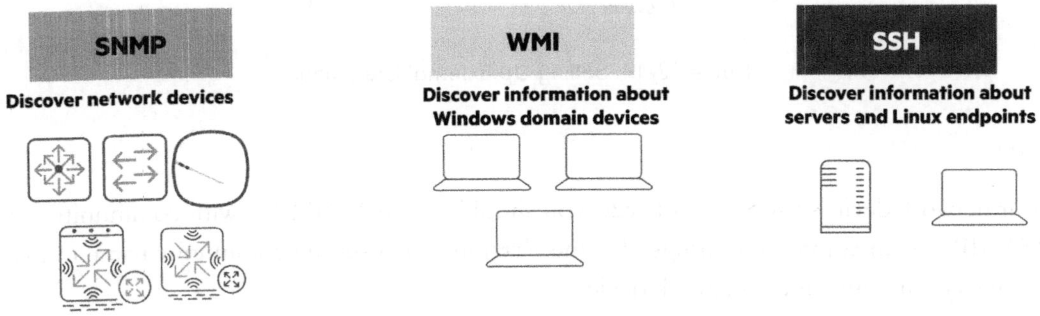

Figure 12-13: Scan Methods

It uses SNMP to discover network devices. Specifically, it collects each device's system name, system location, system contact, and system description. As you will learn in more detail later, CPPM also uses SNMP to collect various information about the device's neighbors so that it can expand the scan.

CPPM uses WMI to discover information about Windows domain computers.

And it uses SSH to discover information about servers and Linux endpoints.

Setting Up External Credentials

To prepare for active scans, you must set up the external credentials that CPPM will use to perform the scan.

To create an external account, navigate to **Administration > External Accounts**. Then click **Add** to create a single account or **Import** to import many accounts from a spreadsheet.

For each account, you specify the type as SNMP, SSH, or Domain/WMI. You can create multiple accounts of each type to hold the specific credentials for different types of devices or devices in different locations, as shown in Figure 12-14.

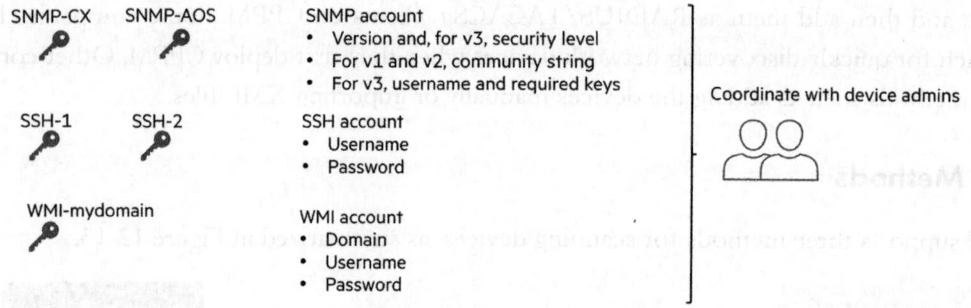

Figure 12-14: Setting Up External Credentials

SNMP

If your network devices use SNMPv1/v2c, you should specify "SNMPv1 with community strings" or "SNMPv2 with community strings" for the Version. Then set the community name to match a read-only community on the network device.

If the network devices require SNMPv3, you must choose of the SNMPv3 options, matching the requirements and algorithms with those used on the network device. The network device admins should tell you the credentials, security levels, and algorithms used for an account that has read-only (or read-write) access. Under the version, you specify whether authentication is required, the algorithm used for authentication if required, and whether privacy is required in addition to authentication. If you select privacy, you choose the privacy protocol in a separate field. After choosing an SNMPv3 version, you specify the username and keys for the read-only account. You will also see key fields for the particular types of security that are required. (In other words, if you choose a version that requires authentication, you must set an authentication key. If the version further requires privacy, you must set a privacy key.) The network admins should tell you the key values as part of the credentials for the account.

SSH

When you choose "SSH" for the Type, you must specify a valid username and password for SSH access to the device.

Domain/WMI

For "Domain/WMI" accounts, you specify the domain name, the username, and the password. CPPM will use this format when logging into a device with these credentials: domain_name\username. Choose a user account that has administrative access to the domain computers.

Setting Up Profiler Subnet Mappings

After you create the external accounts, you must map them to the subnets that contain devices using those accounts. Still within the **Administration > External Accounts** page, select **Profiler Subnet Mappings**.

You can then add multiple mappings. For each mapping, define one or more subnets. Then choose the account type. You can then select one or more accounts of that type.

CPPM uses these mappings when it needs to scan devices. The mappings determine which protocols CPPM can use for scanning devices within that subnet. In the example shown in Figure 12-15, CPPM cannot use WMI to scan an IP address at 10.2.4.4 because it has no WMI credentials mapped to that subnet. Instead CPPM would use SSH to scan this device. CPPM can use WMI to scan 10.1.11.28, however, because it has a WMI mapping that applies to this device's subnet. More specifically, CPPM uses the "WMI-mydomain" account to run the WMI scan.

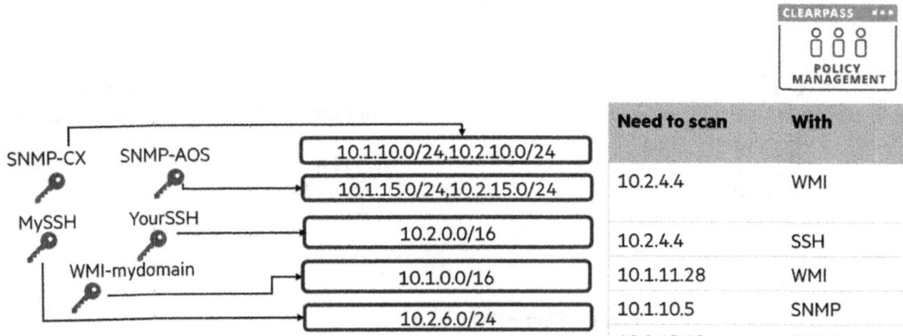

Need to scan	With	Use
10.2.4.4	WMI	Skip...no credentials
10.2.4.4	SSH	YourSSH
10.1.11.28	WMI	WMI-domain
10.1.10.5	SNMP	SNMP-CX
10.2.15.12	SNMP	SNMP-AOS
10.2.6.6	SSH	MySSH

Figure 12-15: Setting Up Profiler Subnet Mappings

This is because the mappings further determine which account or accounts to use for the scan.

For example, you can create a mapping that includes the subnets for your AOS-CX switches—10.1.10.0/24 and 10.2.10.0/24 in this example. You can then select SNMP for the scan type and the particular SNMP account with the credentials used by your AOS-CX switches. In this scenario the AOS devices use different SNMP credentials, so you create a different subnet mapping for the AOS network device subnets and select the "SNMP-AOS" account for that mapping.

You can create multiple mappings for the same subnets, or for overlapping subnets, with different types of accounts, if you want to apply multiple scan methods to the same subnet.

You can also create a subnet mapping with a more specific subnet that overrides a mapping to a larger subnet for that specific subnet only. In the example configuration shown in Figure 12-15, CPPM uses the "MySSH" account to scan devices in 10.2.6.0/24, but "YourSSH" to scan all other devices in 10.2.0.0/16.

Network (Discovery) Scan

A network, or discovery, scan uses seed devices. As shown in Figure 12-16, CPPM begins by sending an SNMP Get message to each seed, asking for three pieces of information: the LLDP neighbors, the CDP neighbors, and the ARP table. You should generally select the default router or routers in your network as the seed. Selecting multiple seeds can be a good idea too. For example, you might specify multiple routing switches. A default router maintains an ARP entry for every device for which it is routing traffic, so its ARP table can yield the IP address and MAC address for every active device in its subnets (as long as the device is sending and receiving routed traffic).

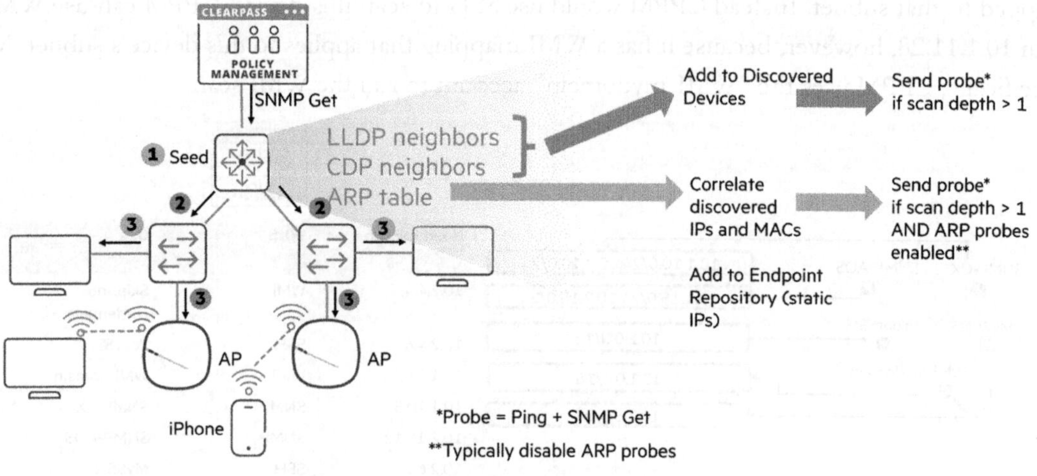

Figure 12-16: Network (Discovery) Scan

The CDP and LLDP tables provide CPPM with a list of devices connected to the seed at Layer 2. CPPM adds these devices to its list of discovered devices. If you want CPPM to send SNMP Get messages to discovered devices, you must set the scan depth greater than 1. The scan depth dictates how many times CPPM repeats the scanning process. If the scan depth is 1, CPPM scans the seeds and then stops. If you set the depth to 2, CPPM scans the seeds and the devices discovered in the seeds' CDP and LLDP tables. If you set the depth to 3, it scans the seeds, the devices discovered in the seeds' CDP and LLDP tables, and the devices discovered in the second devices' CDP and LLDP tables, and so on.

CPPM scans devices with probes. A probe begins with a ping; then CPPM sends an SNMP Get message if the ping receives a response.

As mentioned earlier, CPPM also obtains the ARP table from every device that it scans with SNMP. CPPM uses the ARP table to correlate discovered IP addresses and MAC addresses. You can also enable probing ARP entries, which means that, when conducting the next layer of the scan (as dictated by the scan depth), CPPM sends out probes to devices discovered in the ARP table, as well as the LLDP and CDP tables. Enabling this option could make the scan much longer because the ARP table is typically so much larger than the LLDP and CDP tables, as it includes many client devices as well as network infrastructure devices. If you leave the ARP probe setting disabled, CPPM still uses the ARP table information to correlate IP and MAC addresses. However, it does not send probes to the IP addresses in the table.

Subnet Scan

CPPM can also run subnet scans, illustrated in Figure 12-17. In this case, CPPM scans every device within the subnet or subnets defined in the scan. For example, if you specify 10.1.11.0/24 in the scan, CPPM scans devices 10.1.11.1 through 10.1.11.255.

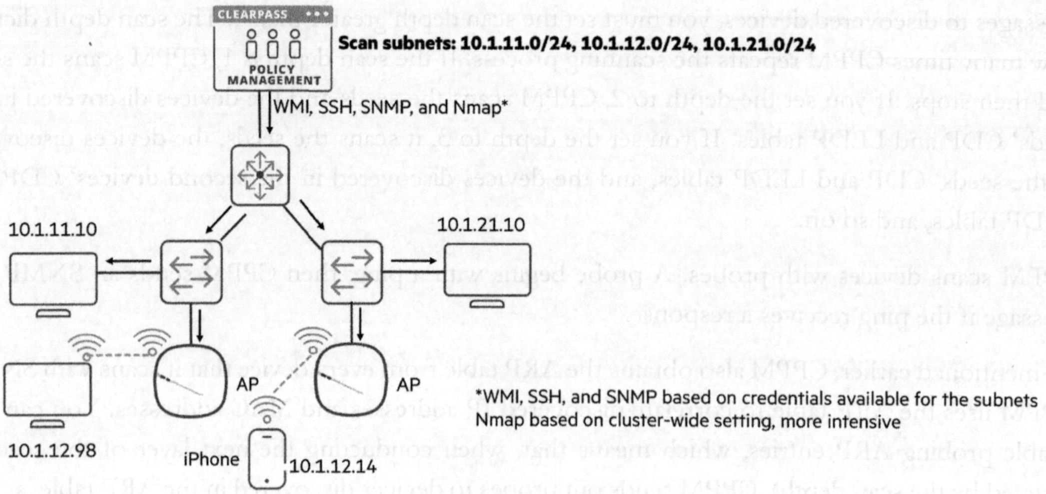

Figure 12-17: Subnet Scan

CPPM uses WMI, SSH, and SNMP to scan the subnets, depending on the external accounts mapped to the subnet. For example, if you have mapped a WMI account to 10.1.11.0/24, or a larger subnet such as 10.1.0.0/16, CPPM uses WMI and the credentials in that account to scan devices in 10.1.11.0/24. Port 135 must be open for the WMI scan to succeed. If 10.1.11.0/24 also has SSH or SNMP credentials associated with it, CPPM uses those protocols for the scan too. But if it does not have credentials of a particular type for the IP addresses, it simply skips scanning those addresses with that protocol.

If you enable Nmap scanning in the cluster-wide settings, CPPM scans devices with Nmap as well. Nmap detects endpoints and scans which application ports are open on those endpoints. It can also figure out endpoints' OS by sending TCP traffic and analyzing how the endpoints respond. Nmap scans are more intensive. You might choose to use this tool in smaller subnets (such as /24), rather than across large subnets (such as /16). You should also be aware that, because hackers also use Nmap for network reconnaissance, the company's security appliances might flag the Nmap scans as a threat. You should discuss the use of this method with the company's security team members, and, if they decide you should use it, they should allow list CPPM.

Guidelines for Running Scans

You can run a scan "on-demand," which means that CPPM runs the scan directly after you create it. You might run an initial on-demand network (discovery) scan after you first install CPPM to obtain a list of network devices. You can then add those network devices as RADIUS/TACACS+ clients. The initial network scan can also discover endpoints and information about them if you have enabled WMI probes.

You can also schedule periodic scans to enable CPPM to discover new endpoints and collect refreshed information. HPE Aruba Networking recommends that you schedule the periodic scans at quiet times during which the network is relatively inactive. Subnet scans can be particularly intensive, so take care to schedule them when they will not interfere with normal operations.

Also note that CPPM might discover devices during subnets scans, but not have the MAC address information for the device. CPPM needs to be able to read network devices' ARP tables to learn the MAC address that correlates to each IP address. CPPM sends out periodic SNMP polls to network devices to collect this information.

To ensure CPPM can run the polls on network devices, edit network device entries (find the list in **Configuration > Network > Devices**). Click the **SNMP Read Settings** tab. Select the **Allow SNMP Read** check box, as shown in Figure 12-18. Configure the credentials to match those on the network devices in question. Then select the **Read ARP table from this device** check box and save the settings.

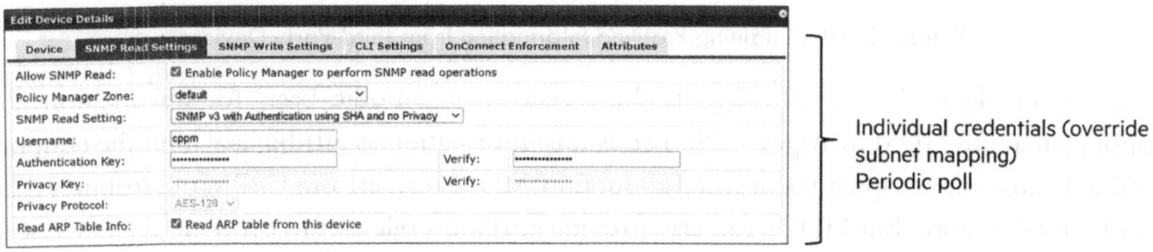

Figure 12-18: Configuration > Network > Devices > *Device*

Obtaining Profiling Information from Third-Party Devices

You can also set up third-party MDM solutions as Endpoint Context Servers. Some examples include Citrix XenMobile and MobileIron Core.

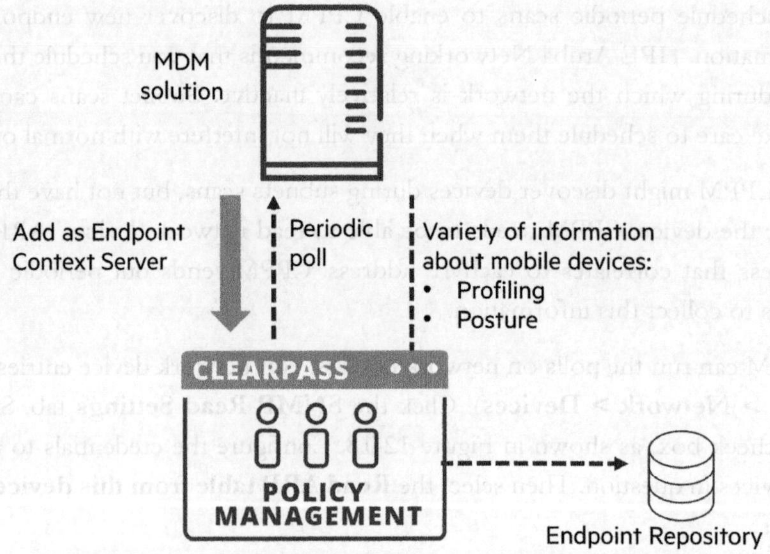

Figure 12-19: Obtaining Profiling Information from Third-Party Devices

As shown in Figure 12-19, CPPM polls the server on a periodic basis for information about endpoints that the MDM manages. CPPM adds this information as attributes within the endpoints' profiles. In an earlier chapter, you learned about how MDMs can inform CPPM of attributes related to endpoints' posture. But MDMs can also provide attributes relevant to profiling the devices such as the device type and OS version.

Using Device Information

Next, you will learn how to use device information collected by CPPM.

CPPM Endpoint Profiler Dashboard

CPPM consolidates the information that it has collected about endpoints in the Endpoint Profiler Dashboard, available in **Monitoring > Profile and Network Scan > Endpoint Profiler**.

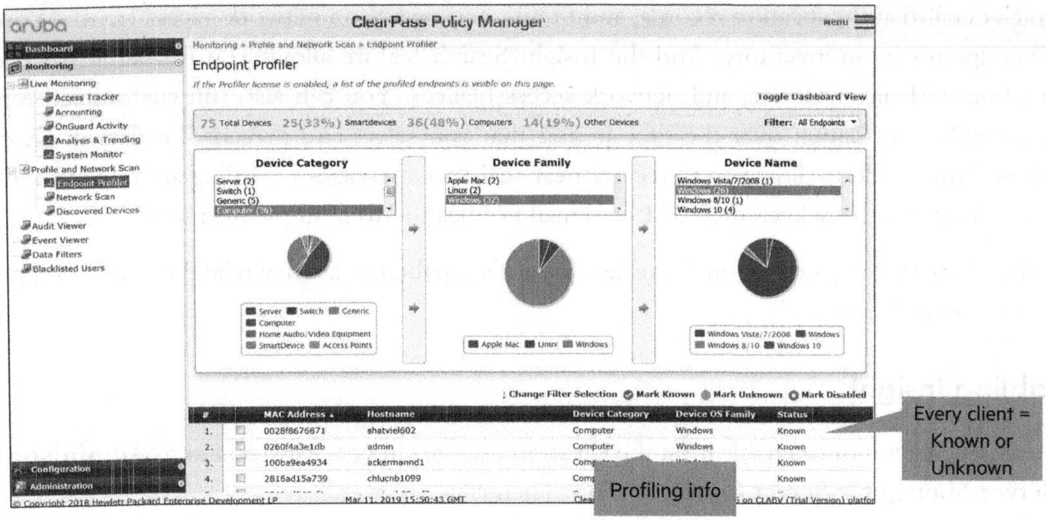

Figure 12-20: CPPM Endpoint Profiler Dashboard

In this dashboard, illustrated in Figure 12-20, you can see a breakdown of devices by category, family, and name. You can find a list of all endpoints, which you can filter based on Device Category, OS Family, and Status.

ClearPass Insight

ClearPass Insight (not to be confused with CPDI) compiles information across ClearPass servers, analyzes the information, and makes that information available to admins in a Dashboard (see Figure 12-21).

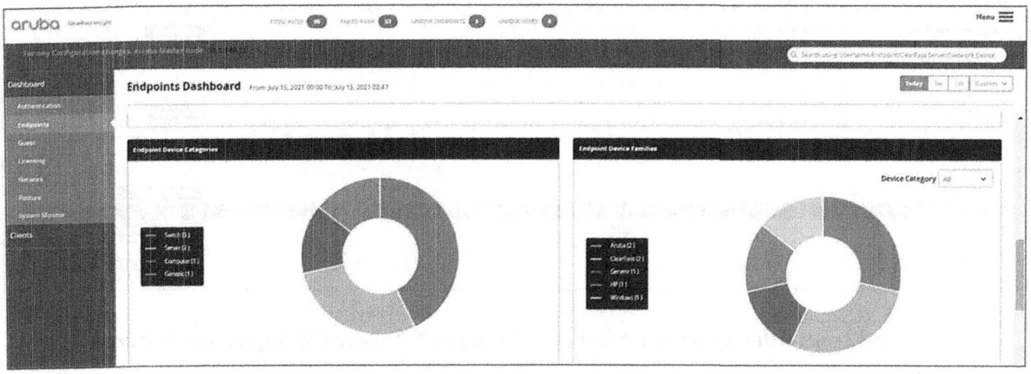

Figure 12-21: ClearPass Insight Endpoints Dashboard

Customizable dashboard "widgets" allow you to view the specific kinds of information you need to monitor and understand what is occurring on the network. You can create customized dashboards to track detailed authentication records, audit trails, and details on network-access trends. You can view endpoints in an inventory. And the Insight Search feature allows you to search for clients, users, Policy Manager servers, and network access devices. You can also run customized reports that compile information over the time period that you select and present the data in an easily readable format. Alerts allow you to receive near-real-time messages regarding anomalous network activity. Alerts can be delivered via SMS or email notification to multiple recipients.

For this chapter, your focus is on ClearPass Insight's capabilities as they relate to inventorying and reporting on endpoints.

Enabling Insight

To use Insight, you must enable it on the ClearPass servers in your cluster. Go to **Administration > Server Manager > Server Configuration**. Then click a server.

In the System tab select the **Enable Insight** check box. Choose one server to act as the primary server and select the **Enable as Insight primary server** check box for that server. (See Figure 12-22.) Note that an Insight primary server is always required, even if you do not have multiple servers in a cluster running Insight.

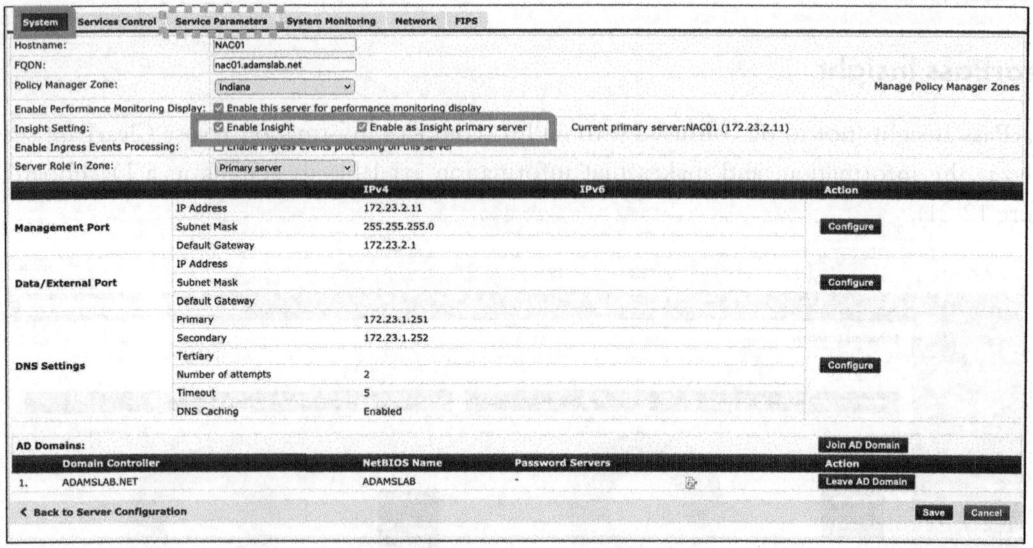

Figure 12-22: **Administration > Server Manager > Server Configuration >** ***Server***

You should also enable logging of accounting interim update packets. This setting allows the server to log the accounting information that it receives from network devices, making that information available for Insight. Of course, you must also set up the network devices to send the accounting interim updates in the first place, which you learned how to do in an earlier chapter. Go to the **Service Parameters** tab to enable this setting (see the dotted line box in Figure 12-22).

Save the settings.

Then repeat these steps on other servers in the cluster. But only make one server act as the Insight Conductor.

Configuring Insight Database

You can alter the database retention settings at the Insight Administration page. This parameter defines how long Insight retains data for the endpoint inventory and reports. As shown in Figure 12-23, the default retention period is 30 days. HPE Aruba Networking recommends increasing this size to at least 90 days. If possible, increase the size even more. Scope the size of the ClearPass server's disk and retain as much data as possible.

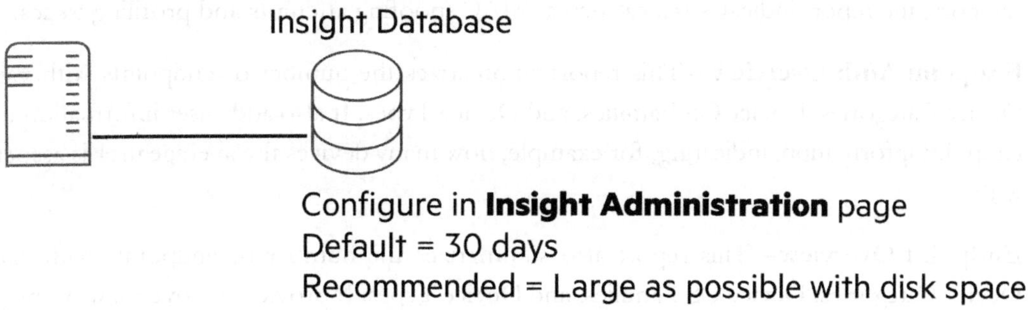

Figure 12-23: Configuring Insight Database

Using Insight Endpoint Reports

Insight offers reports in many categories, which you can run to analyze information, archive that information, and present the information in a useful format to other teams and stakeholders.

To create a report, select **Reports** in Insight's left menu. Click **Create a Report**. A wizard will guide you through creating the report.

As one option, you will choose the report type. For this chapter, you are focusing on reports within the Endpoint Category, as shown in Figure 12-24.

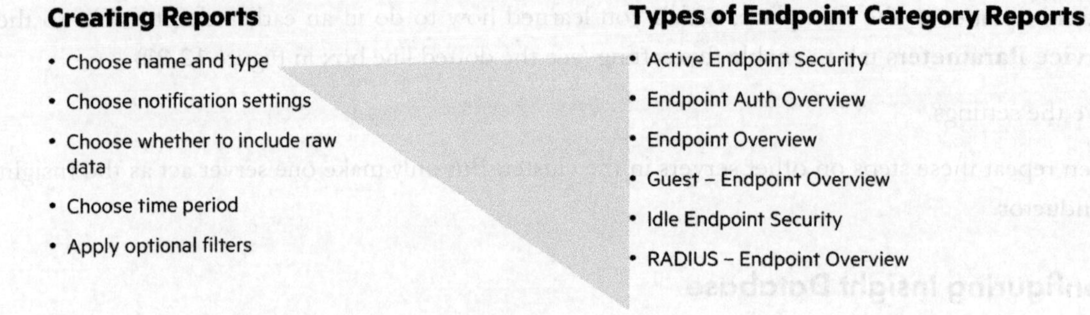

Figure 12-24: Using Insight Endpoint Reports

This category includes these report types:

- **Active Endpoint Security**—This report provides information about endpoints detected by CPPM with an emphasis on potential security issues. In addition to endpoints' status and category, the report indicates issues such as MAC spoofing attempts and profiling issues.
- **Endpoint Auth Overview**—This report summarizes the number of endpoints with various Device Categories, Device OS Families, and Device Types. It also adds user information to the endpoint information, indicating, for example, how many devices the average user is associated with.
- **Endpoint Overview**—This report also summarizes the number of endpoints with various Device Categories, Device OS Families, and Device Types. It provides an overview of the most common categories, families, and names.
- **Guest – Endpoint Overview**—This report is similar to the Endpoint Overview report, but reports on endpoints associated with guest users.
- **Idle Endpoint Security**—This report shows how many known endpoints have not had their profile updated within the time period of the report.
- **RADIUS Endpoint Overview**—This report is similar to the Endpoint Overview, but reports on endpoints that have authenticated with some form of RADIUS (including 802.1X, MAC-Auth, and others).

The wizard will guide you through further options. You can choose to send the report by email, SMS, or both. This setting requires CPPM to have the corresponding notification settings already configured on it.

You can also choose whether to include a CSV file with the raw data used to generate the report. This file can be useful when you want to dive deeper. As you will see, for example, the Active

Endpoint Security report indicates how many endpoints have the "Generic" Device Category. But if you want a complete list of those endpoints, you need the raw data in a CSV file.

You select the time period that the report analyzes. This can be a relative period such as the previous day, week, or month. Or it can be an absolute period, which you set.

The report type itself filters the output. You can also apply optional filters of your own to further limit the output.

Analyzing an Active Endpoint Security Report

After you create the report, you can view and download it. In the following sections, you will look at some of the information that you can obtain from an Active endpoint Security Report and how you can use that information.

Endpoint Device Status

In the Endpoint Device Status section, look for devices with a Disabled status, as shown in Figure 12-25.

If they should be disabled, why are they still connected to the network? The main report simply indicates the number of endpoints with this status, but you can get the full list of devices from the report's raw data in the CSV file.

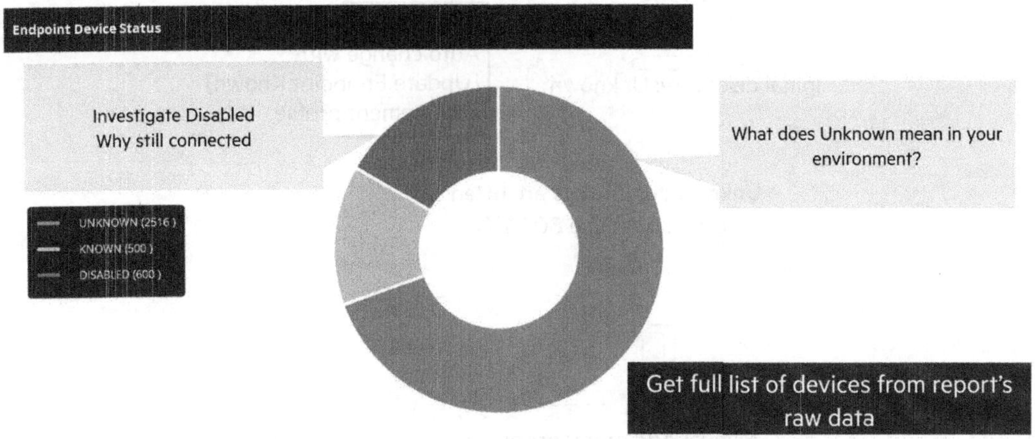

Figure 12-25: Endpoint Device Status

Sort the columns for status, and you can quickly get a list of clients with the "Disabled" status. You can use other columns in the row to track down the endpoint. By checking the Network Access Server (NAS) to which the endpoint is connected, for example, you can start to track down the endpoint's location. If your investigation reveals illegitimate endpoints, you should take steps to remove them. If there is a legitimate reason for the endpoints to be connected, you can leave them.

This report also shows how many clients have a "Known" status or an "Unknown" status. Before you draw conclusions about whether the Unknown endpoints pose a security risk, stop to consider what "Unknown" means in your environment.

More on Unknown versus Known Status

CPPM tracks a status for every endpoint, which is either "Unknown" or "Known."

As shown in Figure 12-26, when CPPM first discovers the endpoint, the status is "Unknown." The status can become "Known" in several ways. You can manually change the status to "Known." Or you use a default enforcement profile, "[Update Endpoint Known]." You can add that enforcement profile to the profile list of an enforcement policy rule to have CPPM automatically update the status for clients to which that rule applies.

You can also add or import endpoints on CPPM, rather than rely on CPPM to discover the endpoints itself. In that case, you can set the status to "Known" in the records that you add or import. Similarly, if CPPM learns about endpoints through third-party integrations, those integrations can add the endpoints with the Known status.

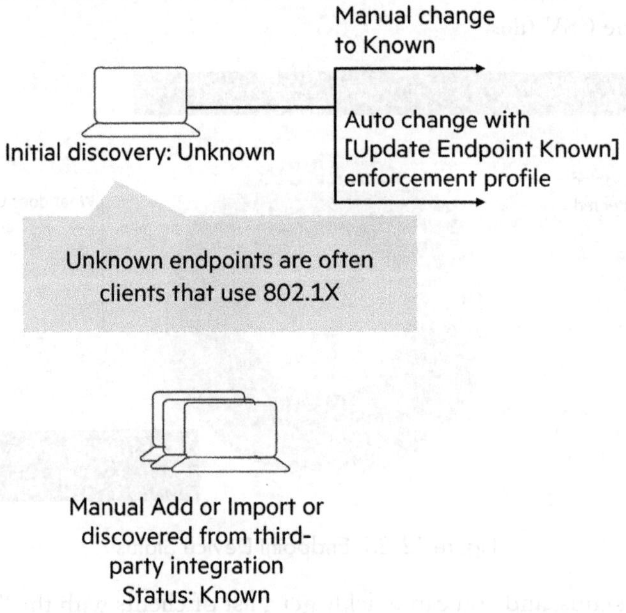

Figure 12-26: Analyzing an Active Endpoint Security Report

Your company can use the "Unknown" versus "Known" status in whatever way makes sense to your company. One common use is for MAC-Auth, which is a less secure form of authentication typically reserved for clients that are incapable of 802.1X. You give clients that need to pass MAC-Auth a Known status, which you might do by importing the clients to the Endpoints Repository with a Known status.

Or many companies use MAC-Auth together with captive portal authentication to enable guests to log into a captive portal once and then become automatically authenticated for a certain period of time. In that case, the guest client first fails MAC-Auth and undergoes captive portal authentication. The enforcement policy rules that grant guests access through the captive portal also include the [Update Status Known] enforcement profile, which enables the client to pass MAC-Auth on the next attempt. You will learn more about using MAC-Auth later in this chapter. This study guide does not cover guest access.

For now, you should be aware of how your company wants to use the Unknown versus Known status so that you can explain the report to the recipients. If you are using the Known status for MAC-Auth, you *want* clients that should undergo the more secure 802.1X to have the Unknown status. In that case, it would be perfectly normal for most endpoints in your network to have the Unknown status.

Endpoints MAC Spoofing

The Active Endpoint Security report also shows you how many spoofing attempts have occurred, as shown in Figure 12-27. Spoofs are indicated when a single MAC address has an initial device fingerprint, which then changes to another fingerprint. A MAC address is typically tied to hardware, so normally the device fingerprint for a MAC address should never change. If it does, hackers are likely spoofing the MAC address with their own devices.

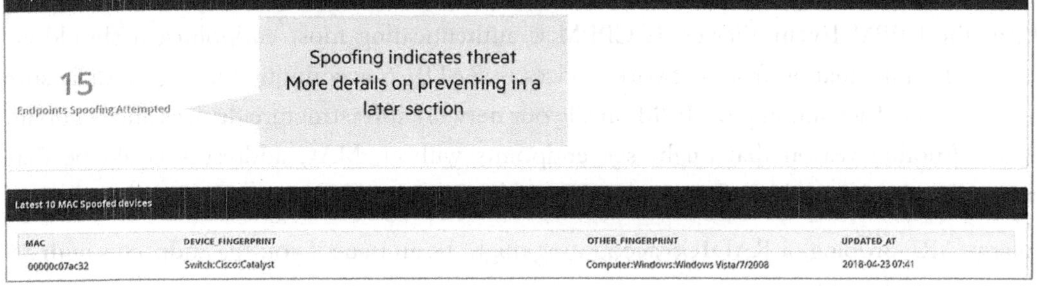

Figure 12-27: Endpoints MAC Spoofing

You should document this threat and check into it. For example, you can find the current NAS to which the endpoint, likely a spoofing endpoint, is connected and track down the threat that way.

You will learn more about preventing MAC spoofing in a later section.

Address Schemes

Another section of the report shows how many endpoints have dynamic IP addresses and how many have static IP addresses (see Figure 12-28). If you see endpoints with static IP addresses, investigate further. Are these truly endpoints with static IP addresses? Or is CPPM missing some DHCP packets? Perhaps network admins missed adding CPPM's IP address to the helper address list in one of the VLANs. Without DHCP packets, CPPM is missing valuable information for fingerprinting, so follow up on this potential clue to an issue.

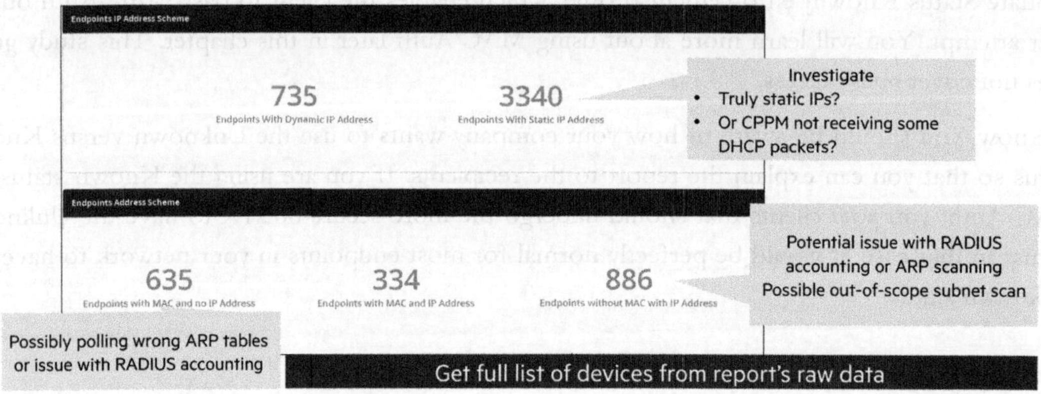

Figure 12-28: Address Schemes

Typically, all endpoints should have an IP address and a MAC address. If you see endpoints listed with IP address only, CPPM discovered the device during a subnet scan, but could not find an MAC address related to the IP address. CPPM can obtain the MAC-to-IP binding by probing ARP tables, so you should investigate whether your network device entries. Have you enabled SNMP and ARP probes on network devices? Are the SNMP credentials correct? You can see errors with SNMP polls in the CPPM Event Viewer. If CPPM is authenticating most endpoints, it should also be receiving this information from network devices in RADIUS accounting messages. Make sure that you have enabled accounting to CPPM on all your network infrastructure devices and set up interim updates. Another reason that might see endpoints without MAC addresses could be that you accidentally scanned a subnet that is not really under the company's control.

Similarly, ARP probing or RADIUS accounting might be an issue if you see endpoints with a MAC address and no IP address.

Profile Issues

You will also see a section labeled "Endpoints with Profile Issues." An example is shown in Figure 12-29.

Unprofiled endpoints indicate that CPPM has learned about the endpoint, typically through the endpoint authenticating, but CPPM has not learned any profiling information about it. Often this means that CPPM is not receiving these endpoints' DHCP Discover packets. Again, you should check the network's DHCP relay settings. If DHCP relay is not used, make sure that you set up CPPM to receive DHCP packets on a Span Port as discussed earlier. If the endpoints have static IP addresses, of course, CPPM cannot profile the devices with DHCP. Instead it should be running active scans to collect this information. In this case, check the scanning settings.

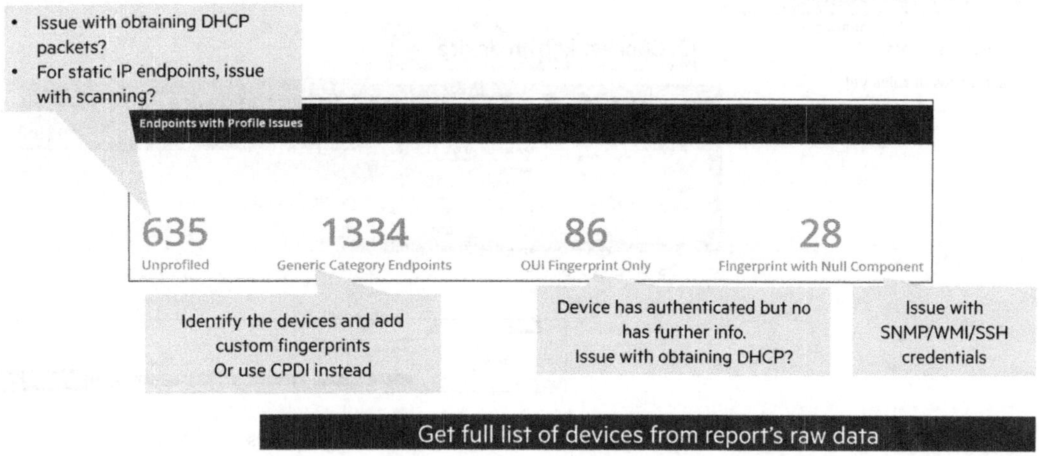

Figure 12-29: Profile Issues

If CPPM has collected profiling attributes about a client, but cannot determine a device category for it, CPPM categorizes the device as "Generic." You should identify these devices and add custom fingerprints for them. (You will learn how in a moment.) If the company has many generic category endpoints, it might be best for the company to deploy CPDI instead of using CPPM's device profiling capabilities, as CPDI has more extensive capabilities in this area.

Typically you should not see endpoints with an OUI Fingerprint Only status. This status means that CPPM has scanned the endpoint, but received no fingerprint during the scan. You will need to check into the scanning and credentials setting for an error. Similarly, if you see the "Fingerprint with Null Component" error, you probably have an issue with the SNMP, WMI, or SSH credentials used to scan that endpoint. Find the IP address for the endpoint in question in the full CSV file for the report. Then look in CPPM's Profiler Subnet Mappings for that endpoint's subnet. Find the associated accounts and double-check the credentials.

Creating Custom Fingerprints

As just mentioned, if you have some devices with a generic category, you should figure out yourself what the device type is and create a custom fingerprint that can apply to that device and all similar devices.

First, as shown in Figure 12-30, add a new fingerprint in **Administration > Dictionaries > Device Fingerprints**. You can choose any meaningful values to you and other administrators for the Category, Family, and Name. At this point, the dictionary has no attributes or rules associated with it yet. You must add rules from a device that CPPM has discovered; you cannot add the rules manually.

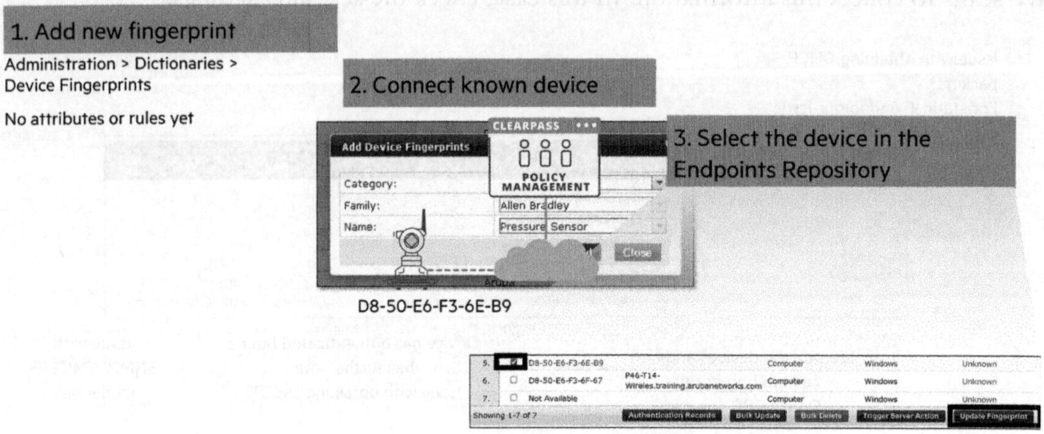

Figure 12-30: Starting to Create Custom Fingerprints

To add the rules, make sure that a representative device for this type is connected to the network and detected by CPPM.

Then find the device's MAC address in CPPM's Endpoint Repository (**Configuration > Identity > Endpoints**). Select the endpoint and click **Update Fingerprint**.

You will see a window such as that shown in Figure 12-31. Manually adjust the Device Category, Device OS Family, and Device Name to match those in the new fingerprint that you previously added. Make sure that the Update Type is set to **Add fingerprint** rule. Select all of the device fingerprints and click **Save**.

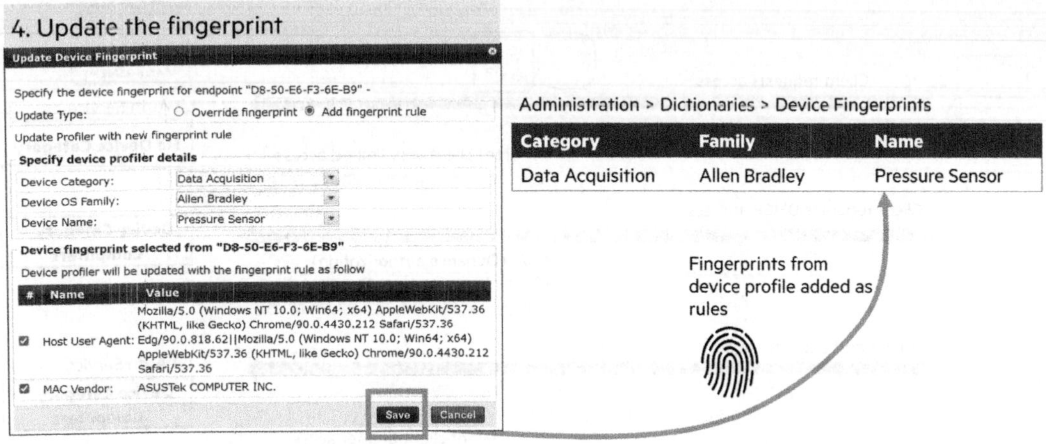

Figure 12-31: Completing the Custom Fingerprint

CPPM then takes the fingerprints from this endpoint's device profile and adds them to the Device Fingerprint that you added previously. Now CPPM can profile other devices of this type with the custom device fingerprint.

Use Profiling in CPPM Services

In this section, you will learn how to set up CPPM services that use profiling.

Example L2 Authentication Process with Device Profiling

You will now learn how to use device profiling information as additional context that CPPM can use in access control decisions. When you are using Layer 2 authentication (802.1X or MAC-Auth) to authenticate clients, the high-level process illustrated in Figure 12-32 occurs.

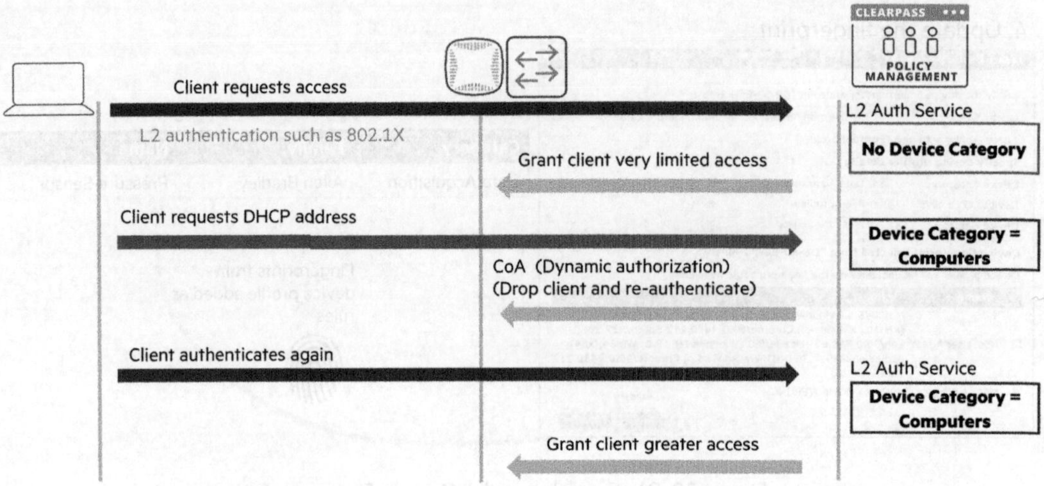

Figure 12-32: Example L2 Authentication Process with Device Profiling

The first time that the client authenticates, CPPM does not have a profile for it and does not know the device category. The enforcement policy within the L2 authentication service has a rule for devices without a device category. This rule assigns the client a profile that provides very limited access.

The client is now connected and requests a DHCP address. CPPM receives the DHCP Discover and profiles the device. Even if CPPM does not have all the information that it might eventually collect for the profile, it should be able to pinpoint the device category. In this example, CPPM profiles the client as a computer.

The change in profile signals CPPM to send a CoA to the network device to which the client is connected.

The client reauthenticates. The L2 authentication service's policies have rules that assign a particular AOS role to users on computers. CPPM informs the network device of the client's new role, and the client is now fully connected with appropriate privileges.

Adding the Profiler to the Service

For this use case you need to enable the **Profile Endpoints** option on the L2 authentication service.

As shown in Figure 12-33, you can find the **Profile Endpoints** option in the **Service** tab of the service. After you select it, CPPM will monitor the profile status for all clients that authenticate through the service, even after the clients have completed authentication. CPPM can then modify the client's access level after the client's device profile changes using the settings that you configure in the **Profile** tab.

Selecting the **Profile Endpoints** option adds the **Profiler** tab to the service.

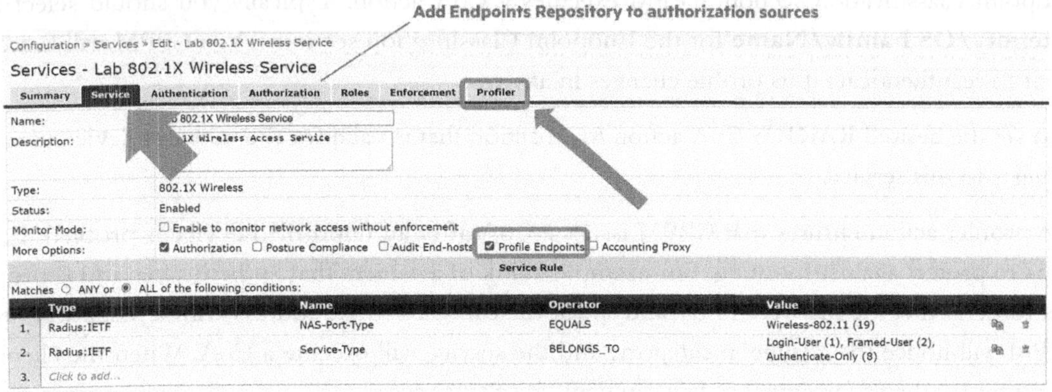

Figure 12-33: Adding the Profiler to the Service

Note that you also often need to select **Authorization** for the service. Then you can add the Endpoints Repository to the Authorization sources. The Endpoints Repository must be either an authentication source or an authorization source for you to include device category, family, or name as conditions in rules.

The profiler functionality is extremely helpful in weaker authentication methods such as MAC authentication where IoT devices could possibly be spoofed on the network. A MAC address is easy to spoof but hardware profiles are much harder to fake.

Configure the Profiler Action

CPPM watches the profile data status for all clients that authenticate with this service. As shown in Figure 12-34, go to the **Profiler** tab to configure how CPPM responds to changes in the status.

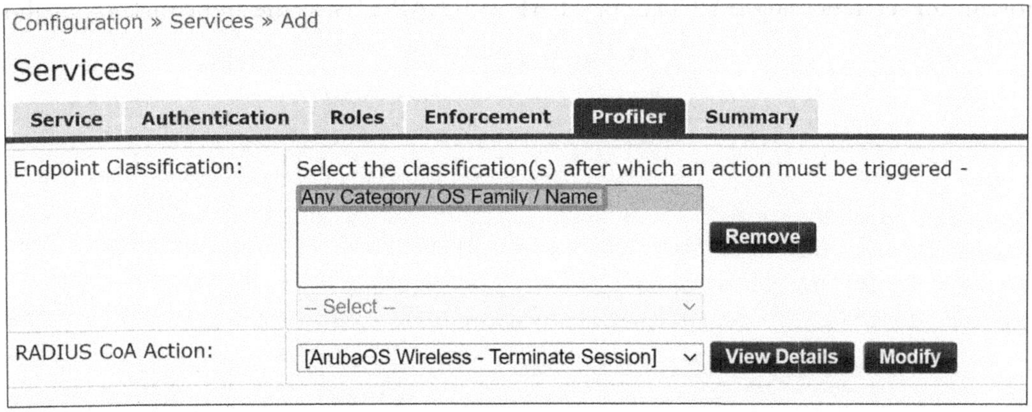

Figure 12-34: Configure the Profiler Action

If a client's classification changes to one of the categories, OS Families, or Names listed in the Endpoint Classification section, CPPM executes a CoA action. Typically you should select **Any Category/OS Family/Name** for the Endpoint Classification section. Then CPPM will force the client to reauthenticate if its profile changes in any way.

Also set the desired RADIUS CoA action to an action that is valid for the network devices sending requests to this service.

The profiler action ensures that CPPM issues a CoA after an unprofiled device is profiled. It also helps to protect against spoofing. For example, think of a camera that authenticates and is profiled as a camera. If someone were to try and spoof the camera's MAC address with a Macbook Pro®, CPPM will notice the change in category, and the service will execute a CoA. When the client re-authenticates the enforcement policy can deny the client access.

You can alternatively configure CPPM to watch for specific changes, such as any change related to the camera category. This means if a device is profiled as a camera and the profile changes to something else, or if a device that is not a camera changes to a camera, both conditions will trigger the CoA action. However, if a device is profiled as a smartphone and changes to profiled as a computer no CoA action will happen.

Example of Incorporating Device Profile Information in a Role Mapping Policy

You need to determine how the company wants to incorporate device profile information in enforcement policies.

In this first example, illustrated in Figure 12-35, a company has a role mapping policy that assigns clients roles based on their AD security group. It also checks the client's account status, which is important for services using EAP-TLS, or TEAP with EAP-TLS, as the authentication method.

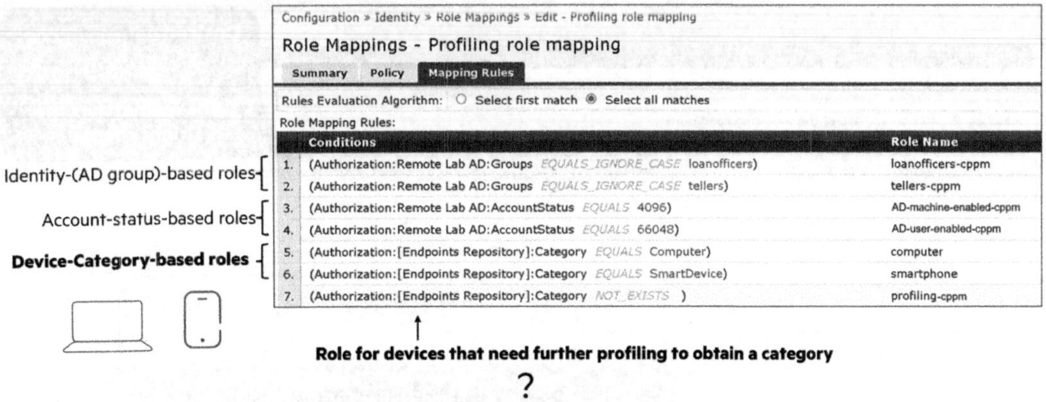

Figure 12-35: Example of Incorporating Device Profile Information in a Role Mapping Policy

Now the company also wants to include device profile context. So you add a couple more rules, which assign roles based on device category. In this example, you assign clients with a category of Computer to a "computer" role and clients with a category of SmartDevice to a "smartphone" role. If you wanted to allow other types of devices to connect, you would simply add more rules and roles for those devices.

You also add a rule for devices that do not have a category. These devices require further profiling so that CPPM can assign them to a category.

More Detailed Guidelines on the Rule for Devices That Need Further Profiling

Take a moment to consider in more detail why the example role mapping policy used the following condition to select devices that need further profiling: "Authorization:[Endpoint Repository] Category NOT_EXISTS."

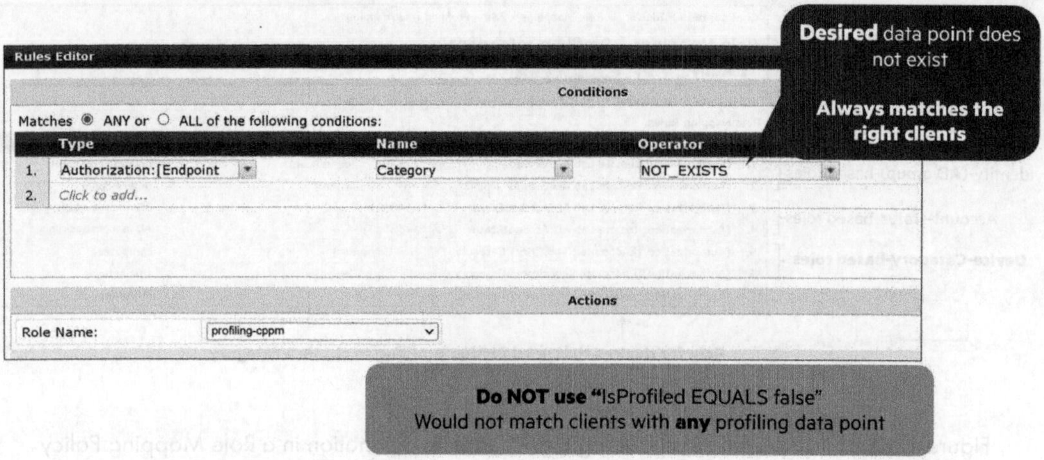

Figure 12-36: More Detailed Guidelines on the Rule for Devices That Need Further Profiling

The best way to select devices that need further profiling is to determine which data point you are using in other rules and evaluate whether it exists. In this scenario you are assigning clients to roles based on Category. So, as shown in Figure 12-36, you should create a rule that selects clients that have no Category and assign them to the "profiling-cppm" role.

Avoid using "IsProfiled EQUALS false" for your rule. Different profiling collectors expose different data points. As long as CPPM knows one data point about the client, it classifies the client as profiled. But CPPM might not actually know the data point that you want to use in rules (Category in this example). To avoid issues, use a rule that evaluates whether the specific data point that you want to use exists.

Example Enforcement Policy That Incorporates Device Profiles

You will now explore how you can use the Device-Category-based roles in an enforcement policy.

The example enforcement policy, illustrated in Figure 12-37, is already distinguishing between employees on domain computers and employees on BYOD devices. The first three rules apply to domain computers, as identified by the fact that the client has passed machine authentication. (It might have passed user authentication too.) You might decide not to include the device-category-based roles as conditions in these rules, as you already know that these clients are domain computers.

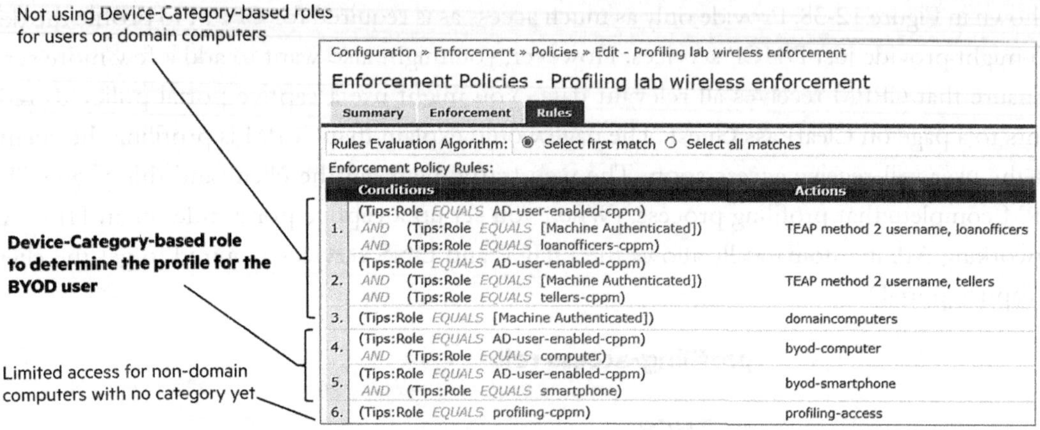

Figure 12-37: Example Enforcement Policy That Incorporates Device Profiles

However, the company might want to distinguish between employees on BYOD laptops and employees on BYOD smartphones. Rules 4 and 5 show how you can achieve this goal. The conditions include the "AD-user-enabled-cppm" roles, verifying that the user is legitimate. The rule further specifies a role associated with a particular device category. Users on computers are assigned to the "byod-computer" enforcement profile and users on smartphones to the "byod-smartphone" enforcement profile. These enforcement profiles use the Aruba-User-Role VSA to assign clients to roles named "byod-computer" and "byod-smartphone." As always, the network devices need matching roles that are configured with the proper firewall policies to enforce the company's security policies.

The final rule applies to clients that receive the "profiling-cppm" role. It assigns a profile named "profiling-access" to the client. This profile sends an Aruba-User-Role VSA for a "profiling-access" AOS role, and that role provides very limited access, as discussed in more detail in a moment.

Note that users on domain computers could receive the "profiling-cppm" role, in addition to other roles, the first time that the computer authenticates to the network. However, because the "Tips:Role EQUALS profiling-cppm" rule is at the bottom of the list, users on domain computers will be initially assigned to their full access profile, even before CPPM finishes profiling them. You might decide to take this approach to give the clients faster initial access, relying on the machine authentication to prove that the clients are domain computers. If, however, you wanted to profile all devices before permitting full access, you would need to move this rule to the top of the list.

Services for the Clients that Need to Be Profiled

As always, the AOS or AOS-CX devices must have roles with the names of those assigned in the Aruba-User-Role VSA. You should lockdown the role assigned to clients that need to be profiled, as shown in Figure 12-38. Provide only as much access as is required for CPPM to profile the device. You might provide just DHCP services. However, you might also want to add a few more services to ensure that CPPM receives all relevant data. You might use a captive portal policy to redirect clients to a page on ClearPass Guest. The page would explain that CPPM is profiling the client and that the user will receive access soon. The web traffic between the client and this page will help CPPM complete that profiling process. (When you set up a captive portal rule on an HPE Aruba Networking AP, it automatically allows the DHCP and DNS services required to get the client to the captive portal.)

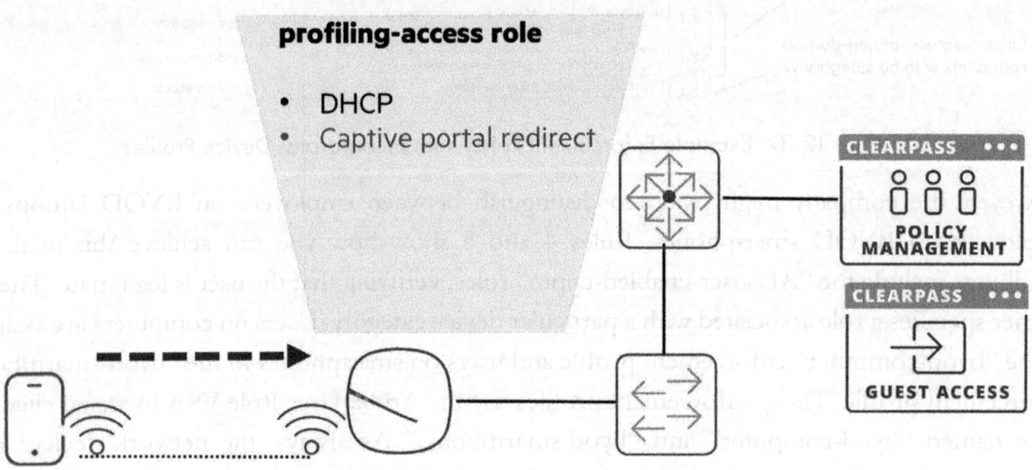

Figure 12-38: Services for the Clients that Need to Be Profiled

Use Device Profiling to Protect Against Spoofing

This section explains how to use device profiling to help protect your networking against spoofing.

MAC-Auth

Up to this point you have focused on 802.1X authentication on WLANs and switch ports and for good reason—802.1X is the most secure option. However, some clients are not capable of 802.1X. "Headless" devices such as cameras, printers, and other devices with limited user interfaces also cannot perform web authentication. MAC authentication (MAC-Auth) provides the network a way to authenticate these clients.

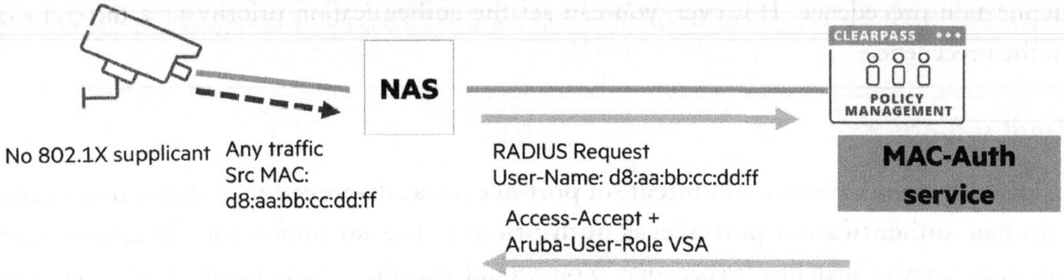

Figure 12-39: MAC-Auth

A client does not need to be configured with any special settings for the network to authenticate it with MAC-Auth. It simply sends its normal traffic, which includes its MAC address. As shown in Figure 12-39, the NAS (which might be a switch, an AP, gateway, or MC) sends a RADIUS request to its RADIUS server. This request uses the client's MAC address as the User name and password.

CPPM uses a MAC-Auth service to authenticate the client. It simply checks whether the MAC address is authorized using the selected authentication source. No further credentials are required. At the end of the process CPPM sends a response to the NAS, indicating whether the client is permitted access.

How MAC-Auth and 802.1X Interact on AOS-CX Switches

Typically, you want to keep 802.1X, as the most secure option, enabled on all of a switch's edge ports. However, you also want MAC-Auth as a failsafe for clients incapable of 802.1X. You can meet both requirements by enabling both port-access methods on all edge ports.

Authentication Priority and Authentication Precedence

Two port-access settings dictate how the methods work together on the port when both are enabled.

Authentication precedence determines which method the switch implements first. By default, 802.1X takes precedence over MAC-Auth. So, the switch first attempts to authenticate a client with 802.1X. Only if the 802.1X authentication times out, the switch implements MAC-Auth. If you change the precedence to MAC-Auth, then 802.1X, the switch first implements MAC-Auth. Because all devices are capable of MAC-Auth, the client either passes or fails MAC-Auth. The client can initiate 802.1X later, though, and achieve an 802.1X authentication status as well.

Authentication priority determines which method's status takes priority over the other when the client has undergone both methods. In other words, if the client has a MAC-Auth authentication status and role, and an 802.1X authentication status and role, which does the switch use as the client's actual status and role on the port? By default, the authentication priority is taken from the authentication precedence. However, you can set the authentication priority to a different order from the precedence.

Default settings

The default settings are **aaa authentication port-access auth-precedence dot1x mac-auth** and the **no aaa authentication port-access auth-priority**. The **no** option for the latter command means that priority matches precedence. You would use this combination when you want to authenticate most clients with 802.1X, but a few clients are entirely incapable of 802.1X, so you will use MAC-Auth for them. As shown in Figure 12-40, when the port comes up, the switch attempts 802.1X. If the client completes 802.1X, and either succeeds or fails, the switch uses the 802.1X status and role to determine the client's access and does not use MAC-Auth. However, if 802.1X times out, the switch implements MAC-Auth. It then uses the role assigned by that method to control the client.

```
aaa authentication port-access auth-precedence dot1x mac-auth
no aaa authentication port-access auth-priority
```

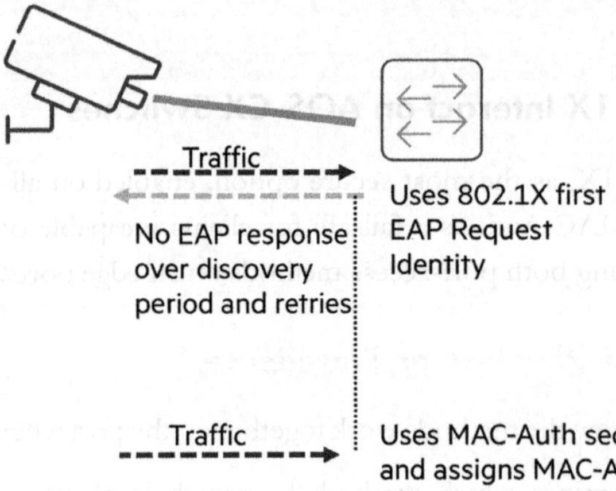

Figure 12-40: Default settings

Timers and Retries

The length of time that it takes for 802.1X to time out depends on three settings, which are configured under the port's **aaa port-access dot1x authenticator** context:

- **discovery-period**—The amount of time that the switch waits between sending EAP/Request Identity packets when no clients are authenticated on the port. The default is 30 seconds.

- **max-eapol-requests**—The number of EAP/Request Identity packets that the switch sends before considering the authentication session over. The default is 5.
- **max-retries**—The number of retries to authenticate a client that is not responding, before considering 802.1X to have timed out. The default is 2.
- **eapol-timeout**—The amount of time that the switch waits for a response after sending an EAP packet in an ongoing EAP transaction before retransmitting the packet. By default, the switch uses the RFC standard of 3 seconds, then doubling for each timeout up to a maximum of 20 seconds.

Sometimes a company might have a mix of devices that are and are not capable of 802.1X. The company might also have some devices that are 802.1X capable, but initially start without the 802.1X configuration. They must connect, download that configuration, and then undergo 802.1X. Examples of such devices are APs. In these cases, the company might not want devices to have to wait several minutes for 802.1X to time out before MAC-Auth succeeds. The company also wants to ensure that devices receive the correct roles. If MAC-Auth succeeds, the device should receive a role from the MAC-Auth process. If 802.1X succeeds, the device should receive a role from 802.1X. And, if both succeeds, generally the 802.1X role should take priority. You can meet these requirements in one of two ways.

Alternative 1

You can configure this command on a switch port: **port-access onboarding-method concurrent enable**. The port then implements all enabled authentication methods at the same time (regardless of the **aaa authentication port-access auth-precedence** setting). A client receives access if it succeeds at any of those methods. It receives the role from the method that succeeds. If multiple methods succeed, by default, the 802.1X role takes priority.

In Figure 12-41, you see the effects of this setting for a device such as an AP. A new AP connects for the first time. The switch port attempts to perform both MAC-Auth and 802.1X. MAC-Auth succeeds, but 802.1X times out because the AP is not set up as an 802.1X client. However, because MAC-Auth succeeded, the AP is permitted to connect with the role assigned through MAC-Auth. (For example, CPPM might have sent an Aruba-User-Role VSA with the RADIUS Access-Accept message.) Using the privileges permitted by that role, the AP connects and receives its configuration, including its 802.1X supplicant settings. Now the AP initiates 802.1X, which succeeds. The client has now passed both MAC-Auth and 802.1X. The new role sent by CPPM during 802.1X takes priority.

```
port-access onboarding-method concurrent enable
```

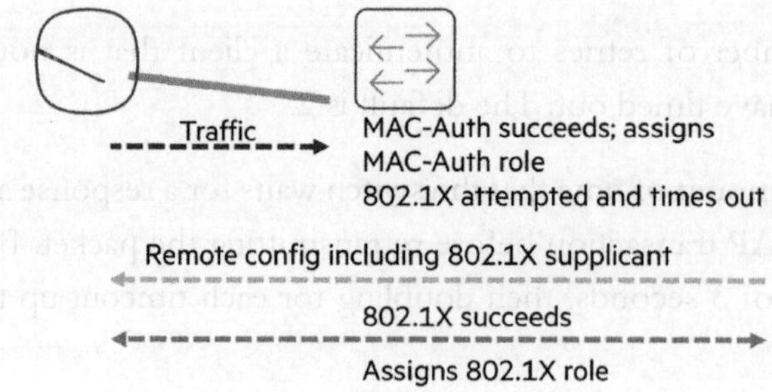

Figure 12-41: Alternative 1

You could use the same setting on other ports that connect to a variety of clients. Typical desktops and laptops, which are capable of 802.1X, fail MAC-Auth, but they succeed at 802.1X and receive the appropriate roles. Non-802.1X capable devices such as printers or some cameras fail at 802.1X, but succeed at MAC-Auth. All devices quickly connect and receive their roles.

You learned about various port-access roles in an earlier chapter. Consider how these roles work when you use **port-access onboarding-method concurrent enable**. If you apply a pre-auth role, the port remains in that role until at least one method has succeeded or all methods have failed. (MAC-Auth typically succeeds or fails quickly.) If all methods fail, and you have applied a reject role or critical role, the switch chooses whether to apply the reject or critical role based on the reason that 802.1X failed.

Alternative 2

Using concurrent methods typically provides the simplest option. However, some third-party RADIUS servers might not accept both 802.1X and MAC-Auth requests. In that case, leave concurrent onboarding disabled. Instead, specify these two commands:

aaa authentication port-access auth-precedence mac-auth dot1x

aaa authentication port-access auth-priority dot1x mac-auth

Now the devices can undergo MAC-Auth immediately, as shown in Figure 12-42. They receive a role that provides sufficient connectivity to receive their 802.1X settings. Then the devices can complete 802.1X authentication. Due to the auth-priority setting, the 802.1X role takes precedence over the MAC-Auth role (even though MAC-Auth occurred first).

```
aaa authentication port-access auth-precedence mac-auth dot1x
aaa authentication port-access auth-priority dot1x mac-auth
```

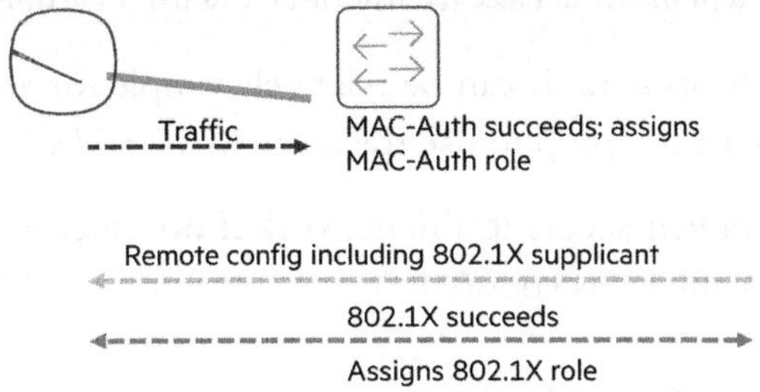

Figure 12-42: Alternative 2

MAC-Auth Vulnerabilities

MAC-Auth does not provide rigorous security because it is easily spoofed, as illustrated in Figure 23-43. Hackers can often discover valid MAC addresses relatively easily. If the hacker is trying to connect to an Ethernet port, the hacker has access to the site. This means that the hacker is probably able to examine devices like printers, and many of those devices have their MAC addresses labeled on them, or will print out their MAC addresses. If the company has a WLAN that uses MAC-Auth, finding a valid MAC address is even easier. The hacker simply needs to sniff the 802.11 frames.

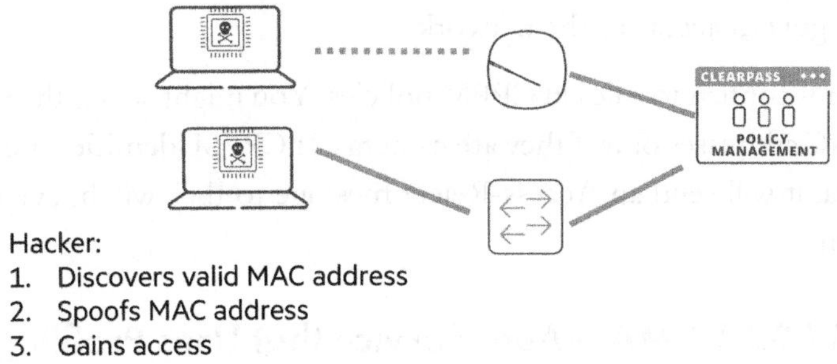

Figure 12-43: MAC-Auth Vulnerabilities

Once hackers know a valid MAC address, it is trivial for them to change their own device's MAC address to match. They can then pass MAC-Auth.

Using Device Profiling to Enhance MAC-Auth

Using endpoint classification and device categories within CPPM policies can help you to enhance the security of the MAC-Auth solution.

In Figure 12-44, you see an example scenario, in which wired IP cameras authenticate by MAC address alone. Without some added form of intelligent access control to identify the device as a valid camera, your network will grant access to any device with a camera MAC address.

An attacker might remove the camera. It can be relatively simple for attackers to find the camera's MAC address and configure their laptop to use the same MAC address. What would happen then?

The laptop would also be granted access to the network if no other way existed to determine that the MAC address being presented was spoofed.

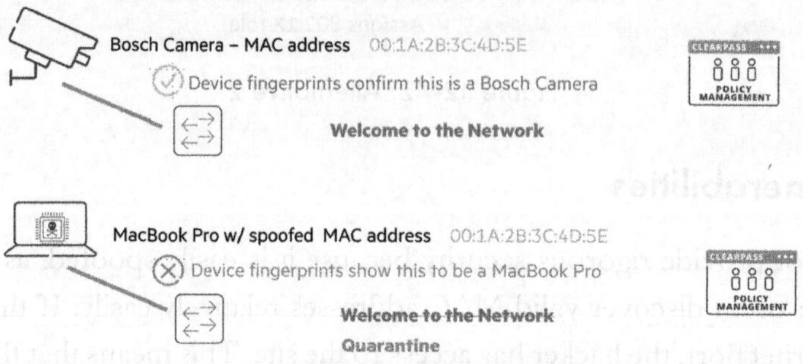

Figure 12-44: Using Device Profiling to Enhance MAC-Auth

However, what if you had a way of telling which type of device was presenting the MAC address? Using the CPPM profiler, you can get a deeper understanding of the real client device type and use that information to govern access to the network.

You can add endpoint context to rules in CPPM policies. You might set up the rules to allow devices with the proper MAC addresses only if they are cameras. If CPPM identifies the device as something other than a camera, it will send an Access-Reject message to the switch, even if the client passed MAC authentication.

Example CPPM 802.1X MAC-Auth Service that Uses Profiling

To provide MAC-Auth with profiling, you would set up a service using the MAC authentication template. This template creates service rules that select MAC-Auth requests (the username is the same as the called station ID). You can choose between two default methods for MAC-Auth:

- [MAC-Auth]—CPPM authenticates a client if its MAC addresses matches an entry in the Endpoints Repository that has a Known status. Admins would need to import the authorized clients' MAC addresses into the Endpoints Repository as Known clients in advance.

- [MAC-Auth Allow All]—CPPM authenticates all clients and uses rules in the enforcement policy exclusively to decide how to handle each client.

You should use the Endpoints Repository for the authentication source for both types of MAC-Auth. For [MAC-Auth Allow All], you do not technically need an authentication source, but you are still required to specify one. Specifying the Endpoints Repository lets you use it as an authorization source.

As shown in Figure 12-45, the role mapping policy and enforcement profile assign clients to CPPM roles and enforcement profiles based on their category. This company has a couple categories of headless devices that need to use MAC-Auth including printers and cameras. The role mapping policy assigns clients profiled as printers to the "printers-cppm" role and clients profiled as cameras to the "cameras-cppm" role. Clients without a device category receive a "profiling-cppm" role. All other clients receive the [Other] role.

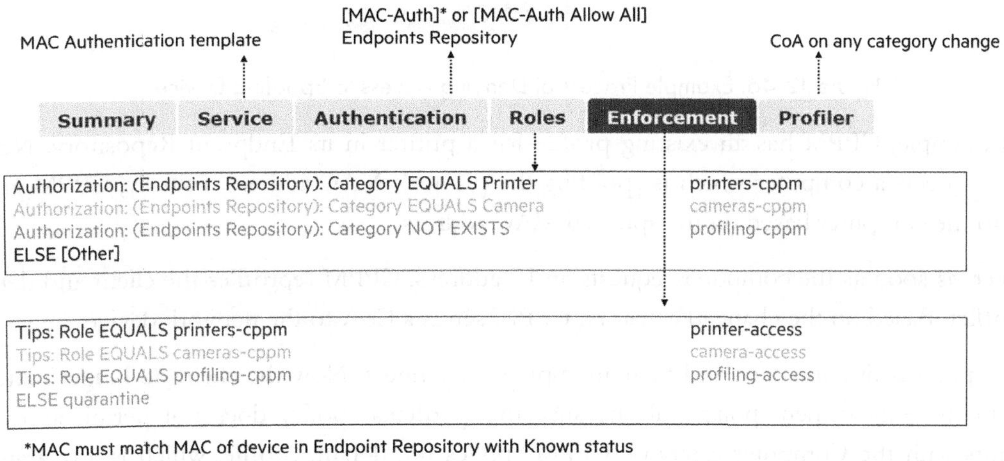

Figure 12-45: Example CPPM 802.1X MAC-Auth Service that Uses Profiling

The enforcement policy maps those CPPM roles to appropriate enforcement profiles. Clients that do not match a rule are assigned to a quarantine role, which essentially locks the client out but stops MAC-Auth from running again and again.

Example Process of Denying Access to Spoofing Device

In Figure 12-46, you see the more detailed process of how CPPM locks the spoofing device out.

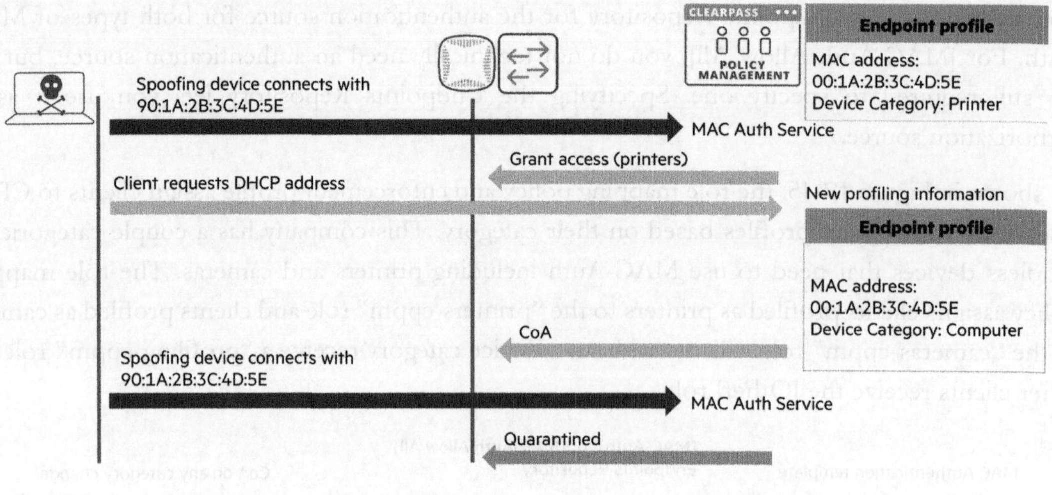

Figure 12-46: Example Process of Denying Access to Spoofing Device

In this example, CPPM has an existing profile for a printer in its Endpoint Repository. Now a hacker connects a computer which is spoofing the printer's MAC address. Initially, CPPM grants access to the computer based on the spoofed MAC address.

However, as soon as the computer requests an IP address, CPPM reprofiles the client and detects the conflict. Based on the changed category, CPPM sends a CoA to the network device.

The computer is disconnected and then attempts to reconnect. Now the client no longer matches a role in the enforcement policy rule because this particular policy does not permit access to endpoints with the Computer category. CPPM applies the default profile, which is a quarantine profile in this example.

If you want to demonstrate this feature to a customer, be aware that CPPM caches the MAC-Auth client's enforcement profile for five minutes after the client's initial connection and automatically reassigns the profile if the client re-connects during that period. After you connect the legitimate device, you should wait for the profile to time out before you connect the spoofing device. (Typically, a device like a printer will have been connected for a significant period of time before a hacker tries to spoof the device.) Also, in the real world, a hacker would connect their own spoofing device. However, in a POC, you might be using the same endpoint to simulate the legitimate device and the hacker device. To re-profile an existing device, CPPM needs to receive a DHCP Discover packet from the device, rather than a DHCP renewal request. Therefore, you should make sure that you force your demonstration device to fully drop its IP address and send out a new DHCP Discover. You might use Access Tracker to send a "bounce port" CoA for the client, which holds the connection down for several seconds.

Endpoints Repository: Conflict Attribute

You can add additional layers of security to the service using the Conflict attribute in the endpoint profile. You should be aware that CPPM continues to profile a device after it has already profiled it, using new information. This helps CPPM to keep track of changes to an endpoint's Device Category and to flag potential issues with spoofing.

The first time that CPPM profiles an endpoint, it sets the Conflict attribute to false. In the example shown in Figure 12-47, a printer connects to the network, completes MAC authentication, and is profiled as a printer. The Conflict attribute is currently false. In the normal course of events, an endpoint's Device Category should not change. The Conflict attribute will only be set to true when CPPM recognizes a change in Device Category. It will *not* be set when an endpoint changes from not profiled to profiled.

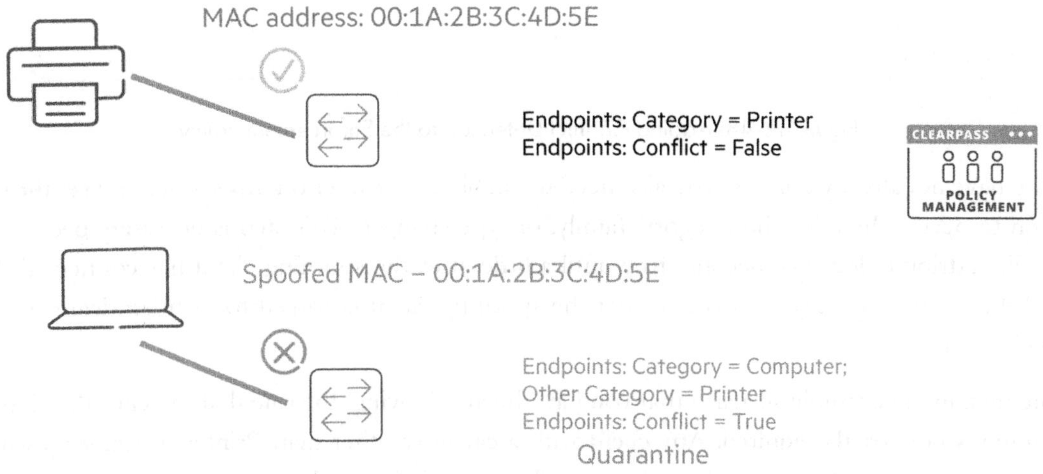

Figure 12-47: Endpoints Repository: Conflict Attribute

But say that hackers spoof the printer's MAC address on their computer. As shown in Figure 12-47, CPPM detects the "same" MAC address, but now profiles it as a Computer. Because the Device Category changed, the Conflict attribute becomes true. This alerts you to the spoofing. You can configure enforcement policies to select any clients with a "Conflict" true flag and apply appropriate actions, such as deny access to the network or sending the helpdesk a message with the endpoint's information.

When demonstrating this feature to customers, keep the same guidelines mentioned earlier in mind.

Adding Conflict Detection to the Enforcement Policy

To identify clients that are spoofing in enforcement policy rules, you should use the "[Endpoints Repository]: Conflict" attribute. Make sure that the service using the policy enables Authorization and has the Endpoints Repository in the Authentication or Authorization sources list.

As shown in Figure 12-48, add a rule to the enforcement policy rules with this condition: [Authorization: Endpoint Repository] Conflict EQUALS true. Set the enforcement profile to an appropriate profile such as one that assigns the client to a quarantine role on the infrastructure device. Make sure to move the rule to the top of the list. Then CPPM will quarantine any client for which the endpoint profile has a Conflict attribute set to true.

Add this rule to the **top**

[Authorization: Endpoints Repository]: Conflict EQUALS true	**quarantine-conflict**
Tips: Role EQUALS printers-cppm	printer-access
Tips: Role EQUALS cameras-cppm	camera-access
Tips: Role EQUALS profiling-cppm	profiling-access
ELSE quarantine	

Figure 12-48: Adding Conflict Detection to the Enforcement Policy

If you have not already done so, you also need to enable the Profiler on this service and set the CoA action to occur whenever the category, family, or type changes. This step is necessary because the Conflict attribute does not become true until a little after the spoofing client has connected. You need the Profiler to trigger a CoA so that the spoofing client is forced to reauthenticate and can then be denied.

Note that, in the example scenario that you have been following, the rule that matches the Conflict attribute is not strictly required. Any client with a category other than Printer or Cameras will be quarantined anyway. However, this rule can still be useful. First, admins might want to distinguish between devices attempting to spoof their MAC address, which indicate a more critical security issue, and devices that simply are not allowed access because they are in the wrong category. Second, some customers might want to permit devices in the Computers category, so the rule that denies access, or quarantines, based on the Conflict attribute is necessary to secure the network.

Ignoring Conflict Flags on Devices Using PXE

Some devices use Preboot Execution Environment (PXE), which allows them to boot an OS image from a location on the network rather than a local disk. VMs often use PXE, but IoT devices and other clients can as well. When PXE is running and before the device has received its OS, CPPM profiles the client for Device OS Family and Device Name. As shown in Figure 12-49, after the client finishes booting, CPPM reclassifies it as whatever OS type it ends up running.

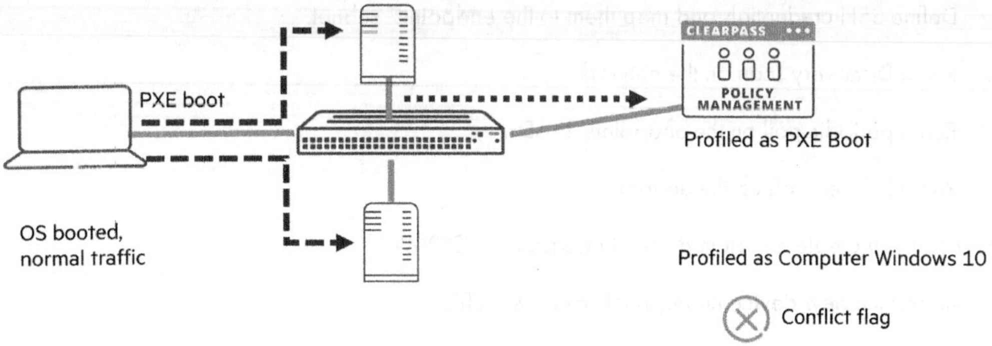

Figure 12-49: Profiling Devices Using PXE

This change in category triggers a conflict flag. However, typically you do not want this situation to trigger that flag because it does not represent a security issue, but simply normal operation. Under **Administration > Server Manager > Server Configuration > Cluster-Wide Parameters > Profiler**, you can set the value of **Ignore Conflict (Network Boot Agents)** as **TRUE**. In such cases, when a device is imaged by using PXE boot, no profiler conflict is triggered.

Summary

Congratulations! You now know all about setting up and using endpoint classification on CPPM. You can collect information to classify devices. You can analyze the collected information and take steps to audit the environment.

You can also use device categories in role mapping and enforcement policies to ensure that each type of device receives the correct level of access. And you can use device profiling as part of a system for enhancing the security of MAC-Auth and preventing MAC spoofing.

Learning checks

1. How do you enable CPPM to receive device fingerprint updates from HPE Aruba Networking?
   a. Integrate CPPM with CPDI.
   b. Integrate CPPM with HPE Aruba Networking Central.
   c. Enable auto endpoint profiler fingerprint updates in the cluster-wide parameters.
   d. Import an XML file from the Administration > Dictionaries > Device Fingerprints page.
2. What steps do you complete to scan endpoints with SSH? (Select two.)

a. Define SSH credentials and map them to the endpoints' subnet.

b. Run a Discovery scan on the network.

c. Run a periodic poll on the endpoints' NADs.

d. Run a Subnet scan on the endpoints.

3. How can you create a custom device fingerprint on CPPM?

a. Import the new device fingerprint from an XML file.

b. Create the new device fingerprint and then import rules from a discovered device.

c. Create the new device fingerprint and then define rules for it manually.

d. Import the new device fingerprint from a third-party solution.

4. You want to configure the profiling action for a CPPM service, but you do not see the Profiler tab. What should you do?

a. Recreate the service, selecting a template that includes "Profiling" in the name.

b. In the Service tab, change the template to one that includes "Profiling" in the name.

c. In the Service tab, select the Profile Endpoints check box.

d. Add a rule that uses the "IsProfiled" attribute to a role mapping or enforcement policy.

5. In which of these circumstances does CPPM set an endpoint's Conflict attribute to true?

a. When the endpoint first connects before it is profiled

b. When the endpoint category changes from unknown to "Computer"

c. When the endpoint category changes from "Cameras" to "Computer"

d. When the endpoint has fingerprints that match multiple device fingerprint entries

---

Note

For answers to all learning checks, refer to the Appendix.

---

13 Device Profiling with HPE Aruba Networking

EXAM OBJECTIVES

✓ Define HPE Aruba Networking ClearPass Device Insight (CPDI)

✓ Understand client profiling integrated in HPE Aruba Networking Central Network Operations

Assumed Knowledge

- Definition of device profiling and basic understanding of profiling methods

Overview

With the proliferation of Internet of Things (IoT) companies need a specialized solution that can identify the many different types of devices that users are adding to the network. CPDI delivers the sophisticated device profiling companies need to gain insight and control over their ever-evolving device ecosystem.

You will begin this chapter by examining the CPDI solution and architecture. You will learn how CPDI supports some of the same profiling methods as HPE Aruba Networking ClearPass Policy Manager (CPPM), but it augments profiling with additional methods such as machine learning.

You will then look at another way customers can obtain client profiling information: client profiles integrated directly in HPE Aruba Networking Central.

Introduction to the Solutions

This section introduces you to HPE Aruba Networking cloud-based solutions for device profiling.

Why Device Profiling from HPE Aruba Networking Cloud Solutions

Device profiling solutions should discover every device that touches the network and then identify the device type and OS. But most solutions struggle with IoT devices, calling many different types of smart devices generic Windows or Linux devices, for example.

Not with HPE Aruba Networking. As indicated in Figure 13-1, HPE Aruba Networking Central and HPE Aruba Networking ClearPass Device Insight (CPDI) provide automatic, comprehensive device discovery and profiling for the many types of devices that exist out in the wild today. They use a broad range of techniques, including machine learning, to characterize device traffic and match characteristics to fingerprints of known device types. In this way, they can tell admins that our example generic Windows device is actually a specific vendor's security camera.

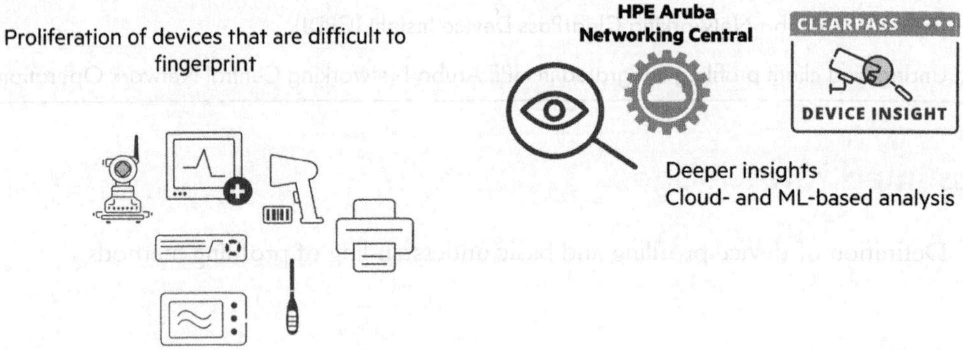

Figure 13-1: Why Device Profiling from HPE Aruba Networking Cloud Solutions

Two Main Options

Organizations have two main ways to obtain this advanced device profiling from HPE Aruba Networking, as summarized in Figure 13-2.

HPE Aruba Networking Central
Client Profiles

- Included in HPE Aruba Networking Central Network Operations subscription
- Uses telemetry from HPE Aruba Networking devices
- Key features:
  - Device profiling with machine learning
  - Collection of multiple attributes about the client and its traffic flows
  - Integration with CPPM

CLEARPASS
DEVICE INSIGHT

- Separate subscription
- Uses active and passive scanning from Data Collectors (supports heterogeneous networks)
- Key features:
  - Device profiling with machine learning
  - Collection of multiple attributes about the client and its traffic flows
  - Integration with CPPM
  - Security scanning (posture and vulnerability assessment)

Figure 13-2: Two Main Options

Client Profiles built into HPE Aruba Networking Central

Client profiles are included within the HPE Aruba Networking Central Network Operations subscription. HPE Aruba Networking Central constructs these profiles using telemetry pulled from HPE Aruba Networking devices, such as APs, switches, and gateways. It uses machine learning (ML)-enhanced techniques to analyze that data and determine the client type, family, and category. The client profile also includes rich context about the client such as the connections that it is establishing and the applications that it is using. As you will learn in more detail in Chapter 15, you can integrate with HPE Aruba Networking ClearPass Policy Manager (CPPM) to feed client profile information to that solution.

HPE Aruba Networking ClearPass Device Insight (CPDI)

Organizations can alternatively obtain CPDI as a separate subscription. CPDI uses Data Collectors, which run active and passive scanning methods to collect information about clients. CPDI can therefore profile clients in heterogenous networks that incorporate third-party network infrastructure while the client profiles in HPE Aruba Networking Central work only for clients connected to HPE Aruba Networking devices.

CPDI provides rich device profiling using a variety of ML techniques. CPDI can also integrate with CPPM and exchange information with it. In addition, CPDI supports security scanning, including posture and vulnerability assessments.

Device Profiling with CPPM vs HPE Aruba Networking Central/CPDI

As you learned earlier, companies can get started with device profiling using CPPM. However, organizations obtain a richer feature set using HPE Aruba Networking Central, either with the CPDI license or the built-in client profiles.

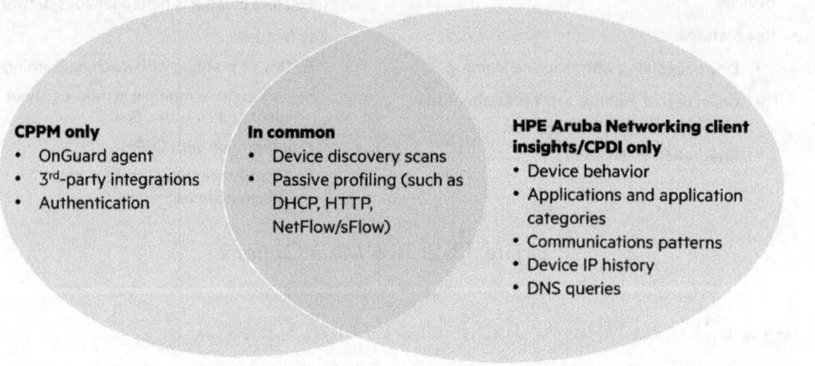

Figure 13-3: Device Profiling with CPPM vs HPE Aruba Networking Central/CPDI

As shown in Figure 13-3, both CPPM and CPDI support device discovery scans, as well as passive fingerprinting methods such as DHCP, HTTP User-Agent strings, NetFlow, and sFlow. (CPPM and CPDI obtain the HTTP User-Agent string in different ways, but both can find it.)

However, CPPM is optimized as an authentication server and policy engine while CPDI is dedicated and optimized to device profiling. It can use DPI to inspect traffic at a deep level. It further applies machine learning to look for patterns and cluster devices together based on these characteristics. It can detect patterns in device behavior and the types of applications and application categories that the device uses. It can also analyze communication patterns such as where the device tends to send traffic and at what bandwidth levels. CPDI can also track a device's IP history, creating a consistent profile for the same client regardless of the IP changes. And it can look at the device's DNS queries to see the hostnames it is accessing, not just the IP addresses. By analyzing all of this information and applying rules and machine learning to it, CPDI can accurately categorize endpoints and create rich endpoint profiles.

CPPM does have a few capabilities that CPDI lacks. CPPM can draw on information from the OnGuard agent (if used), as well as third-party servers. Because CPPM also authenticates clients, it can combine endpoint information with information about the user on that endpoint. Often companies will see the best results from integrating CPPM and CPDI. CPPM can then provide authentication information from CPDI, as well as draw on information delivered by CPDI in its policies.

Automatic Device Classification

CPDI and HPE Aruba Networking Central automatically classify many devices.

First, as shown in Figure 13-4, they check device attributes against system rules to see if it finds a match. System rules are built into the application without any action required from users. If the solution finds a match with a rule, the device is added to the classified devices.

If no match exists for the device within the system rules, the solutions check their MAC range classifiers. If they finds a match there, the device again becomes classified.

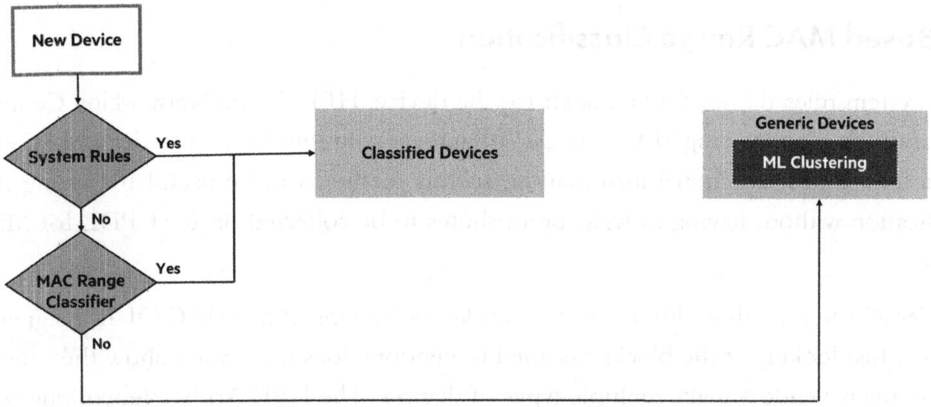

Figure 13-4: Automatic Device Classification

All devices that cannot be classified with system rules or MAC range classifiers become generic devices. In many solutions, users would be left with a pool of unclassified generic devices and not much help moving on from that point. CPDI distinguishes itself by taking a further step. CPDI Analyzer applies ML clustering to all of the generic devices. Instead of seeing a vast list of jumbled "generic" devices, users see a few clusters of similar devices, which they can more easily understand and address. The discussion of the ML clustering in this chapter focuses on CPDI, but you should be aware that HPE Aruba Networking Central client profiles features many of the same benefits.

What Data Are System Rules Based On?

System rules can draw on many sophisticated device attributes. In addition to using DHCP fingerprints, system rules might also reference the destinations and destination ports that the client is using. System rules can also reference a device's open ports, SNMP information, and OS information—all of which can be discovered by Data Collector scans.

These criteria are summarized in Figure 13-5.

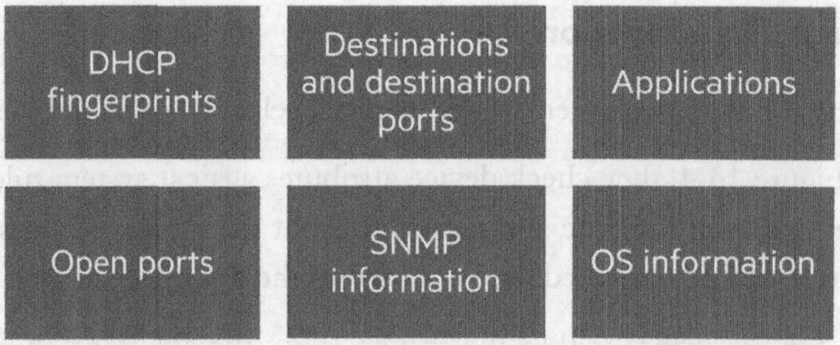

Figure 13-5: What Data Are System Rules Based On?

ML-Based MAC Range Classification

If the system rules do not find a match for the device, HPE Aruba Networking Central or CPDI can classify the device using MAC ranges. Often these solutions know a device's MAC address even before they collect additional information, so this method can be useful for giving the device a classification without having to wait for attributes to be collected or, for CPDI, for ML clustering to run.

This classification method differs from a simple rule looking at the MAC OUIs assigned to various vendors. Just looking at the blocks assigned to vendors does not always show the true device type because many vendors make multiple types of devices. The HPE Aruba Networking solutions, on the other hand, use crowdsourcing and ML to make the MAC-based classification more useful and accurate. On the order of every few weeks, they look across all tenants at rule-based classifications that meet various criteria (such as applying to at least 20 devices). It uses those rules to train itself to understand the MAC address ranges that are associated with various device types. As you see in Figure 13-6, the ML algorithm filters out the "noise" of outlying MAC address-to-type correspondences. It can zero in on the range that indicates the device type with confidence.

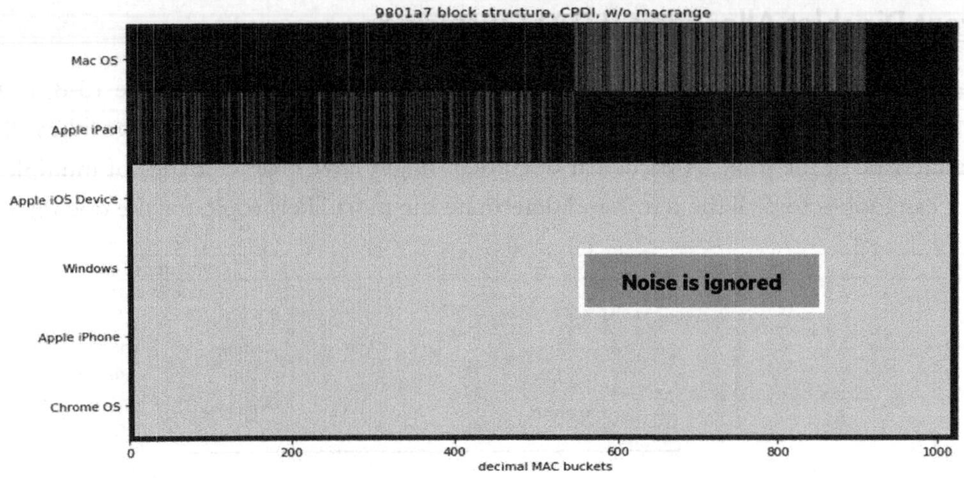

Figure 13-6: ML-Based MAC Range Classification

ML Clustering

As shown in Figure 13-7, CPDI applies ML clustering to generic devices with no classification every 12 hours at 0 UTC and 12 UTC. It runs for all tenants at the same time. If you check your generic devices in the time between when they are discovered and the next clustering process, you will see the generic devices labeled as clustering pending.

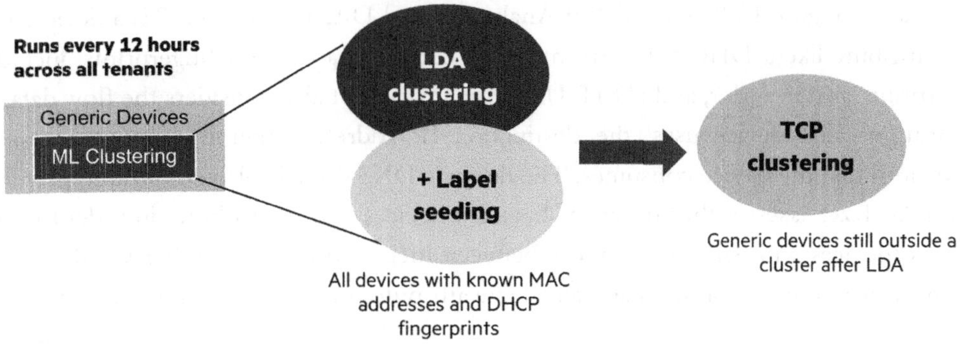

Figure 13-7: ML Clustering

CPDI 2.0.1 supports three mechanisms for clustering. It first applies Latent Dirichlet Allocation (LDA) clustering to devices that have known MAC addresses and DHCP fingerprints. LDA clustering can also use label seeding. CPDI then applies TCP clustering to the devices that are still left unclustered, including ones without MAC addresses.

Latent Dirichlet Allocation (LDA) Clustering

LDA is an ML algorithm for topic modeling. As shown at the top of Figure 13-8, LDA analyzes documents and places them in topic-based sets, based on terms that the algorithm discovers are characteristic of the topic. A particular document might have characteristics of multiple topics, but LDA can look across all the terms and determine the most likely topic for the document.

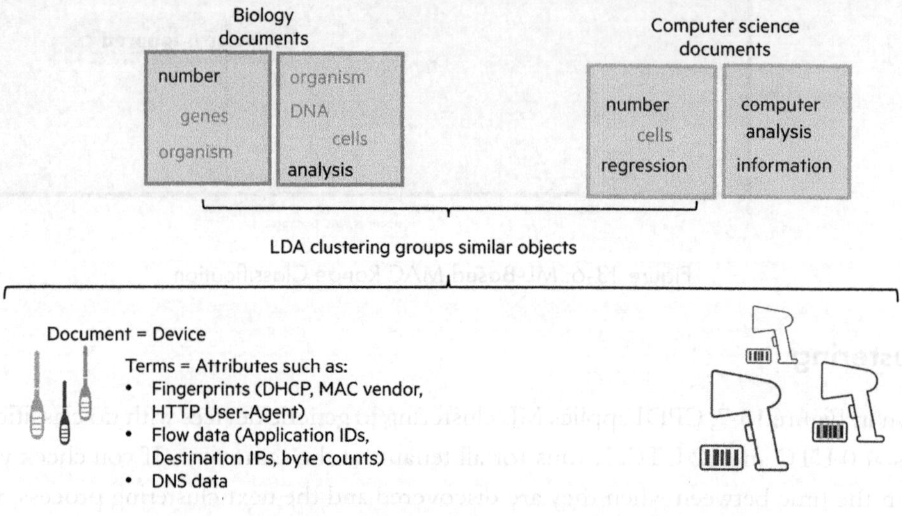

Figure 13-8: Latent Dirichlet Allocation (LDA) Clustering

As shown in Figure 13-8, when CPDI Analyzer uses LDA, a "document" is a device, and a "term" is an attribute like a DHCP fingerprint. LDA considers a device's fingerprints, including DHCP fingerprint, MAC vendor, and HTTP User-Agent string. It also considers the flow data, such as the applications that device uses, the destination IP addresses that the device accesses, and the bandwidth that the device consumes. The device's DNS domain also comes into play. Using LDA, CPDI Analyzer assesses the "distance" between devices' sets of attributes in order to cluster similar devices (devices with smaller distances between their attributes). In other words, it finds devices that have attributes in common and also have attributes that distinguish them from other clusters.

Cluster Recommendations

While CPDI only shows users the generic devices in a cluster, it actually applies the clustering algorithm across all of a tenant's devices, including both classified and generic devices. In this way CPDI can discover classified devices with similar attributes to a generic device cluster. CPDI can then use the classified devices to provide users with a recommendation for classifying the generic device cluster.

For example, a retailer might have several Symbol barcode scanners. As shown in Figure 13-9, CPDI might have succeeded in classifying some of these devices. Other devices do not quite match the rules or ML-based MAC classification and remain unclassified. However, when CPDI Analyzer runs the clustering, it discovers that the classified scanners and the unclassified devices have many attributes in common. It creates a cluster of the unclassified devices and presents a recommendation that these might be Symbol Scanner devices.

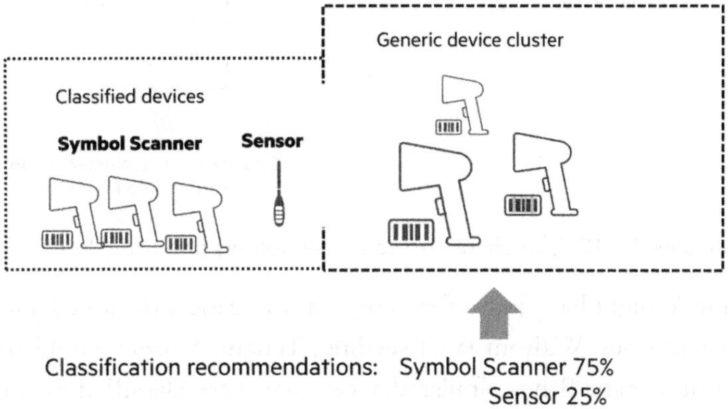

Figure 13-9: Cluster Recommendations

If the cluster includes classified devices of multiple types, CPDI will present multiple recommendations. It will accompany each recommendation with a percentage, based on the percentage of classified devices in the cluster of that type.

Classification Recommendations and Label Seeding

But sometimes LDA produces a cluster without any classified devices, which means that CPDI cannot provide a recommendation, as shown on the left of Figure 13-10. This is most likely to occur with smaller customers or with less popular and less frequently used devices. To address this issue, HPE Aruba Networking adds label seeding to LDA, as shown on the right of Figure 13-10. This feature establishes a "seed set," which consists of the LDA output of the less popular classified devices from across all tenants. CPDI Analyzer can then use that seed set to maximize the chances that every cluster will have at least one classified device in it.

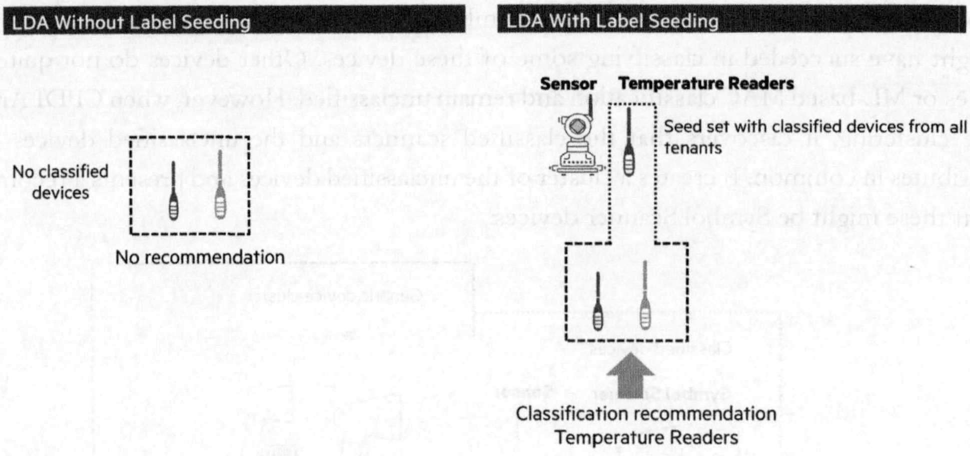

Figure 13-10: Classification Recommendations and Label Seeding

For example, Tenant A might have just a few temperature readers deployed. None of these devices are classified by system rules. Without label seeding, Tenant A users would see a cluster with no recommendation. But Tenant B has similar devices that were classified as a temperature reader. Using the seed set, CPDI Analyzer recognizes that Tenant A's device cluster has attributes similar to Tenant B's readers. It recommends to Tenant A that its device cluster might be temperature readers.

Note that this process occurs in the background, and Tenant A users do not receive any information about Tenant B's devices.

Auto-Classification through Cluster Recommendations

Typically, CPDI Analyzer leaves the generic devices in the generic device cluster and waits for a user to accept the recommended classification or apply a different classification. However, CPDI automatically classifies the generic devices itself if the top recommendation meets two criteria. First, the percentage must be 95% or above. Second, the number of classified devices associated with this recommendation must be 10 or more.

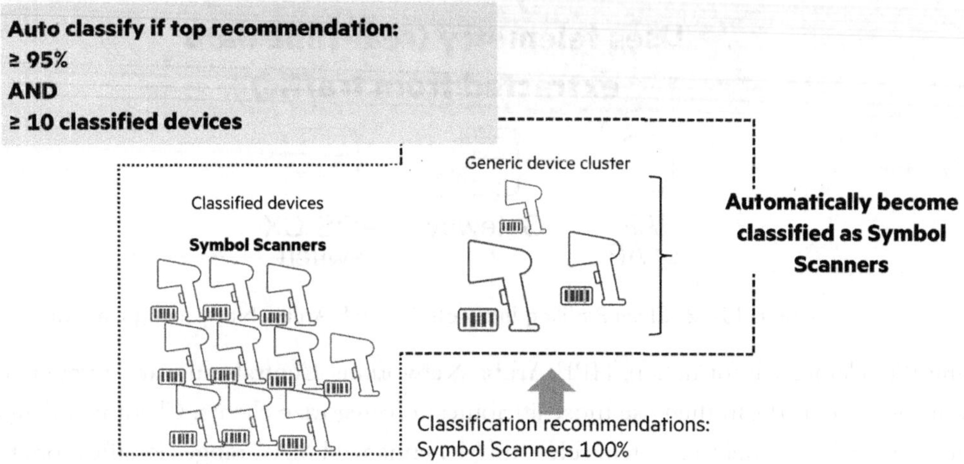

Figure 13-11: Auto-Classification through Cluster Recommendations

In the example shown in Figure 13-11, LDA creates a cluster that includes several generic devices, as well as 10 devices classified as point of sale (PoS) devices. Because all of the classified devices are the same type, CPDI automatically classifies the generic devices as the same type.

Use HPE Aruba Networking Central Client Insight

For this rest of this chapter, you will focus on client profile information integrated in HPE Aruba Networking Central. (The next chapter covers features specific to the standalone CPDI application.)

Client Profiling Integrated in HPE Aruba Networking Central

HPE Aruba Networking Central collects client attributes using telemetry from network devices that it manages, as indicated in Figure 13-12. Network telemetry consists of data that the network devices extract from traffic flowing through them and that they stream in real time to HPE Aruba Networking Central. For example, network infrastructure devices can track the MAC addresses in traffic and inform HPE Aruba Networking Central of those MAC addresses.

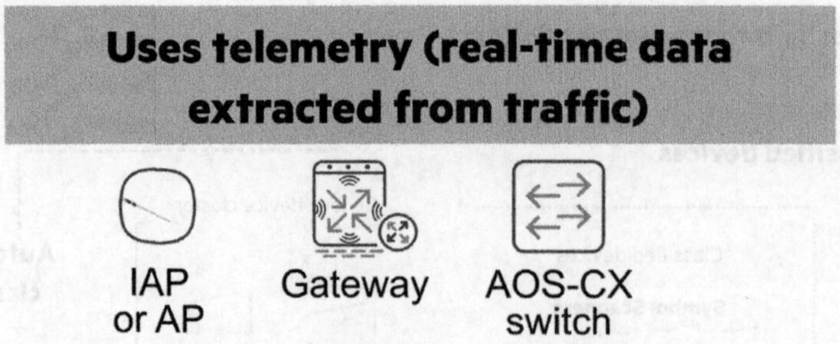

Figure 13-12: Client Profiling Integrated in HPE Aruba Networking Central

Using the telemetry information, HPE Aruba Networking Central can learn a variety of attributes about each client. It can then use those attributes to categorize clients. Client profiling in Network Operations relies on built-in system rules and ML-enhanced MAC range classification. For example, system rules might look at DHCP fingerprints, open ports, and other client attributes to figure out the client type. HPE Aruba Networking client profiles even supports auto-rules, which use ML to automatically create new rules to classify devices, focusing on the most important features and filtering out the noise.

In addition to viewing a client's category, family, and type within the HPE Aruba Networking Central client profile, admins can see all of the client's attributes, as well as track the client's network activity.

Data Collected for Client Profiling

You will now examine the information provided by HPE Aruba Networking devices managed by HPE Aruba Networking Central.

To deliver telemetry data, an HPE Aruba Networking InstantOS version must be 8.7.0.0 or above. APs can provide MAC information, DHCP information, and HTTP User-Agents. (See Figure 13-13.)

Data	Switch*	IAP or AP	Gateway
Static attributes			
MAC attributes	X	X	X
DHCP info		X	X
HTTP User-Agents		X	X
OS via SMB			X
Flow attributes			
Applications			X
Protocols			X
Destination connections			X
Domain names			X

*Additional telemetry available in newer software; this guide focuses on 10.08 and below

Figure 13-13: Data Collected for Client Profiling

As shown in Figure 13-13, gateways submit the same telemetry information as APs, but also much more. They can discover the device OS. Additionally, they can collect information about traffic flows, including applications, protocols, destination connections, and domain names. An EdgeConnect SD-Branch Gateway must run AOS version 8.7.0.0-2.3.0.0 or above. Gateways of any type support telemetry with AOS 10.2 or above.

HPE Aruba Networking Central 2.5.4 and above can also receive MAC addresses from AOS-CX switches; the switches must run AOS-CX version 10.08 or above. The switches can also report additional telemetry, including information about flows, if they support HPE Aruba Networking CX Edge Insights. However, this study guide focuses on feature 10.08 and below, and CX Edge Insights feature was added in version 10.11; the feature might also require an Advanced Feature Pack. For more information, refer to HPE Aruba Networking CX switch documentation.

Also note that these particular capabilities might change with software updates, so check the latest release notes.

Recommendations for Collecting All Attributes on APs and Gateways

HPE Aruba Networking' built-in client profiling can automatically begin collecting information about all clients in gateways' and APs' user tables. However, to ensure that HPE Aruba Networking Central can collect all relevant attributes, you should check a couple settings. Make sure that the gateways or APs apply DPI and application visibility to the clients. You must enable these settings globally. For example, on gateways, go to the **Security > Applications** settings. Then enable firewall visibility and DPI. When you enable these features, you must reboot the devices. For DPI to apply to a client, it must be enabled globally and on the client's role. However, DPI is enabled on the role by default, so it should take effect on all clients after you enable it globally (unless you have manually disabled the feature on the role).

View Client Profiles

You can see the client profiles and classifications within the Network Operations application. Go to the **Clients** page within the Global or Sites view. Then click the **Clients Profile**. Note that the Clients Profile tab is not available for the **Clients** pages in a Group view.

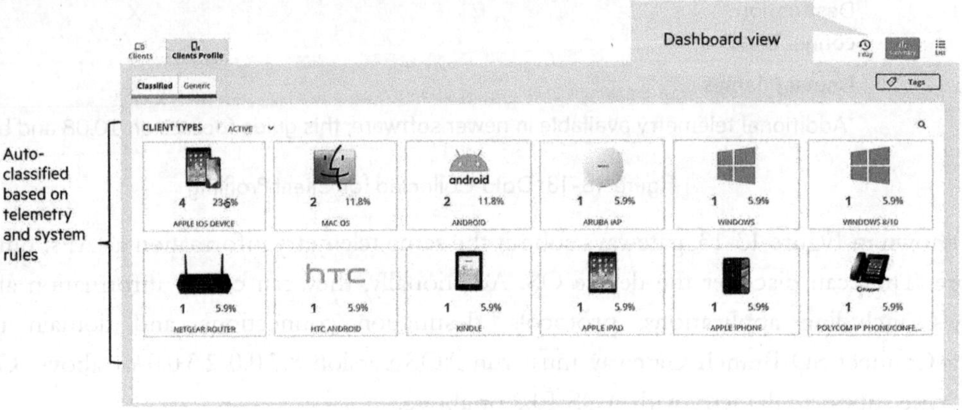

Figure 13-14: View Client Profiles

The **Clients Profile** page shows all of the clients known by HPE Aruba Networking Central, for the Global view, or all the clients at the site, for the Sites view. Clients are grouped by type. HPE Aruba Networking Central automatically assigns this type based on the network device telemetry and system rules. You do not need to take any further steps. You can click any of the client type cards and see the clients assigned to that type.

You can see an example in Figure 13-14. The view and capabilities might change in the future. The same holds true for the rest of this chapter.

View Client Profile Details

When you click a client in the client list, you will drill down into a view of that client. HPE Aruba Networking Central collects a variety of information about clients, including client profiling data, available in a **Profile** tab (see Figure 13-15). Here you see the device's category, family, and type (or OS). You can also see all the attributes collected through telemetry.

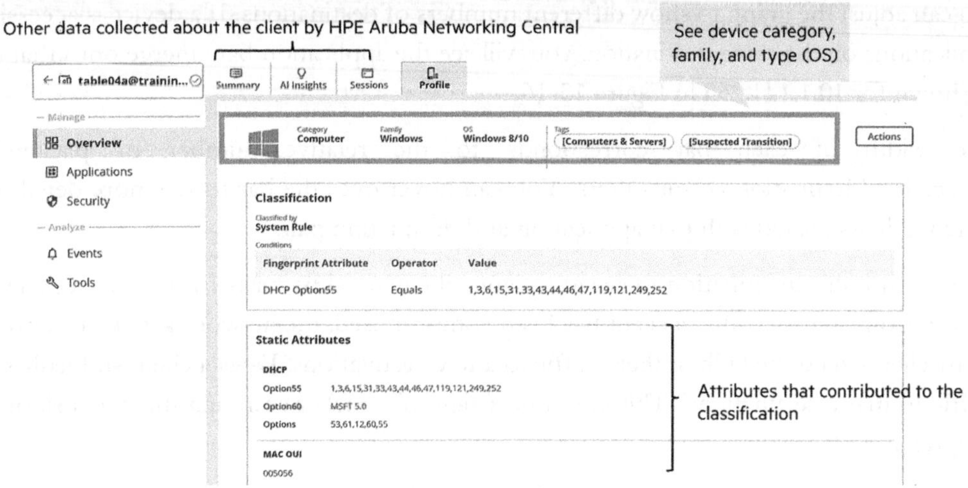

Figure 13-15: View Client Profile Details

View Device Network Activity

Scroll down and examine the Network Activity section to answer questions about the device's network activity and behavior. See Figure 13-16 for an example.

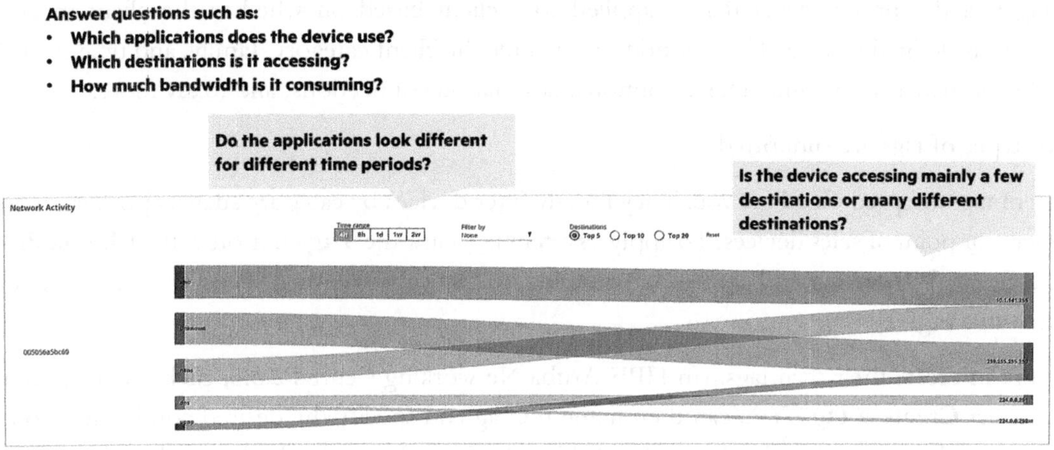

Figure 13-16: View Device Network Activity

The left of the display shows each application that the device is using while the right shows the destinations. A bar forms between each application and the destination on which the device is accessing that application. If a device is accessing that application on more than one destination, the application bar splits into multiple smaller bars, as you see for "dns" in the example. By default the application splits into at most five destinations, which are the top five for that application. But you can adjust the graph to show different numbers of destinations. If a device is accessing multiple applications on the same destination, you will see the application bars merge onto that destination, as shown for 10.1.141.255 in Figure 13-16.

The width of each bar corresponds to the relative number of packets for that application/destination versus others. You can hover over the bar to see more details such as the bandwidth associated with that application and destination pair.

You can use this information to see at a glance the general patterns for this device's traffic. If you have reason to suspect that a client has been compromised, the network activity can provide a clue. Many clients send the bulk of their traffic to a few destinations. Does a client suddenly start to send traffic to many destinations? This could be a sign of a hacker scanning the network or attempting to spread.

You can also adjust the time range for the Network Activity graph. You might use this feature to investigate potential issues. For example, if you see that the application/destination list looks quite different over the past 4 hours as opposed to the past two weeks, this might be a clue that a device has been compromised.

Tags

A tag is a descriptive string that is applied to a client based on whether the client meets the conditions defined in a tag. These conditions include the client category, family, and type, as well as attributes such as user name, DHCP options, destination connections, and many more.

Two types of tags are supported.

System tags are predefined for you. They mostly filter devices by category such as printers, medical devices, or point of sales devices. To apply a system tag, click the **Tags** button in the **Clients Profile** page. Select **System tag** and choose the desired tag. The page will then show only the devices that match that tag.

Users can create their own tags. (In HPE Aruba Networking Central 2.5.5, click the **Tags** button and select **Create a tag**.) You can then name the tag and specify the desired conditions. You can then apply the tag to filter the display by clicking the **Tags** button and selecting User tag. You can also see that tag in the detailed client profile for any client that matches the tag.

System tags in HPE Aruba Networking Central 2.5.5 include:

- Expected transition
- Suspected transition

- Library update transition
- Point of sale devices
- Collaboration and communication
- Networking and switching
- Computers and servers
- Audio and video
- Mobile and gadgets
- Facilities and building automation
- Printers
- Medical devices
- IOT

Example Use Cases for Tags

You can use tags however you want to mark important information about devices.

You might define tags that identify sub-groups of devices within a device type. For example, a healthcare organization might have many medical devices of the same type. As shown in Figure 13-17, you can use tags to identify the specific medical devices at a particular location or accessing specific destination hosts.

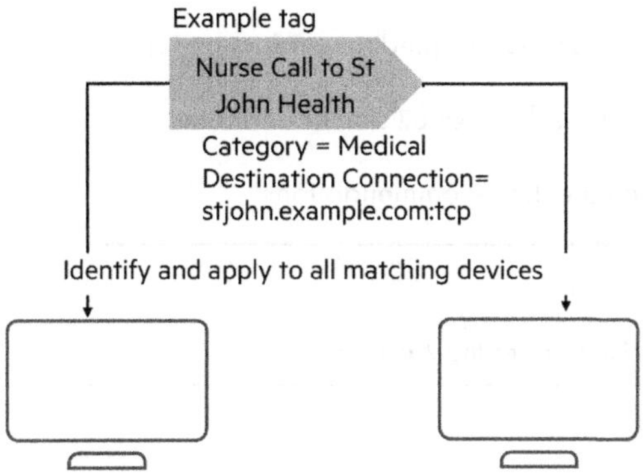

Figure 13-17: Example Use Case for Tags

You can also use tags to flag potentially vulnerable devices. You might want to track which devices have a particular open port. Then if a vulnerability is discovered for that port, it will be easy for admins to find all the potentially affected devices. Similarly, the company might want to track devices with legacy OS versions. In a moment, you will learn more about CPDI security context capabilities, which produce a risk score for devices. You can use tags to track devices with a particular score or higher.

These are just some use cases. You can come up with other ones based on the company's particular needs. No matter what the use case, within CPDI, tags help you to find and track devices that meet particular criteria. When you integrate CPDI with CPPM, the tags become even more powerful. CPPM can use the CPDI tags in role mapping and enforcement policies to affect a device's access rights. You will learn more about this use case in the next chapter.

Summary

Congratulations! You should now understand how CPDI provides companies with richer visibility into their network devices. You also took a look at client profile information that is integrated within HPE Aruba Networking Central. In the next chapter, you will dive into more detail about the standalone CPDI application, learning more about deploying and using it.

Learning check

1. When CPDI applies LDA clustering to generic devices, how does it produce recommendations for the cluster classification?
    a. It sends the cluster to HPE Aruba Networking experts who make the recommendation.
    b. It uses the MAC address ranges assigned to vendors by IEEE.
    c. It looks for classified devices that exhibit similar attributes.
    d. It uses system rules to create the recommendations.

Note

For answers to all learning checks, refer to the Appendix.

---

14 Deploy ClearPass Device Insight

EXAM OBJECTIVES

✓ Define and deploy HPE Aruba Networking ClearPass Device Insight (CPDI)

✓ Analyze endpoint classification data on CPDI

Assumed Knowledge

- Definition of device profiling and basic understanding of profiling methods
- Fundamental ClearPass Device Insight concepts

Overview

CPDI delivers the sophisticated device profiling companies need to gain insight and control over their ever-evolving device ecosystem. In this chapter, you will focus on deploying this application. You will first receive an overview of the application architecture, licensing, and data collectors.

You will then learn how to provision the CPDI application within HPE Aruba Networking Central and add Data Collectors. You will further learn how to set up data collection mechanisms to get your solution running.

Then you will be ready to view and use the information presented by CPDI. You will learn how to navigate through the interface provided for the licensed application. You will also learn how to complete common tasks such as classifying device clusters, creating user rules, and apply tags. And you will explore reports and strategies for furnishing stakeholders with relevant information.

You will then learn how to use CPDI security, a feature unique to the licensed application.

CPDI Overview

This section gives you an overview of CPDI.

CPDI Architecture

As shown in Figure 14-1, CPDI consists of a hybrid on-prem and cloud architecture, which combines the advantages of on-prem data collection with cloud intelligence.

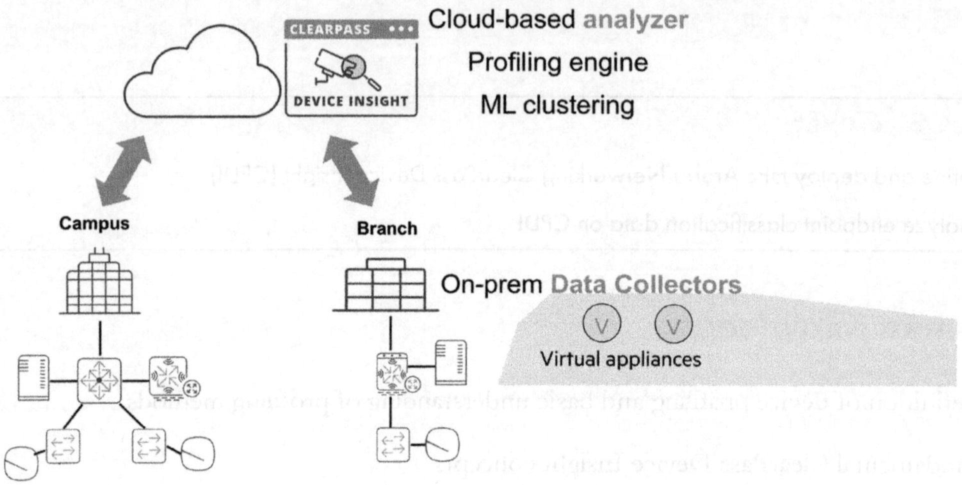

Figure 14-1: CPDI Architecture

On-prem Data Collectors collect data about a company's devices. CPDI Data Collectors are VMs that are dedicated to collecting network data. They support several passive and active methods for gathering data. They start to process that data, extracting, for example, DHCP fingerprints from DHCP Messages and TCP fingerprints from traffic received on a SPAN port. The Collectors send metadata about what they have found to the CPDI Analyzer.

The Analyzer takes the information submitted by the collectors to categorize devices using machine learning. Customers receive a complete inventory of their environment with rich information about each device. Each company has its own account on the CPDI Analyzer, connected to the company's HPE Aruba Networking Central account. Multi-tenancy maintains boundaries between customers' deployments so that each organization can only view information about its own devices.

CPDI Licensing

To use the CPDI application, a company requires a CPDI subscription. Customers can choose from 1-, 3-, and 5-year subscriptions. The subscription entitles CPDI to profile a specific number of endpoint devices. HPE Aruba Networking offers subscriptions for many different numbers of devices: 100, 500, 1000, 2500, 5000, 10,000, 25,000, 50,000, and 100,000. Note, however, that the CPDI standalone application is not supported in MSP mode as of the release of this study guide.

A valid CPDI subscription entitles the organization to download as many copies of the virtual Data Collector image as desired. IT staff members can then deploy the virtual Data Collectors on their choice of hardware that meets the requirements specified by HPE Aruba Networking.

Data Collector Collection Mechanisms

You will now examine in more detail the Data Collectors, which are deployed together with the CPDI application. Data Collectors support several mechanisms for collecting information about devices.

DHCP

The DHCP mechanism should be familiar to you, as it resembles the fingerprinting used by CPPM.

At each site you should add that site's Data Collector as an additional IP helper on DHCP relays. As shown in Figure 14-2, then Data Collectors can receive clients' DHCP Discover messages and extract the DHCP fingerprints to send to the Analyzer. DHCP options collected include options 55, 60, and others.

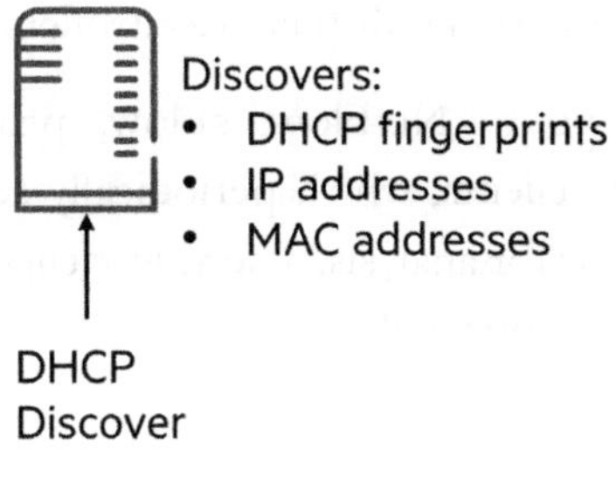

Figure 14-2: DHCP

SPAN

You should connect each Data Collector's SPAN port to a port on a core network infrastructure. That switch should mirror traffic to the SPAN port. The Data Collector analyzes packets to discover information about clients' conversations and applications. The information includes static attributes about devices, their IP addresses, and HTTP User-Agents. Data Collectors can also analyze traffic flows and learn the destination hosts and ports that each device is accessing. Based on that and other information, the Data Collector can determine which applications and protocols a device is using. By listening to DNS traffic on the SPAN port, the Data Collector can also discover domain names associated with IP addresses.

In short, collecting and analyzing SPAN gives CPDI insight into device behavior, as shown in Figure 14-3.

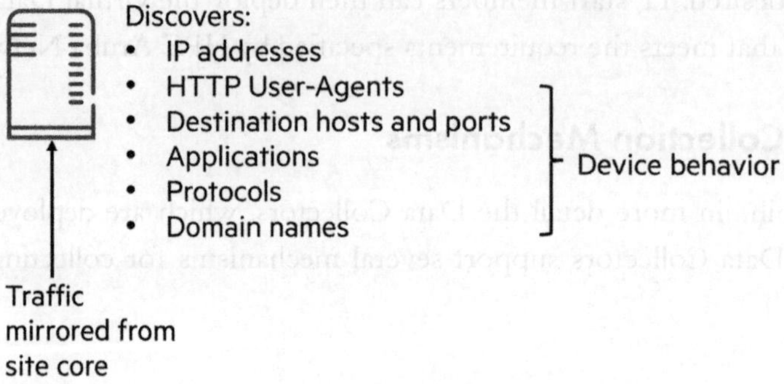

Figure 14-3: SPAN

NetFlow/IPFIX/sFlow

NetFlow is Cisco traffic collection technology. NetFlow clients, which are typically Cisco switches, gather data about conversations, or flows, passing through them. They send the flow information to a collector for analysis. IP Flow Information Export (IPFIX) is based on NetFlow v9, but it is an open standard. HPE Aruba Networking Mobility Controllers (MCs) support IPFIX.

Another open standard alternative to NetFlow, sFlow, provides traffic sampling. Network infrastructure devices act as sFlow clients, which periodically select packets from the traffic flow and send those packets to a collector for analysis. The sFlow clients can also send interface statistics to the collector. AOS-CX switches support sFlow.

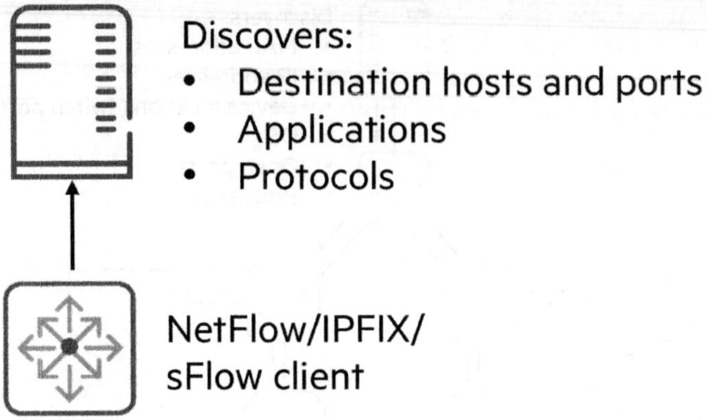

Figure 14-4: NetFlow/IPFIX/sFlow

As shown in Figure 14-4, an HPE Aruba Networking Data Collector can act as a collector for all of these technologies. In this way, CPDI can gain insight into the destination hosts and ports with which each device is communicating, as well as the applications and protocols that the device is using. These technologies give CPDI some insight into device behavior, although less so than with SPAN.

Scans

Data Collectors support active scanning with SNMP, WMI, SSH, and Nmap. As you learned with CPPM, and as shown in Figure 14-5, these scans give insight into devices' MAC addresses, IP addresses, and device location. The Data Collector might also be able to determine the devices' OS, open ports, and processes running on it.

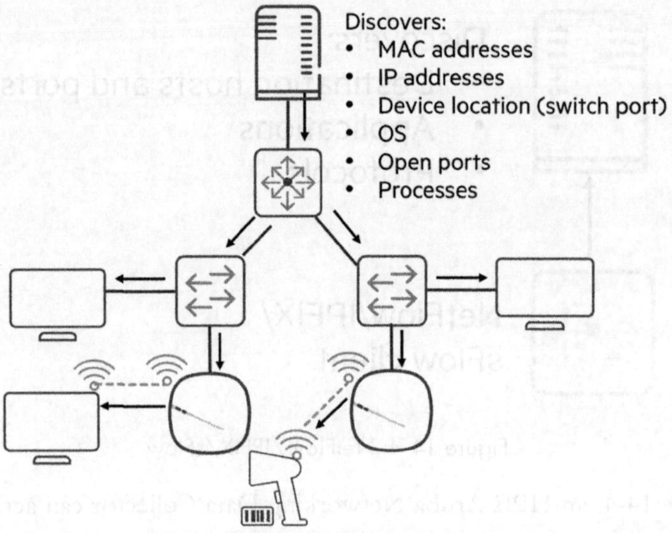

Figure 14-5: Scans

Data Collector Mechanisms Used by System Rules

As you learned in the previous chapter, CPDI has many built-in system rules. These rules draw on multiple device attributes obtained from various collection mechanisms. You should understand where CPDI obtains these attributes so that you can make sure that your solution furnishes the necessary information.

System rules often use DHCP fingerprints, which CPDI can receive from Data Collectors as long as those Data Collectors are properly set up as DHCP helpers.

System rules might also reference the destinations and destination ports that the client is using. CPDI Analyzer can receive this information from Data Collectors using scanning, the SPAN port, or Netflow/IPFIX. BGW telemetry can also furnish this information to the client profiling feature in Network Operations. Scanning, SPAN, and Netflow/IPFIX can also provide application information, which system rules can also use.

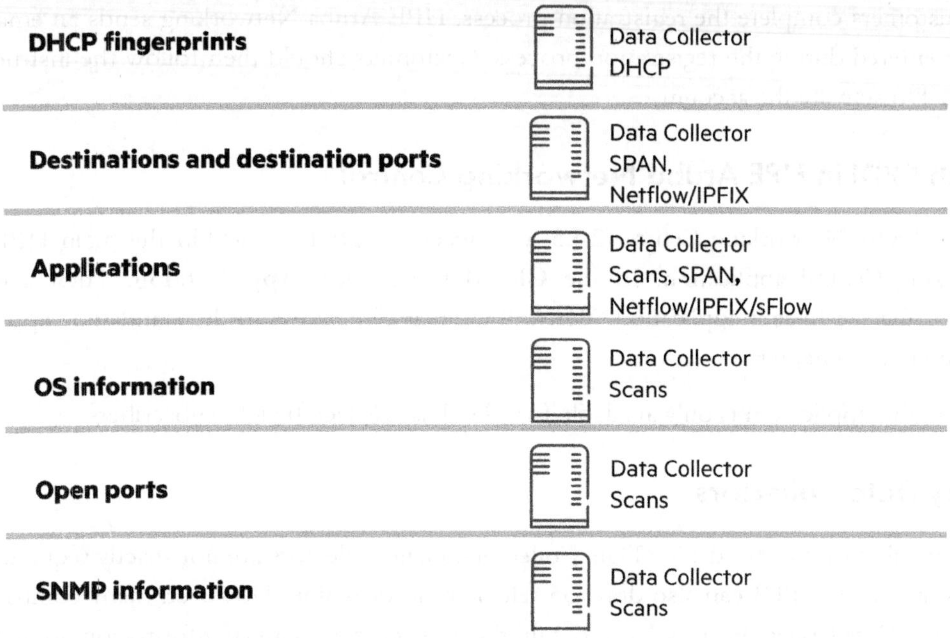

Figure 14-6: Data Collector Mechanisms Used by System Rules

System rules can also reference a device's open ports, SNMP information, and OS information—all of which can be discovered by Data Collector scans.

These capabilities are summarized in Figure 14-6.

Deploy the CPDI Application and Data Collectors

In this section, you will learn how to deploy the CPDI application and data collectors.

Select CPDI for the HPE Aruba Networking Central Account

You will now learn how to deploy the CPDI application.

First the company requires an HPE Aruba Networking Central account. Customers can access the arubanetworks.com site and click a link to register for an account; this link automatically redirects customers to the correct signup URL for the customers' region.

Customers should follow the instructions for registering for the account. When they reach the Interested Apps section, they should make sure to select ClearPass Device Insight. (They can select both Network Operations and ClearPass Device Insight if they want to use HPE Aruba Networking Central to manage the network too.)

After customers complete the registration process, HPE Aruba Networking sends an email to the account entered during the registration process. Customers should then follow the instructions in that email to activate the account.

Launch CPDI in HPE Aruba Networking Central

In HPE Aruba Networking Central 2.5.5, you access CPDI from within the main HPE Aruba Networking Central application. In the **Global** view, select **App Catalog**. Then launch the ClearPass Device Insight application. You will then receive access to the standalone app features described in this chapter.

Note that this application is only available for ClearPass Device Insight subscribers.

Deploy Data Collectors

Next you will learn how to deploy Data Collectors. Data Collectors are not strictly required for the CPDI solution, as CPDI can also draw on telemetry information. However, many companies will want to use Data Collectors in order to obtain the maximum amount of information possible about devices. If the company chooses to use Data Collectors, you should generally deploy one Data Collector at each campus, site, or data center with endpoints you want to profile. Following this guideline prevents too much traffic from flowing over WAN links.

The deployment process is summarized in Figure 14-7.

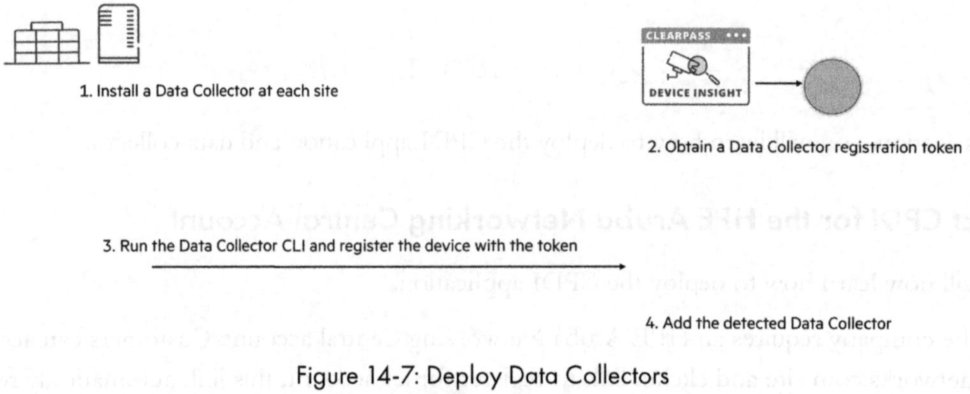

Figure 14-7: Deploy Data Collectors

You begin by installing the physical or virtual Data Collectors. Next you must obtain a Data Collector Registration Token from within the CPDI application. You will reference that token when you run the Data Collector CLI on your Data Collectors. You will also configure several other settings in the CLI to get the Data Collector connected to the network and detected in CPDI.

Finally, you will add the detected Data Collector to CPDI's list of Data Collectors.

Over the sections that follow, you will learn about each step in the process in more detail.

Data Collector Specifications

As mentioned earlier, customers can download the virtual appliance and run as many instances as they want. A virtual Data Collector appliance must be deployed on VMware vSphere 6.0 or later. You should then size the VM for the appliance based on the size of the site that the appliance will support. HPE Aruba Networking also recommends an Intel Ivy Bridge CPU or later. Specification recommendations are summarized in Figure 14-8.

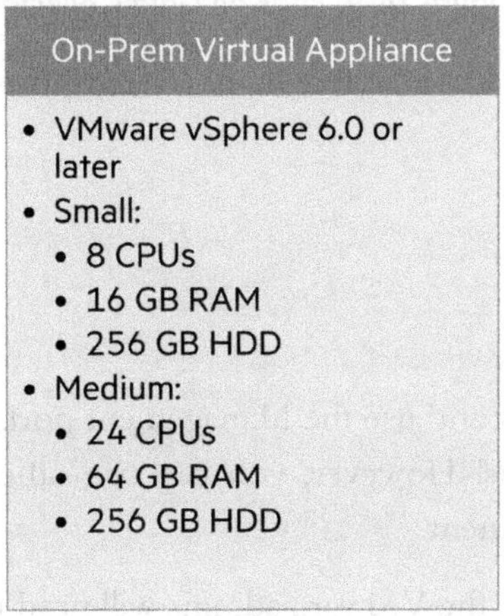

Figure 14-8: Data Collector Specifications

The Data Collectors do *not* require licensing separate from the HPE Aruba Networking ClearPass Device Insight subscription.

Ports on Data Collectors

To understand how to install the Data Collector, you need to understand a bit about its ports.

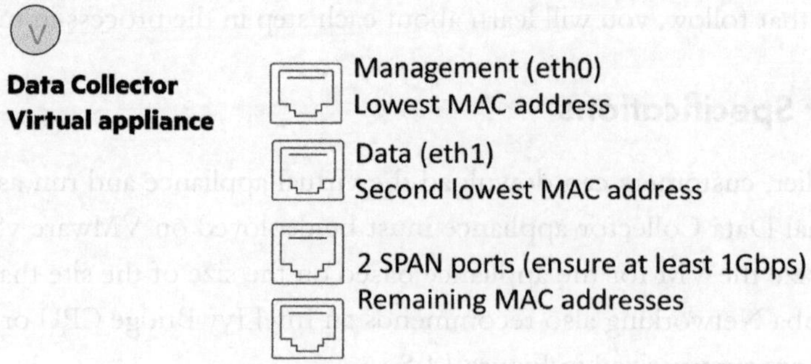

Figure 14-9: Ports on Data Collectors

As shown in Figure 14-9, Virtual Data Collectors have four ports:

- One Management port
- One Data port
- Two Span ports

You are only required to connect and use the Management port, which communicates with CPDI and can perform other functions. However, you can use other ports for specific functions, as explained in more depth in a moment.

If you want to use multiple ports, the VMware admins will need to create multiple virtual NICs for the VM and connect each to the correct network. Inside the VM, the Management Port corresponds with the eth0 interface, and the Data Port corresponds with the eth1 interface. However, this numbering does not necessarily correspond with the numbering for the connected virtual network adapter. The virtualization platform admins can determine the virtual network adapter for the Management Port by finding the adapter with the lowest numbered MAC address. The adapter for the Data Port will have the second lowest MAC address. The two SPAN ports have the highest MAC addresses.

In practice, these NICs might be sharing the ESXi host's physical NICs with other VMs. If you are using the SPAN ports, ensure that they can send and receive at least 1Gbps.

Guidelines for Data Collector Management and Data Ports

You must always connect the Data Collector's Management Port and ensure that it receives an IP address. You can set the IP address statically, or the port can obtain an address with DHCP. The Data Collector uses this port to communicate with HPE Aruba Networking Central and CPDI. Typically, the collector also uses this port to receive passive profiling information, including DHCP traffic and NetFlow/IPFIX/sFlow packets. As shown in Figure 14-10, it also uses this port to run active scans.

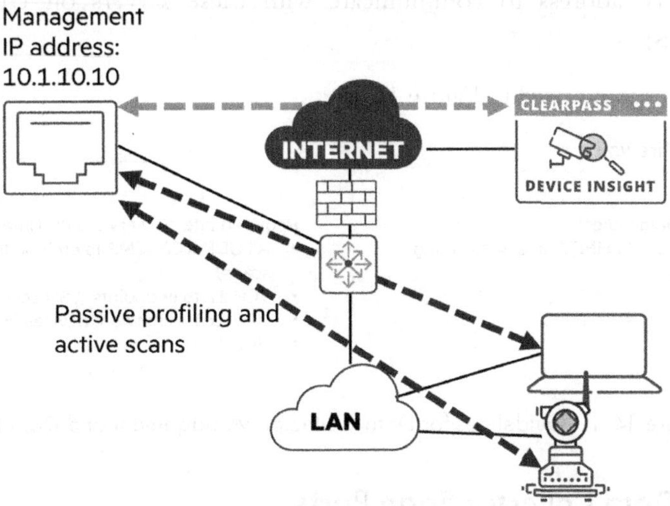

Figure 14-10: Guidelines for Data Collector Management and Data Ports

The Data Collector also supports an optional Data Port. If you choose to connect the Data Port, it requires its own IP address. The Data Collector then uses the Data Port to receive passive profiling information and run active scans. You generally only use this port if the corporate policy mandates separating the management and profiling data.

Summary of Network Services Required for Data Collectors

You should make sure that the perimeter firewall permits the Data Collector to access HPE Aruba Networking Central/CPDI. It must permit outbound TCP 443. If you want the Data Collector to use an external NTP or DNS server, the firewall must also permit the Data Collector to access these servers.

If the internal network infrastructure is implementing ACLs to restrict devices' access, make sure to permit the Data Collector sufficient access for it to perform its responsibilities. You should allow Data Collector to communicate with endpoints on all UDP and TCP ports so that it can perform WMI and Nmap scanning. ICMP is also required. Remember that the Data Collector uses its Data Port IP address for these scans unless its Data port is unconnected, in which case it uses the Management Port IP address. TCP 22 is required for SSH scans; however, if you are enabling full TCP access for other scans, you do not need to permit this port specifically. You should also permit the Data Port IP address (or, if none, Management Port) to send traffic to network devices on UDP

161 and to receive traffic on UDP 162. Then the Data Collector can scan network devices with SNMP. If the Data Collector is using internal NTP and DNS services, make sure to permit the Management Port IP address to communicate with these servers on UDP 123 (NTP) and TCP/UDP 53 (DNS).

These guidelines are summarized in Figure 14-11.

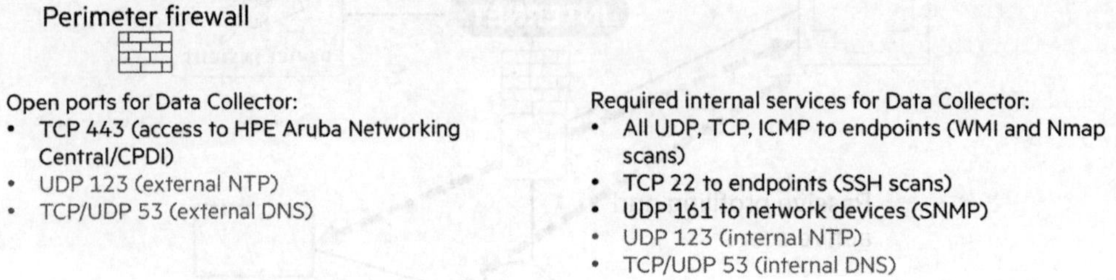

Figure 14-11: Guidelines for Data Collector Management and Data Ports

Guidelines for Data Collector Span Ports

The Span Ports are optional, but recommended. These ports have no IP addresses. The Data Collector listens for traffic on these ports and analyzes the traffic to determine device behavior.

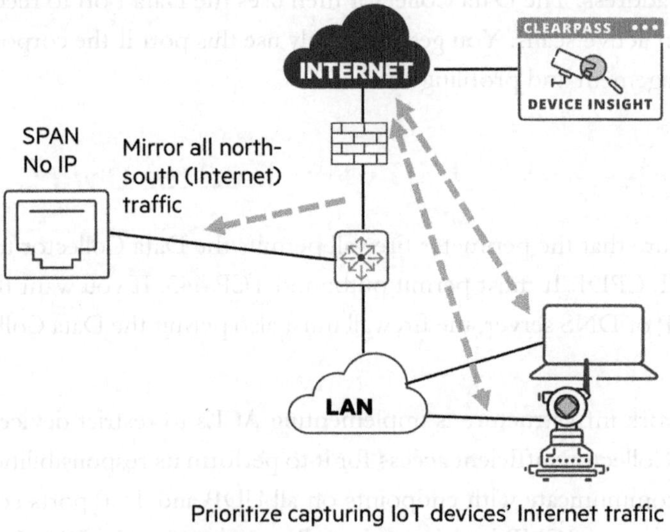

Figure 14-12: Guidelines for Data Collector Span Ports

As shown in Figure 14-12, wherever possible, choose a SPAN location where you can capture all north/south traffic between the endpoints and the internet. In this example, Internet traffic for all clients passes through the core switch. So you can connect the Data Collector SPAN port to a port on the core routing switch. You then create a mirror on this switch, which monitors the port that connects to the Internet and has as its destination the port that connects to the Data Collector SPAN port.

If a site is architected such that you cannot mirror all Internet traffic to the Data Collectors SPAN port, give priority to ensuring that the port receives IoT devices' Internet traffic.

Installing a Virtual Data Collector

Keeping these guidelines in mind, you are almost ready to install the Data Collector.

To install virtual Data Collectors, first download the virtual appliance, which is available from HPE Aruba Networking Central.

In HPE Aruba Networking Central 2.5.5, you go to the Global view. Select **Organization** and then the **Platform Integration** tab. Then click the **Data Collectors** card. Find and click the **get virtual appliance** link. The OVA file for the appliance downloads. (Note that these instructions might change in later versions; refer to the latest documentation for details.)

Give the file to VMware admins to use to deploy the Data Collector as a VM. Communicate a few guidelines. They should use a thick-provisioned, lazy-zeroed disk. They should also reserve resources specifically for the VM. The VM will have four network adapters. They should connect the network adapters based on the plan and guidelines discussed earlier. For example, if they are using a Management Port but no Data Port and one SPAN port, they can use the design shown in Figure 14-13.

- Thick provision lazy zero
- Reserved resources
- Follow previous guidelines to connect network adapters
  Example:
  - Adapter with lowest MAC
    -> VLAN for Management Port
  - Adapter with third lowest MAC
    -> Network for SPAN Port (4095 for VLAN tag pass-through)
  - Other adapters disconnected

Figure 14-13: Guidelines for Deploying the Virtual Data Collector VM

The network adapter with the lowest MAC address is placed in the VLAN in which the company wants to place the controller's Management Port. The adapter with the third lowest MAC address is placed in a network dedicated to SPAN. The network infrastructure often mirrors untagged traffic and tagged traffic in multiple VLANs to this port. VMware admins should configure the network for the SPAN port to use VLAN 4095, which means that the virtual switch simply passes all untagged and tagged traffic.

Obtain Registration Tokens

You are almost ready to set up the installed Data Collector. Part of the setup process involves inputting a registration token on the Data Collector.

You obtain this token from the Data Collectors page in HPE Aruba Networking Central. (In version 2.5.5, go to the Global view. Navigate to the **Organization** page and select the **Platform Integration** tab. Then click the **Data collectors** card.)

If the account does not have any Data Collectors yet, you should use the **Get started** card. Otherwise, will see existing Data Collectors.

In either case, as shown in Figure 14-14, click **Registration Token** and copy the token. Paste the output into a file so that you are ready to upload it on your Data Collectors.

Figure 14-14: Obtain Registration Tokens

Perform the Initial Setup in the Data Collector CLI

Power on the Data Collectors and connect to its console (see Figure 14-15). You will be presented with the Collector CLI. Enter **aruba** as the default username and password.

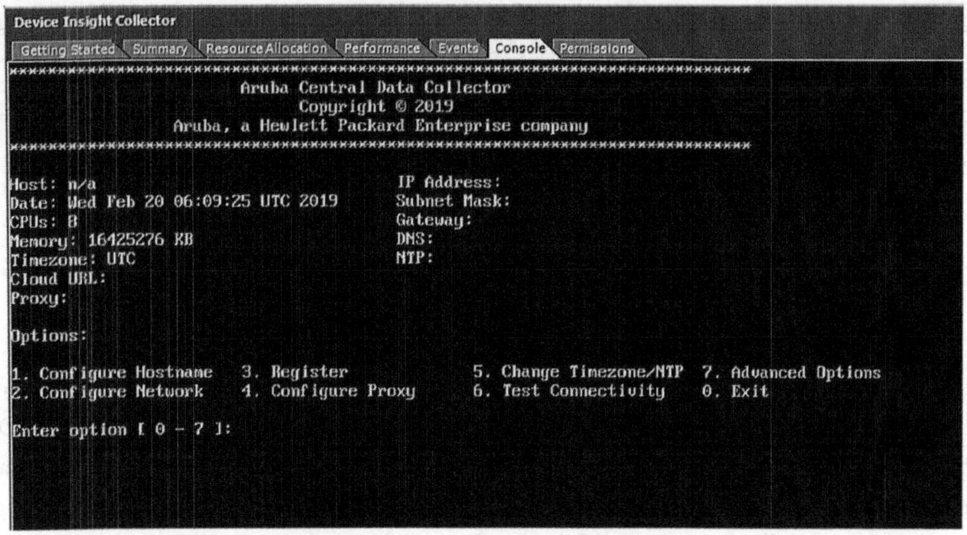

Figure 14-15: Perform the Initial Setup in the Data Collector CLI

You will be presented with a menu of options. Use these options to compete the initial configuration for the Data Collector.

Set a hostname. Then configure the networking settings. The eth0 interface is the Management Port. Configure the IP settings for this port, including the DNS server and any routes and proxy settings required to enable the Data Collector to reach the Internet on this port.

Again, using the menu, set the timezone and point the Data Collector to an NTP server.

Test connectivity with the HPE Aruba Networking cloud using the menu option. You can then register the appliance using the code that you previously collected. After you see the message that the registration was successful, it will take about 20 to 30 minutes for the Data Collector to establish a connection to the HPE Aruba Networking cloud and update its software.

Add Data Collectors in HPE Aruba Networking Central

Registering the appliance automatically directs it to your account in HPE Aruba Networking Central. The appliance is added to a list of appliances that can be created as Data Collectors in your CPDI deployment.

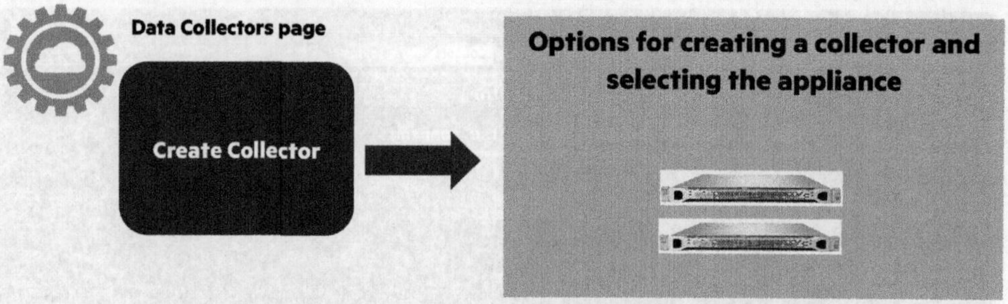

Figure 14-16: Add Data Collectors in HPE Aruba Networking Central

Return to the **Data Collectors** page. Then, as shown in Figure 14-16, select the option to create the collector. You will see options for naming your collector and choosing to install CPDI on it. You will also see a list of detected appliances. Select all the appliances that you want to add and complete the creation dialog. The Data Collectors will be added to the **Data Collectors** page. You can now manage the collectors and configure settings such as active scans on them.

Set Up CPDI Users

You need to create users who are allowed to manage the standalone CPDI application. Starting in HPE Aruba Networking Central 2.5.5, this application has been moved under the HPE GreenLake cloud. You manage user rights for all applications under the HPE GreenLake cloud within that cloud account's **Identity & Access** settings.

From those settings, as shown in Figure 14-17,you can invite new users based on their email addresses. You can then manage the user and assign the ClearPass Device Insight Administrator role.

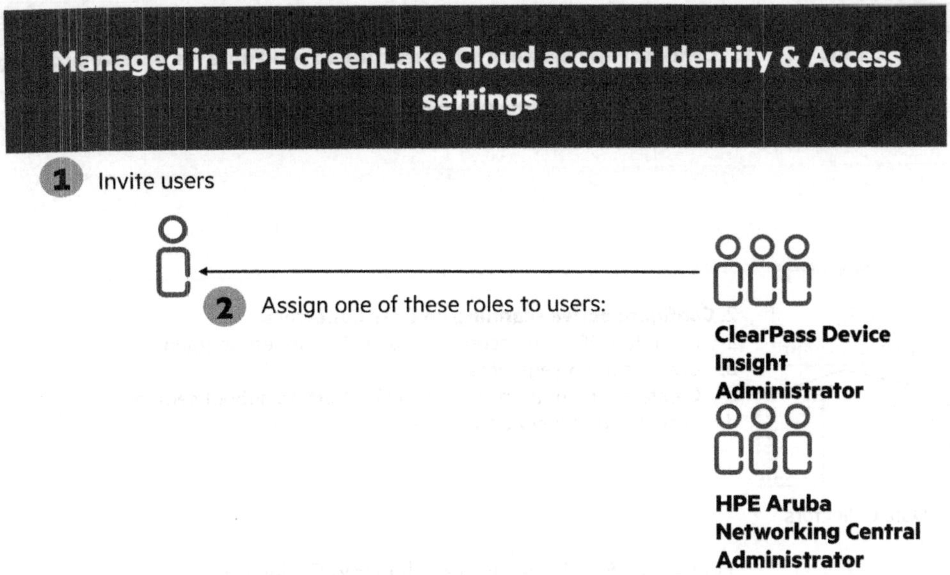

Figure 14-17: Set Up CPDI Users

Users with the HPE Aruba Networking Central Administrator role can also access and manage ClearPass Device Insight, as well as all of the HPE Aruba Networking Central Network Operations features.

Obtain Data on the Data Collectors

You will next learn how to set up the network infrastructure and the CPDI application to ensure that the Data Collectors obtain data about devices.

First, you will learn about setting up the network infrastructure to send the data necessary for passive profiling, including configuring the necessary DHCP relay settings, as well as mirroring traffic to Data Collectors SPAN ports. You can optionally set up NetFlow and sFlow as well.

You will then learn how to configure active scanning on Data Collectors. When you first log into CPDI, you will be prompted to accept or edit the Catch All Subnets segment configuration. After you do so, you can create your own subnet segments. You can then create augmentation methods and attach those methods to subnet segments. As you will learn in more detail in a moment, the subnet segments and augmentation methods control what scanning methods Data Collectors apply to what IP addresses. After you have configured these components, you can schedule discovery and subnet scans.

These steps are summarized in Figure 14-18. However, you will learn about each step in more detail in the sections that follow.

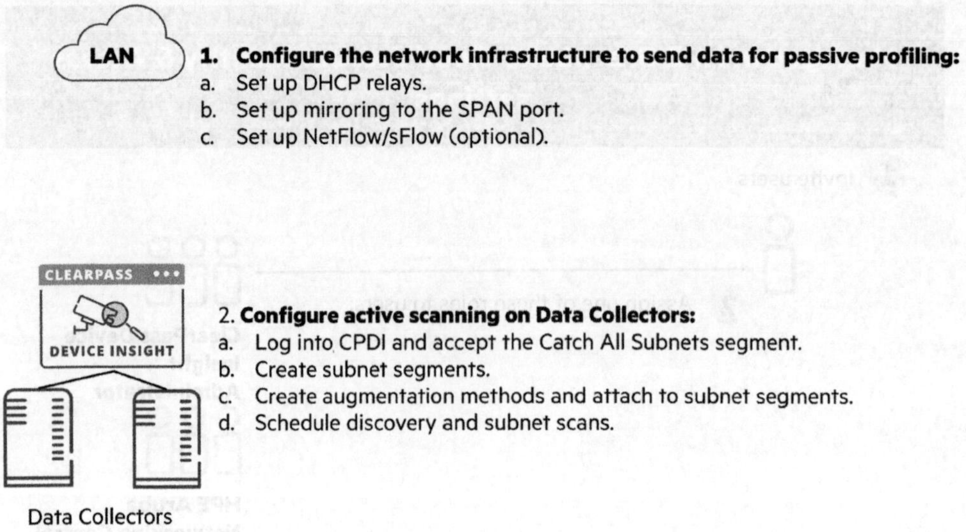

Figure 14-18: Obtain Data on the Data Collectors

Configure DHCP Helper Settings

Most corporate LANs use DHCP relay to direct clients' DHCP requests to a "helper address," which is the corporation's centralized DHCP server. The default router for a subnet typically acts as the DHCP relay for that subnet. In most two-tier networks, a core routing switch (or VSX pair) acts as the default router; in most three-tier networks, aggregation switches (or VSX pairs) act as the default router.

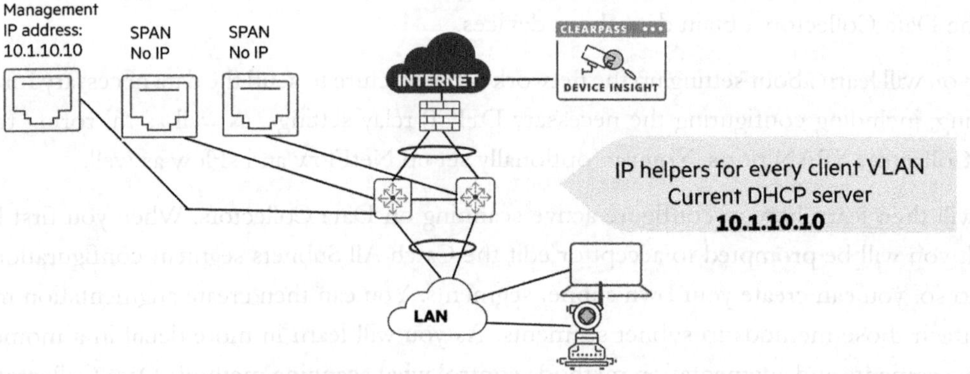

Figure 14-19: Configure DHCP Helper Settings

To copy DHCP information to the Data Collector, configure whatever device is already acting as DHCP relay. As shown in Figure 14-19, add the Data Collector's Data Port IP address to the list of IP helper addresses for every subnet on which you want to enable device profiling. Or if the Data

Collector is not using the Data Port, as in this example, add the collector's Management Port IP address as an IP helper. Be careful to leave the current IP helper address as well, so that clients continue to receive dynamic addresses from the DHCP server.

Usually APs forward traffic into the LAN at Layer 2, so they do not need to act as DHCP relays. If your APs are acting as the DHCP relay for any subnets, though, you would add the Data Collector as an IP helper in those APs' settings.

Set Up Mirroring to Span Ports

As discussed earlier, it is important for the Data Collector to be able to gather a broad range of traffic from clients across the site. Typically, you should deploy the Data Collector at a point where it can receive all inbound and outbound traffic on the site's Internet connection.

As shown in Figure 14-20, often the core network infrastructure device is actually a pair of switches using a virtualization technology such as VSX and the uplink to the Internet is a multi-chassis link aggregation group (M-LAG). To minimize traffic flowing over the inter-switch link (ISL) between the VSX switches, you can connect both switches to one of the SPAN ports on the Data Collector. Then configure both switches to mirror traffic from the M-LAG (lag1 in this example) to the port connected to the Data Collector. From the point of view of the Data Collector, it simply receives some conversations on one port and some on another. (Or for a virtual Data Collector, you might connect the two VSX switches to two ports on an ESXi host that hosts the collector VM. These ports would be uplinks on a virtual switch, which has a network to which the collector's SPAN network adapter is connected.)

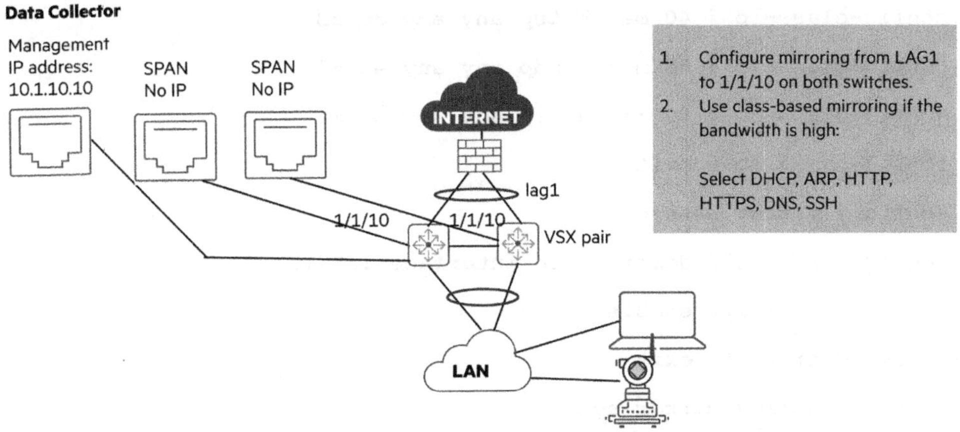

Figure 14-20: Set Up Mirroring to Span Ports

Also pay attention to the bandwidth on the Internet link. While it is possible to connect multiple 10Gb SPAN interfaces on the Data Collector, the Data Collector cannot process the full 20Gbps.

If the physical SPAN interfaces will be receiving more than the supported throughput for your appliance, you should use class-based mirroring to select only specific traffic for mirroring. The data collector's throughput depends on the sizing for the ESXi host and VM resources. The most helpful protocols to include in the class are: DHCP, ARP, HTTP, HTTPS, DNS and SSH. The same applies for sending traffic from a packet broker such as Gigamon.

Use these commands to configure a mirror for this scenario on AOS-CX switches in a VSX pair:

```
VSX-1(config)# mirror session 1
VSX-1(config-mirror-1)# source interface lag1 both
VSX-1(config-mirror-1)# destination interface 1/1/10
VSX-1(config-mirror-1)# enable
VSX-2(config)# mirror session 1
VSX-2(config-mirror-1)# source interface lag1 both
VSX-2(config-mirror-1)# destination interface 1/1/10
VSX-2(config-mirror-1)# enable
```

If you need to use class-based mirroring and the network uses IPv4, use these commands:

```
VSX-1(config)# class ip span
VSX-1(config-class-ip)# 10 match udp any any eq 67
VSX-1(config-class-ip)# 20 match tcp any any eq 80
VSX-1(config-class-ip)# 30 match tcp any any eq 443
VSX-1(config-class-ip)# 40 match tcp any any eq 53
VSX-1(config-class-ip)# 50 match udp any any eq 53
VSX-1(config-class-ip)# 60 match tcp any any eq 22
VSX-1(config-class-ip)# exit
VSX-1(config)# mirror session 1
VSX-1(config-mirror-1)# destination interface 1/1/10
VSX-1(config-mirror-1)# enable
VSX-1(config-mirror-1)# exit
VSX-1(config)# policy mirror_span
VSX-1(config-policy)# 10 class ip span action mirror 1
VSX-1(config-policy)# exit
VSX-1(config)# interface lag1
```

```
VSX-1(config-if-lag1)# apply policy mirror_span out
VSX-1(config-if-lag1)# apply policy mirror_span in
VSX-1(config-if-lag1)# exit
```

You would repeat the same steps on the other switch in the VSX pair.

Initial Log Into CPDI

After you set up the network infrastructure, you are ready to log into CPDI and set up active scanning.

When you first log in, you will be prompted to access or edit the configuration for the "Catch All Subnets" segment. By default, this segment includes all private subnets:

- 10.0.0.0/8
- 192.168.0.0/16
- 172.16.0.0/12

While you can technically have Data Collectors run subnet scans on "Catch All Subnets," such a scan would take much longer to run than a targeted scan on the subnets that you know the company is actually using. As shown in Figure 14-21, typically, you should accept the default "Catch All Subnets" config and simply plan to create your own subnet segments, which you will actually use, later.

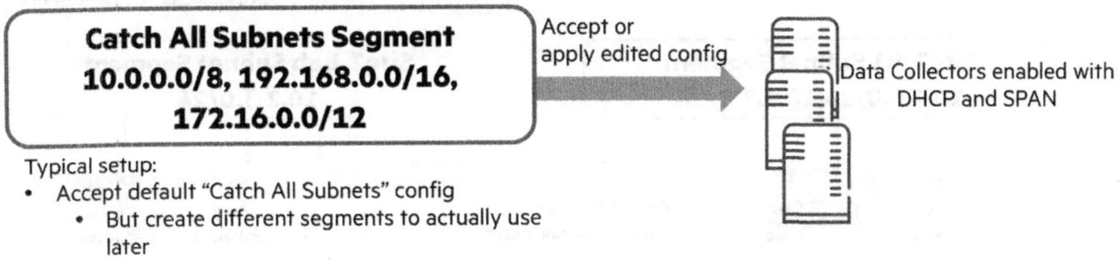

Figure 14-21: Initial Log Into CPDI

After you accept the config, you will be placed in the **Classified Devices - Dashboard** page. The Data Collectors that you earlier added will be activated with DHCP and SPAN enabled on them.

Understand Subnet Segments

When you run a subnet scan, you must select a segment or segments on which to run the scan. The segment definition specifies the IP addresses to be scanned, as well as controls the augmentation methods for the scanning.

When you create a segment, you define these settings for it:

- **Which Data Collector or Collectors are responsible for that subnet**—If a company has multiple sites, HPE Aruba Networking recommends that each site has at least one local Data Collector. You should setup at least one segment for each site and then associate the segment or segments with the Data Collector deployed at that site.
- **Which augmentation methods are enabled on that subnet**—An augmentation method indicates a scanning type and credentials for using that type of scanning. Rather than create one segment subnet per site, you might create multiple segments per site, as shown in this example. You can then set up different augmentation methods on different types of devices. For example, you might create segments just for your network devices, which use the SNMP augmentation method. As another example, the company has main building, B1, at site 2 with many printers, cameras, and domain computers. It also has a lab with Linux devices. You could create separate subnet segments for each type of device and assign different augmentations to each segment. Specifying different subnet segments with different augmentation methods also enables you to specify different credentials for different subnets. In short, you can flexibly assign whatever augmentation methods make sense for any particular use case.

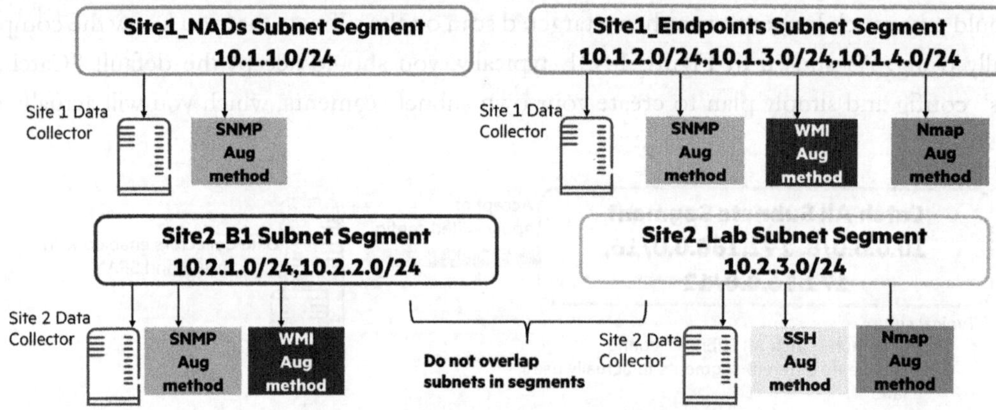

Figure 14-22: Understand Subnet Segments

As indicated in Figure 14-22, you should not specify overlapping subnets in different segments. If you do, you will see a message about the overlap and be encouraged to fix it.

Create Subnet Segments

As shown in Figure 14-23, to create a subnet, go to the **Discovery Settings** page. Click **New Subnet Segment**.

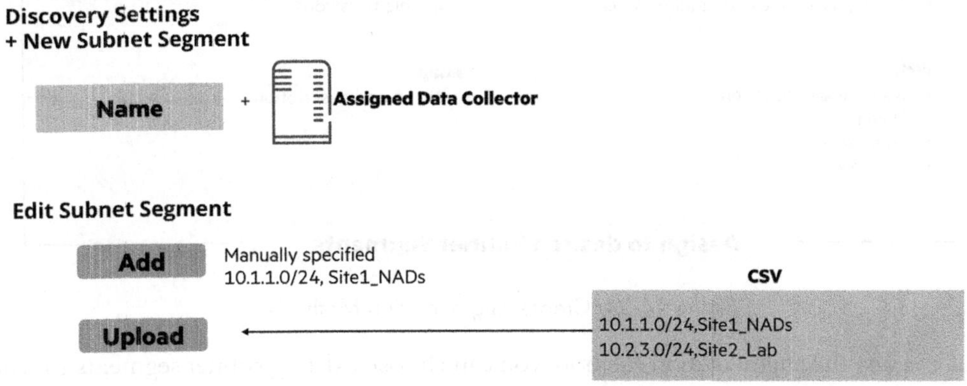

Figure 14-23: Create Subnet Segments

Name the segment and assign it to the Data Collector at the site with that subnet. You must then edit the subnet segment and add one or more subnets. You can add a subnet manually, typing in the subnet address and description manually. Or you can upload a CSV file with the subnet addresses and descriptions.

Create Augmentation Methods

Also in the **Discovery Settings** page you can add the augmentation methods. Add the new aug method and choose the method type. Then specify the settings particular to that type, as shown in Figure 14-24.

The settings for SNMP, SSH, and WMI are very similar to those about which you learned for CPPM External Accounts. Nmap does not require any particular credentials. You simply create the method so that you can attach it to a subnet segment and indicate that you want to run Nmap on that segment.

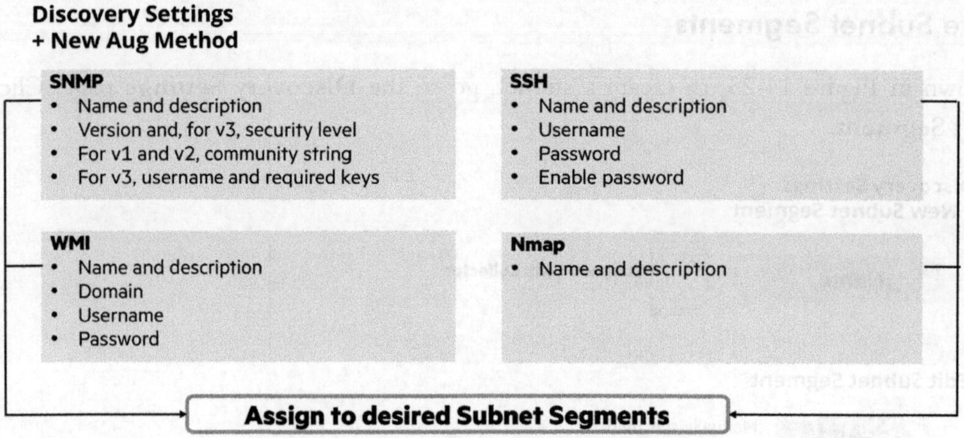

Figure 14-24: Create Augmentation Methods

As part of creating the augmentation method, you can choose existing subnet segments to which to attach that method. Alternatively, after you create the method, you can edit a subnet segment and attach the desired methods to it.

SNMP

If your network devices use SNMPv1/v2c, you should specify "SNMPv1 with community strings" or "SNMPv2 with community strings" for the Version. Then set the community name to match a read-only community on the network device.

If the network devices require SNMPv3, you must choose of the SNMPv3 options, matching the requirements and algorithms with those used on the network device. The network device admins should tell you the credentials, security levels, and algorithms used for an account that has read-only (or read-write) access. Under the version, you specify whether authentication is required, the algorithm used for authentication if required, and whether privacy is required in addition to authentication. If you select privacy, you choose the privacy protocol in a separate field. After choosing an SNMPv3 version, you specify the username and keys for the read-only account. You will also see key fields for the particular types of security that are required. (In other words, if you choose a version that requires authentication, you must set an authentication key. If the version further requires privacy, you must set a privacy key.) The network admins should tell you the key values as part of the credentials for the account.

SSH

When you choose "SSH" for the Type, you must specify a valid username and password for SSH access to the device.

Domain/WMI

For "Domain/WMI" accounts, you specify the domain name, the username, and the password. CPPM will use this format when logging into a device with these credentials: domain_name\username. Choose a user account that has administrative access to the domain computers.

Schedule a Discovery Scan to Discover NADs

Typically, you should run an initial Discovery scan next. This scan, similar to CPPM's Discovery scan, is usually intended to discover network access devices (NADs), such as switches and APs, for Data Collectors to probe. It works with the SNMP augmentation method. Before you schedule the scan, create a subnet segment that includes the NADs' IP addresses and make sure to attach an SNMP augmentation method to that subnet. That method should, of course, specify SNMP credentials that match those on the NADs.

Schedule the scan by going to the **Edit Collector** page for the Data Collector that you want to run the scan (go to **Menu > Discovery Settings > All Collectors** and click the **Edit** icon in the desired collector's card). Then choose to add a scan to the collector.

When you schedule the scan, you specify one or more seed devices. You should generally select the core routing switch or switches at the site as the seed. As shown in Figure 14-25, the Data Collector begins by sending an SNMP Get message to each seed, asking for several pieces of information: the device's sysObjectID (its system name), its bridge MIB, its interface list, its LLDP neighbors, its CDP neighbors, and the ARP table.

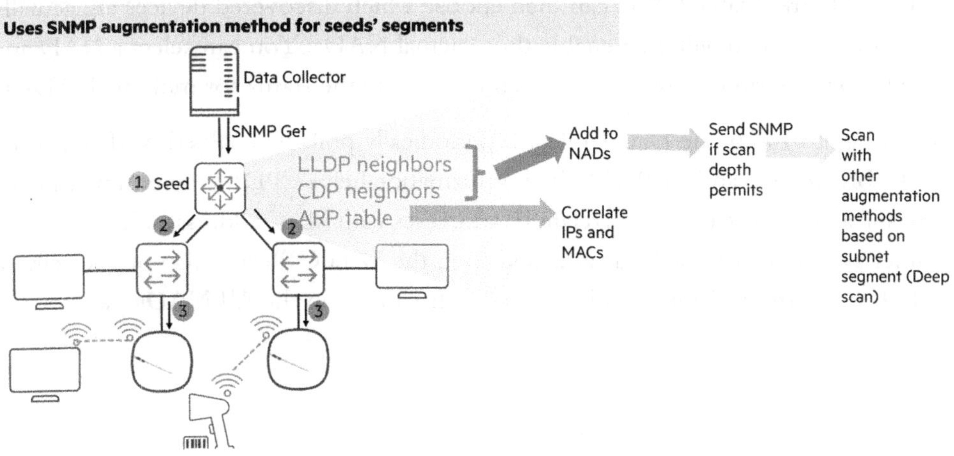

Figure 14-25: Schedule a Discovery Scan to Discover NADs

The CDP and LLDP tables provide the Data Collector with a list of devices connected to the seed at Layer 2. The Data Collector uses the System type (such as bridge) to determine which LLDP and CDP neighbors to add as Network Access Devices (NADs) on this Data Collector.

If you want the Data Collector to send SNMP Get messages to the devices discovered in the seeds' CDP and LLDP tables, you must set a scan depth greater than 1. The scan depth dictates how many times the Data Collector repeats the scanning process. In this example network, a scan depth of 3 takes the scan to the AP level.

As mentioned earlier, the Data Collector also collects information about device's ARP tables. It uses this information to correlate discovered MAC addresses and IP addresses. By default, the Data Collector does not scan the IP addresses that it discovers in the ARP table. But if you enable probing ARP entries, the Data Collector sends SNMP Get requests to all of the devices in the ARP table. However, enabling this option could make the scan much longer as the ARP table typically includes many client devices as well as network infrastructure devices. In addition, you usually do not want client devices potentially getting added as NADs if they respond to the SNMP messages.

After scanning with SNMP out to the scan depth, the Data Collector scans all discovered devices with any other augmentation methods that you configured for the subnet associated with the device.

NAD Management

The discovery scan adds the network infrastructure devices in your network to CPDI. You can view them in the **Discovery Settings** > **All NADs** page. Each NAD is associated with the Data Collector that discovered it.

Initially, all NADs are disabled. You can then choose which discovered devices are actually NADs that you want to poll. You will then enable those actual NADs. You can select a NAD and enable it individually. Or you can use the Actions menu to update the status for multiple NADs at once.

As shown in Figure 14-26, the Data Collector periodically polls its enabled NADs. It uses SNMP to ask for the Bridge MIB and ARP table. This information helps CPDI to determine which detected devices are still active, as well as to correlate IP addresses with MAC addresses. If the NAD's subnet has any other augmentation methods attached to it, the Data Collector also periodically scans the NAD with those methods. You can adjust the poll frequency in the **All NADs** page.

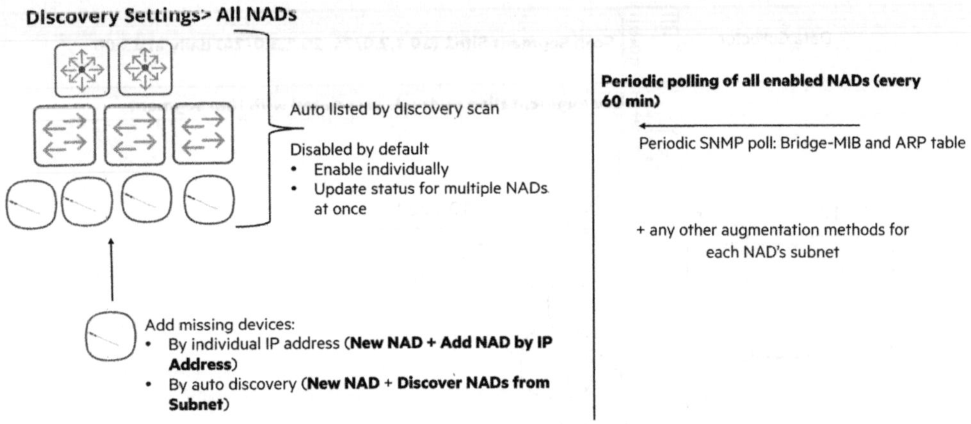

Figure 14-26: NAD Management

Sometimes the Discovery scan might miss some NADs. As indicated in Figure 14-26, you can add these devices manually with their individual IP addresses. Select **New NAD** in the **Discovery Settings** page. You will be given an option to add the NAD by IP address. You will then specify the NAD IP address as well as the Data Collector to which the NAD is assigned. The NAD is then added, and the Data Collector uses SNMP to scan the device and collect its sysObjectID, Bridge-MIB, interface table, LLDP table, CDP table, and ARP table.

Alternatively, you can add all the switches in a subnet through auto discovery. This method lets you add a subnet if your discovery scan failed to add some of the devices that you expected. As a prerequisite for using it, configure SNMP augmentation for the subnet on which you want to discover the NADs. Then select **New NAD** in the **Discovery Settings** page and then select **Discover NADs from Subnet** and select a subnet segment. Then the Data Collector scans that subnet and adds any device on which UDP 161, the port for SNMP, is open.

Schedule Subnet Scans for Endpoints

You should generally use subnet scans to collect information about endpoints.

You schedule subnet scans on Data Collectors in much the same way that you schedule discovery scans. From the **Discovery Settings** > **All Collectors** page, select a Data Collector. Then add a Device Discovery. Select Subnet for the scan type. Then select one of the segments attached to that Data Collector. As shown in Figure 14-27, when the scan runs, the Data Collector scans each device in the subnet segment using the augmentation methods associated with the selected segment.

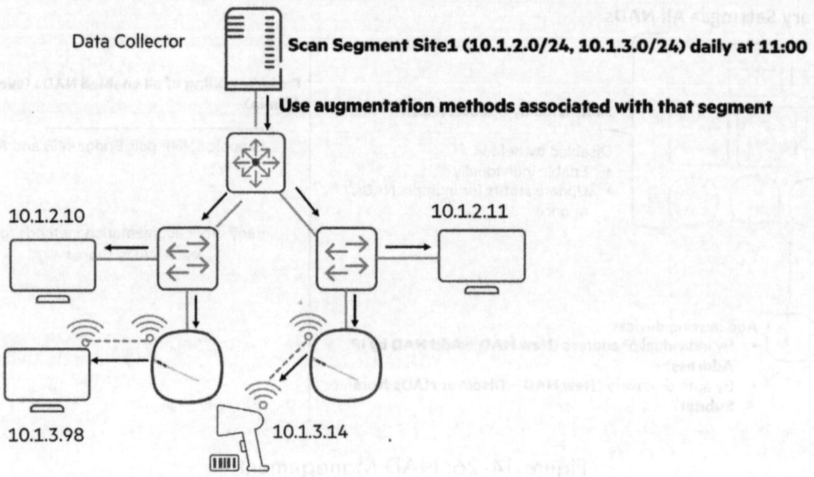

Figure 14-27: Schedule Subnet Scans for Endpoints

As part of the scan configuration, specify a schedule. For example, you might choose to run a daily scan in the middle of the night. However, you need to keep in mind when devices are present, as well as the particular requirements of the organization.

Analyze CPDI Data

In this section, you will learn how to analyze data collected by CPDI. This information resembles that in client profiles provided by HPE Aruba Networking, but the licensed application has some extra features. For example, you will learn how to create rules to reclassify devices.

CPDI Dashboard

When you access the CPDI application, you see the dashboard (see Figure 14-28).

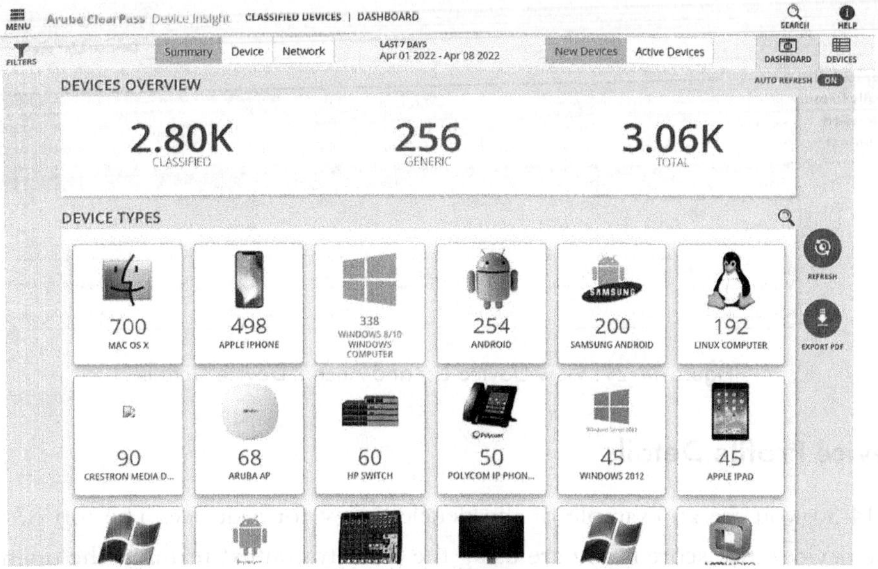

Figure 14-28: CPDI Dashboard

This highly interactive dashboard is designed to provide a modern, easy-to-use view of all devices discovered by CPDI. You can click devices to drill down into more details about them and easily move from one view to another. You can easily search through the devices or apply filters to view just the devices you want. To make the view simpler to understand at a glance, you can upload icons for your custom devices. You can export information about the devices in CSV files. You can also run reports to analyze the data and present it to stakeholders. By storing data over the long term, CPDI ensures that companies can access all relevant data about their devices and their history. However, CPDI might clean up data for devices that have been inactive for 30 or more days.

View Device List and Access Device Details

In the Classified Devices or User-Classified Devices dashboards, you can click the **Devices** tab to see a list of the devices (see Figure 14-29). From this view, you can filter and sort the devices according to various attributes. You can also perform bulk actions such as deleting attributes from all devices. CPDI gathers detailed context about each device. You can access this profile by hovering over a device and clicking the details icon that appears.

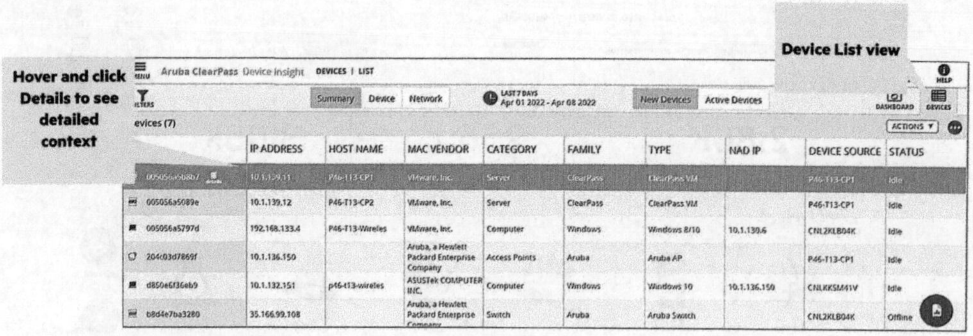

Figure 14-29: View Device List and Access Device Details

View Device Profile Details

In Figure 14-30, you see an example of the detailed view for a device. The top of the window displays the device's risk score if you are using the security context features, the online or offline status, the hostname (if known), and the device type.

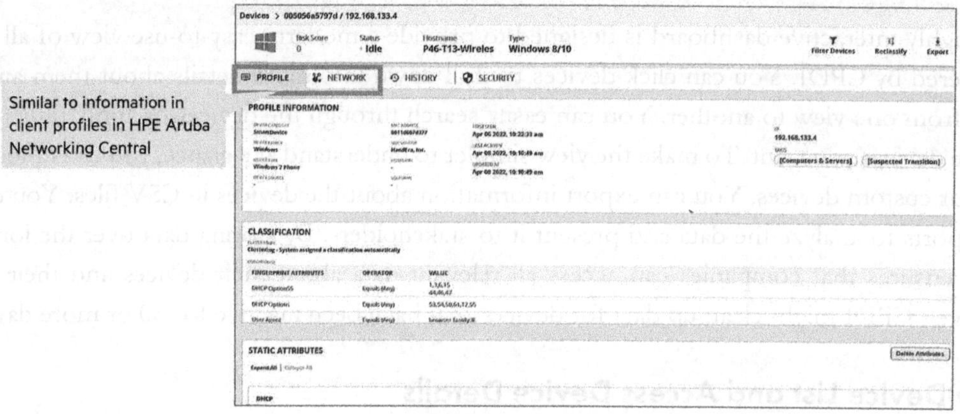

Figure 14-30: View Device Profile Details

The **Profile** tab provides additional information about the device category, family, and type, as well as the device MAC address and vendor. If you integrate CPDI with CPPM, you can further see the username for the user on the device.

Check the Classification section to discover why the device was classified with this category, family, and type. In this example, ML clustering placed the device in cluster with classified devices. The number of classified and percentage of classified devices of the same type met the threshold for auto-classification, so the device was automatically classified. Other classifications include system rules, user rules, and MAC address ranges. The classification might also show that the device was clustered and a user then accepted a recommended classification. The conditions listed in the Classification section show the attributes that contributed to the device being classified in this way. You can scroll down further in the profile to see all of the static attributes and flow attributes collected for this device.

Click the **Network** tab to view more details about the device location and network activity.

The Network Information section can help you to locate the device based on NAD IP and port, as well as the segment to which the device belongs.

Look through the Network Activity to answer questions about the device's network activity and behavior. You can follow similar guidelines to those discussed for client profiles in HPE Aruba Networking Central.

Device History

The Device History tab gives you insight into the device's activity and status over a longer period (see Figure 14-31). A history gives you deeper insight into the client. Instead of just seeing that a device has an unhealthy posture, for example, you can see all of the context around the posture assignment. You can also detect patterns. If client often changes between healthy and unhealthy posture, what is happening? Does the user not understand the security policies? Is the user acting in a way that can expose the company to exploit? You might need to help educate users.

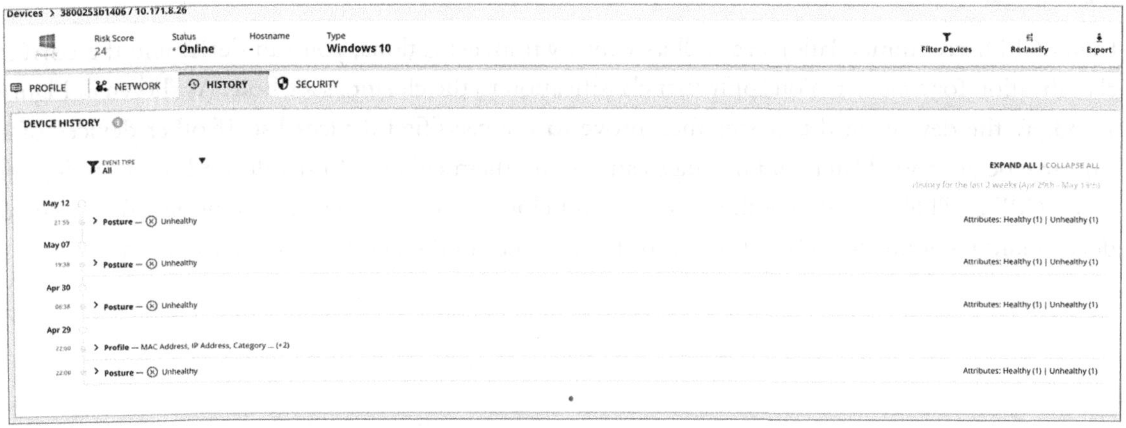

Figure 14-31: Device History

Being able to see a client's IP changes is also useful. Some solutions track devices by IP address, which can make a single device look like multiple devices if the IP address changes. CPDI's approach helps you to detect patterns even if a client's IP address changes.

View Generic Clusters

Now that you have seen how to use some of the information that CPPM collects about devices, you will look at ways to classify devices that are currently in the generic devices list. As you recall, CPDI classifies as many devices as it can and then uses ML to group the remaining devices in clusters. You can find the clusters in the **Generic Devices** page (see Figure 14-32). Click one of the clusters to see more details and classify the cluster.

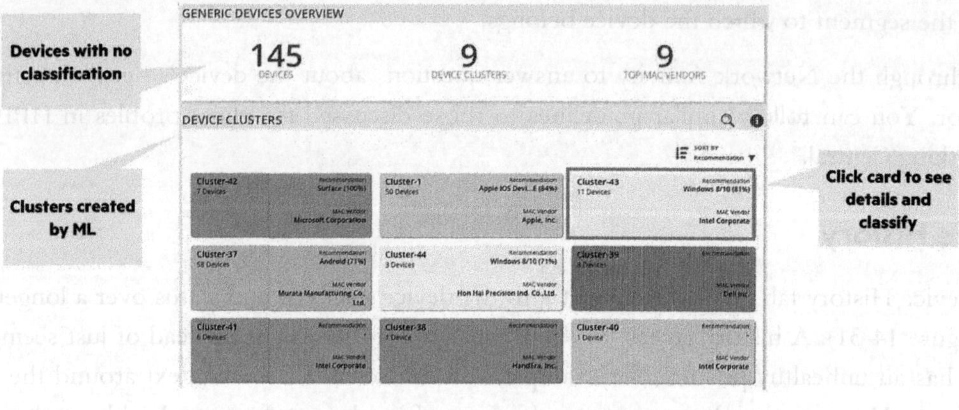

Figure 14-32: View Generic Clusters

Applying User Classifications and User Rules to Generic Clusters

Using CPDI recommendations as well as your own investigations, you can determine the correct classification for a cluster. You apply that classification to the cluster as a whole. As shown in Figure 14-33, all the devices in the cluster then move to the classified devices list. If other devices of a similar type are added later, and no rules can classify them either, CPDI will run ML clustering on them. CPDI will likely find that the new devices belong to the same cluster as the already classified devices and recommend a classification for them. You can then classify the new devices.

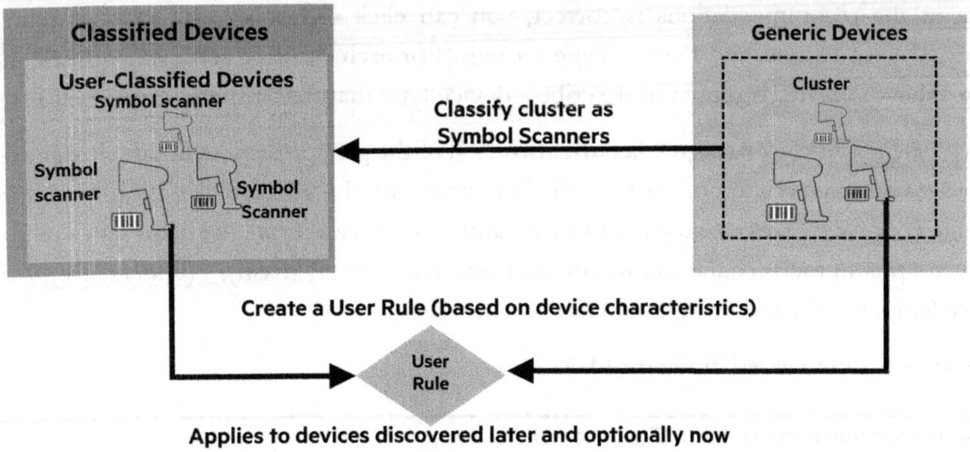

Figure 14-33: Applying User Classifications and User Rules to Generic Clusters

However, as shown in Figure 14-33, you can make this entire process faster and easier by also adding user rules, which will classify the new devices immediately. You can define user rules from scratch based on a broad variety of attributes. But CPDI makes it easier for you to define these rules. You can select a generic device or a user-classified device and choose to create a rule from it. Then simply select the attributes characteristic of the device and make a rule. When you save the user rule, you can optionally apply it to existing devices then. No matter what your choice, the user rule will apply to devices that are discovered later.

You can combine the two forms of classification to easily classify your devices. You can select a cluster and classify it based on the recommendation. Then you can go to the user-classified devices, choose one of the devices, and create a user rule from it. Then new devices will be classified based on that rule without having to go through ML clustering. Or you can first create the rule based on a generic device and then classify the cluster. The choice is yours, but the end result is much the same.

Classifying a Cluster with a Recommendation

Now look in more detail about how you apply the classifications.

When you click a cluster, you see the **Cluster Details** page. If CPDI Analyzer was able to determine any recommendations for the cluster, you will see a Classification Recommendations section. As you learned earlier, recommendations derive from classified devices with similar characteristics. Recommendations include a confidence percentage, which is based on the number of similar classified of the recommended type, as compared to the total number of similar classified devices of any type. Generally, the higher the percentage, the more likely the classification.

If none of the recommendations is correct, you can click **Other** and fill in your own Device Category, Device Family, and Device Type settings. For each of these settings, you can specify an existing value or create your own to describe a device type that does not yet exist in CPDI.

When you are ready, click **Assign Classification**. All of the generic devices in this cluster are moved to the User Classified section of the Classified Devices with the selected classification. Note that, if you create your own Device Category, Device Family, or Device Type, the name can use upper and lower case alpha-numeric characters and these special characters: - (dash), _ (underscore), / (forward slash), & (ampersand), and () (right and left parenthesis).

This process is summarized in Figure 14-34.

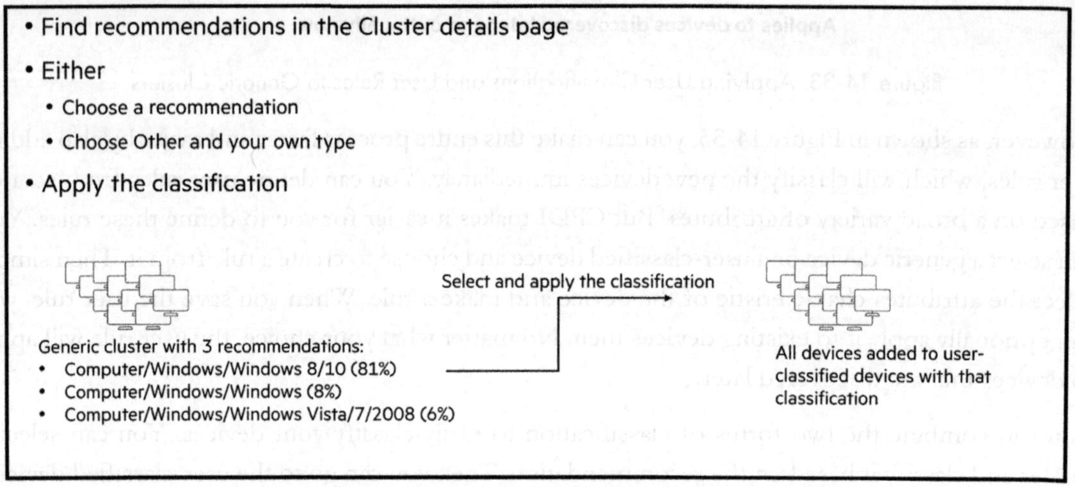

Figure 14-34: Classifying a Cluster with a Recommendation

Creating User Rules

In addition to classifying clusters, you might want to create user rules, which will immediately apply to new devices as they are discovered.

You can create a rule from scratch. Simply click the button for creating a rule. You will see a dialog, in which you can specify conditions and rules.

The conditions are attributes that the device needs to match. The rule is the category, family, and type that you want to assign to matching devices.

It can be very challenging to pinpoint exactly which attributes match a particular device type. Often it is easiest to create the rule based on an existing device of that type. For that option, create the rule from the **Device Details** page for a particular device. This method auto-populates conditions for you.

Defining the User Rule

The dialog for creating the rule first includes an Edit Conditions section, as shown in Figure 14-35.

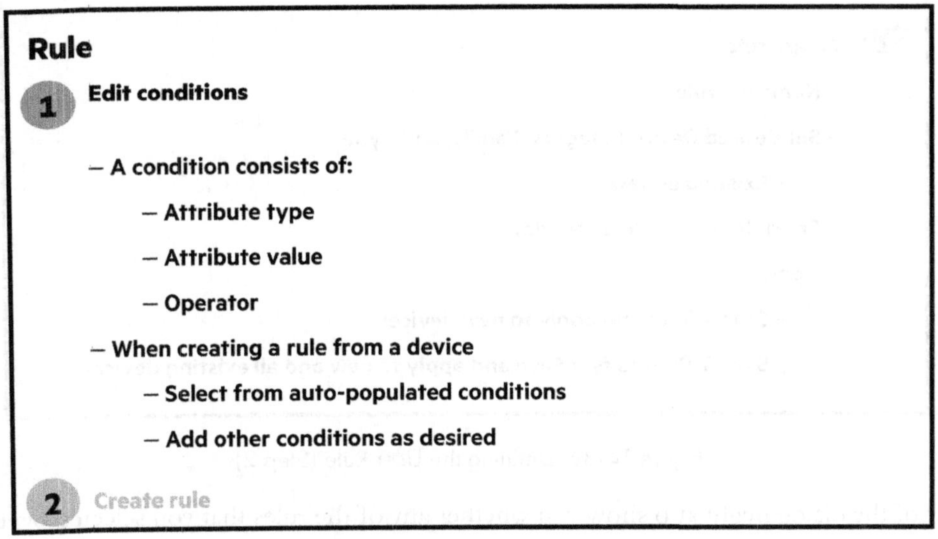

Figure 14-35: Defining the User Rule (Step 1)

Each condition includes an attribute type, such as device category, MAC vendor, open port, application, or destination host. CPDI supports many other static and flow attributes. It also includes the desired value for that type. You also specify an operator, which dictates how CPDI matches devices' attributes to the specified values. If you select "Contains (All)" for example, the device's attributes of that type must contain all of the values listed. "Contains or Equals (Any)," on the other hand, establishes an "or" logic; the device only needs to match one of the values.

You can include multiple types of conditions, allowing you to create a sophisticated rule.

As mentioned earlier, when you create the rule from a device, that section is auto-populated with some of the device's attributes. You can then select check boxes to choose which attributes to use as conditions in your rule. You can also add more conditions. When you create the rule from a device, the dialog automatically presents you with possible values for the attributes based on the device's values for those attributes.

Next you move to the section for creating rules. As shown in Figure 14-36, you set the rule name. Then choose the Device Category, Device Family, and Device Type. You can use an existing category, family, and type, or specify your own. If you specify a new type, you will be prompted to browse for an image for that type. That image will then appear in the dashboard for devices of this type. The image should be a .png, .jpg, or .jpeg file.

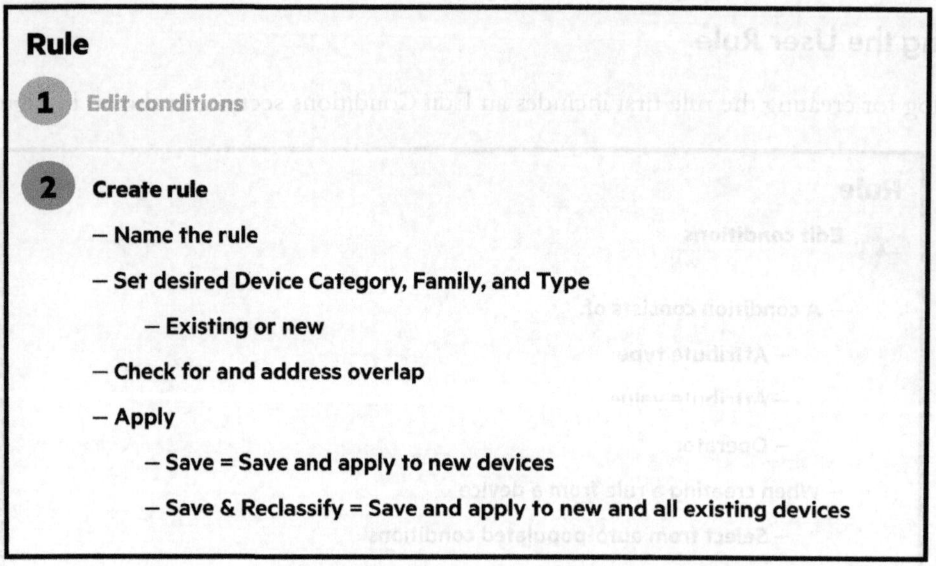

Figure 14-36: Defining the User Rule (Step 2)

This part of the dialog might also show you whether any of the rules that you selected overlap with other rules. If they do, you should typically adjust the rules to avoid overlap. For example, if you are creating a rule from a particular device, rather than select just one open port on the device, you might select the complete port list used by the device.

When you have finished defining the rule, you can choose between clicking **Save** or **Save & Reclassify**. Click Save if you only want the new rule to apply to new devices discovered later. Click **Save & Reclassify** if you want the rule to apply to existing devices, as well as new devices.

The device category, family, and type and its image will be emailed to HPE Aruba Networking. HPE Aruba Networking might replace the image with one that it has that better reflects that type. If so, the image in your tenant's dashboard will be updated with HPE Aruba Networking's image.

Summary of the Classification Workflow With User Input

In Figure 14-37, you see a summary of how CPDI classifies devices after users have classified clusters and created rules. As you see, CPDI applies the user rules first before any system rules or MAC Range Classifiers.

Users can delete rules or cluster assignments. This action affects already classified devices as well as devices discovered later. The devices that were earlier classified by that rule or cluster assignment are returned to the generic devices pool, and CPDI applies ML clustering to them.

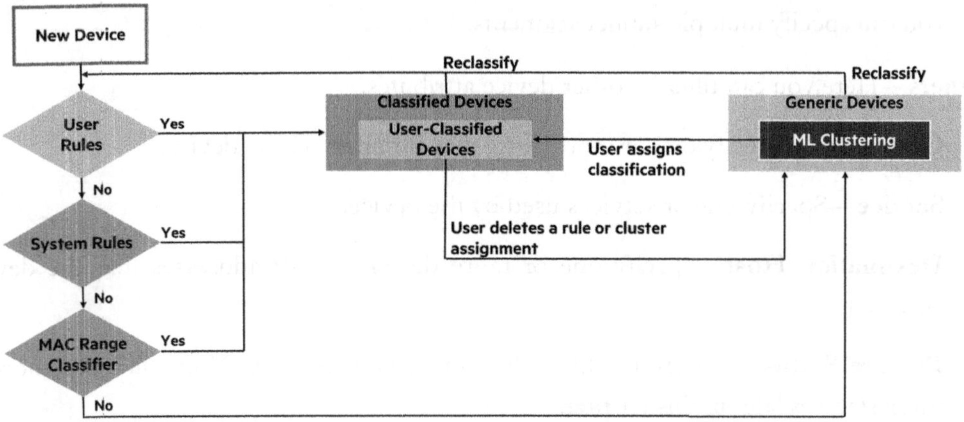

Figure 14-37: Summary of the Classification Workflow With User Input

Using Filters

Next you will look at ways that you can use the CPDI data. From the Classified Devices, User-Classified Devices, and Generic Devices dashboards, you can apply filters to search for devices that meet specific criteria. You can search for all the devices of a category, family, or type. You can find devices that are using specific applications or accessing specific destinations. You can see all the devices at a particular location. And you can fulfill many other use cases.

To create the filter, click the filter icon in the dashboard. In this example, you are filtering the classified devices, but you can apply similar filters in the User-Classified Devices and Generic Devices dashboards.

As shown in Figure 14-38, filters have three main categories of criteria. For any of this criteria, you can choose to include or exclude the specified value.

- **What**—Use the What category to filter devices by category, family, or type.
- **Where**—You have a few different options for how you specify the device's location. You can specify multiple options:
  - **NAD IP**—Specify the IP address of the NAD to which the device is connected. You can specify multiple IP addresses to find devices connected to any of those NADs.

- **Subnet**—Specify the subnet to which the device's IP address belongs. You can specify multiple subnets.
- **Segment**—Specify the subnet segment to which the device's IP address belongs. Again, you can specify multiple subnet segments.

- **Others**—Here you can filter by other device attributes:
  - **Open Port**—Specify one or more ports that are open on the device.
  - **Service**—Specify one or services used by the device.
  - **Destination Host**—Specify one or more destination IP addresses that the device has accessed.
  - **Posture Status**—Specify Healthy, Unhealthy, or Unknown. You will learn more about these statuses later in this chapter.
  - **Risk Score**—Specify a score, and the filter will show devices with that score or higher. You will learn more about risk scores later is this chapter.

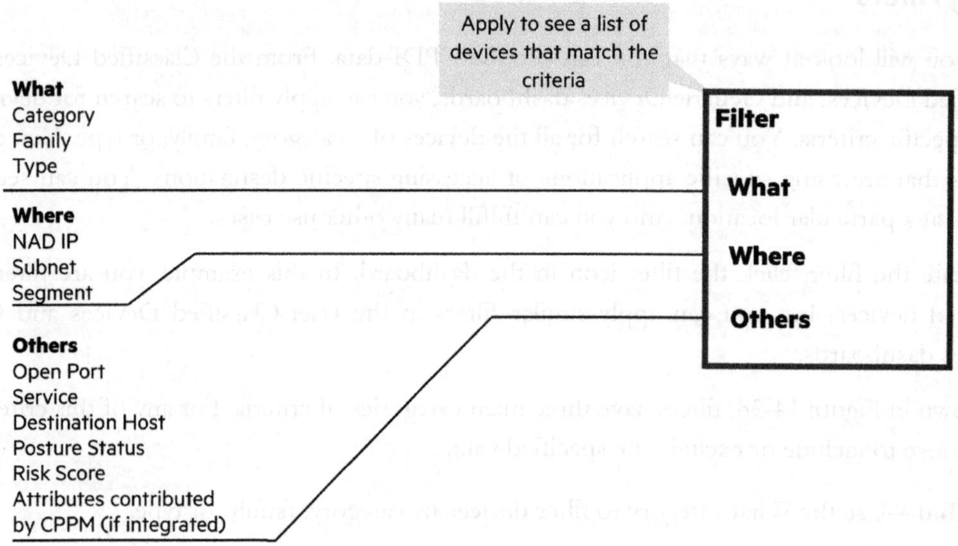

Figure 14-38: Using Filters

There are several other attributes which you can use if CPDI is integrated with CPPM. These include Authentication Method, Authentication Source, Access Type, Username, MDM Server Name, and MDM Server Type.

You can specify as many criteria as you want. Devices must match all criteria of different types for the filter to apply to them. This example filter finds all Windows 8/10 devices that are connected to the NAD at 10.10.22.251 and have port 137 open. However, if you include more than one option for the same condition type, those options are applied as "or." For example, if you include 10.10.22.251 and 10.10.22.252 for NAD IP, the filter finds devices that are connected to 10.10.22.251 or 10.10.22.252.

Creating Tags

You can save any filter that you create as a tag by applying the **Save as tag** action.

Figure 14-39: Creating Tags

As indicated in Figure 14-39, saving the tag makes it easy to apply commonly used filters whenever required. When users select the Filter icon, they see a list of User Tags and can select the tag that they want to use to filter the device list.

In addition, CPDI dynamically adds the tag to all devices that meet the filter criteria. As soon as you save the tag, it applies the tag to any existing devices that meet the criteria. And if a device is later discovered to meet the criteria, CPDI adds the tag to it then. Users can see the tags applied to a device in the **Device Details** page at the **Profile** tab.

You can use tags however you want to mark important information about devices, much as you learned for client profile tags in HPE Aruba Networking Central. To review the use cases, refer to Chapter 13.

Creating Reports

You can create reports to export and archive information that CPDI has collected. To begin creating the report, click the **Create Report** button, which is available from many pages within CPDI. For example, you can find this button on any of the device dashboards.

When you click the button, you will see a dialog for creating the report. The template for the report is auto-filled in based on where you clicked the button. For example, if you clicked the button in the Classified Devices dashboard, the report is automatically set to report on classified devices. If you clicked the button from the Classified Devices list view, though, the template would be for

Classified Devices List. Similarly the templates will be different if you click the button in the User-Classified Devices dashboard or Devices list, or the Generic Devices dashboard or Devices list.

For the report frequency, choose whether you want to run the report immediately and only once (on demand) or periodically. If you choose a periodic option, you will be prompted to specify the time, weekday, or date on which the report should run.

Specify the recipients for the report by email address. You can include multiple email addresses, separated by commas.

The valid format for the report depends on the type. Dashboard reports require the PDF format and List reports require the CSV format. If you want to change the format, exit out of the dialog, change your view, and then create the report from the new view.

Finish the report by exporting (for an On Demand report) or saving (for a periodic report).

Use CPDI Security

CPDI supports security assessments. In this section, you will learn about implementing and using these features.

CPDI Security Context Capabilities

The CPDI application, as of version 2.0.1, provides security context capabilities.

Using agentless scanning CPDI collects information about a device's vulnerabilities and posture. Based on these assessments, CPDI generates a risk score for the device. An example of the security details for a device in shown in Figure 14-40.

Figure 14-40: Example Device Security Details

You will learn more about the security features over the rest of this section.

CPDI Agentless Scanning

CPDI uses agentless scanning to assess devices' vulnerability and posture. CPDI supports vulnerability and posture scanning for 64-bit Windows OS of the following versions:

- Windows 8, 8.1, 10
- Windows Server 2012, 2016, 2019

Agentless scanning makes use of WMI, so you must attach a WMI augmentation method to the segment or segments with the devices that you want to scan. This method must have, as its credentials, a domain account that is part of the local Administrators' group on the Windows devices.

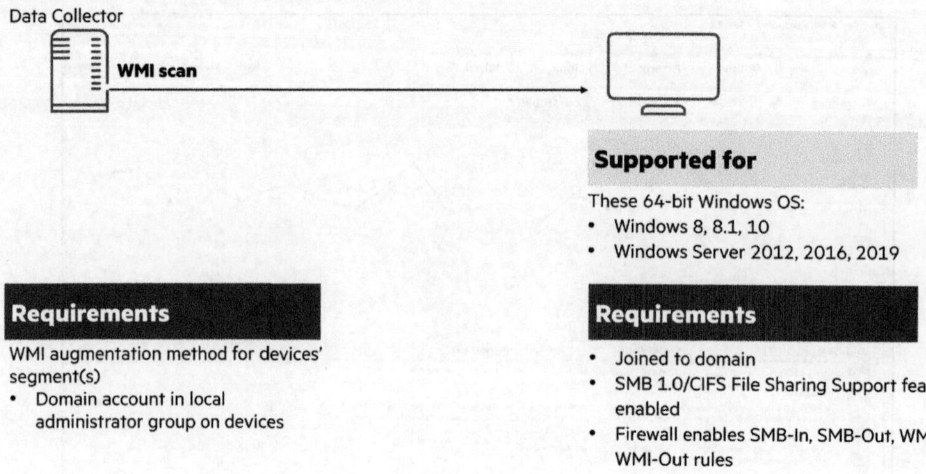

Figure 14-41: CPDI Agentless Scanning

As shown in Figure 14-41, you must also make sure that the devices to be scanned meet several requirements. They must be joined to the domain. They must have SMBv2 and Administrative Shares enabled on them (check the Windows features). Their firewalls must permit TCP ports 135, 139, and 445. You can configure the Windows Firewall to permit these ports by enabling these rules:

- File and Print Sharing (SMB-In)
- File and Print Sharing (SMB-Out)
- Windows Management Instrumentation (WMI-In)
- Windows Management Instrumentation (WMI-Out)

The WMI augmentation method can alternatively reference a local user account on the Windows devices. However, this setup is less commonly used. If you do use a local user account, it must be an Administrator on the devices. It must also be permitted to run psexec scripts, which could require a registry edit.

- Edit HKEY_LOCAL_MACHINE\SOFTWARE\Microsoft\Windows\CurrentVersion\Policies\System
- If the LocalAccountTokenFilterPolicy registry entry does not exist, follow these steps:
  - Select New, and then click DWORD Value.
  - Type LocalAccountTokenFilterPolicy, and then press ENTER.

- Right-click LocalAccountTokenFilterPolicy, and then click Modify.
- Set the Value data to 1.

Enabling CPDI Vulnerability and Posture Assessments

To configure the enable CPDI vulnerability and posture assessments, select **Security Settings** in the CPDI menu.

First, toggle the Security Analysis setting to **On**. Enabling this setting tells CPDI to run agentless scanning and scan for vulnerabilities.

Next choose the Data Source. If you have integrated CPPM with CPDI, which is covered in the next chapter, you can select **ClearPass Device Insight** or **ClearPass Policy Manager (CPPM)**. However, the focus for this chapter is using CPDI.

You then choose whether to enable posture assessments or not. You must enable security analysis and vulnerabilities in order to enable posture assessments.

These guidelines are summarized in Figure 14-42.

Assess vulnerabilities only	Assess vulnerabilities and posture
Security Analysis = On	Security Analysis = On
Data Source = ClearPass Device Insight	Data Source = ClearPass Device Insight
Enable Posture Assessment = Off	Enable Posture Assessment = On

Figure 14-42: Options for Security Settings

When Agentless Scanning Runs

When you have enabled security analysis and set up a WMI augmentation method for devices' subnets, CPDI initiates agentless scanning on those devices automatically. The Data Collector responsible for the device runs the scan when the device is first detected—for example, when its DHCP fingerprint is received. After that, the Data Collector runs the scan every four hours on active devices, as shown in Figure 14-43. It also runs when a device first becomes active again after more than four hours of inactivity.

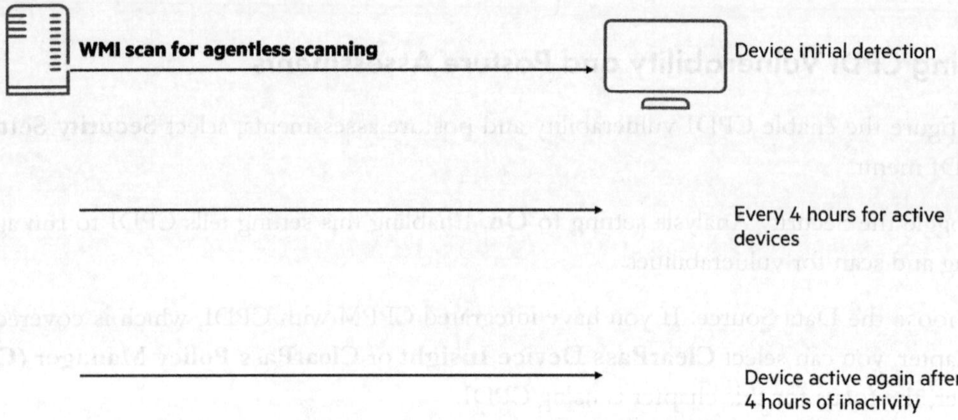

Figure 14-43: When Agentless Scanning Runs

You do not need to schedule WMI scans for the agentless scanning. However, if you do schedule a WMI scan, the agentless scan occurs.

Vulnerability Assessment

When Security Analysis is enabled, CPDI checks supported devices for the vulnerabilities defined by the National Vulnerability Database (NVD). It looks for vulnerabilities in all major products from Microsoft, Apple, Adobe, Google, and Mozilla.

For Microsoft products, only vulnerabilities from 2016 and later are reported. In addition, CPDI only reports vulnerabilities for which Microsoft has provided a hotfix. For non-Microsoft products, CPDI reports all vulnerabilities.

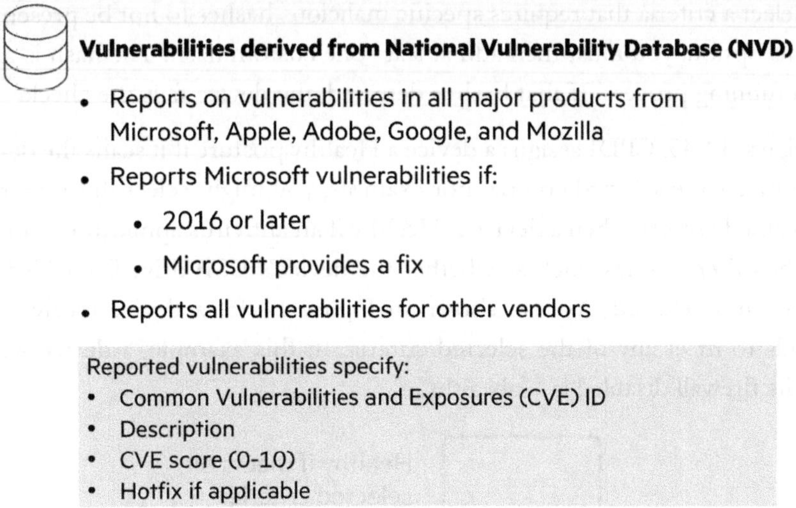

Figure 14-44: Vulnerability Assessment

CPDI lists the discovered vulnerabilities by Common Vulnerabilities and Exposures (CVE) ID and provides a brief description of each vulnerability. It also shows the CVE score associated with each vulnerability. A CVE score is based on a variety of factors such as how easy it is for hackers to exploit the vulnerability and how much damage hackers could do if they succeed in exploiting the vulnerability. The higher the score, the worse the vulnerability. If the vulnerability has a hotfix associated with it, CPDI reports that as well.

These capabilities are summarized in Figure 14-44.

Posture Assessment

During agentless scanning, CPDI also runs posture assessments if you have enabled posture assessment and if the Data Source is ClearPass Device Insight. In this case, you configure the posture rules in the **Security Settings** page. In CPDI 2.0.1, you can have CPDI check for these settings:

- Antivirus enabled
- Host firewall enabled
- Encryption enabled
- USB drive disabled
- Dual NICs disabled

You can also select a criteria that requires specific malicious hashes to not be present on the device. If you select this option, you must then add at least one custom hash. The hash is an MD5 hash of the binary of a running process. If that hash is detected, the device fails the check.

As shown in Figure 14-45, CPDI assigns a device a Healthy posture if it scans the device and detects that the device meets the selected criteria. For example, you might select the antivirus enabled and host firewall enabled criteria. Then a device is Healthy if an antivirus application and its host firewall are enabled. The other settings, such as whether the device's USB drive is enabled, do not matter because they are not selected. As also shown in Figure 14-45, a device receives the Unhealthy posture if it fails to meet any of the selected criteria. In this example, a device with no antivirus application or its firewall disabled is Unhealthy.

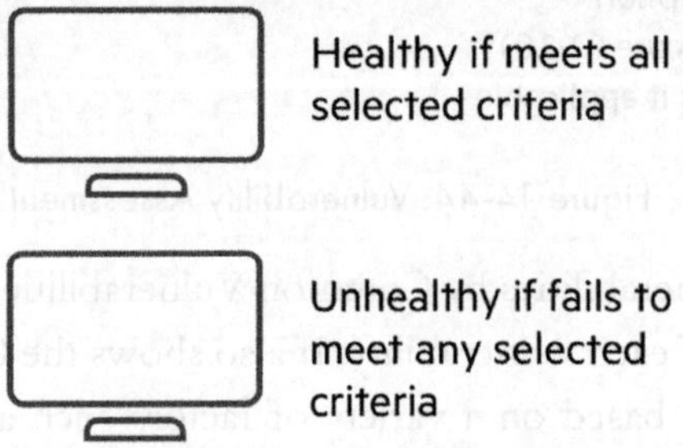

Figure 14-45: Posture Assessment

Controlling Posture for Devices That Cannot Be Scanned

By default, CPDI assigns devices that it cannot scan an Unknown posture. However, in the **Security Settings** page, you can adjust this setting such that CPDI assigns these devices a Healthy or Unhealthy posture instead. You can configure the setting separately for Windows devices that cannot be checked and for non-Windows devices. For example, as shown in Figure 14-46, you might want to tell CPDI to configure non-Windows device's posture as Healthy. Agentless scanning is not supported on these devices, so assigning a Healthy posture here is the only way for those devices to ever receive a Healthy posture. In this way, you can avoid adding points to the risk score on all of your non-Windows devices.

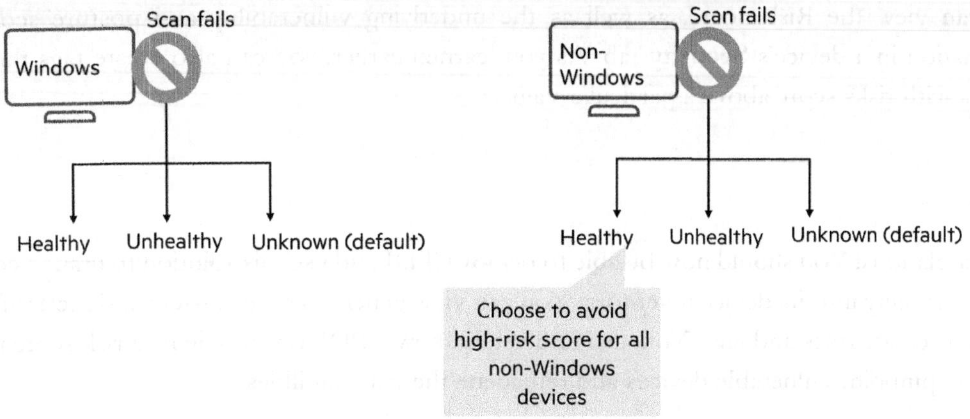

Figure 14-46: Controlling Posture for Devices That Cannot Be Scanned

Understanding the Risk Score

When CPDI assesses vulnerabilities, it assigns each supported device a vulnerability score. This score ranges from 0 to 100.

As shown in Figure 14-47, CPDI derives the score by first multiplying the CVE associated with each detected vulnerability, which range from 0 to 10, by 10. It then sums these values. Finally, it applies a logistic function, which is not quite an average of the values, but a representative value between 0 to 100.

If CPDI is also assessing posture, it assigns the device a posture score. The score is 0 if the device is healthy, 50 if its posture is unknown, and 100 if the posture is unhealthy.

CPDI then multiplies the vulnerability score by .7 to produce the vulnerability contribution to the risk score. It multiplies the posture score by .3 to produce the posture score contribution to the risk score. It then adds these two scores together.

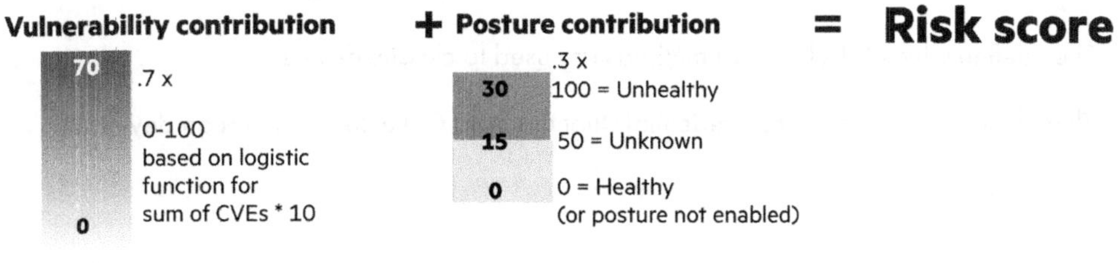

Figure 14-47: Understanding the Risk Score

You can view the Risk Score, as well as the underlying vulnerability and posture scores and information in a device's **Security** tab. As you learned earlier, you can also create tags that select devices with risks score above a particular value.

Summary

Congratulations! You should now be able to deploy CPDI and use this solution to grant a company deeper visibility into its device inventory. You can view generic device clusters and reclassify them, as well as create rules and tags. You also understand how CPDI assigns clients a risk score that can help you pinpoint vulnerable devices and remediate the vulnerabilities.

In the next chapter, you will learn how to integrate CPPM with CPDI or with the device insight capabilities built into HPE Aruba Networking Central. In this way, you will enhance the capabilities of both solutions.

Learning checks

1. What port must be open between a CPDI Data Collector and HPE Aruba Networking Central?
   a. TCP 8080
   b. TCP 443
   c. UDP 22
   d. UDP 4422
2. What do the augmentation methods for a segment define?
   a. They provide expanded MAC address ranges to supplement classification.
   b. They define which additional user rules to apply to classify devices.
   c. They define which ML clustering methods are used to classify devices.
   d. They define the methods and credentials that the Data Collector uses to scan devices.

3. Which correctly describes a tag?
    a. A custom rule that reclassifies device types based on a variety of attributes
    b. A description that applies to devices based on various criteria
    c. A device attribute that is collected by a Data Collector
    d. A flow attribute that is collected by a Data Collector
4. What is a requirement for CPDI to collect vulnerability and posture information on Windows devices?
    a. The devices must have 32-bit operating systems.
    b. The devices must use Windows 7.
    c. CPDI agents must be installed on the devices.
    d. A WMI augmentation method must be attached to the segment for the devices.

Note

For answers to all learning checks, refer to the Appendix.

---

15 Integrate Device Insight with CPPM

EXAM OBJECTIVES

✓ Integrate HPE Aruba Networking ClearPass Policy Manager (CPPM) and Device Insight

✓ Mitigate threats by using Device Insight to identify traffic flows and apply tags and CPPM to take actions based on tags

Assumed Knowledge

- Fundamental HPE Aruba Networking ClearPass Device Insight concepts
- Fundamental HPE Aruba Networking ClearPass Policy Manager (CPPM) components, such as services, roles, enforcement profiles, and enforcement policies

Overview

You can integrate HPE Aruba Networking Central's client profile information with HPE Aruba Networking ClearPass Policy Manager (CPPM). Similarly, you can integrate HPE Aruba Networking ClearPass Device Insight with CPPM. This study guide will refer to both forms of integration as Device Insight integration. By integrating Device Insight and CPPM, customers can more effectively identify and remediate security vulnerabilities on the network.

This chapter begins by explaining exactly how to integrate these solutions. You will see what prerequisites must be met for integration to avoid roadblocks during implementation. And you will cover what implications this integration has for the network to ensure that no unintended consequences result from integration.

In the second section, you will learn about putting these integrated solutions to work to enhance network security. You will see how admins can use Device Insight tags to enrich CPPM's understanding of clients on the network and guide CPPM's response to any security incidents.

Integrate Device Insight with CPPM

You will first learn how to integrate Device Insight with CPPM.

Why Integrate CPPM and Device Insight?

Why should admins take the effort to integrate CPPM and Device Insight? Because integrating these solutions allows them to feed each other information and boost overall visibility to strengthen network security, as summarized in Figure 15-1.

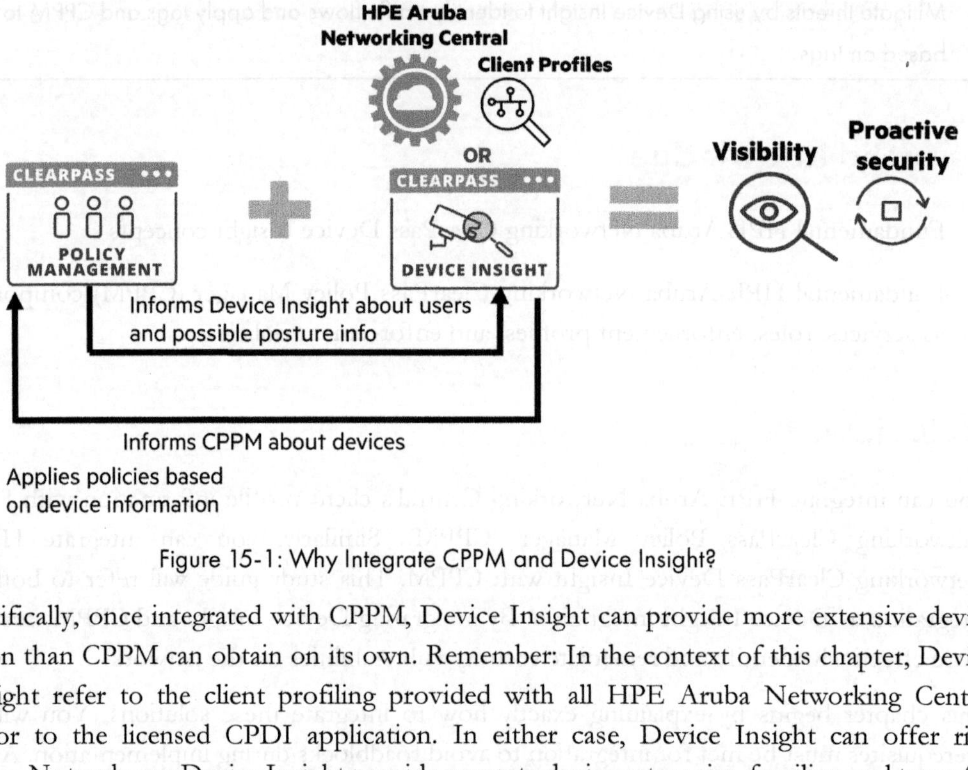

Figure 15-1: Why Integrate CPPM and Device Insight?

More specifically, once integrated with CPPM, Device Insight can provide more extensive device information than CPPM can obtain on its own. Remember: In the context of this chapter, Device Insight might refer to the client profiling provided with all HPE Aruba Networking Central accounts or to the licensed CPDI application. In either case, Device Insight can offer rich information. Not only can Device Insight provide accurate device categories, families, and types for a broad range of devices, including IoT, but it can also give CPPM additional context through tags. HPE Aruba Networking Central or CPDI admins can apply tags to devices based on very flexible criteria, including behavior-like criteria such as the applications that the device is using and the destinations that it is contacting. CPPM can then use this information in policies to control client access.

CPPM, in turn, can provide HPE Aruba Networking Central or CPDI with information about users and authentication to expand their endpoint profiles. CPPM can also provide posture information to these solutions.

CPPM and Device Insight Integration Architecture

When you integrate CPPM with Device Insight, CPPM creates a collector service. The collector service receives profile requests from various services in CPPM and acquires device data. As shown in Figure 15-2, the collector service then forwards this information on to the Analyzer used by HPE Aruba Networking Central client profiles or CPDI.

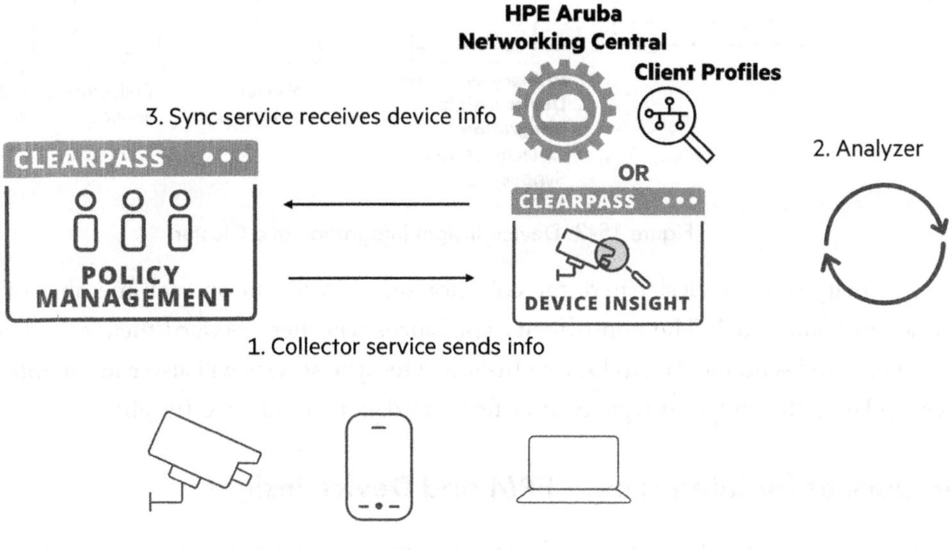

Figure 15-2: CPPM and Device Insight Integration Architecture

This Analyzer processes device characteristics as well as communication and behavior patterns. Using rules and ML, it matches each device to a set of known fingerprints and classifies the device.

The sync service, another subsystem of CPPM, then fetches the endpoint's classification from Analyzer and updates CPPM's endpoint repository.

These updates happen in real time, and to back them up, CPPM also polls Analyzer periodically.

Device Insight Integration in a Cluster

When CPPM is running on a cluster of servers, admins must designate a primary server for Device Insight integration. Typically, the Publisher is a good candidate to serve as the primary server.

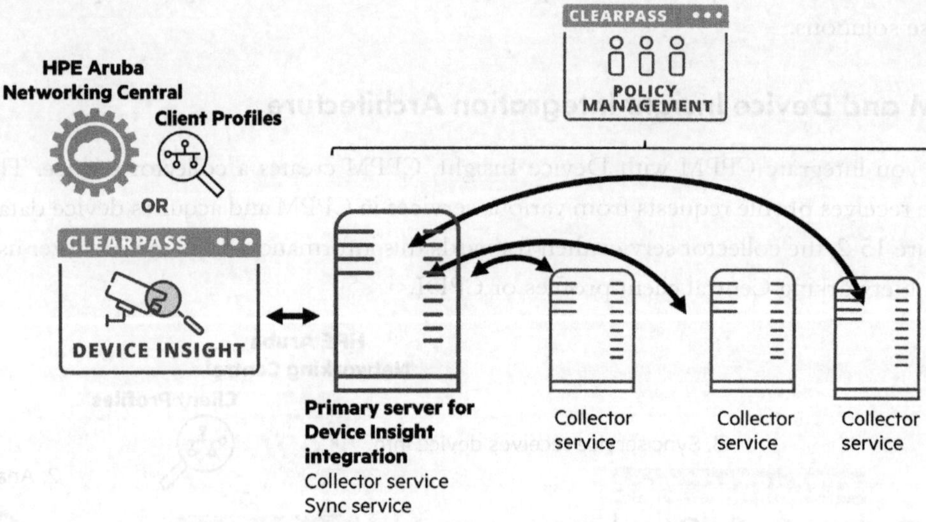

Figure 15-3: Device Insight Integration in a Cluster

Once a primary server is designated, the collector service will run on each member of the cluster, as shown in Figure 15-3. The non-primary appliances will then forward their data to the primary server, which will send the data to Device Insight. The sync service will also run on only the primary server, updating the endpoint registry after fetching data from Device Insight.

Prerequisites for Integrating CPPM and Device Insight

Before you enable CPPM and Device Insight integration, ensure that you meet the prerequisites, which are summarized in Figure 15-4.

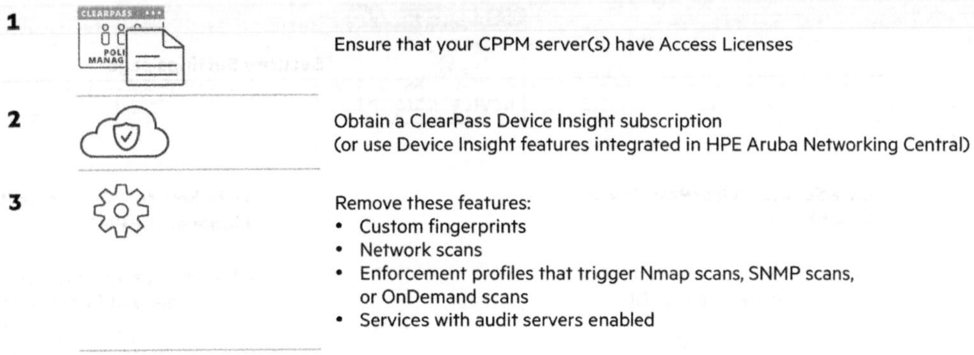

Figure 15-4: Prerequisites for Integrating CPPM and Device Insight

The CPPM server or cluster requires Access Licenses.

Next, you must have a subscription for CPDI. Or you can alternatively use the Device Insight features integrated in HPE Aruba Networking Central.

Lastly, you must disable several features that you may have configured. Specifically, you must remove any custom fingerprints or network scans that you have configured. You must remove any enforcement profiles that are configured with Nmap scans, SNMP scans, or OnDemand scans. And you must disable or remove any services that have audit servers enabled.

Understand Implications of Integration

When you integrate CPPM and Device Insight, CPPM has access to the rich profiling information collected by CPDI and HPE Aruba Networking Central. Therefore, you often do not need to configure CPPM to collect profiling information. CPPM can still use device profile information, such as endpoint category, within its policies, relying on the information provided by the cloud solution.

Depending on the CPPM version, CPPM's device profiler and related features, such as scanning, might be disabled when you enable Device Insight integration. You should check the documentation for your CPPM version to understand how CPDI and CPPM interact.

Security Context Integration between CPPM and CPDI

Before you integrate CPPM and the licensed CPDI application, you should also understand which solution's vulnerability and posture assessment information is used, based on the settings in CPDI.

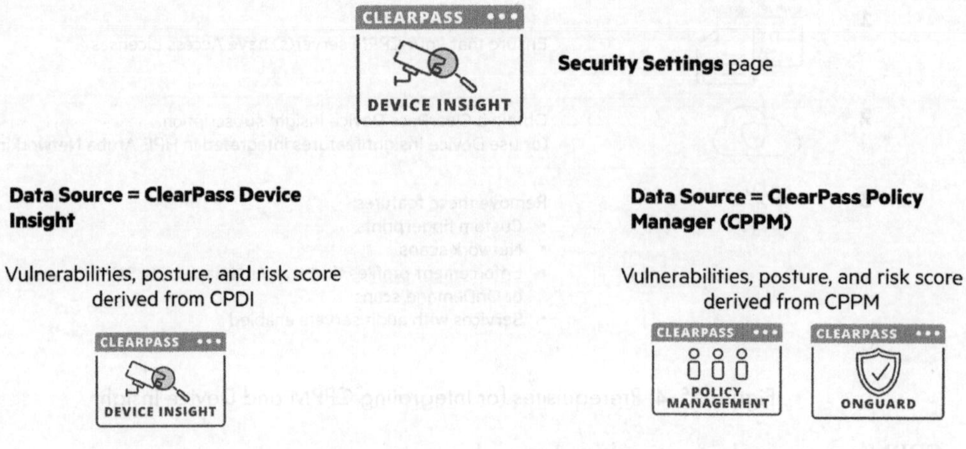

Figure 15-5: Security Context Integration between CPPM and CPDI

In the **Security Settings** page, the Data Source indicates whether CPDI takes the vulnerabilities, posture, and score from its own agentless scanning or from CPPM. As shown on the left of Figure 15-5, if you select **ClearPass Device Insight** as the Data Source, CPDI will use its own scanning to derive the vulnerability, posture, and risk score shown for clients. However, it will not overwrite posture information on CPPM. As shown on the right of Figure 15-5, if you select **ClearPass Policy Manager (CPPM)** as the Data Source, on the other hand, CPPM sends the vulnerabilities, posture, and risk scores to CPDI.

Obtaining Registration Tokens

To integrate CPPM with Device Insight, you need a registration token. You use the same token that you do for Data Collectors.

If you need a refresher on how to find this token, follow these directions for HPE Aruba Networking Central 2.5.5. From the Global view, go to the **Organization** page. Then select the **Platform Integration** tab. Click the **Data collectors** card. From there, you will either see a Get Started wizard or a list of Data Collectors. In either case, you should see a **Registration Token** button. Click **Registration Token** and copy the token, as indicated in Figure 15-6. Paste the output into a file so that you are ready to upload it on your Data Collectors.

Figure 15-6: Obtaining Registration Tokens

Detailed steps might change in later HPE Aruba Networking Central versions; refer to the latest documentation.

Configuring Device Insight Integration on CPPM

Configuring Device Insight Integration on CPPM is simple. Navigate to **Administration > Server Manager > Device Insight**, and click the **Enable** option for Device Insight Integration (see Figure 15-7).

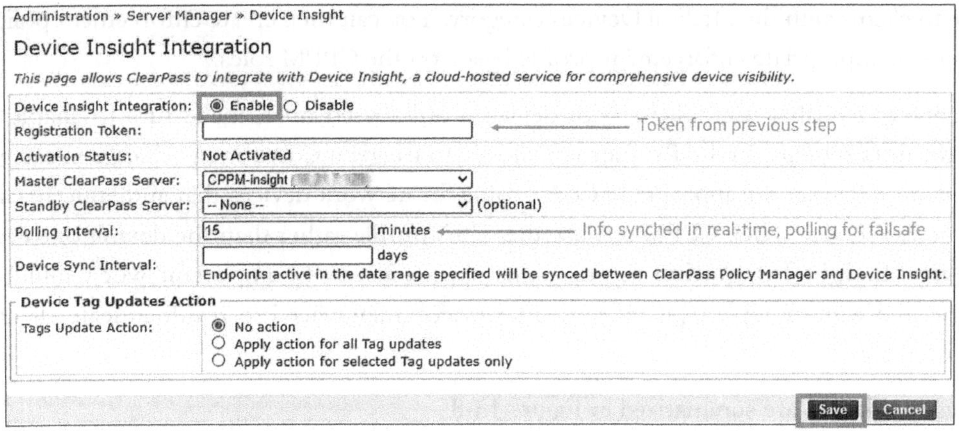

Figure 15-7: Configuring Device Insight Integration on CPPM

Copy in the registration token that you collected.

If you are integrating a cluster with CPPM, select the primary ClearPass Server. You can also select a standby server.

Device Insight updates CPPM with device information in real time. However, if no communication occurs between CPPM and Device Insight over the course of the polling interval, then CPPM polls Device Insight to ensure that it has the latest devices, classification details, and Device Insight tags. The Device Sync interval, available in v6.9.1 and above, determines for which endpoints CPPM and

Device Insight sync information. They only sync info for devices that have been active in the previous specified number of days. (In earlier versions, CPPM and Device Insight synched information for all endpoints.)

As of v6.9.2, you can configure the settings for tags. You will learn more about using tags a bit later in this chapter.

When you have filled out the settings, click **Save**.

Using Device Insight Information in CPPM Services and Policies

You have already learned how to set up CPPM to use device profile information to assign endpoints to ClearPass roles and eventually to enforcement profiles. The same concepts apply when CPPM is receiving the device profile information from Device Insight instead of its own device profiler.

You can create rules within role mapping policies or enforcement policies that reference the device profile attributes. These attributes are stored within these namespaces: Authorization: [Endpoints Repository] and Endpoint. For example, you can create role mapping policy rules that assign a "smartphone-CPPM" role to clients with the SmartDevice category and a "medical-device-CPPM" role to clients with the Medical Devices category. You can then create enforcement policy rules that assign an appropriate enforcement profile based on the CPPM roles.

In the services that use policies with device profile-based rules, make sure to enable the **Profile Endpoints** settings. In the **Profiler** tab choose to trigger a CoA for any change in category, family, or name. Then set the appropriate CoA profile for network devices that authenticate clients to this particular service. Then CPPM can use that CoA profile rather than the default one defined in the **Device Insight Integration** settings; this approach can be important when multiple services, associated with multiple types of network devices, might need to reauthenticate clients based on updated tags.

These guidelines are summarized in Figure 15-8.

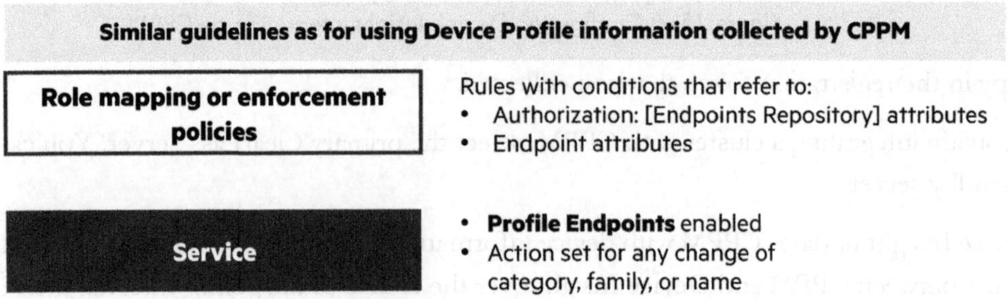

Figure 15-8: Using Device Insight Information in CPPM Services and Policies

Use Device Insight Integration to Detect and Mitigate Threats

In this section, you will explore ways to use CPPM's integration with Device Insight to detect and mitigate threats.

Device Insight Tags and CPPM Integration

As you learned previously, admins can create tags in Device Insight. These tags select devices based on a broad range of conditions. Within CPDI and HPE Aruba Networking Central, they are meant to convey useful information about devices and make it easier for users to find devices that exhibit characteristics of interest.

When you integrate CPPM with Device Insight, the tags can provide valuable information for CPPM to consider when determining what access level to grant to clients. For example, as shown at the top of Figure 15-9, CPDI might tag devices as having a high-risk score. CPPM can use that information to decide to apply a remediation enforcement profile to the client. This profile could limit the client's access, and the impact of a potential attack, until the vulnerabilities can be remediated.

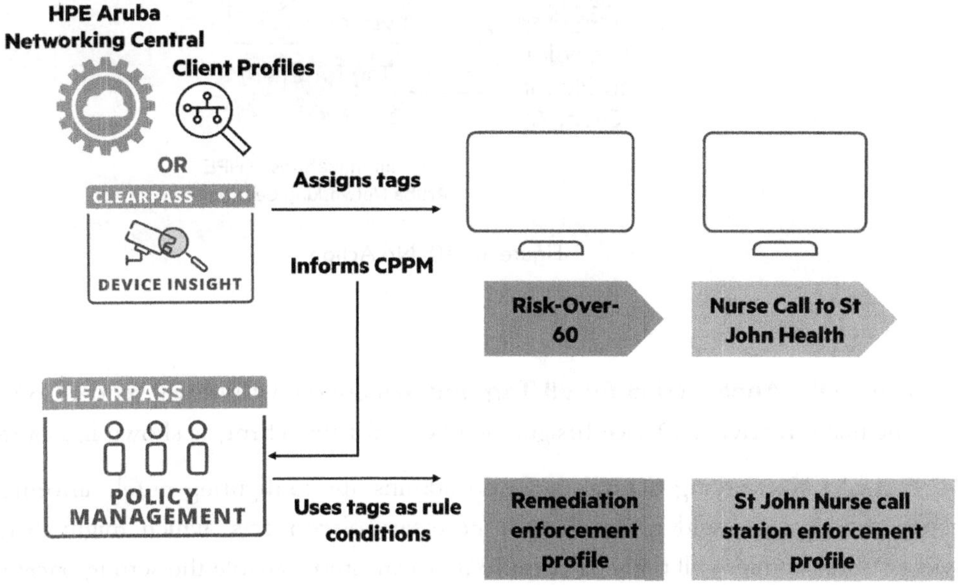

Figure 15-9: Device Insight Tags and CPPM Integration

Or either CPDI or HPE Aruba Networking Central might tag devices based on their location or other context. CPPM can use this information to apply an appropriate enforcement profile with a particular Aruba-User-Role VSA, as shown at the bottom of Figure 15-9.

Understanding Tag Update Action Options

A device's tags might change. If you want CPPM to be able to apply a new enforcement profile based on the latest tags, you must configure CPPM to issue a CoA in response to a tag update.

Configure the **Tags Update Action** in the Device Insight integration settings. Navigate to **Administration** > **Server Manager** > **Device Insight** in the CPPM UI. You have three options for **Tags Update Action**.

No Action

The **No action** option tells CPPM not to issue CoAs, as indicated in Figure 15-10. This action is only suitable if all tags are purely information; CPPM is not using them as conditions in role mapping or enforcement policies.

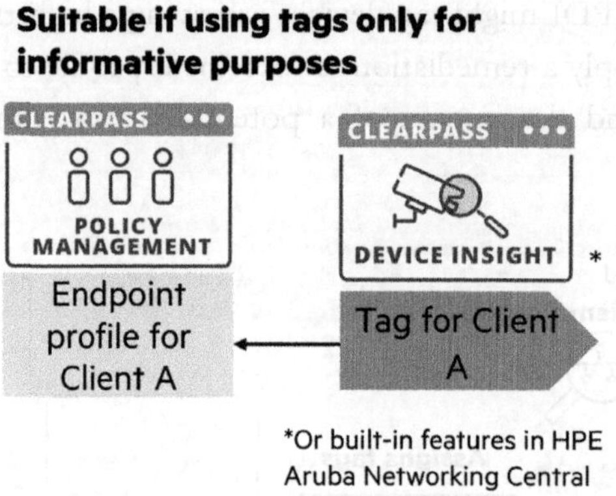

Figure 15-10: No Action

Apply Action for All Tag Updates

If you choose the **Apply action for all Tag updates** option, CPPM will send a CoA for a client every time that it receives a Device Insight tag update for that client, as shown in Figure 15-11.

If HPE Aruba Networking Central or CPDI admins are configuring purely informational tags, selecting this option could cause many unnecessary disconnects, which might disrupt clients. However, some admins still prefer it because they can simply enable the setting once; they do not need to return later and add new tags to the selected list, which can minimize mistakes in the future.

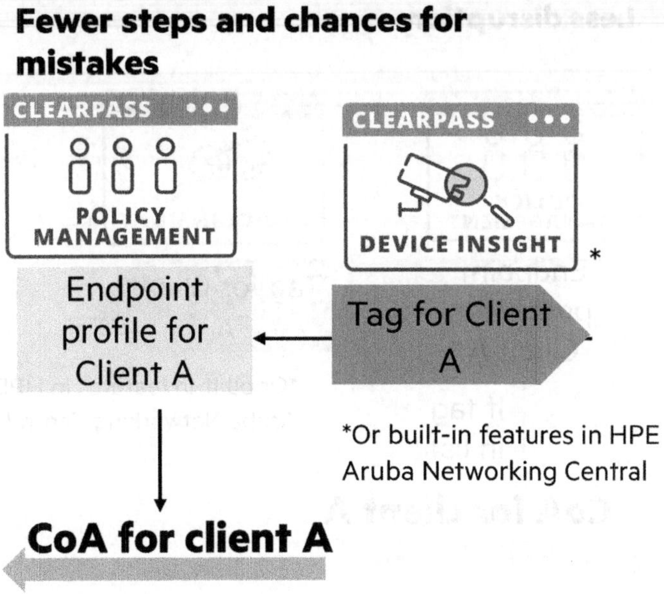

Figure 15-11: Apply Action for All Tag Updates

Apply Action for Selected Tag Updates Only

However, if you are concerned about disruption, you should select **Apply action for selected tag updates only** option, as indicated in Figure 15-12. Then, in the list below, add the specific tags that are used in role mapping policy or enforcement policy rule conditions. These are the tags that might change the client's enforcement profile, so these are the tags that should trigger a CoA.

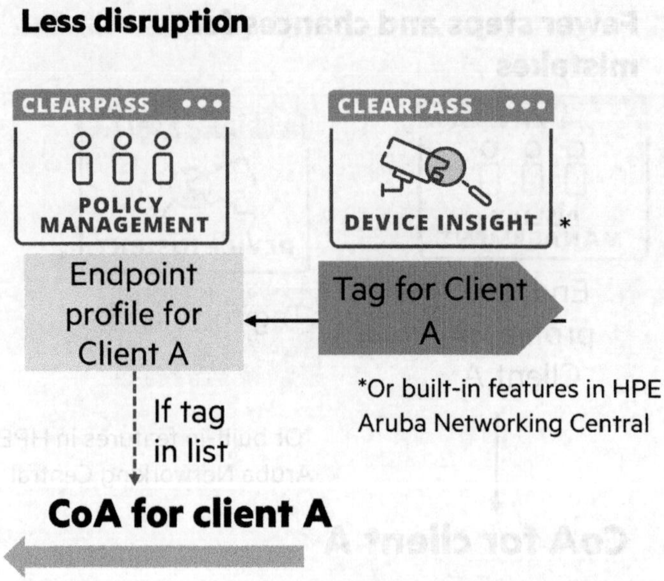

Figure 15-12: Apply Action for Selected Tag Updates Only

Choosing an Action

When CPPM triggers the action (whether for all tag updates or for selecting tag updates), it uses the action defined in the **Profiler** tab for the service that authenticated the client in question. However, if this action is not configured, CPPM uses the RADIUS CoA Action defined in the Device Insight Integration settings.

Overview of Using Tags to Apply Enforcement Profiles

In Figure 15-13, you see a basic overview of the process by which CPPM can use tags to apply a specific enforcement policy to a client.

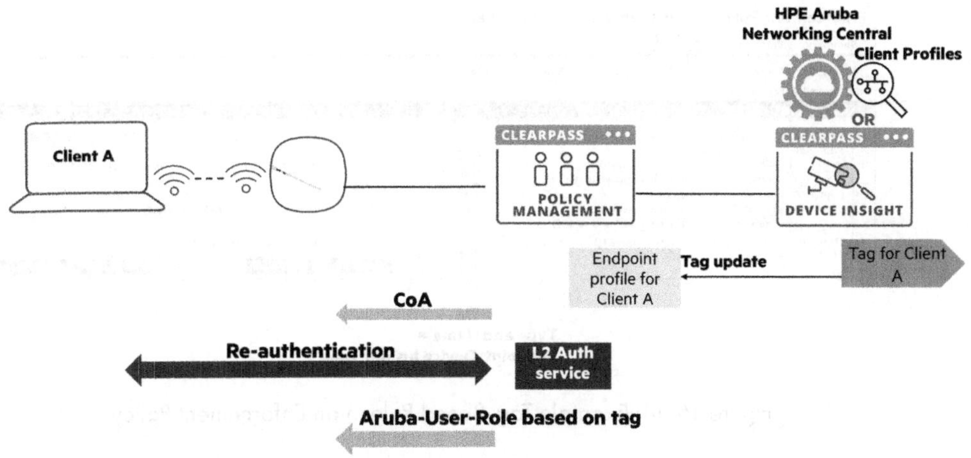

Figure 15-13: Overview of Using Tags to Apply Enforcement Profiles

First, an admin creates a tag in HPE Aruba Networking Central or CPDI, and that tag applies to a device. The cloud solution informs CPPM of the new device tag for that client. CPPM places that tag in the client's endpoint profile.

Assuming the integration settings use the proper option for **Tags Update Action**, CPPM sends a CoA to the network device (an AP or switch) to which the client is connected.

After the network device enforces the CoA, the client will re-authenticate to an 802.1X service, or other L2 authentication service, on CPPM. A rule in that service's enforcement policy uses the tag as a condition. The rule will apply an enforcement profile with an appropriate Aruba-User-Role. For example, if the tag indicates that the device is vulnerable, the rule associated with that tag might place the user in an Aruba-User-Role for remediation.

Example Tag-Based Rule in an Enforcement Policy

You can use Device Insight tags in role mapping policies and enforcement policies, just like you use other attributes in an endpoint profile. If you want to use a tag to adjust clients' access, you will need to add a tag-based rule to the policies for any service that a client could potentially use to authenticate, such as the wired and wireless 802.1X service on CPPM.

Keep in mind that whether you add the rule to a role mapping policy or an enforcement policy depends on what approach you're using in general and your particular use case.

In the example shown in Figure 15-14, you are using an enforcement policy. You add a rule with the Endpoint Type and the Device Insight Tags Name in the condition. You set the condition value equal to "suspicious_dst." Then you select the profiles that you want to apply.

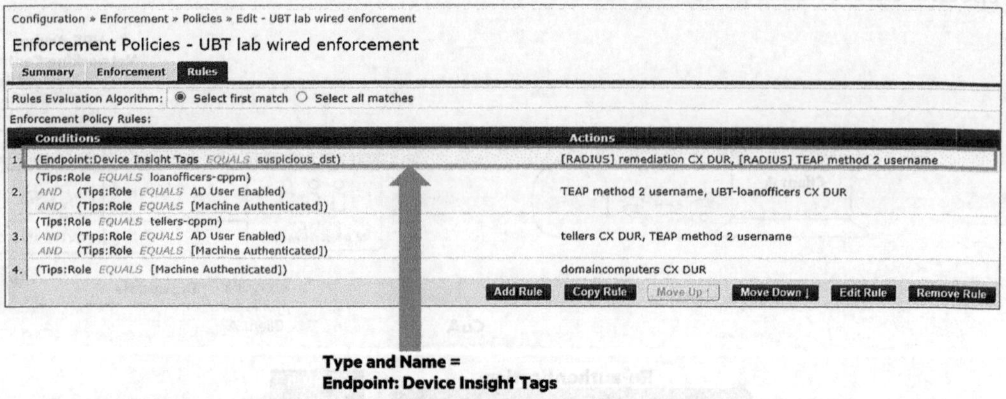

Figure 15-14: Example Tag-Based Rule in an Enforcement Policy

Example Use Case Using Network Activity

Now look at a complete example use case, which involves integration between CPPM and the licensed CPDI application. You could also implement the same use case with HPE Aruba Networking Central's client profiles with no CPDI license required.

In this example, illustrated in Figure 15-15, CPDI admins analyze the network activity, applications, and destination hosts on endpoints. They might discover signs of a compromised device such as an IoT device contacting a known command and control host. The admins can quickly respond by creating a tag from the profile (named "suspicious_dst" in this example). As you recall, the tag comes up with all the attributes of that device and the admins can select just the attributes that indicate the device is compromised.

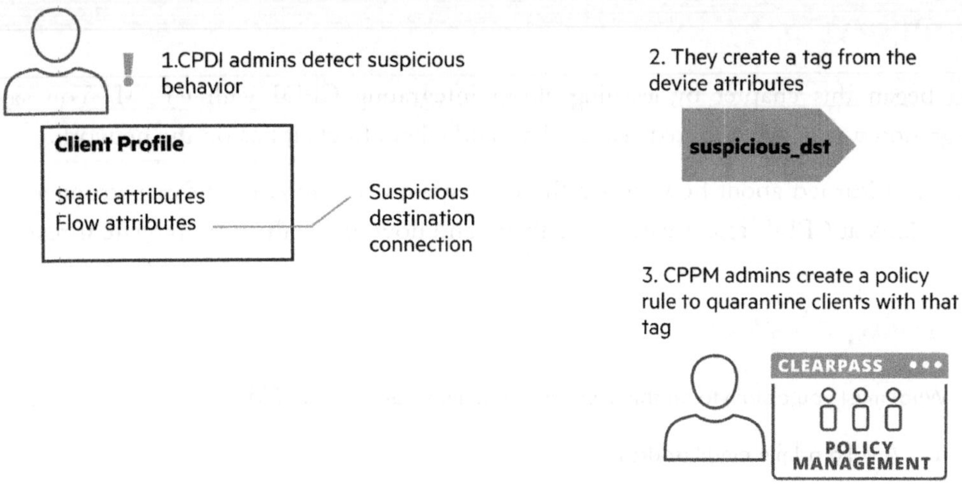

Figure 15-15: Example Use Case Using Network Activity (Steps 1-3)

The CPDI admins inform the CPPM admins about the new tag. The CPPM admins add a rule to each 802.1X service enforcement policy. That rule selects endpoints with that Device Insight tag and applies a quarantine or remediation profile. If CPPM is using the Disconnect for Selected Tags option, the admins add the new tag to the list of selected tags. (CPPM admins can also issue a CoA on the initial device or devices on which admins detected the suspicious behavior. Or admins might have already dealt with those devices.)

Later other devices display the same suspicious behavior, as shown in Figure 15-16. Without admin intervention, CPDI automatically applies the "suspicious_dst" tag. It also updates CPPM to inform CPPM that the devices have this tag. CPPM then automatically issues CoAs for the compromised devices with that tag. The devices reauthenticate and CPPM quarantines them. CPDI and CPPM have worked together to protect the network.

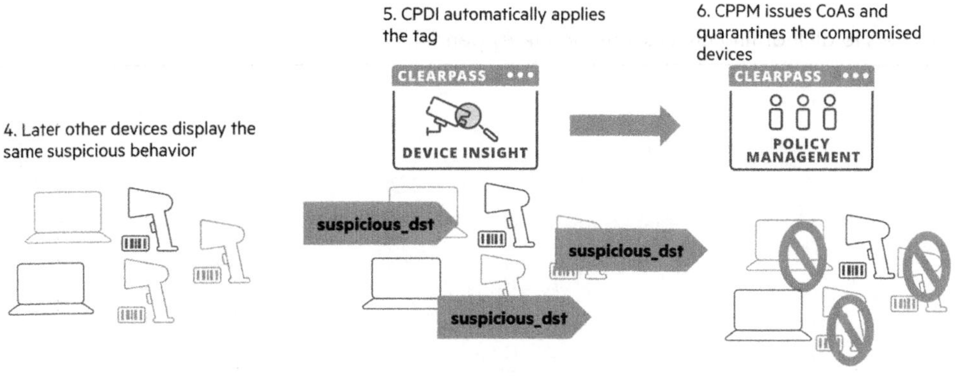

Figure 15-16: Example Use Case Using Network Activity (Steps 4-6)

Summary

You began this chapter by learning about integrating CPDI with CPPM. You saw how this integration works, what prerequisites it has, and what effects it has on the network.

Then you learned about how to use these integrated solutions to enforce stronger security. You took a look at CPDI tags, use cases for them, and how to use them to mitigate threats.

Learning checks

1. What must you obtain to enable Device Insight integration on CPPM?
   a. A valid admin email address
   b. An admin password
   c. A private key
   d. A registration token
2. CPPM is integrated with Device CPDI. What determines whether CPPM sends a CoA when a device's tag is updated by CPDI?
   a. The settings in the Profiler tab for the service that authenticated the device
   b. The settings in the ClearPass Device Integration page
   c. The cluster-wide General settings
   d. The network device settings for the NAD to which the device is connected

---

Note

For answers to all learning checks, refer to the Appendix.

---

16 Use Packet Captures to Investigate Security Issues

EXAM OBJECTIVES

- ✓ Perform packet capture on HPE Aruba Networking infrastructure locally and using HPE Aruba Networking Central
- ✓ Interpret packet captures

Assumed Knowledge

- Basics of deploying HPE Aruba Networking infrastructure products
- Basics of HPE Aruba Networking solution architectures
- Basic 802.11 and Ethernet principles

Overview

When you need to dig deep into an issue, packet captures provide a powerful tool.

In this chapter, you will review how to capture packets on HPE Aruba Networking network infrastructure devices. You will learn about a variety of strategies that you can take to obtain the packets, depending on what traffic you want to capture and your network architecture.

You will then learn how to use Wireshark so that you can analyze the packets that you have discovered. After a brief introduction to basic guidelines for using the tool, the chapter will guide you through a few scenarios for detecting security issues using Wireshark packet captures. Specifically, you will look at detecting port scans and investigating an ARP poisoning attack.

Review Packet Captures

In this section, you will review packet captures.

Review Packet Capture

Packet captures can be useful tools for investigating security issues. As shown in Figure 16-1, when a device "captures" packets, it records traffic that is flowing through it. A packet capturing device might also save either the packets themselves, or it could save an analysis of the packets to a file for later examination. The device might also send the packets to another device, such as an Intrusion Detection System (IDS), for examination.

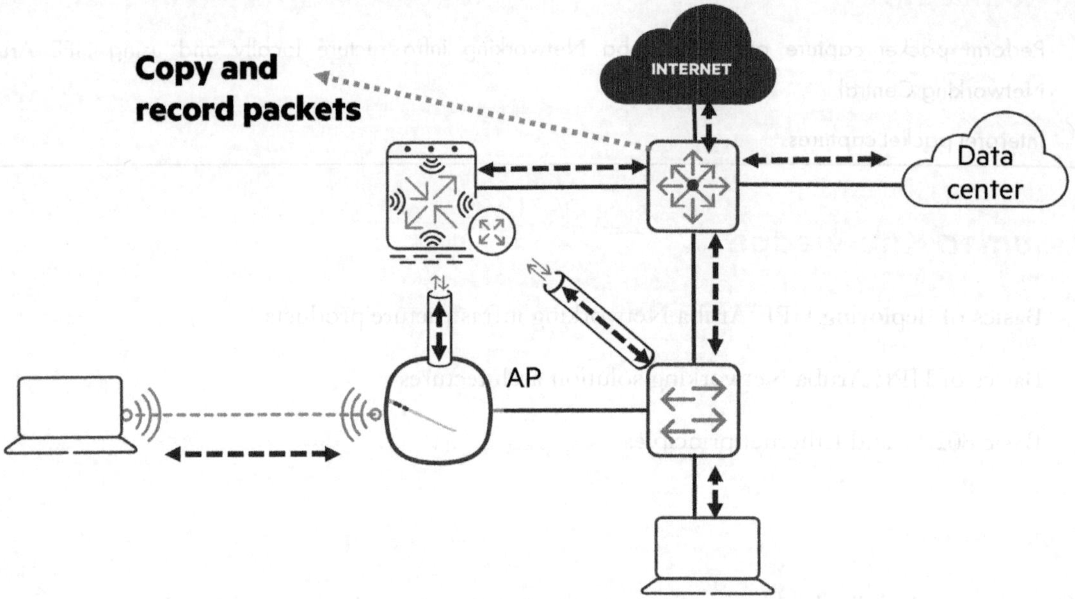

Figure 16-1: Review Packet Capture

Considerations for Implementing the Packet Capture

To use packet captures effectively, you need to carefully consider the goals of your investigation. What traffic are you interested in? Are you attempting to gather traffic across the network as the preliminary step in pinpointing a problem? Or do you already know that a client is behaving suspiciously, and you want to investigate that client further?

When you know your target, consider the traffic flow. Generally you want to capture traffic as close to the source of the issue as possible. However, when you do not know what the source is, you might need to capture a large amount of traffic as a first step in looking for the source. In that case, you could set up a capture at a bottleneck where most traffic flows, such as the link between a core routing switch and the data center.

For wireless traffic you also need to consider decryption. Based on the wireless architecture, an AP, gateway, or MC might be decrypting the wireless traffic. In AOS-10, APs encrypt and decrypt wireless traffic for both bridged and tunneled SSIDs, as shown in Figure 16-2. The AP might then encrypt the traffic with IPsec again; however, you can capture the traffic on the AP before that encryption occurs.

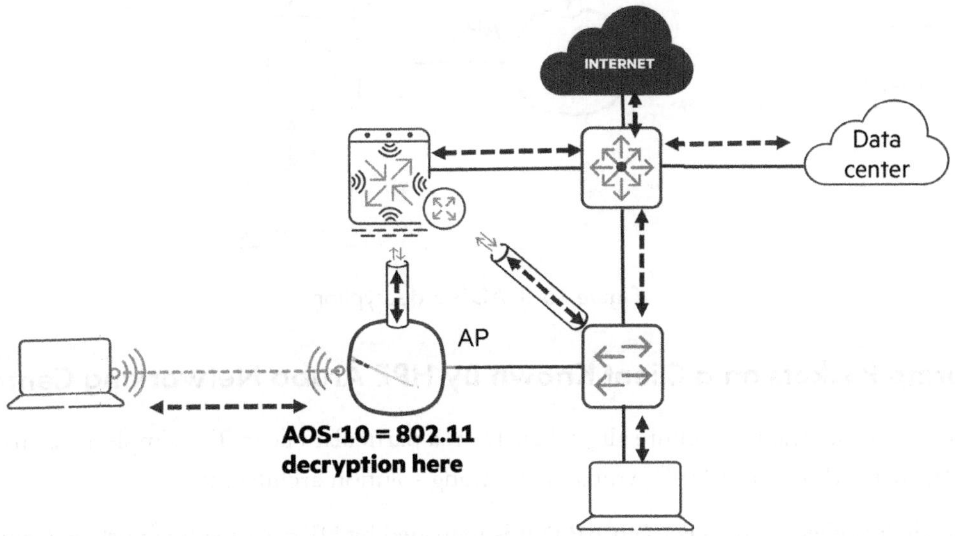

Figure 16-2: AOS-10 decryption

However, in AOS-8 architectures, MCs typically encrypt and decrypt the traffic, as shown in Figure 16-3. Understand the architecture that the company is using so that you can capture the traffic after decryption.

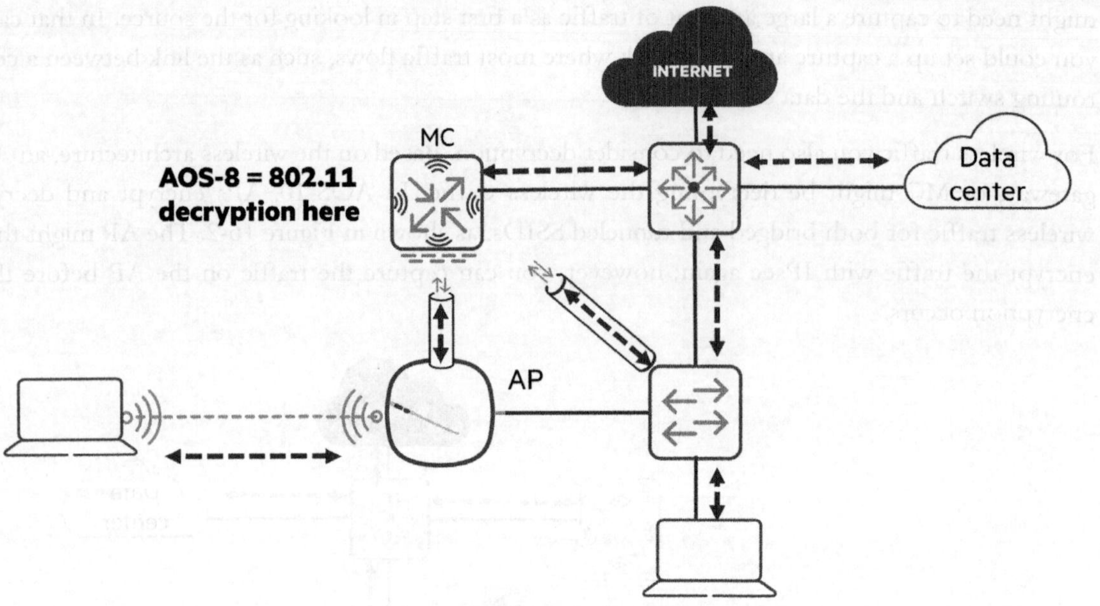

Figure 16-3: AOS-8 decryption

Capturing Packets on a Client Known By HPE Aruba Networking Central

Sometimes you will want to capture all packets from a particular client. The simplest way to capture that traffic depends on your HPE Aruba Networking solution architecture.

When a wireless client connects to an AP that is managed by HPE Aruba Networking Central, you can easily initiate a packet capture from within the cloud console. You can use this feature regardless of whether the AP is bridging or tunneling the client traffic. Similarly, you can initiate packet captures from HPE Aruba Networking Central on wired clients that are connected to an AOS-CX switch or have their traffic tunneled to a gateway.

As shown in Figure 16-4, when you start the packet capture on a client's traffic, HPE Aruba Networking Central tells the device to which the client is connected to execute the packet capture. For up to 15 minutes, the device copies all packets flowing to and from that client and sends the packets to HPE Aruba Networking Central. That solution saves the packets in a Packet Capture (PCAP) file, which you can download to your local station at the end of the packet capturing session.

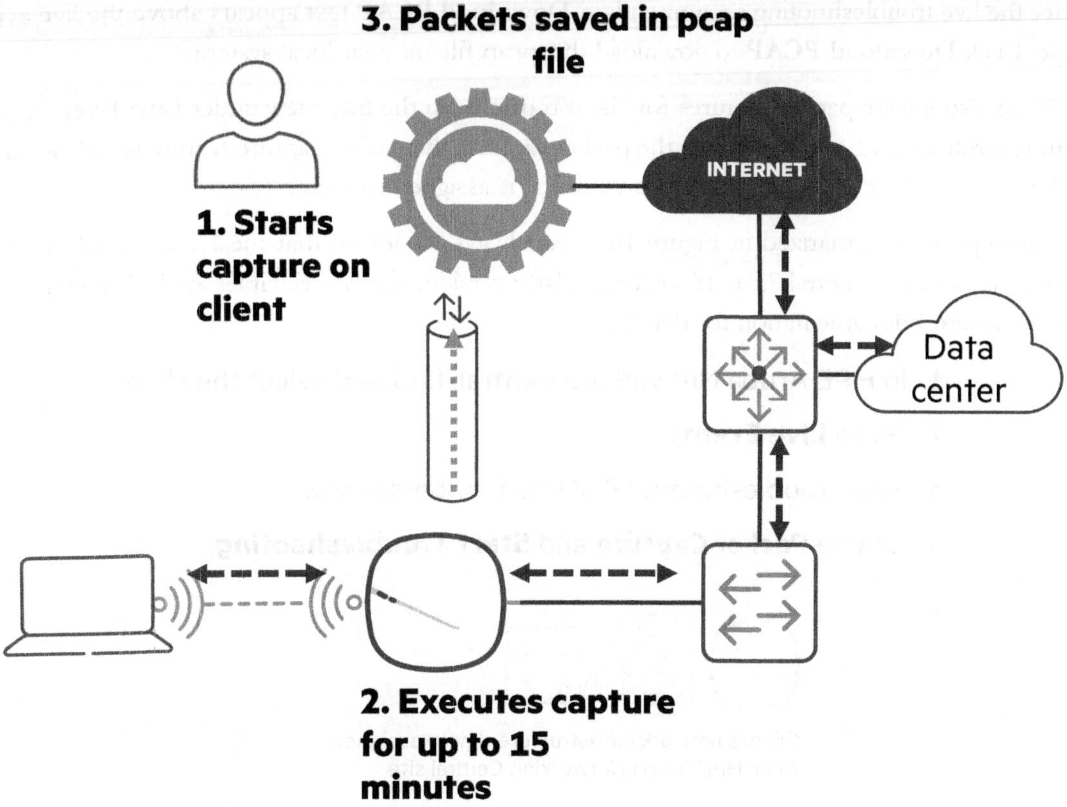

Figure 16-4: Capturing Packets on a Client Known By HPE Aruba Networking Central

Note, however, that all network devices and device software versions do not necessarily support this feature, so you should check the capabilities and requirements for your particular network devices.

Implementing a Packet Capture with HPE Aruba Networking Central

To implement a packet capture with HPE Aruba Networking Central, go to the **Clients** page from any view. Click the name of the client for which you want to capture traffic. You will be taken to the **Clients Details** page.

Select **Live Events**..

Click **Stop Troubleshooting**, as troubleshooting automatically begins. Then enable the **Packet Capture** toggle button and click **Start Troubleshooting**. The live troubleshooting session runs for up to 15 minutes. You can stop the troubleshooting session at any time before that if you choose.

After the live troubleshooting session ends, a **Download PCAP** text appears above the live events table. Click **Download PCAP** to download the .pcap file on your local system.

You can also initiate packet captures for client traffic from the **Site** view under **Live Events**. Note that, regardless of where you initiate the packet capture, the packet capture feature is only available if the network device to which the client connects is assigned to a site.

These steps are summarized in Figure 16-5. As always, be aware that these steps apply to HPE Aruba Networking Central 2.5.6. If you have a later version, similar principles apply, but you should check the latest documentation for details.

1. In HPE Aruba Networking Central, find and select the client.
2. Go to **Live Events**.
3. Stop troubleshooting if started automatically.
4. Enable **Packet Capture** and **Start Troubleshooting**.

Figure 16-5: Implementing a Packet Capture with HPE Aruba Networking Central

Capturing Packets to and from Clients on a Gateway or MC

If the company is using a controller-based architecture, the AP is tunneling the wireless client's traffic to a Mobility Controller (MC), which decrypts the traffic. You can initiate a datapath capture of the client's traffic from the CLI, as indicated in Figure 16-6. Similarly, you can initiate a datapath capture of any client connected in the MC user-table, including UBT clients and clients in a client-to-site VPN. The same holds true for HPE Aruba Networking gateways.

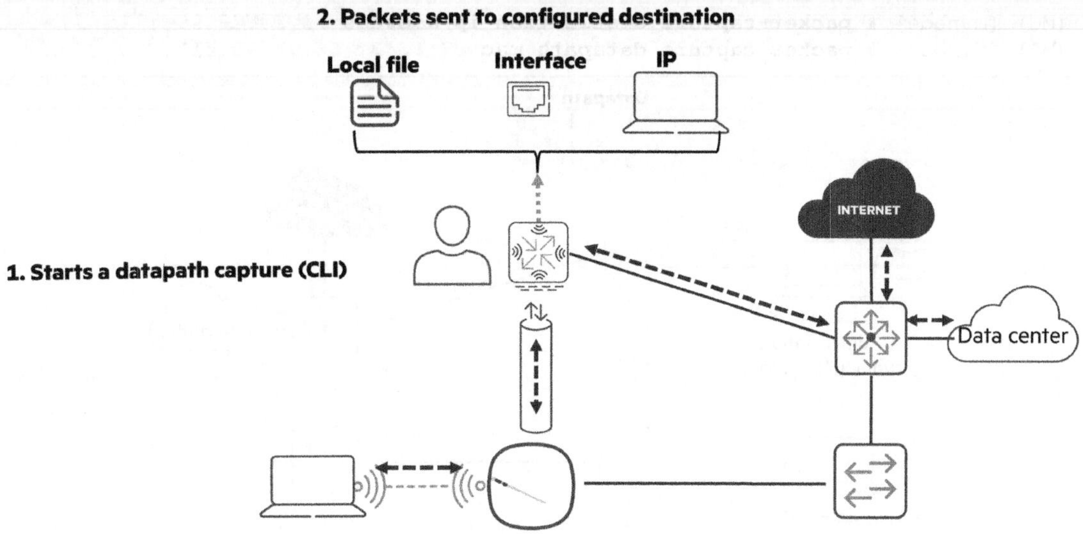

Figure 16-6: Capturing Packets to and from Clients on a Gateway or MC

When you initiate a datapath capture on an MC or gateway, the device copies all of the traffic to and from the client and sends the packet copies to the configured packet capture destination. As shown in Figure 16-6, you can configure this destination as a local file on the MC, an interface on the MC to which you would attach a device running a packet analyzer like Wireshark, or the IP address of a device running a packet analyzer.

Implementing an MC or Gateway Datapath Packet Capture to a Remote IP

You can use the IP address destination option to analyze the captured packets directly on your management station. Start a packet analyzer like Wireshark on your station (you will learn more details about doing this in the next section). Then configure that station as the destination for the packet capture on the MC or gateway. Use this command: **packet-capture destination ip-address <IP_address>**

Find the MAC address of the client whose traffic you want to capture. Then start the packet-capture with this command: **packet-capture datapath mac <MAC_address> <all | decrypted | encrypted>**. You can choose whether to capture all traffic to and from the client, just the decrypted traffic, or just the encrypted traffic. (See Figure 16-7.)

```
(MC) [mynode] # packet-capture destination ip-address 10.1.140.90
(MC) [mynode] # packet-capture datapath mac d8:12:34:56:78:90 all
```

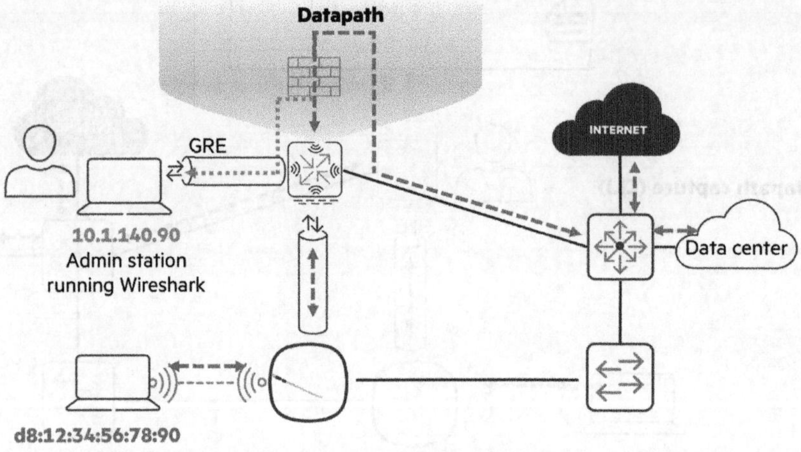

Figure 16-7: Implementing an MC or Gateway Datapath Packet Capture to a Remote IP

The MC or gateway then encapsulates the client traffic with GRE and sends the traffic to the destination IP address.

When you have finished investigating the issue, make sure to turn the packet capture off with the **no** form of the same command: **no packet-capture datapath mac <MAC_address> <all | decrypted | encrypted>**

Implementing an MC or Gateway Datapath Packet Capture to a File

Sometimes you will want to collect the packets in a file, which you can then copy off of the MC or gateway and analyze elsewhere. For this option, set the packet capture destination to the local-filesystem as shown in Figure 16-8. Then initiate the datapath packet capture in the same way explained previously.

```
(MC) [mynode] # packet-capture destination local-filesystem
(MC) [mynode] # packet-capture datapath mac d8:12:34:56:78:90 all
(MC) [mynode] # no packet-capture datapath mac d8:12:34:56:78:90 all
(MC) [mynode] # copy /var/log/oslog/datapath.pcap <ftp: | scp: | tftp: | usb:>
               <dst> <dst filename>
```

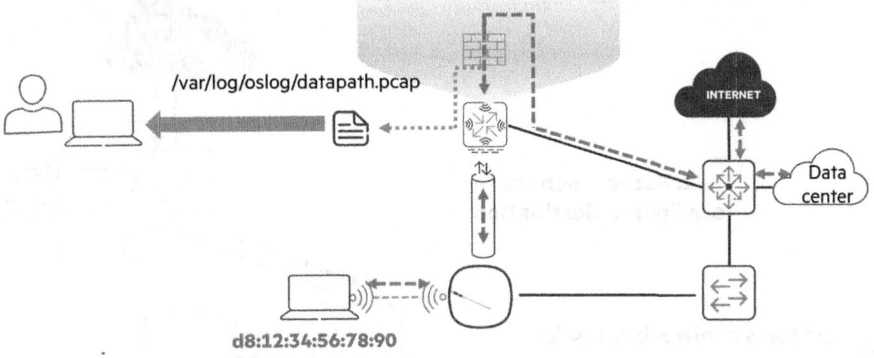

Figure 16-8: Implementing an MC or Gateway Datapath Packet Capture to a File

When you are done with the capture, disable it. The packets are saved in this file: /var/log/oslog/datapath.pcap. Copy the file to an FTP server, SCP server, or TFTP server. Or you can copy the file to a USB drive.

Capturing Packets to and from a Wired Client on an AOS-CX Switch

If you want to capture packets flowing through an AOS-CX switch, use mirroring. A mirroring session copies packets received from or transmitted to one or more source interfaces. It then sends those packets to a destination. As shown in Figure 16-9, options for the destination include an interface or a tunnel to a remote IP address. You can use these options to send the packets to an device that is running a packet analyzer.

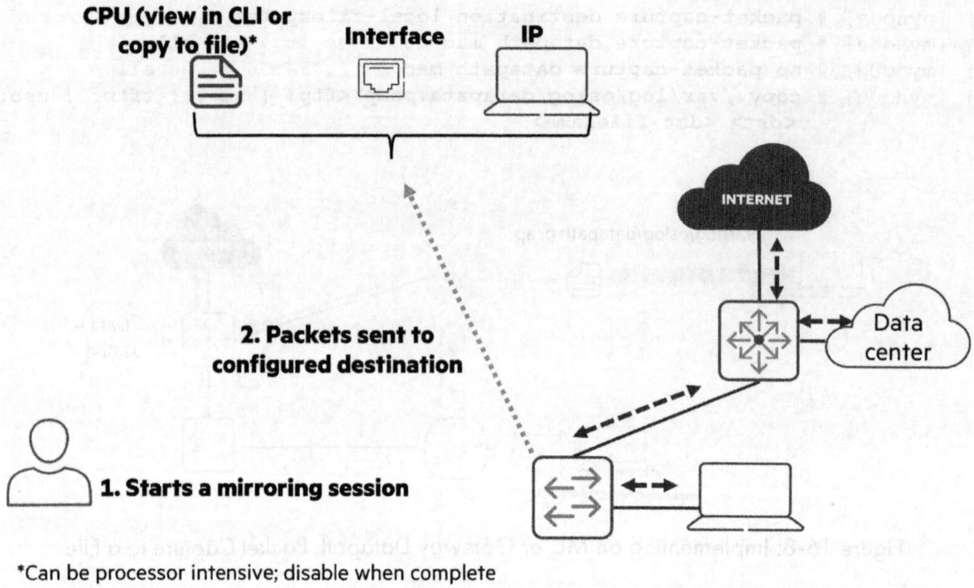

Figure 16-9: Capturing Packets to and from a Wired Client on an AOS-CX Switch

If you want the switch to capture and analyze the packets itself, you can set the destination to "cpu." You can then view the packets on the switch directly by entering these commands: **diagnostic** and then **diag utilities tshark**. However, it is usually easier to interpret the packets in an application like Wireshark. You can place the last 32 bytes of the CPU capture in a file with these commands: **diagnostic** and then **diag utilities tshark file**. Then copy the tshark file to an SFTP server.

However, the CPU capture can be processor intensive. Often the best approach is to set up a client running Wireshark and remotely mirror the traffic to it.

AOS-CX switches support multiple mirroring sessions. However, they only support only one session using the CPU as the destination.

Running a Packet Analysis Directly on the Switch

If you want to capture packets and analyze them directly on the switch, you can do so by initiating a tcpdump in the shell. You might also choose this option when you want to capture just a specific type of traffic, such as RADIUS traffic in the example shown in Figure 16-10. It is possible to use classes to filter a mirror session, but the configuration takes a few extra steps.

```
AOS-CX# start-shell
AOS-CX:~$ sudo ip netns exec swns tcpdump -vv -i any port 1812 or port 1813
or port 3799
16:09:23.780428 ethertype IPv4, IP (tos 0x0, ttl 63, id 14614, offset 0,
flags [DF], proto UDP (17), length 151)
    clearpassxx.training.arubanetworks.com.radius > P46-TXX-6300.50007: [udp
sum ok] RADIUS, length: 123
Access-Challenge (11), id: 0xd3, Authenticator:
c0196620a3ba3671db0f2990c8b11373
  EAP-Message Attribute (79), length: 43, Value: .n
[Ctrl+c]
12 packets captured
102 packets received by filter
90 packets dropped by kernel
AOS-CX:~$ exit
```

Warning: Shell is locked down to protect the switch, but still be cautious

Figure 16-10: Running a Packet Analysis Directly on the Switch

The shell is locked down to prevent users from causing damage to the switch functions. However, whenever you access the shell, you should be cautious and avoid entering commands that you do not understand. To start the shell, enter: **start-shell**

Use this command to start the packet capture: **sudo ip netns exec swns tcpdump**

When you specify **-vv** option, the CLI outputs more complete information about each packet. If you omit this option, you simply see one line about each packet that arrives—which can still be useful if you are simply trying to see if particular types of packets are flowing.

Use the **-i** option to indicate on which interfaces you want to capture the traffic. In this example, you are capturing traffic on all interfaces.

You can use the port options to select the UDP or TCP port for the particular type of traffic that you want to capture.

Mirroring Traffic from a Switch Port to a Packet Analyzer

In Figure 16-11, you see an example of how to mirror traffic to a client running Wireshark.

```
AOS-CX(config)# mirror session 1
AOS-CX(config-mirror-1)# source interface 1/1/4 both
AOS-CX(config-mirror-1)# destination tunnel 10.1.140.90 source 10.1.130.5
AOS-CX(config-mirror-1)# enable
```

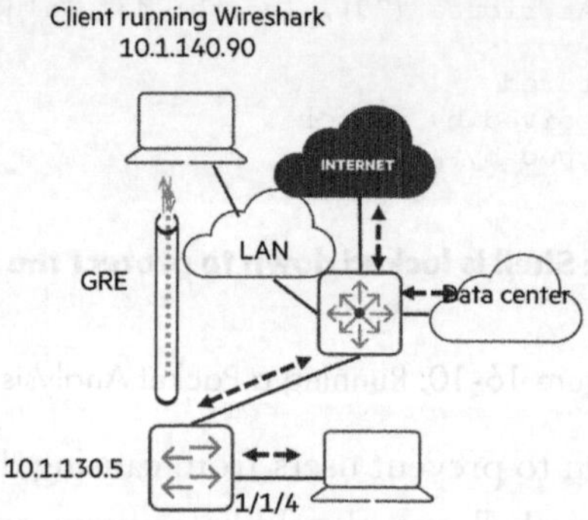

Figure 16-11: Mirroring Traffic from a Switch Port to a Packet Analyzer

Begin by creating the mirror session. Each session has a unique number: **mirror session <*ID*>**

Next set the source interface or interfaces for the session: **source interface <*port_ID*> <rx | tx | both>**. In this example, you want to capture traffic from the client connected to interface 1/1/4, so you set that interface as the source. You can choose to mirror traffic received on the interface (from the client in this example), traffic sent on the interface (to the client in this example), or both. Generally it is most useful to see both sides of the conversation.

Next set the destination. In this scenario, your client running Wireshark is not connected directly to the mirroring switch, so you need to set up remote mirroring. Use this command: **destination tunnel <*dest_ip*> source <*switch_ip*> [vrf <VRF_id> dscp <value>]**. As you see, you set the first IP address in the command to the remote IP address to which you want to send the traffic. The source IP address is an IP address on the switch. Specify the VRF in which the switch reaches the destination if not **default**.

When you are ready to start mirroring packets to the destination, enter **enable**. When you are done, you can disable the mirror: **no enable**.

Use Wireshark

In this section, you will learn tips for using Wireshark to analyze packets captured from the network.

Wireshark

You have reviewed how to capture packets and send them to a device running a packet analyzer. Now you will learn about using Wireshark, one of the most common packet analyzers, to interpret the packet capture. As shown in Figure 16-12, Wireshark can capture packets that arrive on an interface on the device running Wireshark. It decodes the packets and outputs information about them. You can also use Wireshark to open a saved PCAP file and view the packets decoded within that file.

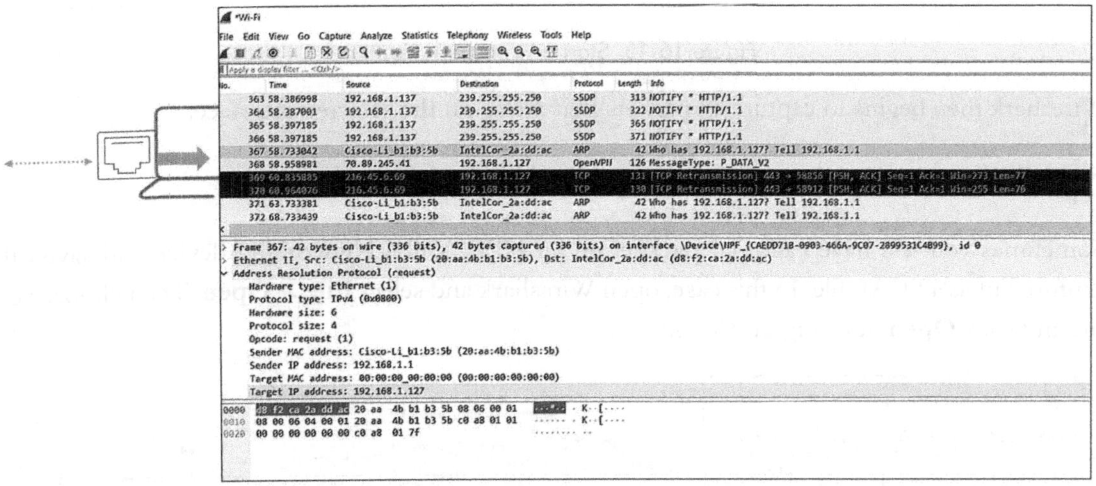

Figure 16-12: Wireshark

You could spend many hours practicing using Wireshark, so this study guide clearly cannot teach you all the aspects of this tool. However, it will give you some basic steps and pointers to get you started. This section features screenshots and instructions for Wireshark v2.6.5. Steps might differ a bit on other versions; however, you can apply many of the same principles.

Start a Wireshark Capture

When you open Wireshark, you will see a section called "Capture," which lists all of the interfaces on the device (see Figure 16-13). You can double-click the interface to immediately start capturing all of the packets on that interface. You will want to use this option when you are mirroring packets from an HPE Aruba Networking network device to the device running Wireshark.

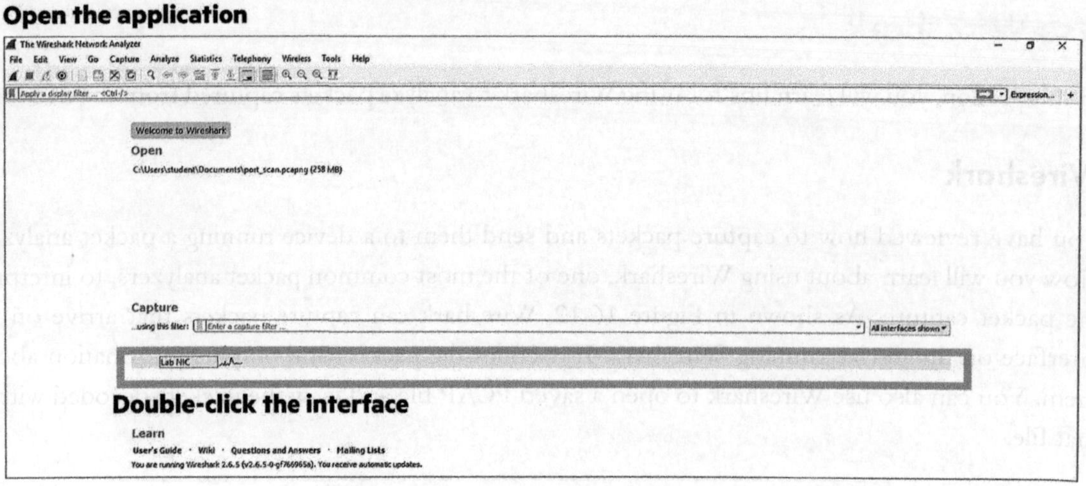

Figure 16-13: Start a Wireshark Capture

Wireshark then begins to capture all packets that arrive on the selected interface.

Open a PCAP File

Sometimes you will have captured packets on an HPE Aruba Networking device and saved the capture out as a PCAP file. In this case, open Wireshark and select **File > Open**. Then choose your file and click **Open**. (See Figure 16-14.)

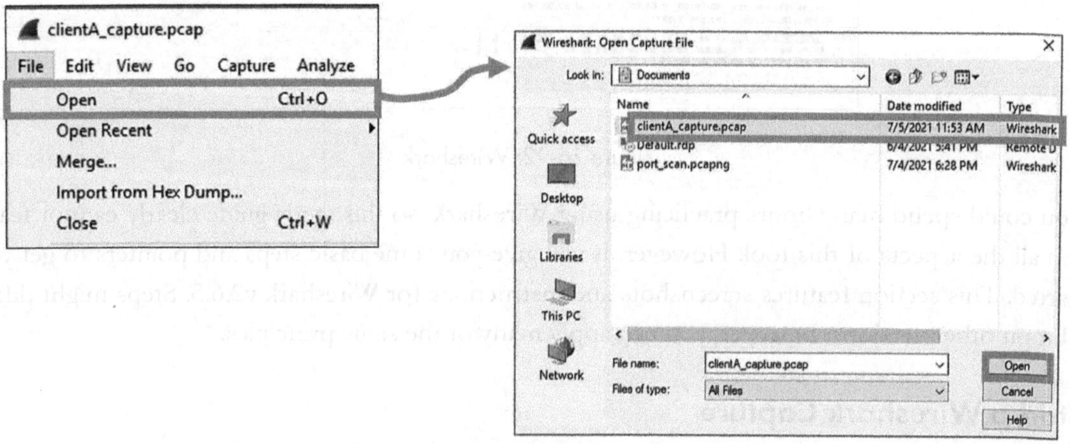

Figure 16-14: Open a PCAP File

Wireshark Output

After you start the packet capture or open the file, the top section of the application shows a list of packets ordered by time (see Figure 16-15). Wireshark provides a brief summary of important information about the packet, including its source IP address, destination IP address, protocol, length, and brief information (**Info**) about the packet contents. For example, Wireshark decodes ARP requests as **"Who has < target IP address>? Tell <sender IP address>."**

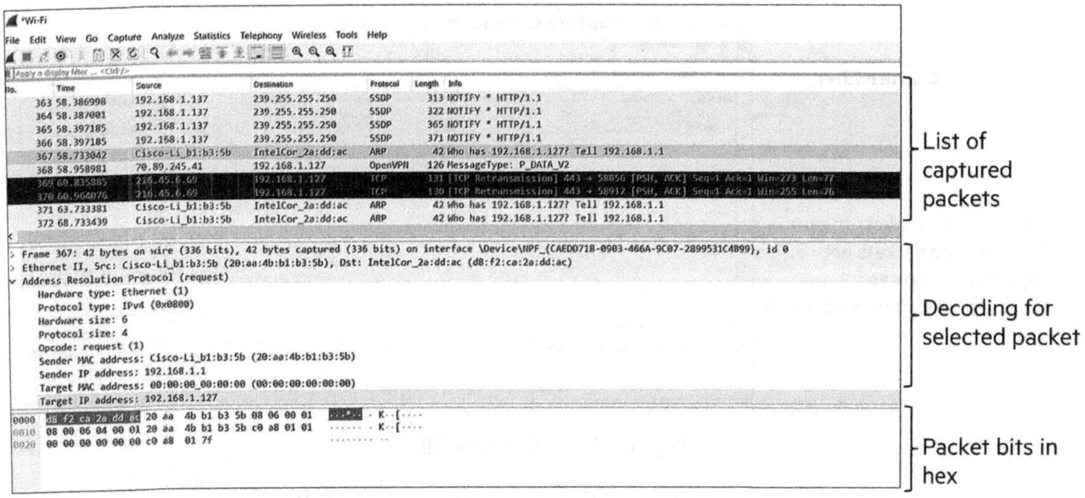

Figure 16-15: Wireshark Output

If you are running a live capture on an interface, as more packets arrive, Wireshark adds them to the bottom of the list. You can select a packet in the list and use the arrow keys to move up and down the packets.

When you select a packet, the middle section of the application shows more detailed decoding of that packet (see Figure 16-15). This section is organized in headers and the payload. You can expand any of the headers to see the fields within that header. In this example, the ARP header is expanded, and you can see the sender MAC address, sender IP address, target MAC address, and target IP address.

The bottom of the interface shows the packet bits in hex (see Figure 16-15). It highlights the bits that correspond to the particular field that you have selected in the middle section.

Capture Filter

Often you will want to narrow the packet output so that you can focus on conversations of interest. Wireshark supports both capture filters and display filters.

As shown in Figure 16-16, you apply a capture filter before you start the capture. Wireshark then only captures and displays packets with the specified characteristics. You can only remove the capture filter by stopping the capture and restarting with a new filter. You cannot remove the filter and see a broader range of earlier packets because Wireshark never captured the packets in the first place. You might use a capture filter when you are mirroring traffic to the Wireshark device, and you want to capture only the mirrored traffic, filtering out any random traffic that the device might be sending or receiving on its own behalf.

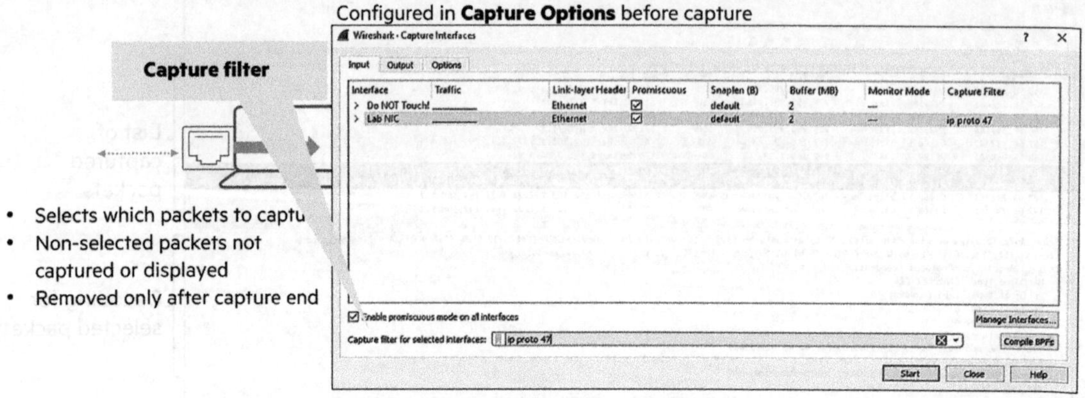

Figure 16-16: Capture Filter

If you are mirroring traffic from an AOS-CX switch, filter for the GRE protocol with this string: **ip proto 47**. To filter for Aruba_ERM traffic, use this string: **udp port 5555**. These examples are shown in Figure 16-17. Capture filters use libpcap filter language. For more information, refer to Wireshark documentation.

Capture Filter	Description
ip proto 47	Selects GRE packets
udp port 5555	Selects to and from UDP port 5555 (Aruba_ERM)

Figure 16-17: Example Capture Filters

Display Filter

A display filter is temporary. As shown in Figure 16-18, the display filter looks for packets that match the filter specifications and removes all other packets from the display. However, Wireshark still retains memory of the removed packets, and it continues to capture all packets permitted by the capture filter. You can remove the display filter and see all the captured packets again. You can then apply a new display filter as you desire.

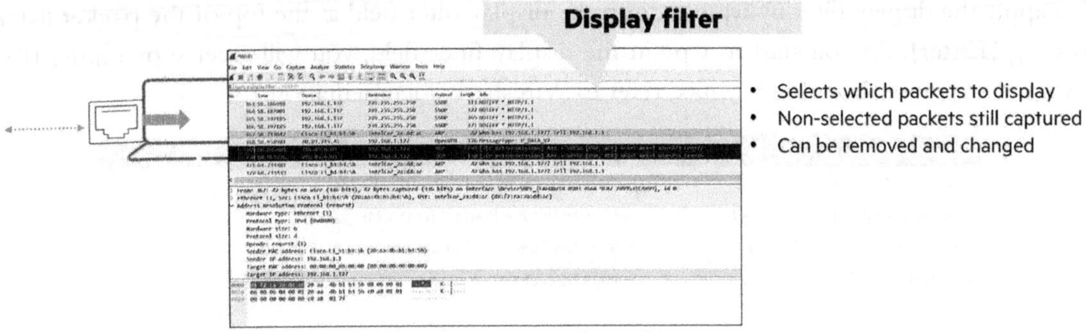

Figure 16-18: Display Filter

Display Filter Notation

Display filters use this basic notation: *parameter operator value*. The Operators table in Figure 16-19 shows some of the operators that you can use in display filters. To find parameters, you can start typing in the display filter field and see suggestions of valid parameters.

Operators	Description
==	Equals
!	Not
or	Match either condition
and &&	Match both conditions

Figure 16-19: Operators

You can apply display filters to help you focus in on a particular issue. Several examples are shown in Figure 16-20.

For example, you might be investigating an attack on a particular server. You can filter for that server's IP address to see traffic to and from it. Or you might be looking for signs of a ping sweep. You can filter for ICMP echoes (type 8) as well as echo replies (type 0). As you see in that example, you can combine multiple statements together with an "or" if you want to select packets that meet either criteria. Or you can use "&&" or "and" to select packets that meet both criteria. Use "!" to negate a statement.

HPE Aruba Networking Central captures include both management and data frames for wireless traffic. The management frames can be useful for troubleshooting, but sometimes you want to focus on the data sent to and received from a client. You can filter for 802.11 data frames with this filter: wlan.fc.type == 2

You apply the display filter by typing it into the display filter field at the top of the packet list and pressing **[Enter]**. As you start to type in the display filter field, you will receive prompts. If you want to learn more about filter syntax, refer to Wireshark documentation.

Display Filter	Description
ip.addr == 10.1.10.10	Selects packets to and from 10.1.10.10
icmp.type == 8 or icmp.type == 0	Selects both type 8 and type 0 ICMP packets
dns and !ip.src == 10.254.1.21	Selects DNS packets that are not from 10.254.1.21
wlan.fc.type == 2	Selects 802.11 data frames

Figure 16-20: Example Display Filters

Decoding GRE Packets From HPE Aruba Networking Devices

In addition to decoding the typical protocols used in network communications, Wireshark can decode the packets sent to it from an AOS-CX mirror session, AOS gateway datapath packet capture, or AOS MC datapath packet capture.

In the example shown in Figure 16-21, an AOS-CX switch at 10.1.140.6 has a remote mirroring session with the device running Wireshark, which is at 10.1.140.90. Wireshark automatically decapsulates the GRE packets and shows the inner packet in the list of packets. In this example, packet 14 is selected. You can see in the middle section that this packet has an outer IP header with a source and destination of the switch and Wireshark device. It also has a GRE header. Inside the GRE header you can see the original Ethernet header, including the VLAN tag. You can then see the inner IP header and protocol header. As you see, the display of packets shows the inner source IP address, inner destination IP address, and inner protocol.

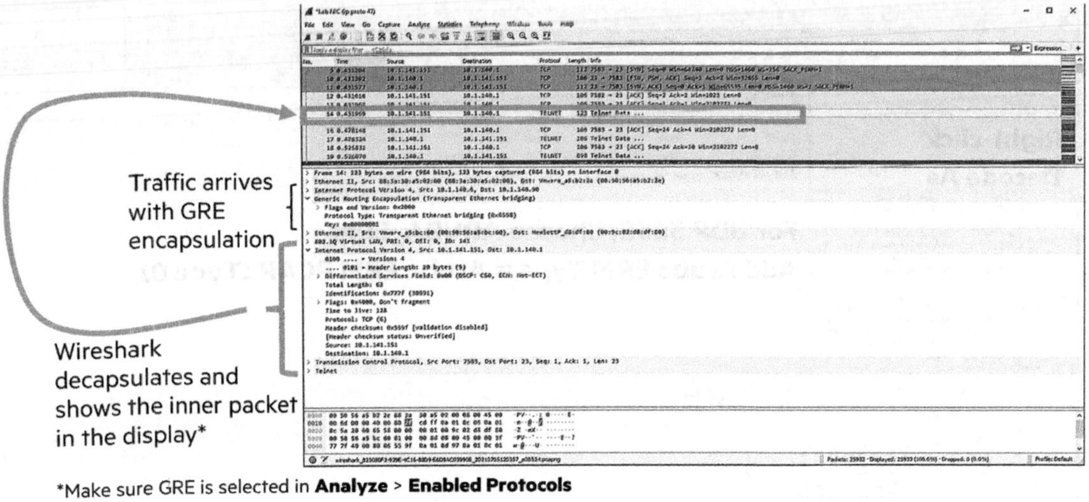

Figure 16-21: Decoding GRE Packets From HPE Aruba Networking Devices

This automatic decapsulation makes it easy for you to focus on the traffic that was mirrored, rather than see a lot of GRE packets.

You can verify that Wireshark is decoding the packets by going to **Analyze > Enabled Protocols**. Then search for **GRE** and verify that the check box is selected.

Decoding Aruba_ERM Packets

HPE Aruba Networking CAPs capture packets with Aruba_ERM. Wireshark does not know how to decode this protocol by default, so you will need to configure it to do so.

As you see in Figure 16-22, Wireshark initially presents the packets from the CAP as UDP packets destined to port 5555. Right-click one of the packets and select **Decode As**. The window appears with a filter already configured to match the packet that you selected. Select **Aruba_ERM** in the Current column.

You must also click the plus (+) icon and add another definition to tell Wireshark that the Aruba_ERM traffic is a packet capture. Select **Aruba ERM Type** and select **Aruba ERM PCAP (Type 0)** in the Current column.

Click **Save**.

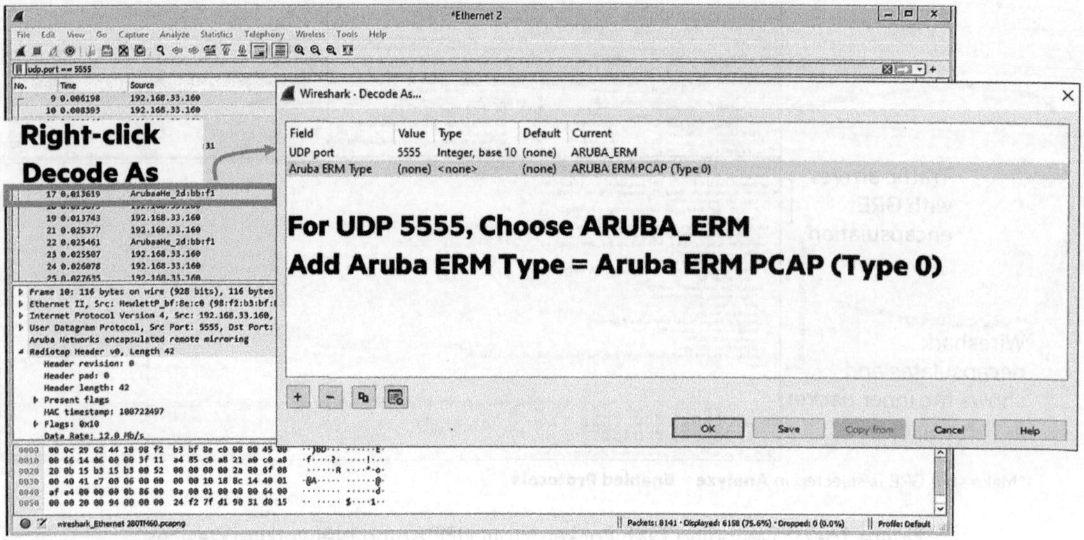

Figure 16-22: Decoding Aruba_ERM Packets

When you perform this task in your Wireshark application once, the configuration is saved for other times that you use the application.

Decoding 802.11 Frames Captured in HPE Aruba Networking Central

When you capture packets in HPE Aruba Networking Central, the AP captures the complete 802.11 frame. At default settings Wireshark cannot properly decode the traffic, so it appears as indistinguishable 802.11 frames.

To view intelligible packets, you must select **Edit** in the Wireshark menu and then preferences (see Figure 16-23). Expand **Protocols** and scroll to **IEEE 802.11**. In Figure 16-23, you see several options for how Wireshark decodes 802.11 frames. You must tell Wireshark to ignore the Protection bit with the IV. Also tell it to validate the FCS checksum if possible. You might need to experiment with toggling the **Assume packets have FCS** settings. Click **OK** to apply the settings. After you tell Wireshark how to decode the IEEE frames correctly, you will then be able to see all of the 802.11 frames properly decoded to show the inner IP headers and deeper protocol layers.

The packet capture includes all of the 802.11 frames, including both management and data frames.

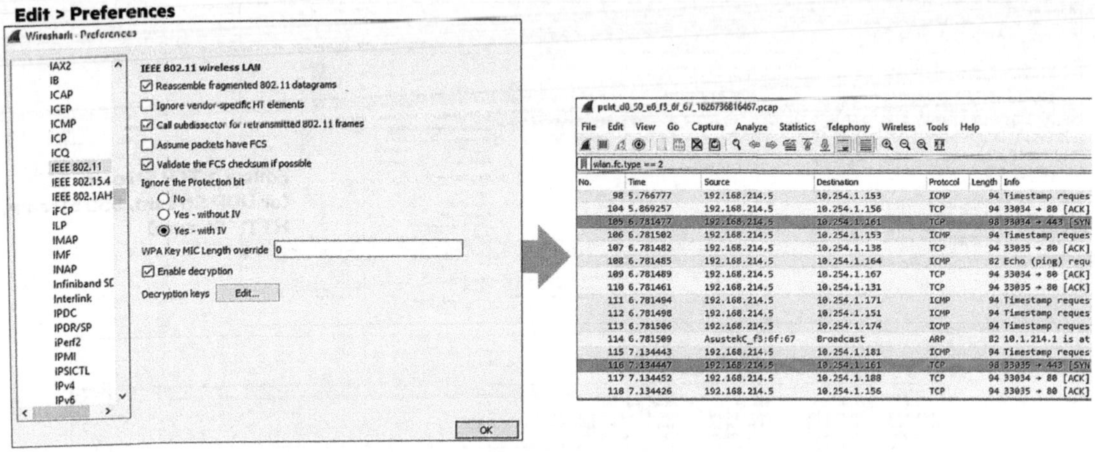

Figure 16-23: Decoding 802.11 Frames Captured in HPE Aruba Networking Central

Tracing a Conversation

Sometimes you will want to trace a particular conversation. You could apply a display filter for the source and destination IP address in question. However, those IP addresses might be participating in multiple conversations with each other. The easiest option can be to trace the conversation.

Select a packet in the conversation of interest, as shown in Figure 16-24. Right-click and select **Follow** and then the type of stream that is appropriate for that conversation. As you see, Wireshark can follow TCP, UDP, SSL, or HTTP streams.

Wireshark then automatically applies a filter for that stream. You will see all the packets associated with it. You can move through the conversation and drill down into details for any of the packets.

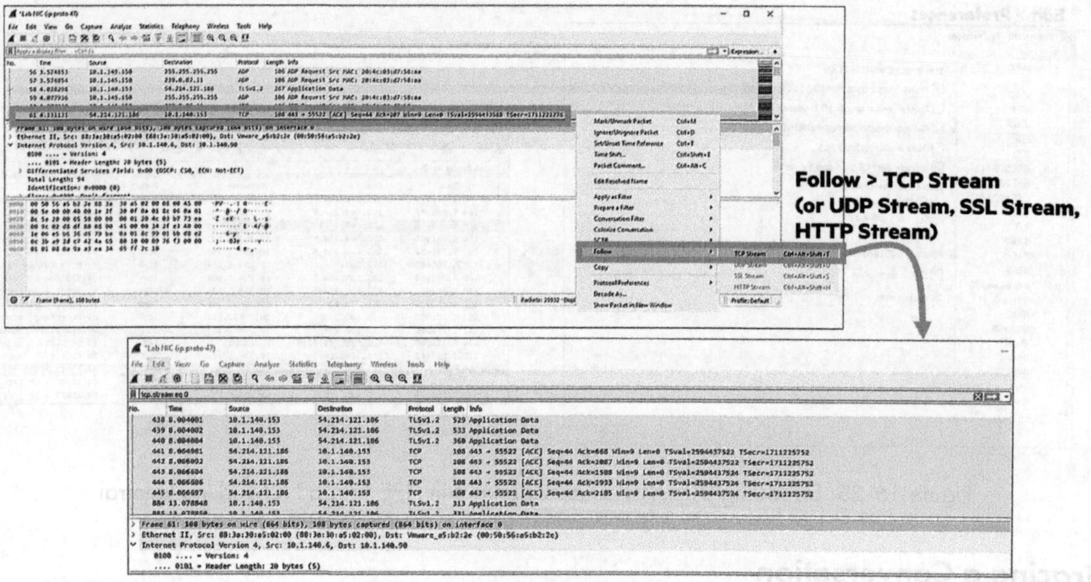

Figure 16-24: Tracing a Conversation

Detect Port Scans with Wireshark

This section explains how you can detect port scans with Wireshark.

Detect Ping Sweeps

You will now look at how you can use Wireshark to detect various types of sweeps and scans. While innocent reasons for such scans exist—for example, CPDI Data Collectors use some types of scans—hackers also use these scans to reconnoiter a network and plan their next steps in the attack.

First you will look at ping sweeps, which are a simple first step in finding active devices. ICMP ping sweeps use ICMP type 8 messages (echo messages), so you should filter for those with this syntax in the display filter: **icmp.type==8**. To see the echo replies, add **or icmp.type==0**. Press **[Enter]** to apply the filter.

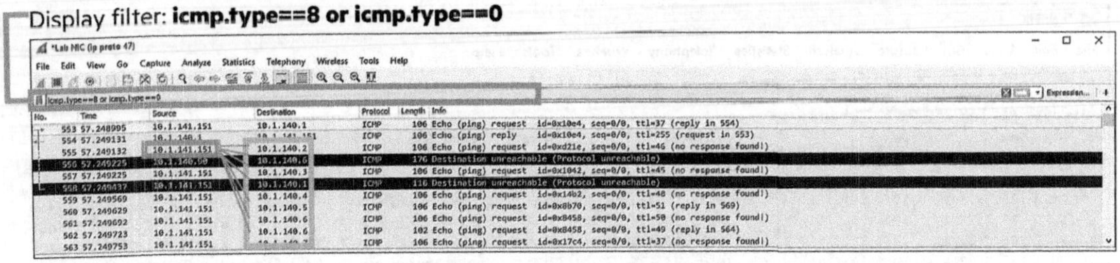

Figure 16-25: Detect Ping Sweeps

Any active network will feature pings, so seeing output after you apply this filter does *not* necessarily mean that you have detected a ping sweep. Instead, look for additional signs of a sweep. First you will need a general sense of your network based on analyzing the network over time. How many pings are normal for this particular environment, and are you seeing more than usual within a specific time period? Next look for many pings from the same source. In the example shown in Figure 16-25, IP address 10.1.141.151 is sending the bulk of the requests within a short period. You can also see how 10.1.141.151 is sending pings to many different incrementing destinations, moving from 10.1.140.1 to 10.1.140.2 to 10.1.140.3 and so on. This methodical approach is a sign of a ping sweep.

Other Types of Sweeps

Because some networks and devices filter ICMP, hackers might continue to sweep the network in other ways. The hackers can then build a more complete picture of the network.

TCP "Ping"

In a TCP "ping," the scanner sends a TCP Syn or TCP Ack message to one arbitrary port on all the IP addresses in a range. Any response indicates that an active device has that IP address. Ports 80 or 443 are often used for the sweep. You should understand that you will often see legitimate packets with these flags. What indicates a scan is the fact that a single source rapidly attempts to connect to many destinations, as shown in Figure 16-26.

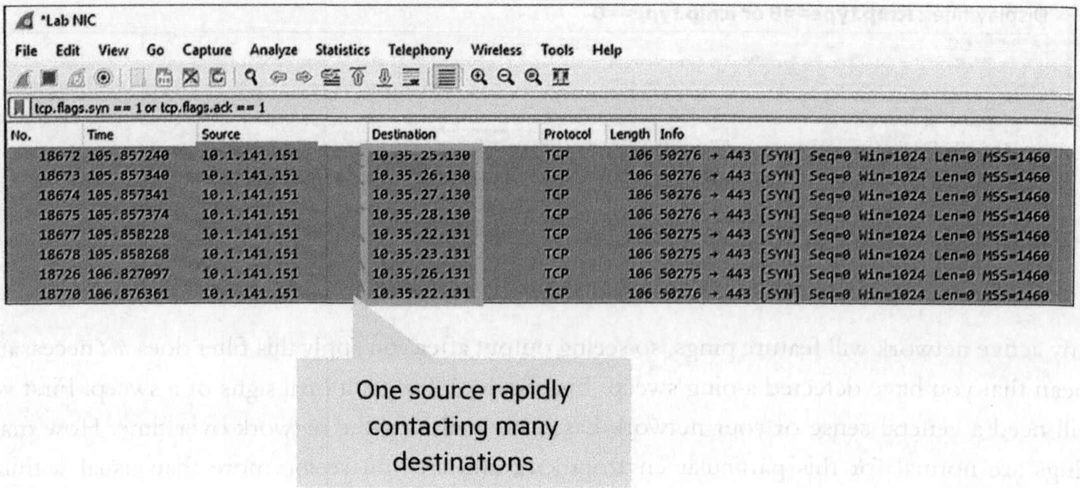

Figure 16-26: TCP "Ping"

UDP "Ping"

Hackers can similarly use UDP sweeps to try to get past firewalls that filter ICMP and TCP. The hacker sends UDP messages to a port that is likely to be closed. If a device exists at that IP address and the port is closed on it, the device responds with an ICMP unreachable. Again, you can detect the sweep based on seeing a chunk of UDP packets from the same source to many different IP addresses on the same port. As shown in Figure 16-27, Nmap uses UDP port 40125 by default for UDP sweeps. This port is rarely used otherwise, so seeing packets to this port is a good sign that someone is performing reconnaissance on the network.

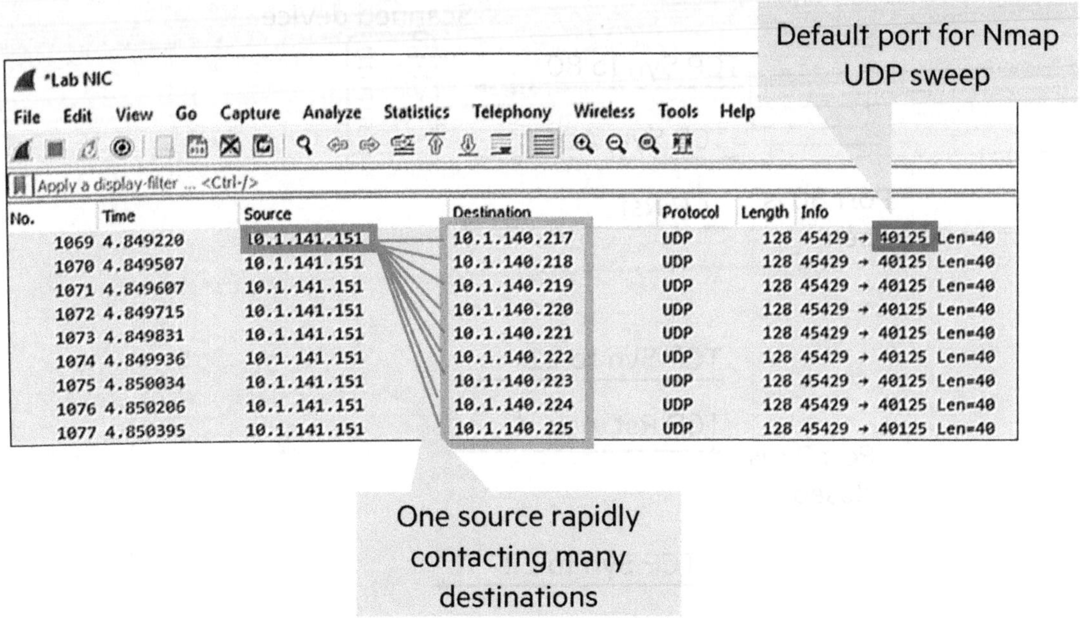

Figure 16-27: UDP "Ping"

TCP Syn Scan

Hackers often run sweeps as preludes to a more intense port scan, in which they attempt to discover open ports on devices detected as active during the sweep.

To discover the TCP ports that are open on a scanned device, hackers often run TCP Syn scans. Sometimes called half-open TCP scans because the scanner does not fully open the TCP session, TCP Syn scans are relatively fast and unobtrusive. The application, such as like Nmap, sends a TCP Syn to each port that it wants to scan. As shown at the top of Figure 16-28, if the port is open on the scanned device, the device returns a TCP packet with the Syn and Ack flags set. The application then knows that the port is open. Because it does not really want to open a session, it sends a TCP Rst to the scanned device.

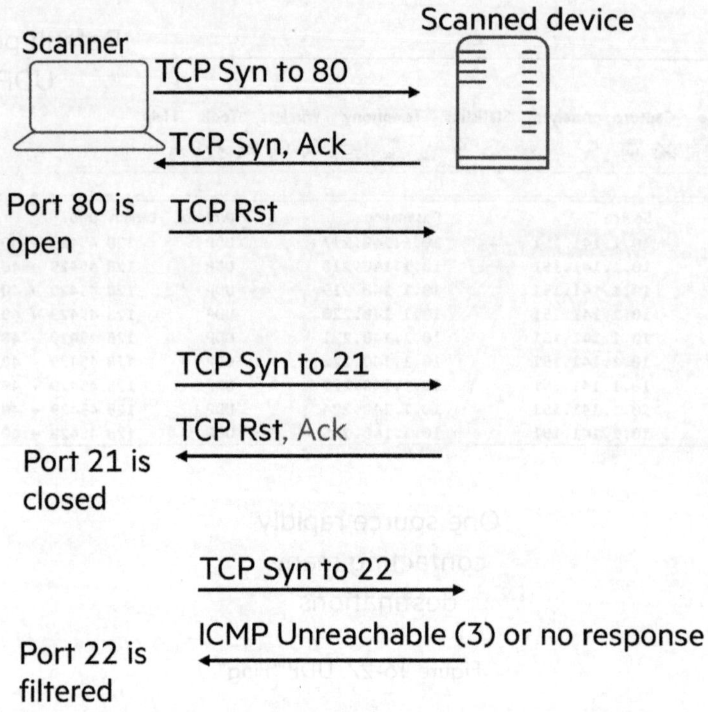

Figure 16-28: TCP Syn Scan

If the port is closed, the scanned device responds with a TCP Rst or possibly a TCP Rst, Ack message, as shown at the middle of Figure 16-28. A firewall might lie between the scanner and the scanned device. If the firewall denies access to a port, the scanner will either receive no response or an ICMP Unreachable (type 3) message, as shown at the bottom of Figure 16-28. The scanner marks those ports as filtered.

Detect TCP Syn Scans

To detect a TCP Syn scan, you can look for an unusually large number of packets with TCP RST flags set. In Figure 16-29, you see a display filter for such packets: **tcp.flags.reset == 1**. (The **1** indicates that the flag must be set.) You might also add or **icmp.type==3** to see the ICMP Unreachable messages sent if a firewall is filtering some ports. These filters detect all the final messages in the TCP Syn scan.

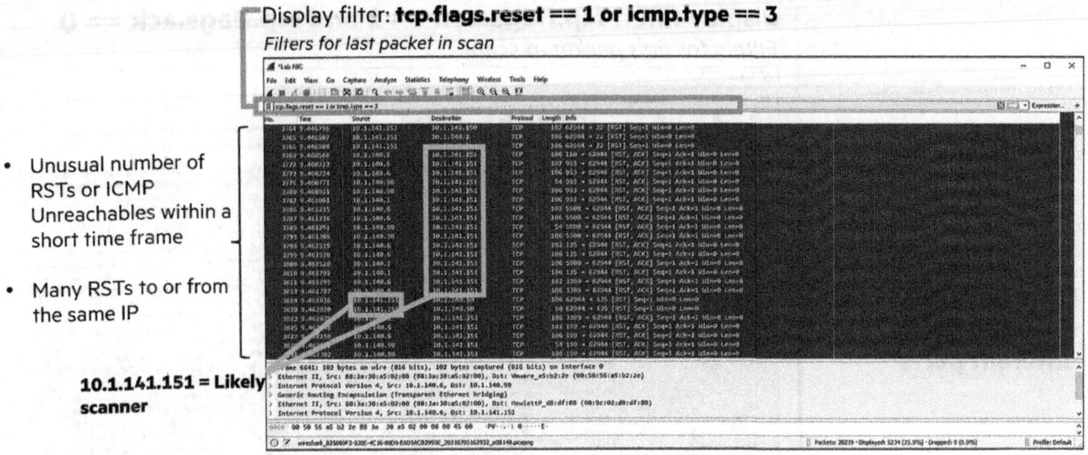

Figure 16-29: Detect TCP Syn Scans

While clients and servers do sometimes reset sessions, the display shown in Figure 16-29 has several signs of a TCP scan. The vast majority of the RSTs within a short time frame are destined to a single IP address, 10.1.141.151. And the RSTs that are destined to another IP address are actually from 10.1.141.151. You can ID 10.1.141.151 as likely to be running a port scan.

Another Way to Search for TCP Syn Scans

Alternatively, you might filter for the first packet in the scans. These packets have the Syn flag set, but not the Ack flag. Use this filter: **tcp.flags.syn == 1 and tcp.flags.ack == 0**. Perfectly legitimate traffic features flags of this type. This packet simply indicates the start of a TCP session. However, in Figure 16-30, you see the signs of a scan. The same IP address is rapidly trying to connect to many different ports. Legitimate clients rarely need to open so many connections in rapid succession. The source of these connection attempts is likely running a port scan.

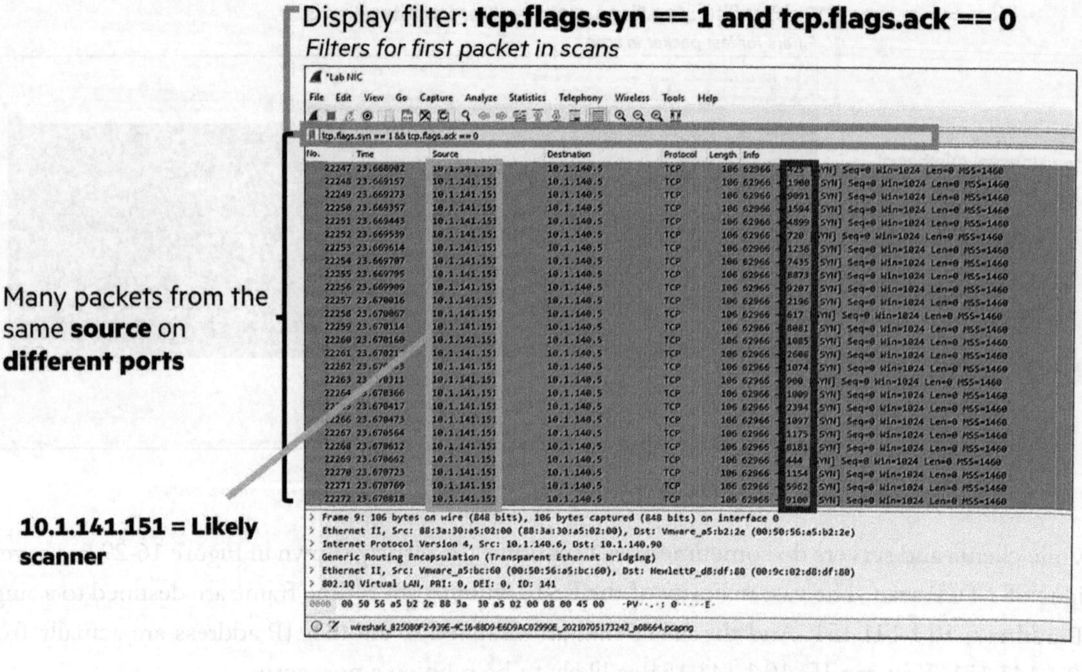

Figure 16-30: Another Way to Search for TCP Syn Scans

Further Investigating the TCP Syn Scan

After you determine that a particular IP address is likely running port scan, you can investigate further by filtering for that IP address—10.1.141.151 in this example. Use **ip.addr** to see packets both to and from this IP address. In the example shown in Figure 16-31, you continue to see signs of a scan. The suspected scanner is sending TCP Syn messages on many different ports to different IP addresses.

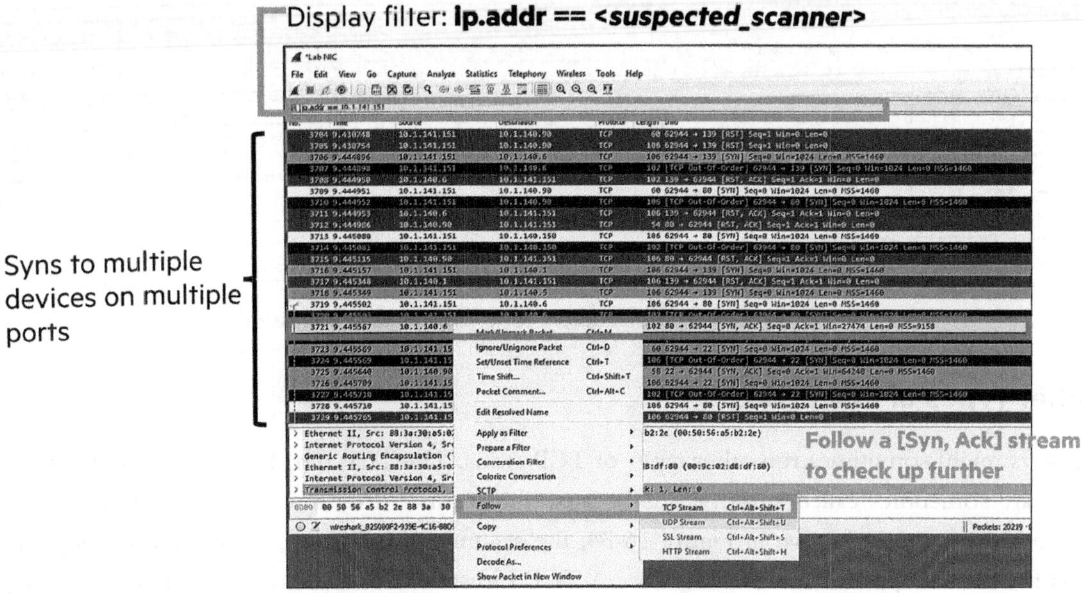

Figure 16-31: Further Investigating the TCP Syn Scan

To check up further, you can follow one of the streams for which the scanned device responded with a SYN, ACK, indicating that the port is open.

TCP Syn Scan versus Legitimate TCP Session

In Figure 16-32, you see the result of following one such stream. As you see, 10.1.141.151 immediately reset the session as soon as it received the Syn, Ack. That behavior indicates a scan.

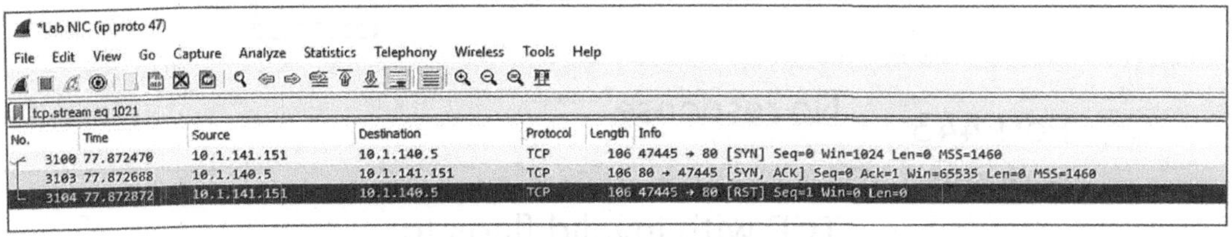

Figure 16-32: Packet Capture for a TCP Syn Scan

A normal TCP session would look more like the one shown in Figure 16-33, in which the client and server exchange many packets with application data.

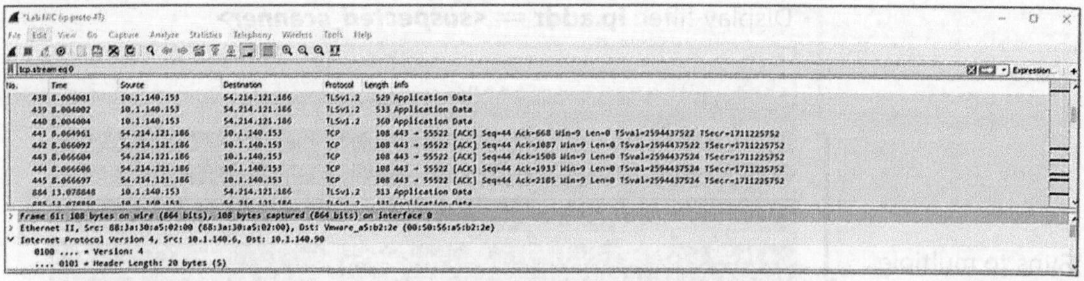

Figure 16-33: Packet Capture for a Legitimate TCP Session

Other Types of TCP Scans

Hackers might sometimes run other types of TCP scans. The TCP Fin, TCP Null, and TCP Xmas scans are sometimes called stealth scans because the scanner does not need to form a session to conduct the scan. As shown in Figure 16-34, the scanner simply sends a packet with illegitimate flags for the first packet in a TCP session. A TCP Fin scan uses a packet with the Fin flag set. A TCP Null scan uses a packet with no flags. And a TCP Xmas scan sets the Fin, Psh, and Urg flags. All of these scans are looking for similar behavior. If the scanned device sends no response, the port is open. If the device sends a TCP Rst, the port is closed.

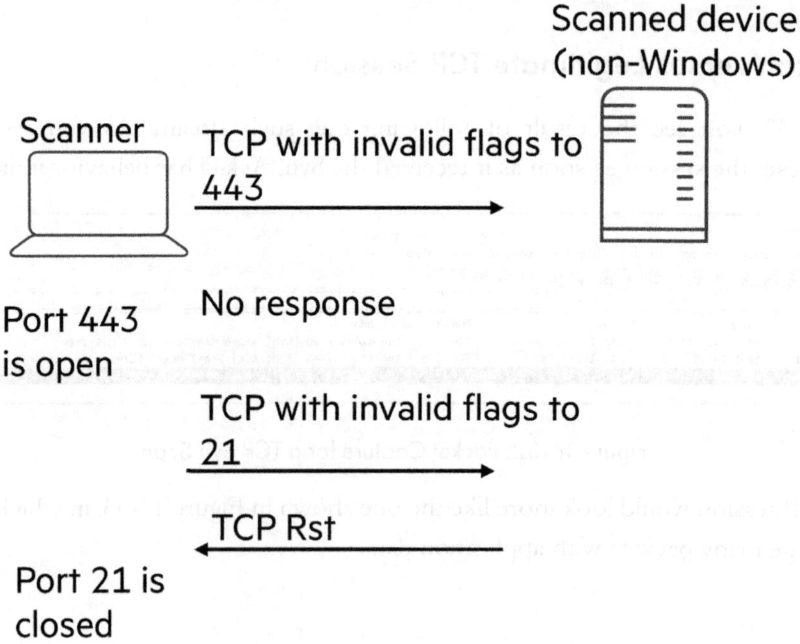

Figure 16-34: Other Types of TCP Scans

Note that these types of scans do not work against Windows OS because Windows devices send a TCP Rst in response to the illegitimate packet whether or not the port is closed. For this reason, hackers might not use these scans as often as the TCP Syn scans. However, you should still be aware of them.

You can use the filters shown in Figure 16-35 to look for these scans in Wireshark. For the TCP Xmas scan, rather than filtering for each flag separately, it can be faster to filter for the Hex value associated with those flags: 0x029. Seeing packets with the Fin flag set does not necessarily indicate a scan, as devices legitimately send such packets at the end of a session. You should look for other signs of a scan such as many such messages from the same source to different ports and destinations.

These packets do not legitimately occur

Scan to Detect	Wireshark Filter
TCP Fin	tcp.flags.fin == 1
TCP Null	tcp.flags == 0x000
TCP Xmas	tcp.flags.fin == 1 and tcp.flags.psh == 1 and tcp.flags.urg == 1 tcp.flags == 0x029

Figure 16-35: Wireshark Filters for Detecting Scans

The combination of flags for TCP Null and TCP Xmas scans should not occur in legitimate traffic. If you filter for those packets and see results, you know that a hacker is running a scan. In Figure 16-36, you can see a Wireshark capture of a TCP Xmas scan.

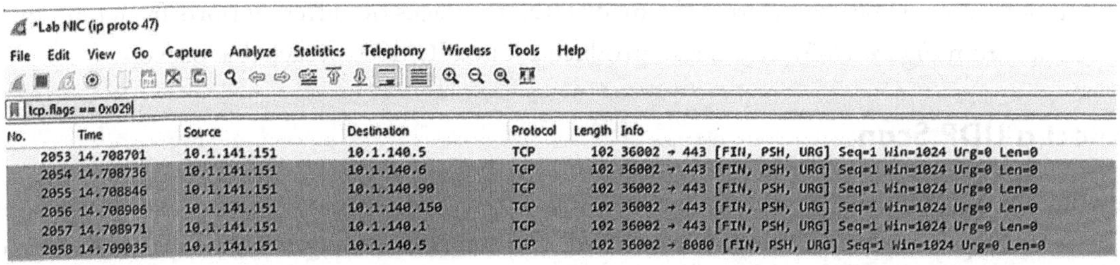

Figure 16-36: Wireshark Capture of a TCP Xmas Scan

UDP Port Scans

Hackers might also conduct UDP port scans to gather information about which UDP ports are open, and potentially exploitable, on scanned devices. The scanner sends a series of UDP messages on different ports. If it receives no response to the message, it considers the port open. (Rarely the scanned device might send a UDP response, in which case the port scanner would also consider the port open.) If the scanner receives an ICMP Unreachable (type 3) response, it considers the port closed if the code is 3. If the code is 1, 2, 9, 10, or 13, it considers the port open.

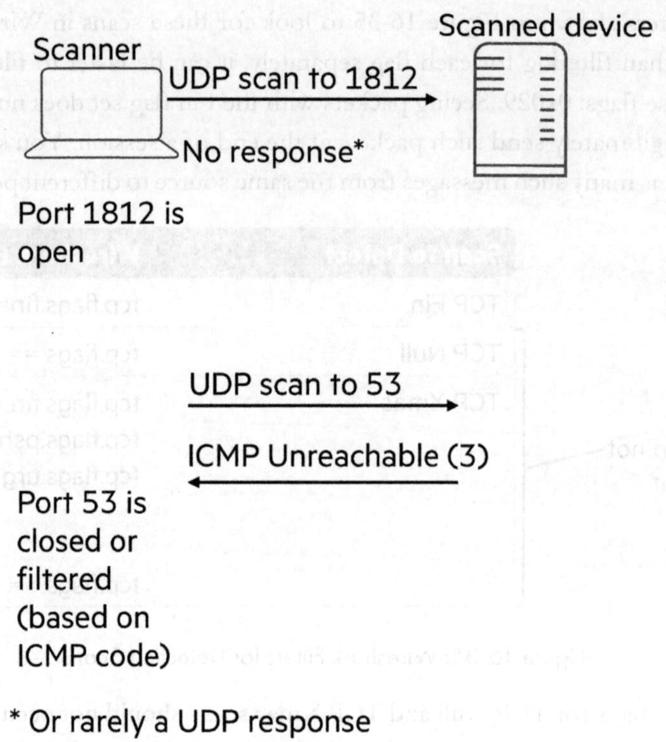

Figure 16-37: UDP Port Scans

A UDP port scan will be characterized by many UDP messages on different ports from one source. As you see from Figure 16-37, it will also produce a lot of ICMP Unreachable packets.

Detect a UDP Scan

Filtering for **icmp.type==3** can help tip you off to UDP scans, just as it can for TCP scans. Again, you are looking for an unusual number of ICMP Unreachable messages within a short time frame and destined to the same IP address.

You can select one of these packets and remove the display filter to see the packets that occurred around the same time. When the display shows many UDP requests from that IP address, you can be reasonably confident that the device at that address is conducting a UDP scan. In the example shown in Figure 16-38, 10.1.140.151 is sending UDP requests to a few IP addresses within a range. The requests are for several different commonly scanned ports.

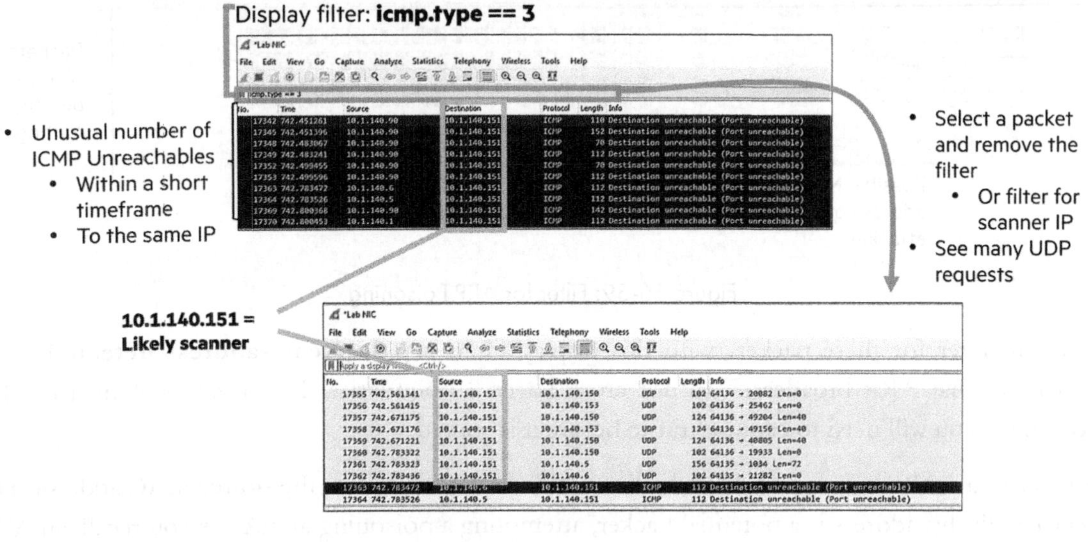

Figure 16-38: Detect a UDP Scan

As with other scan, you can continue to investigate by filtering by the likely scanner's IP address. You can then see if perhaps the scanner is running other types of scans or even other attacks.

Investigating ARP Poisoning with Packet Captures

This section guides you through investigating ARP poisoning with packet captures.

Filter for ARP Poisoning

Wireshark can give you a clue that an ARP poisoning has occurred. If a packet capture includes ARP messages that claim the same IP address at different MAC addresses, Wireshark marks the messages with the second MAC address. As you see in Figure 16-38, the Info field states "duplicate use of <IP address> detected!"

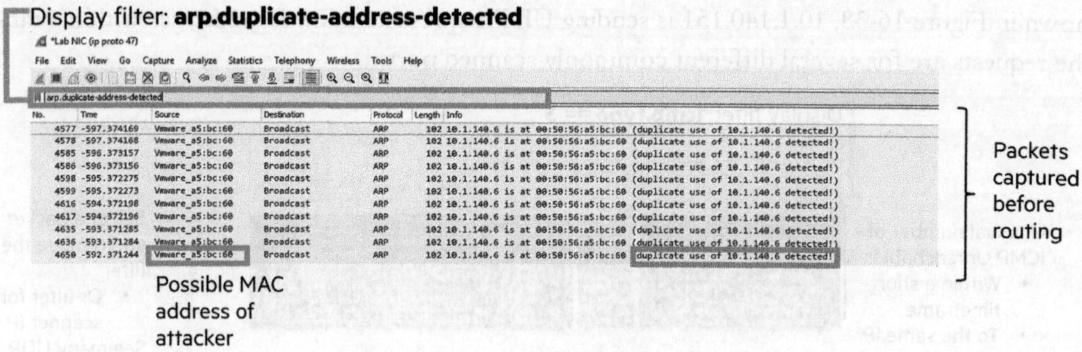

Figure 16-39: Filter for ARP Poisoning

You can filter for these packets with this display filter: **arp.duplicate-address-detected**. Also remember that ARP broadcasts do not cross Layer 2 boundaries. If you are looking for ARP poisoning, you will need to capture traffic before it is routed.

If you see any ARP responses with a duplicate address detected, note the source MAC address. The device with this address is a potential hacker, attempting a poisoning attack. As you recall, an ARP poisoning attack is often a prelude to a man-in-the-middle (MITM) attack.

Also remember that you can prevent such attacks before they occur by setting up ARP inspection on your AOS-CX switches.

Find the Original MAC Address

You can further investigate by looking for the other ARP messages that claim the IP address in question.

As shown in Figure 16-40, use a filter such as this: **arp.dst.proto_ipv4 == <IP_address>**. Note the MAC address associated with those messages. Wireshark does not know the legitimate device that owns the IP address. Depending on the timing of the attack and the packet capture, the attacker might have either address. Sometimes you will be able to tell relatively quickly which MAC address is legitimate. If one MAC address belongs to a network infrastructure device, and the other to a client, the client is the likely attacker. Other times you should investigate both MAC addresses.

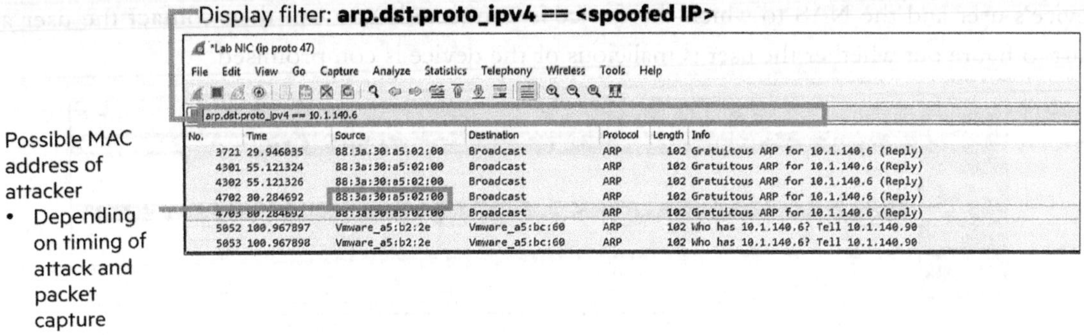

Figure 16-40: Find the Original MAC Address

Collecting Information about Suspicious Clients

You can collect more information about the client MAC address or addresses in question using a variety of HPE Aruba Networking tools.

In ClearPass, you can find this information in several places. You can search for the potential hackers' MAC addresses in Insight Inventory. You can also find endpoint information in the Endpoints Repository. Or you can use Access Tracker to find the user authenticated on those clients. As shown in Figure 16-41, filter based on **Host MAC-Address**. Open up the record to find information such as the NAS to which the client is connected.

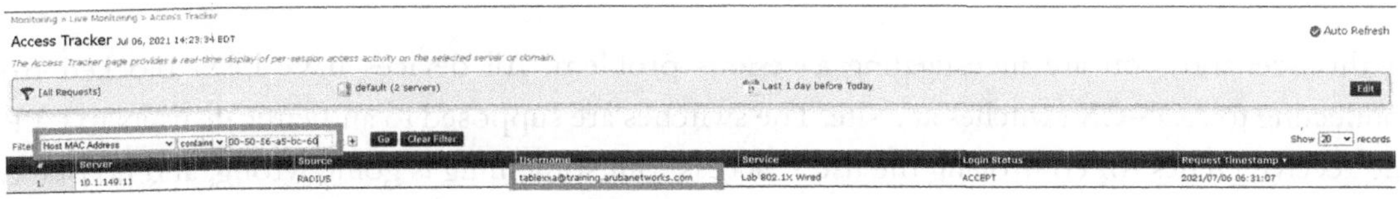

Figure 16-41: Collecting Information about Suspicious Clients

You can also search for the client in HPE Aruba Networking Central. From there you can find the client's username, connected NAS, and initiate a packet capture. If you are using CPDI, that application will have information about the endpoints' profile, network activity, and history.

You can also look in the network infrastructure itself. You can search gateway or MC client lists or even switch MAC forwarding tables.

Example of Finding Information in ClearPass Insight

In Figure 16-42, you see an example of how you can find more information about the potential attacker's client in ClearPass Insight. In the left menu, select **Endpoints**. Find the MAC address or addresses that you recorded earlier. Click the address to view information such as the name of the device's user and the NAS to which the device is connected. You can then contact the user and start to figure out whether the user is malicious or the device is compromised.

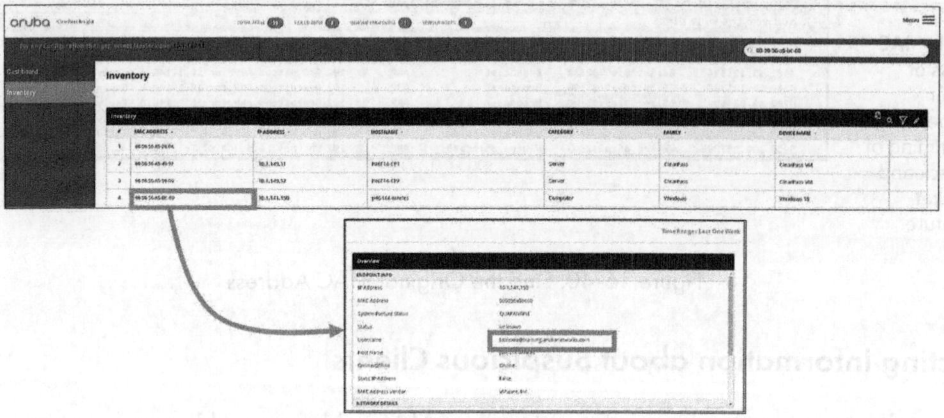

Figure 16-42: Example of Finding Information in ClearPass Insight

Scenario: Use Wireshark to Troubleshoot

This section presents a scenario of using Wireshark to troubleshoot.

The Problem

In this scenario, you are investigating a serious problem. All devices have been blocked from connecting to AOS-CX switches at a site. The switches are supposed to authenticate users to CPPM and receive DURs for controlling the users' access. But something is going wrong, and no one can connect to the network, as indicated in Figure 16-43.

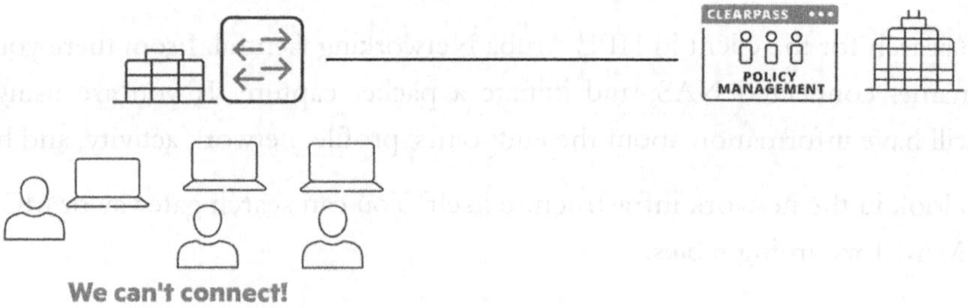

Figure 16-43: The Problem

Initial Investigation

You start investigating the issue by viewing the clients' authentication port-access status. As shown in Figure 16-44, you see that the clients have actually authenticated successfully. However, their roles are not applying. When you look at the roles, you see that the switch failed to download roles because it does not trust CPPM's HTTPS server certificate.

```
P46-TXX-6300# show aaa authentication port-access int all client-status

Port Access Client Status Details

Client 00:50:56:a5:bc:60, tablexxa@training.arubanetworks.com
==============================
<output omitted>
  Authentication Details
  ----------------------
    Status            : dot1x Authenticated
<output omitted>
  Authorization Details
  ----------------------
    Role    :
    Status  : Not Ready
```

```
P46-TXX-6300(config-if)# show port-access role
Role Information:

Name   : loanofficers_CX_DUR-3023-13
Type   : clearpass
Status: Failed, Server Certificate Invalid
---------------------------------------------
```

Figure 16-44: AOS-CX Switch Authentication Port-Access Status

Usual Checks Do Not Show the Problem

You next check the certificate on CPPM to ensure that the switch has the proper root CA certificate installed in a TA profile.

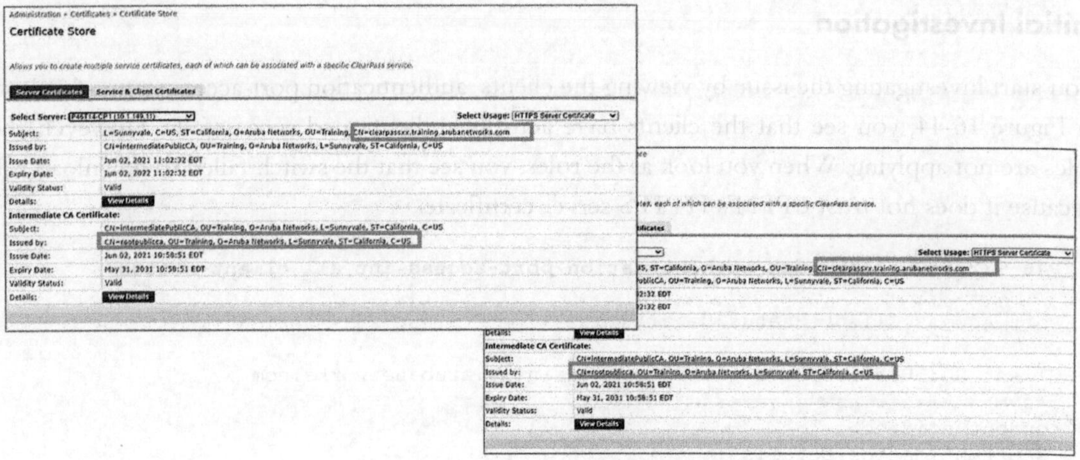

Figure 16-45: CPPM Certificate Store

As shown in Figure 16-45, you view the HTTPS server certificate on *both* servers in the cluster. You see that the root authority is rootpublicCA. You also check the subject name in the certificates, as this must match the RADIUS server name specified on the switch.

You check the TA profile on the switch, as shown in Figure 16-46, and see that the correct root CA certificate is installed and valid. The RADIUS server settings, not shown in Figure 16-46, are also correct.

```
P46-TXX-6300# show crypto pki ta-profile publicCA

  TA Profile Name     : publicCA
  Revocation Check    : disabled
  <-output omitted->
  TA Certificate      : Installed and valid
   Certificate:
    Data:
        Version: 3 (0x2)
        Serial Number:
            68:87:92:bf:5b:76:96:f2:a3:a2:8c:2a:1f:bf:ff:aa:20:28:66:03
    Signature Algorithm: sha256WithRSAEncryption
        Issuer: C=US, ST=California, L=Sunnyvale, O=Aruba Networks,
OU=Training, CN=rootpublicca
        Validity
            Not Before: Jun  2 03:04:13 2021 GMT
            Not After : May 31 03:04:13 2031 GMT
```

Figure 16-46: AOS-CX Switch TA Profile

Capture Packets to Find the Issue

Because the issue is resisting your normal troubleshooting procedures, you decide to run a packet capture to investigate further. As indicated in Figure 16-47, you set up a mirror on the switch uplink for sent and received traffic. You set the mirror destination to your laptop, which is running Wireshark. You then have a user try to connect to the network again.

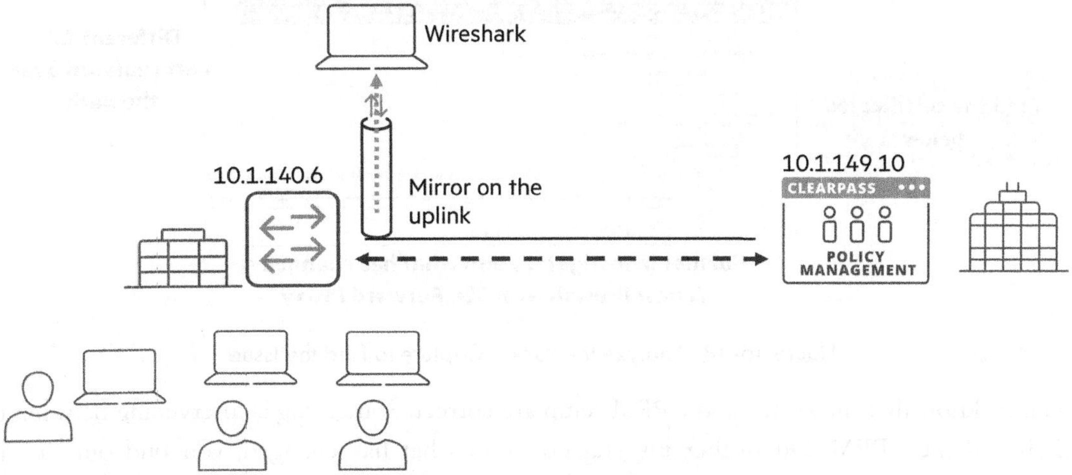

Figure 16-47: Capture Packets to Find the Issue

Analyze the Packet Capture to Find the Issue

You know that an error is occurring in the certificate trust, so you are focusing on that exchange. In the packet capture, you filter for the switch IP address, 10.1.140.6, and CPPM's IP address, 10.1.149.10. You then look for the TLS exchange that occurs when the switch attempts to download the roles. You look for the packet with a certificate after the Server Hello. This packet has the server certificate.

You expand headers in the middle section until you have expanded the Certificate and can view information about it. In Figure 16-48, you see that the switch is *not* receiving the same certificate that you saw listed for CPPM's HTTPS certificate. This certificate is signed by a different CA: intCA.companyx.example.com.

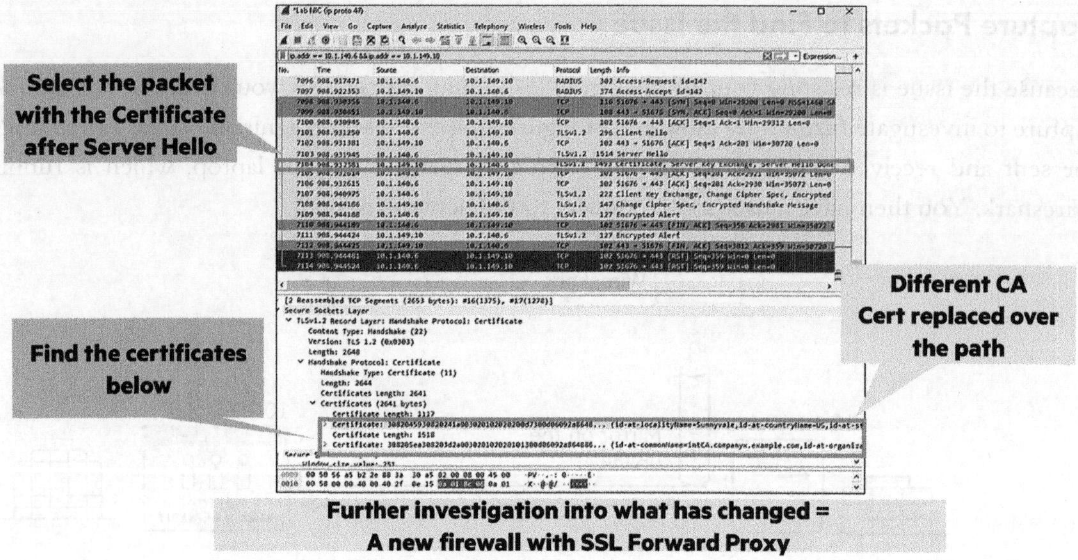

Figure 16-48: Analyze the Packet Capture to Find the Issue

You now know that the switch and CPPM setup are correct. Something is intervening between the switch and the CPPM. On further investigation into what has changed, you find out that the company has implemented a new firewall. This firewall uses SSL Forward Proxy which involves replacing certificates in communications with new certificates signed by the firewall on the fly. This explains how the CA certificate changed enroute. You will need to set up the AOS-CX switch to trust the firewall's root CA certificate to resolve the issue. (Or the company will need to change how it is implementing this security feature to keep the switch and CPPM communications intact.)

Summary

Congratulations! You can now follow your security team's instructions to capture relevant packets for a security investigation. You have also gotten started with analyzing these packet captures yourself using Wireshark. You can navigate the Wireshark interface and apply simple display filters to zero in on relevant data for your investigation. You can now recognize port scans and other potential security issues.

Learning checks

1. What is one difference between Wireshark capture filters and display filters?
   a. Only capture filters can be removed and changed during an active capture.
   b. Only capture filters can filter packets based on protocol.
   c. Only display filters can be removed and changed during an active capture.
   d. Only display filters can filter packets based on protocol.
2. What is a sign of a TCP Syn scan? (Multiple answers can be correct.)
   a. A device that immediately sends TCP RST packets in response to Syn + Ack messages
   b. Any TCP packet that has the Syn flag set without the Ack flag
   c. Any TCP packet that has the Syn flag set with the Ack flag
   d. Many TCP Syn packets from the same source to different destinations and ports
3. Below are listed several different instructions that the security team might give you. For each one consider what steps you would take next.
   a. "We have detected suspicious activity on a wireless client at IP address 10.1.55.36. Please capture its traffic for 15 minutes and give us the results in a PCAP file." (This environment uses an AOS-8 architecture with MCs and CAPs.)
   b. "We have detected suspicious activity on a wireless client at IP address 10.5.32.17. Please capture its traffic for 30 minutes and give us the results in a PCAP file." (This environment uses an AOS-10 architecture with APs and HPE Aruba Networking Central.)
   c. "We think that an intrusion might have occurred in Building 1. We want to capture all traffic coming between the access switch in that building and the data center or Internet. Please send the traffic to 10.2.6.8, where we are running an analyzer." (This switch belongs to the CX 6300 series.)

Note

For answers to all learning checks, refer to the Appendix.

---

17 Secure Remote and Branch Access

EXAM OBJECTIVES

✓ Explain how to use HPE Aruba Networking SSE to secure access from anywhere

✓ Describe how HPE Aruba Networking EdgeConnect solutions secure branch connectivity

Assumed Knowledge

- Basics of HPE Aruba Networking solution architectures
- Encryption and PKI fundamentals
- Unified SASE fundamentals

Overview

This chapter will begin by reviewing how HPE Aruba Networking SSE helps customers surmount their challenges in securing remote access. This section also introduces you to the HPE Aruba Networking SSE management console and explains how to set up Identity Providers (IdPs) for authenticating users for SSE services.

You will then learn how to implement these services, including an overview and configuration guidelines for HPE Aruba Networking ZTNA, HPE Aruba Networking SWG, and HPE Aruba Networking CASB.

Next, you will learn how to secure branch access with HPE Aruba Networking. You will review the HPE Aruba Networking EdgeConnect portfolio and learn how the solutions in the portfolio automate branch security.

Introduction to Securing Remote Access

This section reviews the benefits of HPE Aruba Networking SSE for securing remote access. It also introduces you to managing HPE Aruba Networking SSE and setting up the authentication services used by ZTNA, SWG, and CASB.

Traditional Option for Securing Remote Access

Traditionally companies have granted remote users access to headquarters by implementing a client-to-site virtual private network (VPN). As shown in Figure 17-1, a client-to-site VPN establishes a secure tunnel between the client and a gateway (sometimes called VPN concentrator [VPN-C]) at the headquarters.

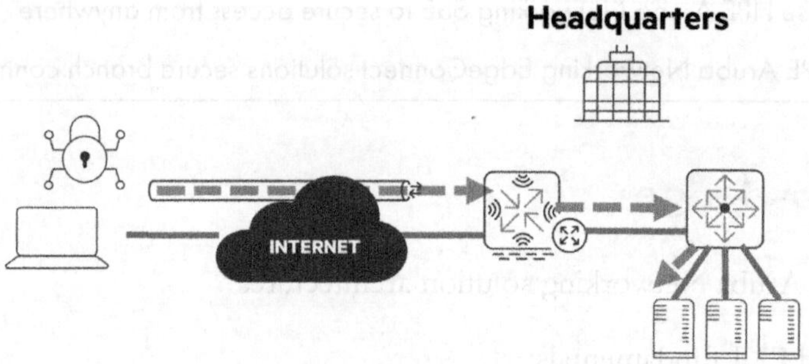

Figure 17-1: Client-to-Site VPN

This VPN often uses IP Security (IPsec) to establish the tunnel. IPsec encapsulates traffic, encrypts the payload, and ensures the traffic's integrity. Client-to-site VPNs typically use Internet Key Exchange (IKE) to authenticate clients, negotiate the encryption and integrity keys, and set up the tunnel.

Problems with the Traditional Client-to-Site VPN

While client-to-site VPNs have served their purpose for many years, they have some drawbacks, as summarized in Figure 17-2.

Many legacy client-to-site VPNs force users to authenticate but then grant excessive privileges to those authenticated users. If the company lacks the ability to apply access controls to internal clients, the company is just as likely to control the remote users. These companies often take a legacy view of security in which they use source IP addresses to determine the device's location, and they grant implicit trust to devices with internal IP address. As you have learned throughout this study guide, this approach no longer grants sufficient security even when the devices in question are truly internal.

The security gaps widen even more when those devices are actually remote devices, which have been granted internal IP addresses by the client-to-site VPN.

With traditional client-to-site VPN solutions, companies often lack visibility into connected clients and how they are behaving on the network.

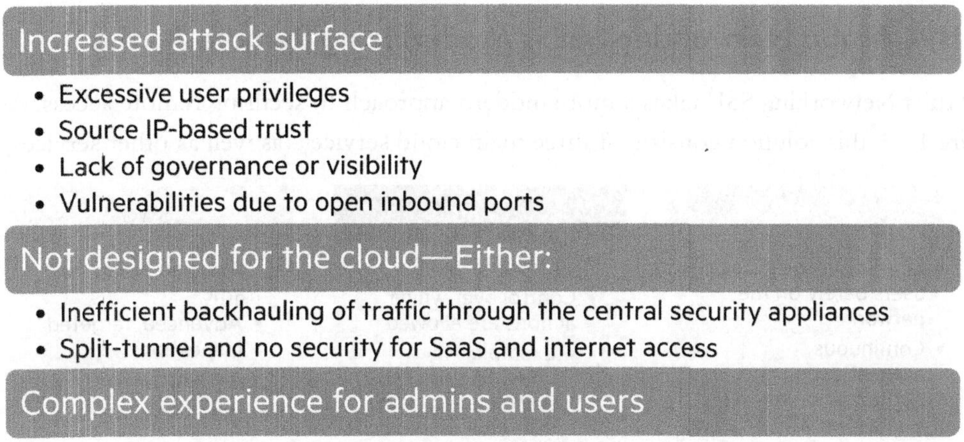

Figure 17-2: Problems with the Traditional Client-to-Site VPN

As another drawback, client-to-site VPNs are not designed for cloud apps. They force organizations to make a choice: backhauling all remote user traffic to the data center or splitting some traffic out and not sending it over the Internet.

Backhauling all remote user traffic to the headquarters allows the organization pass that traffic through central security appliances. However, this option introduces inefficiencies in the traffic flow for cloud apps. Rather than flow directly to the internet, the traffic must go to the headquarters to the internet and then back again. The increased latency can affect user experience. In addition, VPN gateways and central security appliances must handle more traffic, potentially compromising those appliances' performance or forcing the company to deploy more expensive equipment.

To avoid these issues companies might choose to have remote clients use "split-tunnel." That is the clients split out their cloud and internet traffic, sending it directly to the internet rather than over the VPN tunnel. However, with this approach, the organization loses the opportunity to control and secure that traffic.

Finally, client-to-site VPNs are complex to deploy, manage, and troubleshoot, adding to already overburdened network admins' work. Users also often find these solutions difficult to use. The user might need to install a client and set it up. If something goes wrong, the user will likely lack the expertise to correct the issue and will require admins to help. Sometimes users—such as contractors—have multiple VPNs clients running on their machines, and those clients interfere with each other.

In short, client-to-site VPNs no longer meet the needs for modern organizations with many remote workers and cloud apps.

How HPE Aruba Networking SSE is Modernizing Remote Access

HPE Aruba Networking SSE takes a more modern approach to securing remote access. As shown in Figure 17-3, this solution consists of three main cloud services (as well as other services).

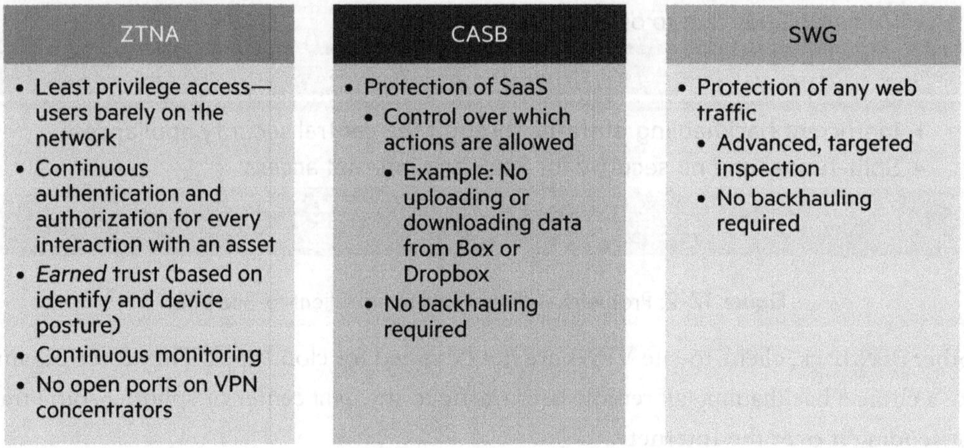

Figure 17-3: How HPE Aruba Networking SSE is Modernizing Remote Access

HPE Aruba Networking Zero Trust Network Access (ZTNA)

ZTNA closes security gaps by taking a new approach to granting users' access to private resources and applications. Rather than placing remote clients on the internal network—and then granting those clients intrinsic trust—ZTNA provides least privilege access. Users are barely on the network. Instead, they receive brokered access to specific authorized applications.

To determine whether to grant that access, HPE Aruba Networking ZTNA enforces continuous authentication and authorization for every interaction with an asset. For example, if a user wants to access an internal SharePoint folder, the user must authenticate to receive that access. Then when the user also wants to run the company's private accounting app, the user must authenticate and be authorized for that app as well. (Note, that this authentication and authorization can sometimes occur in the background from the user perspective.) In short, users continually earn trust based on factors such as their identity and the posture (or health) of their device.

HPE Aruba Networking ZTNA also provides continuous monitoring, helping admins achieve better visibility into who is accessing which applications when.

This solution replaces client-to-site VPNs, but it is not a VPN. HPE Aruba Networking ZTNA does not require VPN concentrators at the organization's main office or data center. Therefore, the organization avoids the potential vulnerability of open ports.

HPE Aruba Networking Cloud Access Security Broker (CASB)

HPE Aruba Networking CASB protects users' access to software as a service (SaaS) applications. This solution can inspect users' traffic and enforce controls to protect the organization. For example, the company might choose to prevent users from uploading or downloading data from Box, Dropbox, and other file sharing SaaS applications. As a cloud service, HPE Aruba Networking CASB enforces this security efficiently without forcing organizations to backhaul traffic to their own site for inspection.

HPE Aruba Networking Secure Web Gateway (SWG)

HPE Aruba Networking SWG protects any web traffic. Organizations can establish rules for permitting or blocking users' access to various web site categories based on their internal policies. They can also implement advanced inspection on traffic destined to particular categories or particular sites, detecting malware. Again, HPE Aruba Networking SWG provides all these benefits without backhauling traffic to a main office or data center.

HPE Aruba Networking SSE Architecture

Over the next several sections of this chapter, you will learn how to set up HPE Aruba Networking SSE to achieve these benefits. Before diving into those sections, take a moment to view the HPE Aruba Networking SSE architecture at a high level (see Figure 17-4).

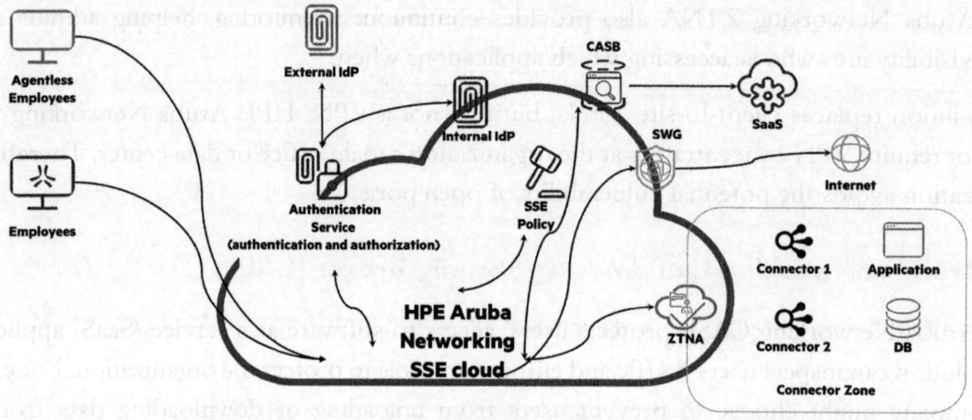

Figure 17-4: HPE Aruba Networking SSE Architecture

HPE Aruba Networking SSE runs in the cloud. It provides three main services.

HPE Aruba Networking ZTNA works with connectors deployed with connector zones at organizations' own sites. It uses these connectors to establish connections between users and resources, such as applications and databases.

HPE Aruba Networking SWG handles users' access to web sites on the Internet.

HPE Aruba Networking CASB brokers connections between users and SaaS applications.

Admins set up the rules for establishing and regulating these connections with the SSE policy.

HPE Aruba Networking SSE also includes an authentication service, which authenticates users, who can then by authorized for access to various applications. This service can use an internal Identity Provider (IdP), included in the solution, or it can integrate with one or more external IdPs.

HPE Aruba SSE Management Console

Each company receives a "workspace" dedicated to them within HPE Aruba Networking SSE cloud.

You access your workspace and set up HPE Aruba Networking ZTNA, CASB, SWG, and other features at the HPE Aruba Networking management console (see Figure 17-5). This console provides role-based access control (RBAC), allowing admins to assign different administrative rights to different users.

Figure 17-5: HPE Aruba SSE Management Console

HPE Aruba Networking SSE and Identity Providers (IdPs)

HPE Aruba Networking SSE uses Identity Providers (IdPs) to authenticate and authorize users for various forms of access. As shown in Figure 17-6, it integrates with several common third-party IdPs, as well as provides its own built-in IdP.

Third-party integration

- Recommendations:
  - Use options already used by the organization
  - Multiple permitted (ideal for mergers and acquisitions)
- Options
  - Microsoft Azure AD (Entra)
  - Okta
  - PingFederate
  - JumpCloud
  - Generic SAML

Built-in IdP

- Recommendation: Use when organization has no IdP

Figure 17-6: HPE Aruba Networking SSE and Identity Providers (IdPs)

Third-Party Integration

Many organizations already use an IdP to authenticate users for various purposes. In this case, HPE Aruba Networking recommends that you the same IdP for HPE Aruba Networking SSE. Then admins can continue to manage identities within a single location, simplifying processes for the customer and minimizing potential issues, such as forgetting to revoke an account when a user leaves the organization. Users also benefit; they can log in with familiar credentials.

HPE Aruba Networking SSE supports integration with multiple IdPs in the same workspace, which might be required if mergers and acquisitions have resulted in the company using more than IdP. You simply attach the desired IdP to each application.

As of the release of this study guide, HPE Aruba Networking SSE supports these IdPs:

- Microsoft Azure AD (renamed Entra)
- Okta
- PingFederate
- JumpCloud

If the organization uses a different IdP, which supports Security Assertion Markup Language (SAML), you can set up a generic SAML integration to accommodate that IdP.

Built-In IdP

For organizations without their own IdP, you can leverage the IdP built into the solution. This option also works well for labs and demos.

Using the Built-In IdP

You will first examine how to use the built-in IdP, as the simpler option.

You must add a user account for each user who needs access to HPE Aruba Networking SSE services. User account settings include username (used to log into the user portal), first name, last name, and email address. After you create the account, the user receive an invitation at the specified email address. You can specify the user's password when you create the account. Or you can leave the password unset, in which case the user is prompted to create their own password.

You should also set up user groups and assign users to them. These user groups will make it easier for you to establish policy rules. Rather than grant access to various applications to individual users, you can grant access to full user groups.

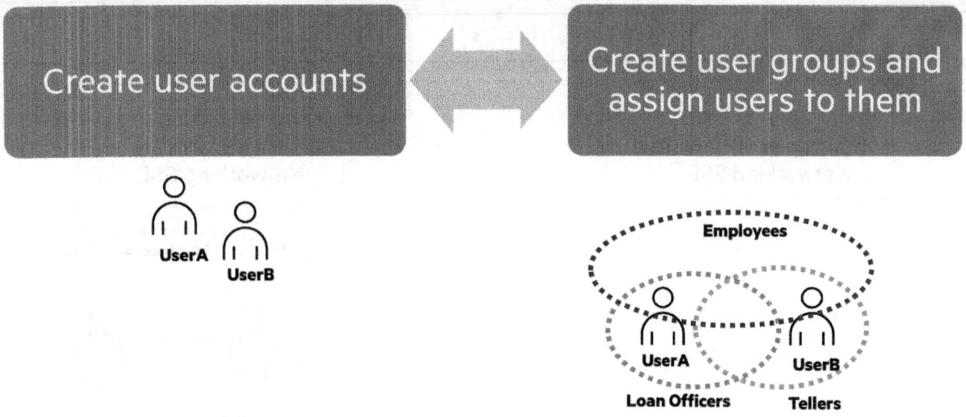

Figure 17-7: Using the Built-In IdP

As shown in Figure 17-7, HPE Aruba Networking SSE gives you great flexibility in setting up user groups, allowing users to belong to multiple groups. For example, UserA might be part of a Loan Officers group, and UserB part of a Tellers group. Both users could be part of an Employees group. You can then easily set up rules. One rule might grant access to generic applications to all employees. Other rules might grant access to specialized applications to Loan Officers only or to Tellers only.

Note that the setup order is also flexible. You can create user groups first and then assign users to groups when you create the user accounts.

Using Third-Party IdPs

To integrate HPE Aruba Networking SSE with a third-party SSE, you must complete some initial setup within the third-party platform, as summarized in Figure 17-8.

This setup typically includes adding an application to the IdP and generating various keys for the application to use to authenticate to the IdP. The application settings might also include redirect URLs on HPE Aruba Networking SSE to which the IdP redirects users after they authenticate to the IdP. You must then add an IdP integration on HPE Aruba Networking SSE, referencing the authentication credentials and other settings established on the IdP. For precise instructions for particular IdPs, refer to HPE Aruba Networking SSE documentation.

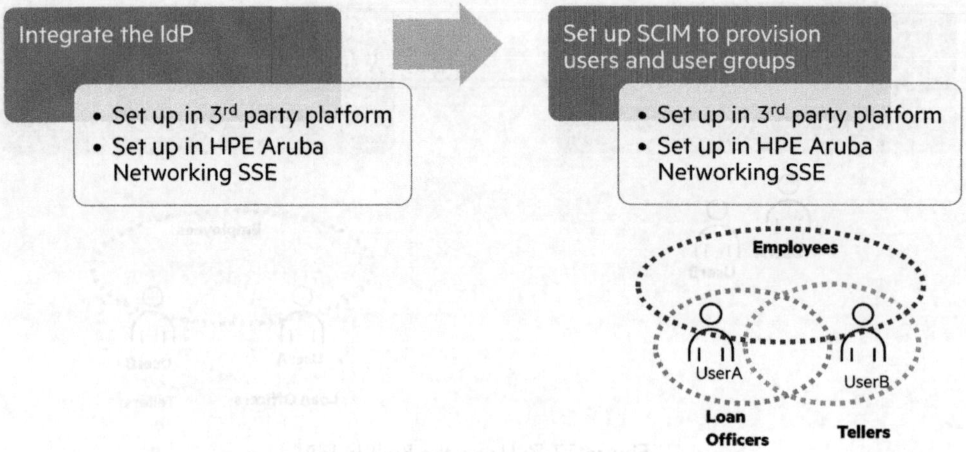

Figure 17-8: Using Third-Party IdPs

The initial IdP integration permits HPE Aruba Networking SSE to authenticate users using the IdP. However, HPE Aruba Networking SSE also needs to know about user accounts and the groups to which users belong so that admins can reference those entities when they set up policy rules. You can reproduce these entities on HPE Aruba Networking SSE manually. However, as a simpler, more scalable solution, you should generally set up SCIM, which allows the IdP to automatically provision users and user groups on HPE Aruba Networking SSE.

Again, you must complete some prerequisite setup within the 3rd party IdP platform. The SCIM setup depends on the specific IdP. You then edit the IdP integration on HPE Aruba Networking SSE and use the advanced settings to enable SCIM. Configure the SCIM settings, syncing values with those established on the 3rd party IdP platform. Refer to HPE Aruba Networking SSE documentation for step-by-step instructions for each supported IdP provider.

After you have configured SCIM on the IdP and HPE Aruba Networking SSE, you should see user accounts and user groups on HPE Aruba Networking SSE. You should also see that the users are assigned to the same groups to which they are assigned within the third-party IdP.

Implementing HPE Aruba Networking ZTNA

In this section, you will learn how to implement HPE Aruba Networking ZTNA.

Overview of the HPE Aruba Networking ZTNA Process (Option A)

Before you dive into setting up HPE Aruba Networking ZTNA, you might find it useful to examine what ZTNA looks like from a user perspective (see Figure 17-9).

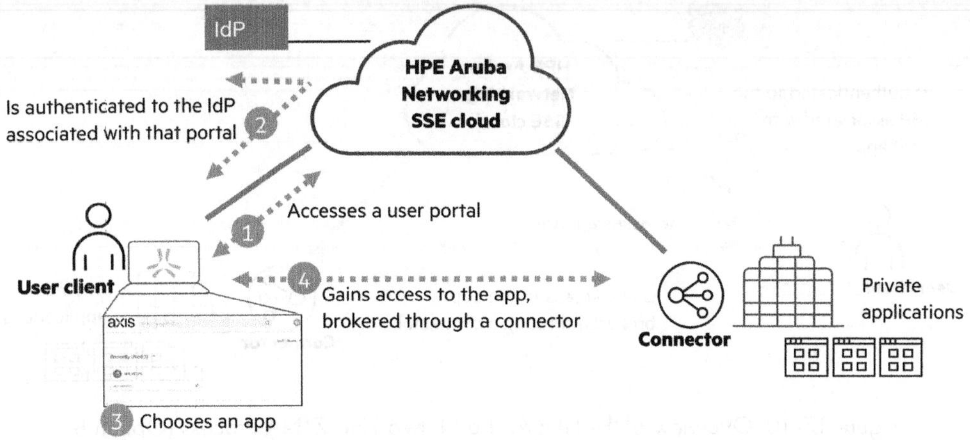

Figure 17-9: Overview of the HPE Aruba Networking ZTNA Process (Option A)

1. A user logs into the HPE Aruba Networking SSE user portal.
2. Each IdP configured on a workspace has its own user portal. HPE Aruba Networking SSE authenticates the user to the IdP in question to log the user in.
3. That portal shows a dashboard of private applications associated with the IdP in question and to which this particular user is permitted access. The user chooses an app and clicks it.
4. The user gain access to the app as easily as that. In the background, HPE Aruba Networking SSE directs the client to a connector in the app's zone. That connector brokers a connection between the client and the app.

Overview of the HPE Aruba Networking ZTNA Process (Option B)

HPE Aruba Networking ZTNA supports another simple option for users, illustrated in Figure 1 7-10. As you set up private applications within HPE Aruba Networking SSE, each application receives a URL, hosted on the HPE Aruba Networking SSE Cloud. Rather than log into the user portal and select the app from a dashboard, a user can follow this process:

1. The user accesses the app URL.
2. HPE Aruba Networking SSE authenticates the user to the IdP associated with the application.
3. The process then continues in the same way you already examined. The connector brokers a connection between the client and application.

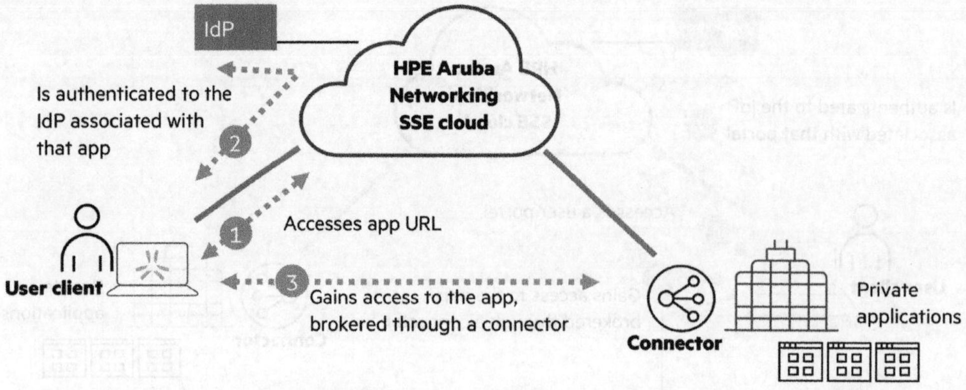

Figure 17-10: Overview of the HPE Aruba Networking ZTNA Process (Option B)

Agent versus Agentless Deployments

HPE Aruba Networking ZTNA supports clients with or without agents installed on them. The differences between these options are summarized in Figure 17-11.

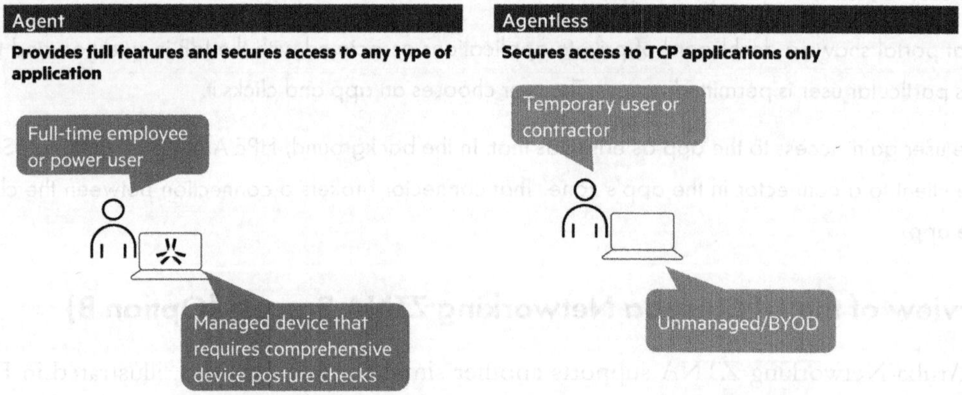

Figure 17-11: Agent versus Agentless Deployments

Agent

Clients with agents receive access to a fuller set of features, so you should generally deploy them on clients used by full-time employees or power users.

For example, agents can access any type of application, including HTTP/HTTPS, Secure Shell (SSH), Remote Desktop Protocol (RDP), and other TCP and UDP applications. Agents can also grant users access to a complete network segment if the company so chooses.

Agents also support device posture checks, so you should deploy them on managed devices that require such checks.

As you will learn in later sections, a client must have an agent to access HPE Aruba Networking SWG and CASB services.

Users can easily download the agent from the HPE Aruba Networking SSE user portal and install it on their own devices. No configuration is required.

Agentless

Agentless deployments provide a great option for temporary users or contractors. Clients can still use HPE Aruba Networking ZTNA services without an agent, but they can only receive access to TCP applications. Because the agent is required for posture checks, companies typically use the agentless option for unmanaged Bring Your Own Device (BYOD) clients that do not require those checks.

Overview of HPE Aruba Networking ZTNA Setup Process

You have seen how simple HPE Aruba Networking ZTNA seems from the user viewpoint. You will now learn how easily you can set up the solution on the admin side.

As for all HPE Aruba Networking SSE features, you must ensure that you have set up the built-in or external IdP that you plan to authenticate users for access to applications. You already learned about this configuration.

You must also create a connector zone for each location in which private applications reside and deploy connectors in the zone.

You must set up the applications to which you want to grant users access within HPE Aruba Networking SSE.

Finally, you create policy rules to control which users are permitted access to various applications under various conditions.

These steps are summarized in Figure 17-12.

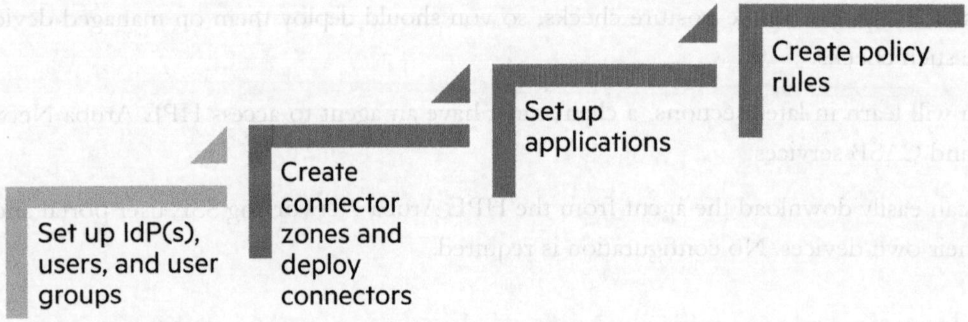

Figure 17-12: Overview of HPE Aruba Networking ZTNA Setup Process

Creating Connector Zones

You will now learn more about creating connector zones.

As indicated in Figure 17-13, a zone defines a destination at which users reach various applications. It might be a data center, granting users access to private applications that run in that data center. Or it might be a public cloud instance, granting users access to infrastructure as a service (IaaS). You can define zones however makes sense for your organization. Typically, admins define at least one zone per private site, such as a data center. However, you could also define multiple zones per site if the company segments resources within the data center. For example, for a multi-tenant data center, you could create a TenantA zone and a TenantB zone.

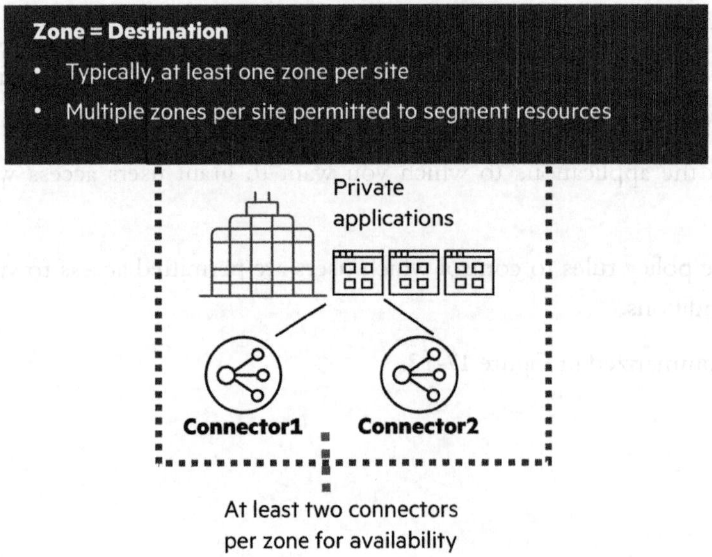

Figure 17-13: Creating Connector Zones

A zone also establishes a container in which you deploy connectors. A connector must be able to access all the private applications within that zone. HPE Aruba Networking recommends deploying at least two connectors per zone for availability purposes.

Creating Connectors

HPE Aruba Networking SSE makes it easy to deploy connectors. As shown in Figure 17-14, when you a connector, you simply assign it a name and select its connector zone.

You can then choose one of three options for deploying the connector:

- **Linux**—You can set up the connector on a supported Linux server, which could be a bare metal server or VM.
- **VM (OVA)**—HPE Aruba Networking provides an OVA, which you can install on a supported hypervisor. The OVA provides a Linux VM, which is ready to act as a connector with just a few simple steps.
- **AWS**—You can deploy the connector as an AWS. You would use this option when you want to establish a connector zone to services running in an organization's AWS account.

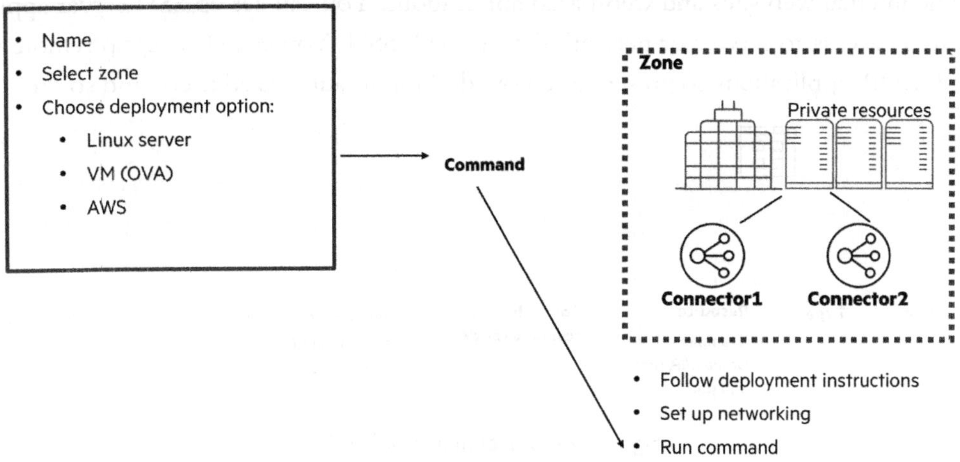

Figure 17-14: Creating Connectors

The connector settings help you deploy the connector. For example, you can download the OVA file by clicking a link in the settings window.

To find an up-to-date list of supported Linux OS and hypervisors for the OVA, as well as to find connectivity requirements, refer to this link: https://docs.axissecurity.com/docs/server-requirements-for-the-connector

After you establish the initial connector settings, the settings window displays a command, which includes a token. You should keep the window open so that you can copy the command to the connector. However, if you accidentally close the window, you can also edit the connector and re-generate the token, which will display the command again.

You should follow the simple deployment instructions for the option that you selected. You should also set up networking on the server, VM, or AWS instance running the connector.

You can then copy in the command from the connector settings. The command runs a script for establishing a connection between the connector and the connector zone in your HPE Aruba Networking workspace. The token included in the command authenticates the connector as valid. If the command executes correctly, you should be able to see the connector in the HPE Aruba Networking SSE management console. The console also reports on the connector's health.

Defining Applications

With the connector zone setup and the connectors deployed, you can add applications that are accessible in that zone. You should add an application for each application or resource that you want to provide to remote users. For example, you can set up web applications to grant users access to specific internal web sites and web-based applications. You can set up SSH server applications to give users access to servers or network devices with SSH. You can also set up remote desktop protocol (RDP) applications so present a remote desktop to authorized users, and so on.

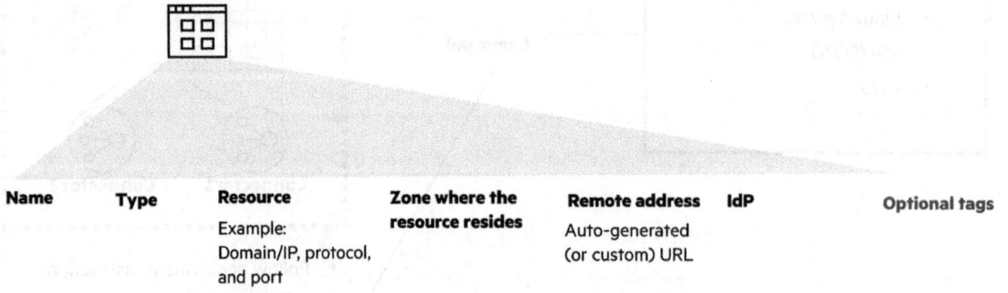

Figure 17-15: Defining Applications

The precise settings for the application depend on the application type. However, at a high level, applications include these settings (see also Figure 17-15):

- **Resource**—The resource settings define the domain name or IP address at which the application is accessed. They might also define a protocol and port number. For example, for a web application, you can define HTTP or HTTPS as the protocol. By default, the application uses the well-known ports for HTTP and HTTPS. However, if your application uses a custom port, you can specify that port.

- **Zone**—You must indicate the zone where the resource resides. All connectors in that zone must be able to reach the application at the domain name or IP address and port indicated.
- **Remote address**—The remote address defines a URL at which remote users access the application. (As you learned earlier, users can either log into the HPE Aruba Networking user portal and select applications or access the applications directly from their URL.) HPE Aruba Networking SSE automatically generates a unique URL for the application; this URL is hosted in its cloud. However, if your company has its own domain, you could specify a custom URL.
- **IdP**—You must specify the IdP that authenticates and authorizes clients for accessing this application. In this way, if your organization uses multiple IdPs, you can easily select the correct IdP for each application.
- **Tags**—You can create application tags, which you then apply to applications. For example, you might create a "loan officers" tag, which you apply to each loan officers' application.

You can accelerate the application definition by creating a spreadsheet, each row of which defines an application. You can then import that spreadsheet in the HPE Aruba Networking SSE management console to configure the applications in bulk. For specifications for that spreadsheet, see the HPE Aruba Networking SSE documentation.

Application Types

In Figure 17-16, you see some of the most common application types supported by HPE Aruba Networking SSE. As you see, in addition to web-based applications, you can provide many options for accessing servers from RDP to SSH and VNC. You can also provide access to a Git server or database server. You can also provide access to a complete range of IP addresses, if necessary. With an Active Directory application, you can permit users to authenticate to AD and receive domain services. A DNS application sends users' DNS requests through the specified DNS servers.

Type	Description	Agent required?
Web Applications		
Self-Hosted Web Application	Defines internal web sites	
SAML	Defines a SAML server	
Server Access		
RDP – Remote Desktop Server	Windows Desktop access	
Remote Desktop Pool	Defines access to a pool of RDP servers	
SSH Server	SSH access to a system	
SSH Range	Defines access to a pool of SSH servers	X
Git Server	Defines a version control system	
Database Server	Defines a database server	
VNC Server	Defines a VNC server	
Network Access		X
Network Range	Defines a range of IP addresses	X
Active Directory	Agent users can be authenticated with Active Directory	X
DNS Application	Agent users' DNS requests can go to specified DNS servers	X

Figure 17-16: Application Types

As shown in the table in Figure 17-16, clients can access most of these applications with or without an agent. However, a client must have an agent to access some application types, including:

- SSH Range
- Network Range
- Active Directory
- DNS Application

Note that this list of application types is not comprehensive. For a complete list, refer to the HPE Aruba Networking documentation.

Establishing Policy Rules

With the applications defined, you are ready to define policy rules to control which users can access which applications under which circumstances.

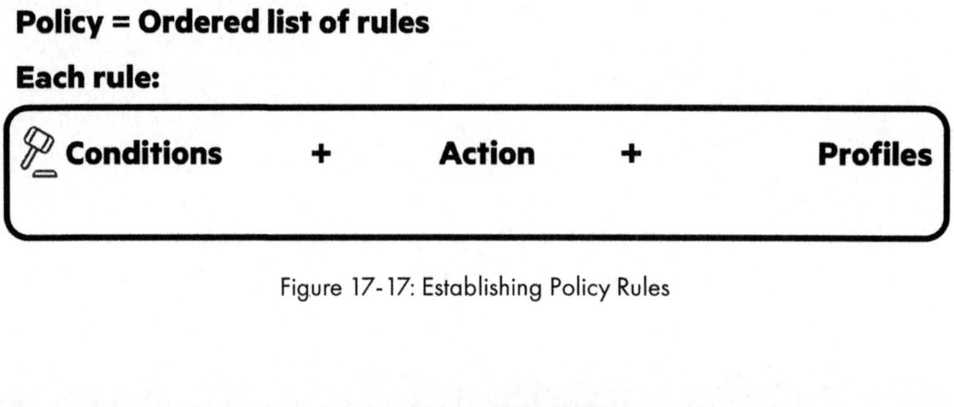

Figure 17-17: Establishing Policy Rules

As shown in Figure 17-17, the HPE Aruba Networking SSE policy is an ordered list of rules. Each rule consists of three broad components: conditions, an action, and profiles.

The policy rule conditions match characteristics of traffic. If traffic matches all the conditions for a rule (and do not match those for any rules higher in the list), HPE Aruba Networking SSE applies the rule action and profiles.

Policy Rule Conditions

As shown in Figure 17-18, HPE Aruba Networking SSE supports several options for conditions, enabling you to create flexible rules that control clients' access in granular ways.

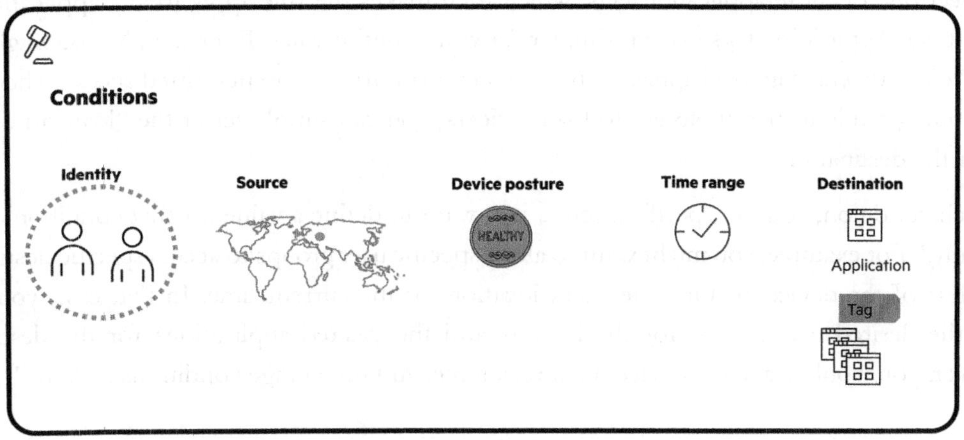

Figure 17-18: Establishing Policy Rules

- **Identity**—For identity, you can select one or more users or user groups.
- **Source**—The source indicates the location from which the user connects. HPE Aruba Networking SSE comes with predefined geolocations. You can also define your own IP ranges to specify for source conditions.
- **Device posture**—You can define device trust rules with the checks that you want to implement on the client. You can then reference one or more of those rules in the policy rule. The rule only applies if the client passes the checks configured in those rules. HPE Aruba Networking SSE supports two types of device trust rules:
    - **Agent**—If the client is running an agent, that agent can analyze the client and check whether the client complies with certain conditions, such as having antivirus enabled. The agent supports checks on MacOS, Windows, Linux, iOS, and Android.

- **Certificate**—You can configure this form of device trust rule for agentless devices. The rule specifies a CA, and HPE Aruba Networking SSE checks whether the client has a valid certificate signed under that CA.
- This study guide does not cover device posture in detail. For more information, take additional training or use the documentation.

- **Time range**—You can specify the times at which this rule applies, such as during working hours or on weekends.
- **Destination**—You can select one or more applications that you defined earlier. You can also select one or more application tags, which then makes the rule apply to any application with that tag. Application tags make it simpler for you to define rules. For example, you might apply a "loan officers" tag to all applications to which loan officers are permitted access. Then, when defining a rule to permit access to loan officers, you can simply select the "loan officers" tag for the destination.

For each condition, you can specify whether you want to define a value for that condition or leave it an "any." For example, you might want to allow specific user groups to access specific destinations regardless of the device posture, the client location, or the current time. In that case, you would select the desired user groups for the identity and the desired applications for the destination. However, you would leave the device posture, source, and time range conditions at "any."

Policy Rule Actions

As shown in Figure 17-19, you can choose to allow the traffic that meets the conditions. In that case, HPE Aruba Networking SSE brokers a connection to the destination application using the zone associated with that application.

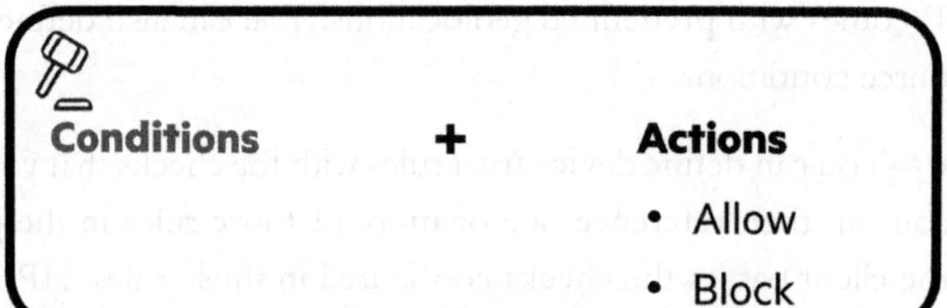

Figure 17-19: Policy Rule Actions

Alternatively, a rule can block the traffic. The user cannot access the application, and HPE Aruba Networking SSE redirects the user to a web page indicating that access to the resource is forbidden.

You should have users install HPE Aruba Networking SSE's CA certificate on their machines to avoid certificate errors when they view the blocking web page.

Policy Rule Profiles

The rule also indicates profiles, as shown in Figure 17-20. The profiles refine the permissions for various types of access.

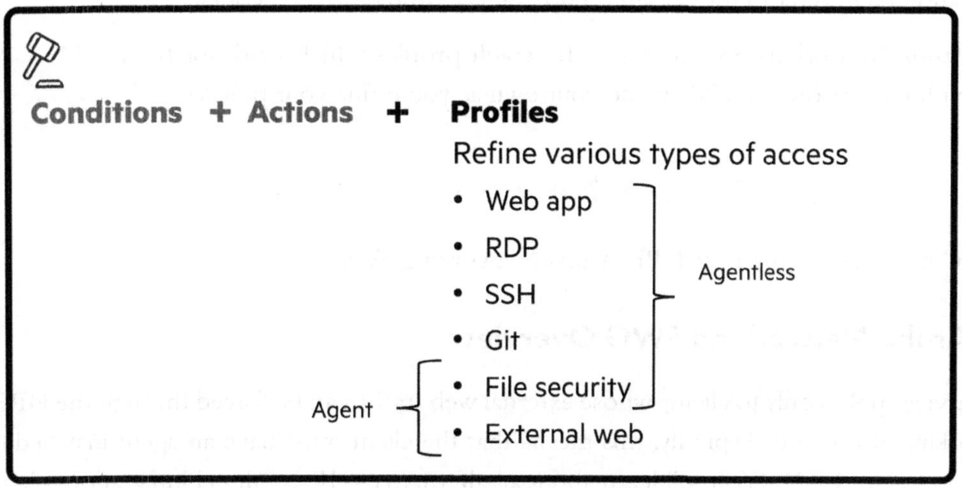

Figure 17-20: Policy Rule Profiles

HPE Aruba Networking SSE supports several profiles for agentless clients:

- **Web app**—This profile affects users' access to the company's self-hosted web applications. For example, the profile can control whether users are allowed to upload and download files from the application.
- **RDP**—With this profile, you can enable or disable various types of RDP access. For example, you can control whether the RDP session occurs over the web or with a native RDP client. You can also enable or disable the remote clipboard, file transfers, and printing, as well as control session lengths.
- **SSH**—Use this profile to control settings for brokered SSH sessions. Among other settings, you control the session length and token expiration time. You can also enable a web SSH client, which gives users access to the SSH session through their Web browser. You can enable or disable file transfers and copying and pasting with the web SSH client. You can also permit access through native SSH (clients on the users' own devices).
- **Git**—This profile controls the token expiration time for Git access.

External web profiles apply to clients with agents. An external web profile enables or disables features such as SSL inspection and threat protection.

HPE Aruba Networking SSE supports file security profiles for both agentless clients (accessing private applications with ZTNA) and clients with agents (accessing SaaS applications with CASB). File security profiles, coupled with external web profiles, empower you to implement data loss prevention (DLP). However, this feature is beyond the scope of this study guide. For more information on it, refer to the *Deploying HPE Aruba Networking SSE* course.

HPE Aruba Networking SSE comes with default profiles which work for basic ZTNA. You can begin with these profiles and then add your own as you refine your policies.

Implementing HPE Aruba Networking SWG

You will next learn how to use HPE Aruba Networking SWG.

HPE Aruba Networking SWG Overview

This service applies only to clients whose external web traffic can be forced through the HPE Aruba Networking SSE cloud. Typically, this means that the client must have an agent installed on it, as shown in Figure 17-21. (You will learn about an alternative—IPsec tunnel integration—later.)

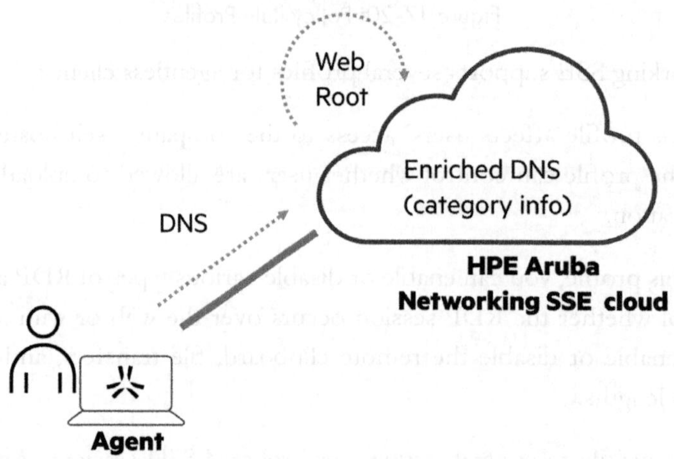

Figure 17-21: HPE Aruba Networking SWG Overview

HPE Aruba Networking SSE inspects the web traffic and blocks or allows that traffic based on policy rules. When it allows the traffic, it can optionally inspect the traffic for threats such as malware.

Often you want to construct your rules based on web categories. For example, you might want to block access to known phishing sites. Or an organization—such as a government entity—might want to block access to gambling or potentially offensive sites. To determine the web category, HPE Aruba Networking SWG relies on DNS monitoring using the HPE Aruba Networking SSE agent. That is, the agent sends information about the client's DNS requests HPE Aruba Networking SSE. The solution integrates with Web Root to determine the web category.

Overview of HPE Aruba Networking SWG Setup

You can get started with HPE Aruba Networking SWG easily, following the process outlined in Figure 17-22.

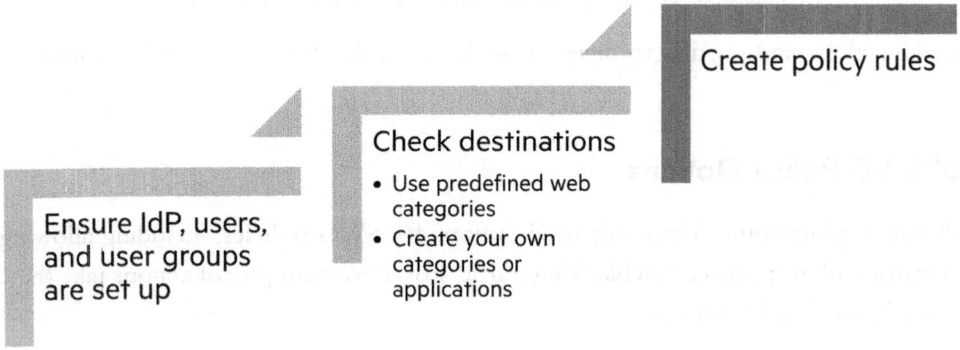

Figure 17-22: Overview of HPE Aruba Networking SWG Setup

Just like HPE Aruba Networking ZTNA, this feature relies on policy rules.

Within those rules, you often include identity information. For example, you might want to block certain web sites for some users, but not others. If you have already set up IdP, you can use that IdP's users and user groups. Users log in through the agent, and HPE Aruba Networking SWG will use that authentication to determine the user identity when applying the web category rules.

Your HPE Aruba Networking SSE workspace comes with many predefined web categories, so you can often jump straight to setting up policy rules. However, if necessary, you can create your own custom categories. In addition, if you want to allow or block access to a specific public website, you can specify that website as a self-hosted web application that uses the public connector zone.

Policy Rules for HPE Aruba Networking SWG

To create a typical HPE Aruba Networking SWG rule, set the destination to one or more web categories, as shown in Figure 17-23. You can also define any of the conditions about which you learned earlier. For example, you can make the rule apply to specific user groups or at specific times.

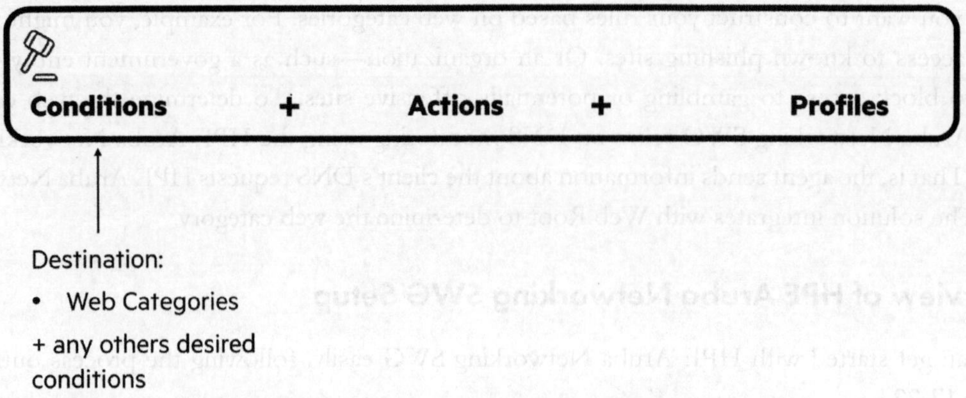

Figure 17-23: Policy Rules for HPE Aruba Networking SWG

You use the actions and profiles to dictate how HPE Aruba Networking SSE should handle the traffic.

Typical SWG Policy Options

You will now explore some commonly used options for SWG policies, including allowing traffic, allowing traffic with inspection, and blocking traffic. For more examples of actions take the *Deploying HPE Aruba Networking SSE* course.

Allow

For the first example, illustrated in Figure 17-24, you want to define a rule that permits users to browse the Internet directly without their traffic flowing through the HPE Aruba Networking SSE cloud. This type of rule works well for low-risk categories that your organization permits.

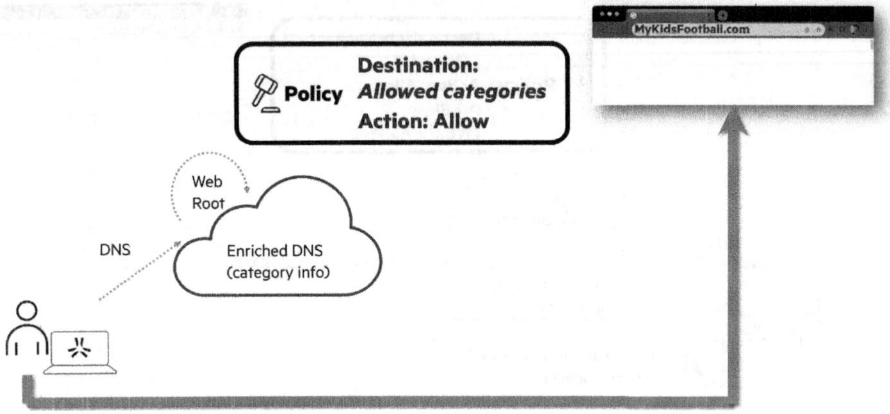

Figure 17-24: Allow

Within the rule, simply select the allowed web categories for the destination and add any other desired conditions. Set the action to Allow.

Note that your workspace comes with a default rule at the end of the policy list that allows all external web traffic. So, if you want to permit certain categories only, make sure to add an explicit rule that blocks other categories. As an alternative approach, if your organization permits most categories, you can create rules that block certain categories and use the default rule to allow everything else.

Allow + Inspection

For the next example, illustrated in Figure 17-25, you want to enhance security for traffic destined to questionable or potentially risky sites. You want to allow traffic but also inspect it for threats. Highly security conscious organizations might inspect traffic for all web sites.

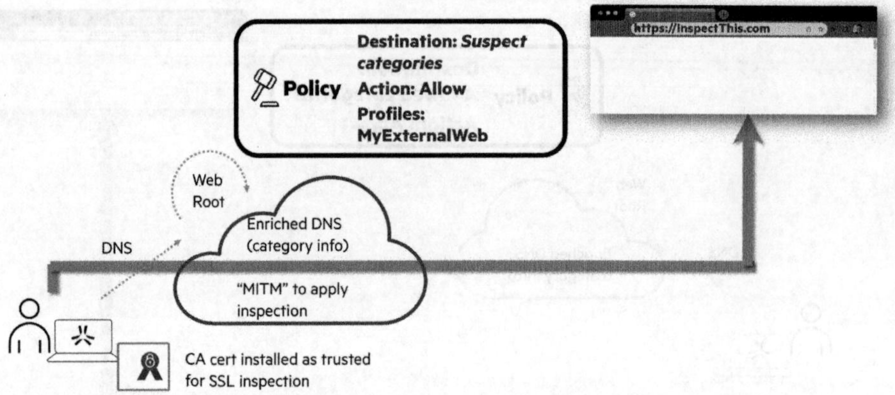

Figure 17-25: Allow + Inspection

To implement this inspection, you set the action to allow and specify a custom external web profile. Within that profile, you can enable these features:

- **SSL inspection**—This feature enables the HPE Aruba Networking SSE to decrypt and inspect SSL traffic. Because most traffic on the web uses HTTPS, you must enable this feature for visibility into the traffic. The HPE Aruba Networking SSE cloud acts as a "man-in-the-middle" (MITM) between the client and the website, establishing an HTTPS session with the client. For this session to establish correctly, the client must have the HPE Aruba Networking SSE CA certificate installed on it as trusted. Users can download the certificate from the user portal.
- **Session visibility**—This feature sends logs about the users' DNS and web browsing activities to SIEM.
- **Threat Intelligence protection**—Enable this feature to detect malware and other threats in the traffic.

Block

Sometimes an organization has categories of sites to which it wants to block access entirely. In this case, set up a rule with those categories set as the destination. Set the action to block, as shown in Figure 17-26. The agent will direct the clients' traffic to a 403 page on the HPE Aruba Networking SSE cloud. Remember to have users download the HPE Aruba Networking SSE CA certificate from the user portal and install that certificate as trusted. Otherwise, they will see a security error on the 403 page.

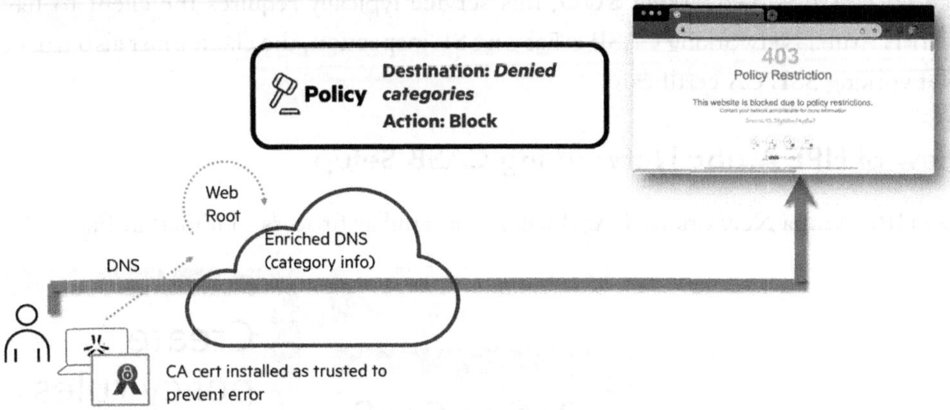

Figure 17-26: Block

Implementing HPE Aruba Networking CASB

You will now learn more about HPE Aruba Networking CASB.

HPE Aruba Networking CASB Overview

As shown in Figure 17-27, HPE Aruba Networking CASB brokers connections between clients and SaaS applications. As part of the brokering process, the service can apply security policies, allowing or restricting various actions. For example, an organization can choose whether to permit uploading and downloading files for certain SaaS applications.

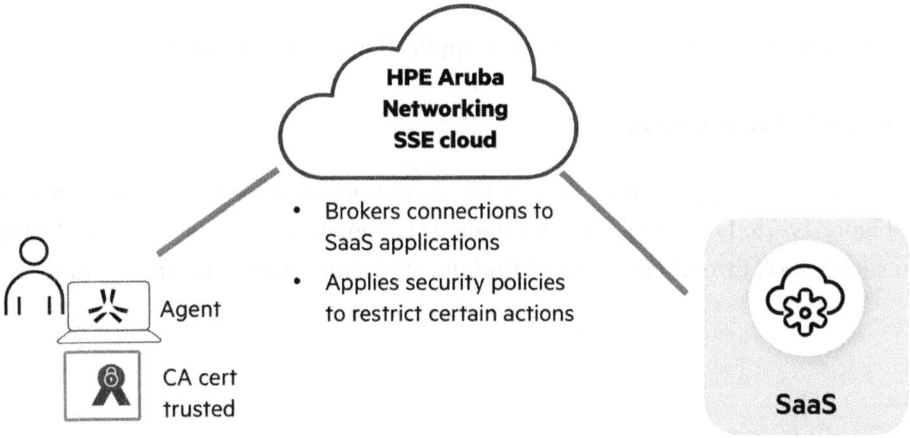

Figure 17-27: HPE Aruba Networking CASB Overview

Much like HPE Aruba Networking SWG, this service typically requires the client to have agent. Because HPE Aruba Networking CASB relies on SSL inspection, the client must also trust the HPE Aruba Networking SSE CA certificate.

Overview of HPE Aruba Networking CASB Setup

Setting up HPE Aruba Networking CASB follows a familiar process, outlined in Figure 17-28.

Figure 17-28: Overview of HPE Aruba Networking CASB Setup

Again, ensure that you have set up the correct IdP or IdPs. You can then reference the IdPs' user and user groups in policy rules.

Next, define SaaS applications within your workspace for the applications to which you want to control users' access.

You can then create policy rules that use those applications as destinations.

Defining SaaS Applications

When you define a SaaS application, you give the definition a name and select the SaaS category, as shown in Figure 17-29. HPE Aruba Networking SSE comes with many predefined categories for SaaS. You simply look through the list and find the application that your organization uses.

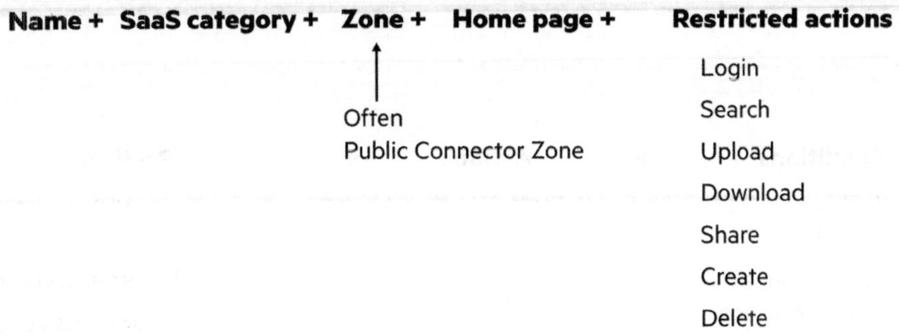

Figure 17-29: Defining SaaS Applications

You must set a zone for the application. Often this is the Public Connector Zone, allowing users' traffic to remain in the cloud, rather than backhauling the traffic through a company site.

You can optionally define a home page for the application. If you do, HPE Aruba Networking CASB will direct clients to this home page when they access the application.

Finally, you select any actions that you want to restrict users from performing with this application. Options include:

- Login
- Search
- Upload
- Download
- Share
- Create
- Delete

Defining Policy Rules for HPE Aruba Networking CASB

When you define a policy rule for HPE Aruba Networking CASB, simply select the SaaS application or applications that you have already defined. Apply other conditions as desired.

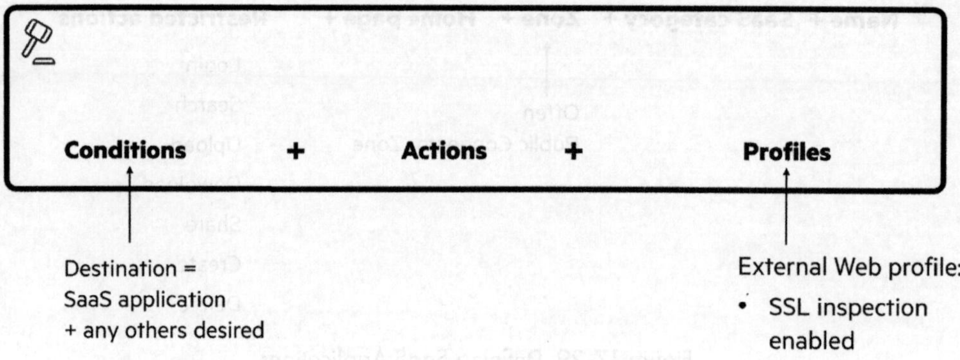

Figure 17-30: Defining Policy Rules for HPE Aruba Networking CASB

Set the action as desired to permit or block access to the application. As shown in Figure 17-30, you *must* also specify an external Web profile with SSL inspection enabled. Only then can HPE Aruba Networking inspect the traffic and control it as you have specified.

HPE Aruba Networking SSE conclusion

This brief section summarizes what you have learned about HPE Aruba Networking SSE and points out some additional features out of scope for this training.

Integrating with SD-WAN Solutions

As you have learned, various features such as HPE Aruba Networking SWG, CASB, and sometimes DLP typically require clients to have agents. For example, without the agent, remote clients can simply access SaaS applications on their own; the CASB service never has a chance to broker the connection instead.

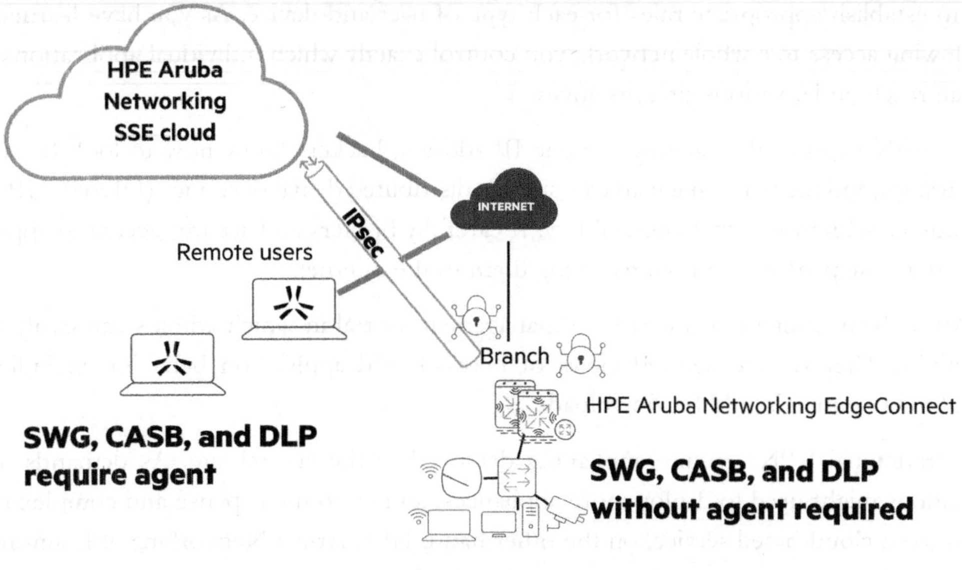

Figure 17-31: Integrating with SD-WAN Solutions

However, an exception applies for clients at branch sites with gateway appliances, such as HPE Aruba Networking EdgeConnect solutions. As shown in Figure 17-31, you can establish an IPsec tunnel between the branch appliance and HPE Aruba Networking SSE cloud. The HPE Aruba Networking SSE can implement SWG, CASB, and DLP for the branch clients even if those clients don't have agents.

You can learn more about this option at the expert-level of the security certification track or in the *Deploying HPE Aruba Networking SSE* course.

Benefits of HPE Aruba Networking SSE Versus Legacy VPN

Throughout this chapter, you have explored ways that HPE Aruba Networking SSE exceeds the security capabilities of a legacy VPN.

Unlike some legacy VPNs, which give authenticated remote clients blanket access to the private network, HPE Aruba Networking SSE controls users' access granularly. Policies can draw on rich context, including device posture checks, destinations, time of day, location, and IP address range.

Rather than authenticate the client once—when establishing a VPN connection—HPE Aruba Networking SSE continuously authenticates and authorizes clients for each resource that they attempt to access.

With legacy VPNs, admins often struggle to set up appropriate restrictions for contractors, temporary employees, and BYOD devices. As you have seen, HPE Aruba Networking SSE makes it easy to establish appropriate rules for each type of user and device. As you have learned, rather than allowing access to a whole network, you control exactly which individual applications various users can reach under various circumstances.

A legacy VPN exposes the gateway's public IP address; hackers know how to look for and scan such gateways, and they can launch attacks such as distributed denial of service (DDoS). HPE Aruba Networking SSE removes that vulnerability. It carefully brokers and secures access to applications published via the platform without exposing them to the internet.

HPE Aruba Networking SSE provides a management portal in which admins can easily track all user activity. They receive detailed views of network and application behavior, including URL history, SSH commands, and file downloads.

Finally, traditional VPNs require appliances deployed at the central site. As demands increase, organizations might need to deploy more appliances, adding to the expense and complexity of the solution. As a cloud-based service, on the other hand, HPE Aruba Networking SSE automatically scales as traffic increases—no reconfiguration required.

These benefits are summarized in Figure 17-32.

Capability	Description	HPE Aruba Networking SSE	VPN
Policy: Granular Access	Granular policy access to specific applications based on rich context such as: • Device posture checks • Conditions such as destinations, time of day, location, and IP address range	Yes	No
Continuous Enforcement	Continuous authentication throughout the session and not only at initial login	Yes	No
3d party Limited Access	• Restricted access for contractors, temporary employees, BYOD devices • Different access for different users/groups based on posture and conditions	Yes	No
Network Isolation	Access restricted to individual applications – NOT allowed to the whole network	Yes	No
Application Isolation	Applications published via the platform NOT exposed to the internet	Yes	No
Visibility	• All user activity tracked • Detailed views of network and application behavior	Yes	No
Auto scaling	Automatic scaling as traffic increases without network reconfiguration	Yes	No

Figure 17-32: Benefits of HPE Aruba Networking SSE Versus Legacy VPN

Securing Branch Access

In this section, you will learn more about how HPE Aruba Networking products help you to secure branch access.

Traditional Option for Securing Branch Access

To connect a branch to headquarters—particularly over the Internet, an inherently insecure network—companies often rely on site-to-site VPNs.

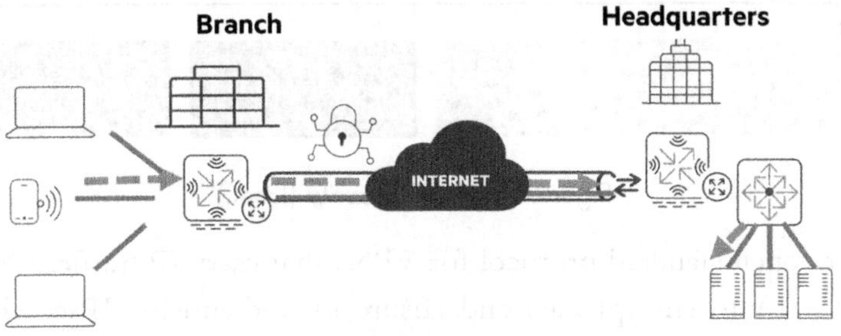

Figure 17-33: Site-to-Site VPN Based on IPsec

A site-to-site VPN securely connects two sites together, as shown in Figure 17-33. For example, a company can use a site-to-site VPN to connect a branch office to a central data center over the Internet—a much more cost-effective option than a private WAN line.

IP Security (IPsec)

Before you dive into HPE Aruba Networking's approach to site-to-site VPNs, you should learn just a bit more about IP Security (IPsec)

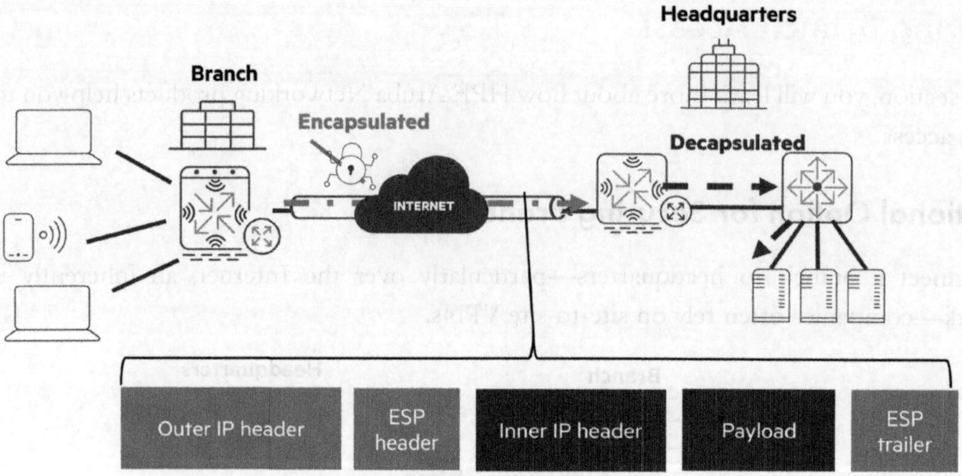

Figure 17-34: IP Security (IPsec)

IPsec is a common, open-standard protocol for VPNs that carry IP traffic. IPsec uses algorithms such as AES and SHA to encrypt data and ensure its authenticity. IPsec also provides replay protection.

For the purposes of this study guide, you are interested in tunnel mode IPsec, which sets up a tunnel between two IPsec hosts or gateways. This tunnel can carry traffic on behalf of clients behind the IPsec gateways in a site-to-site VPN, such as the one shown in Figure 17-34. Each gateway encapsulates clients' IP traffic with an ESP header and outer IP header. The outer IP header includes the source and destination gateways' IP addresses while the inner IP header retains the client's source IP address and the IP address of the resource that the client is accessing.

IPsec Security Associations (SAs)

Each IPsec tunnel features two security associations (SAs), one for carrying traffic in each direction, as shown in Figure 17-35. The SA defines parameters such as the algorithms used to authenticate and encrypt traffic, as well as the actual authentication and encryption keys. The SAs and their keys can be defined manually on each side of the tunnel. However, often the two tunnel endpoints use Internet Key Exchange (IKE) or IKEv2 to negotiate the SA automatically. Using IKE/IKEv2 endpoints negotiate algorithms, authenticate each other, and securely generate shared keys for the SAs.

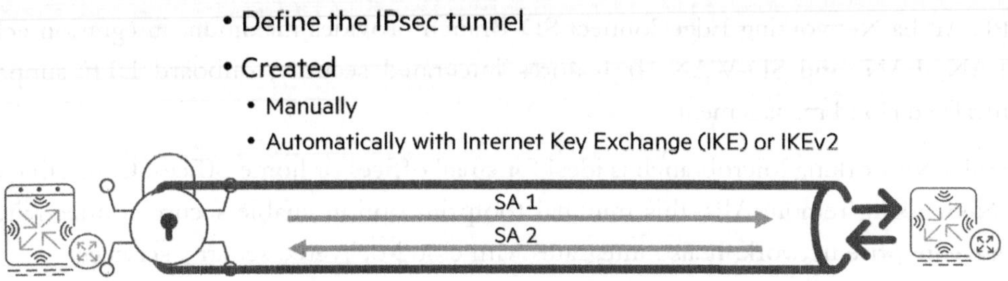

Figure 17-35: IPsec Security Associations (SAs)

Review the HPE Aruba Networking Unified SD-WAN Fabric

With this brief background in VPN technology, you are ready to learn about HPE Aruba Networking options for securing branch access. Take a moment to review the HPE Aruba Networking unified SD-WAN fabric, which is illustrated in Figure 17-36.

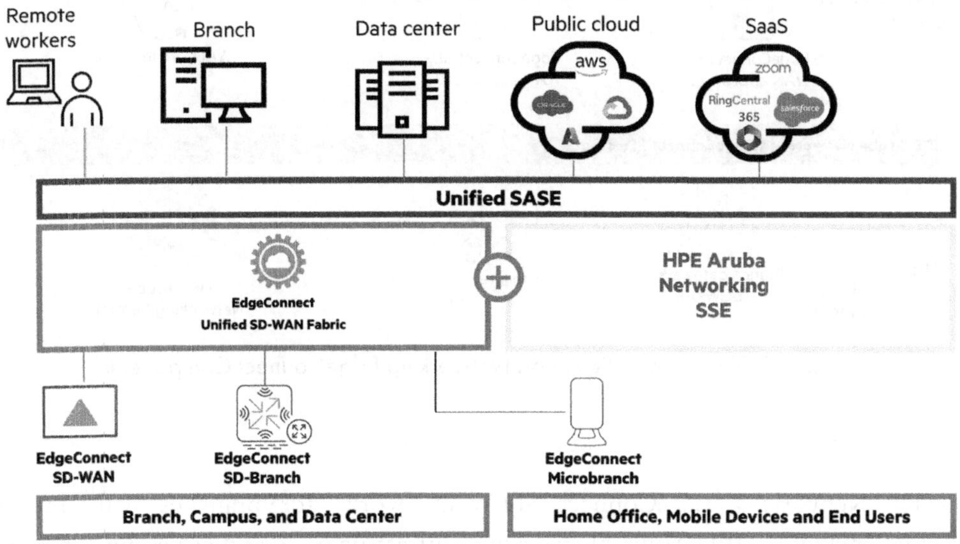

Figure 17-36: Review the HPE Aruba Networking Unified SD-WAN Fabric

As you know, unified SASE solutions consist of two key components: SSE—you have already explored HPE's services in this area—and SD-WAN. For SD-WAN, HPE offers a unified SD-WAN fabric based on one or more HPE Aruba Networking EdgeConnect products:

- HPE Aruba Networking EdgeConnect SD-WAN offers advanced SD-WAN capabilities. It continuously learns and adapts to changing business need, delivering maximum network and application performance from the edge to the cloud.

- HPE Aruba Networking EdgeConnect SD-Branch provides maximum integration across the WLAN, LAN, and SD-WAN. It features integrated security, onboard LTE support, and centralized cloud management.

HPE Aruba Networking Microbranch is ideal for small offices or home offices. Consisting of HPE Aruba Networking remote APs, this minimal footprint option enable secure connectivity to the corporate enterprise network. It also integrates with cloud-delivered security services.

Review HPE Aruba Networking EdgeConnect Components

Also take a moment to review the components of EdgeConnect SD-WAN and EdgeConnect SD-Branch solutions, which are shown in Figure 17-37 and described in more detail below.

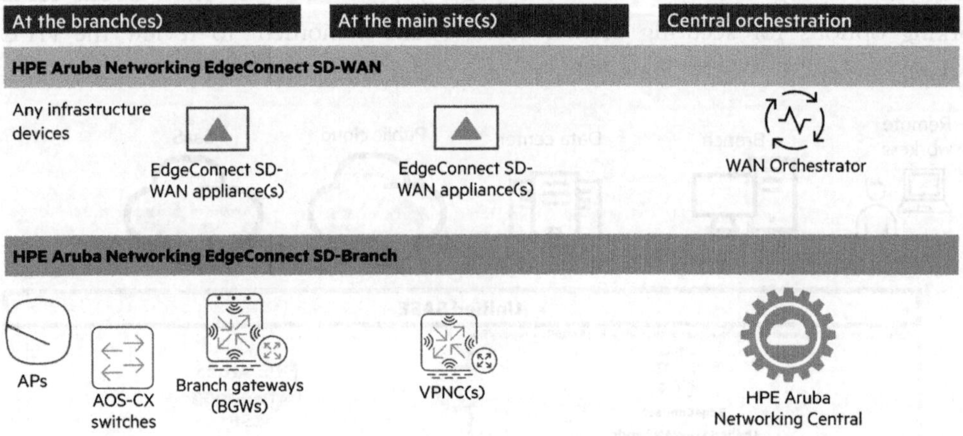

Figure 17-37: Review HPE Aruba Networking EdgeConnect Components

HPE Aruba Networking EdgeConnect SD-WAN

HPE Aruba Networking EdgeConnect SD-WAN works for branches with any network infrastructure, including HPE Aruba Networking's infrastructure or other vendors' infrastructure. You deploy a single EdgeConnect SD-WAN appliance at each branch site or a pair of appliances for high availability. You also deploy EdgeConnect SD-WAN appliances at the main site to act as hubs. Appliances establish connectivity with each other, as orchestrated by WAN Orchestrator. HPE Aruba Networking provides both physical and virtual options for the SD-WAN appliance.

Admins manage the SD-WAN solution with WAN Orchestrator. Companies can deploy WAN Orchestrator as a VM at their main site, or they can purchase a cloud subscription for it.

HPE Aruba Networking EdgeConnect SD-Branch

An Aruba SD-Branch solution includes a branch gateway at each site. Larger sites, or sites that require particularly high availability, can have two branch gateways deployed in a cluster. Typically, companies should deploy HPE Aruba Networking 9x00 Series as the branch gateways. They can also use the 7000 Series for this role; however, only the 9x00 Series Gateways support full Unified Threat Management (UTM) capabilities.

The company also requires headend gateways at the headquarters or data center. These gateways are sometimes called VPN Concentrators (VPNCs). Smaller organizations can use the 7010 Series or 9000s for this role, while larger enterprises should deploy 7210 models or above. (Note that the 7005, 7008, and 7205 models are not supported.) HPE Aruba Networking also offers Virtual Gateways, which companies deploy on hardware of their choice. HPE Aruba Networking 7200 Series models can theoretically operate as branch gateways; however, they support more clients and greater throughput than a branch requires.

An EdgeConnect SD-Branch solution uses the same HPE Aruba Networking APs and AOS-CX switches as any AOS 10 architecture.

HPE Aruba Networking Central provides cloud-based orchestration for the Edge Connect SD-Branch solution, as well as cloud-based management and automation for all the network devices.

As of the release of this study guide, EdgeConnect SD-Branch features are supported with Foundation licenses; however some features require Advanced, Foundation with Security, or Advanced with Security licenses. Refer to the latest datasheets for up-to-date information on licensing.

Example EdgeConnect SD-WAN Architectures

In an EdgeConnect SD-WAN solution, each appliance is a spoke or a hub. Spoke appliances connect to hub appliances, but not to each other. A simple architecture might feature a spoke at several branches and a hub at a main office. Note that each connection, illustrated as a simple line in Figure 17-38, represents a secure IPsec tunnel established over a flexible choice of media, such as internet, MPLS, or LTE. In fact, the line could even represent multiple tunnels established over multiple media. EdgeConnect SD-WAN can flexibly steer traffic over the correct path as dictated by changing circumstances.

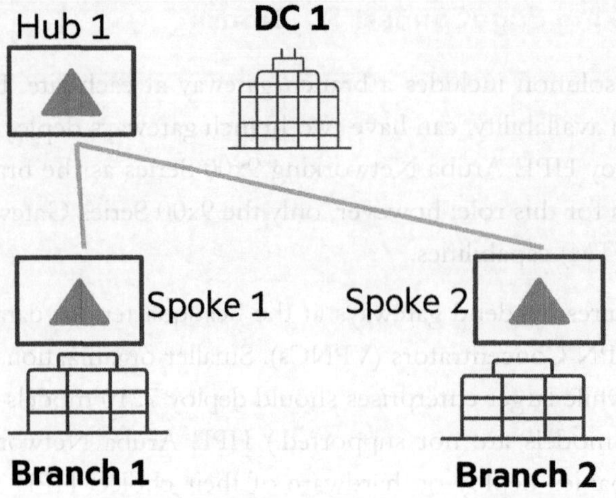

Figure 17-38: Simple Spoke-and-Hub Architecture

EdgeConnect SD-WAN also makes it easy for companies to set up more complex architectures for larger enterprises. You can establish multiple spoke-and-hub regions and connect those regions through hubs. In the example shown in Figure 17-39, the hubs in regions 1, 2, and 3 all establish connections with each other. (Again, each line represents a secure IPsec tunnel or tunnels established over the internet, MPLS, or LTE.)

For high availability, you could have multiple hubs within a region, as illustration for region 2. You can even have pairs of spokes at branch offices., although that option is not illustrated in this example.

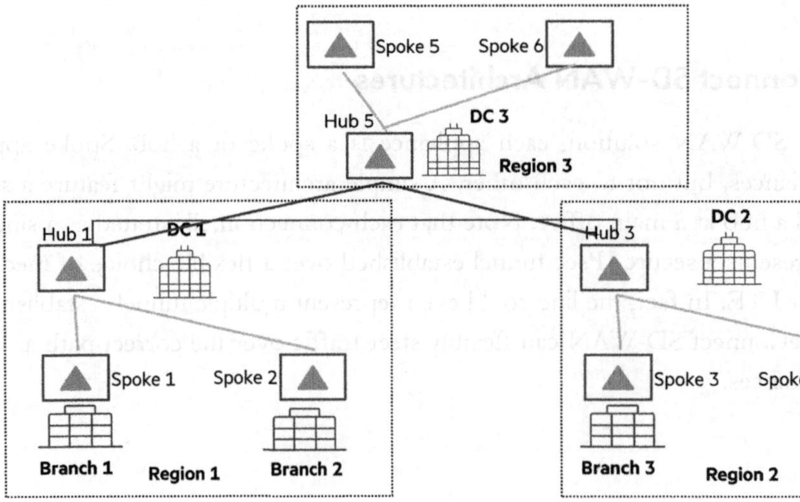

Figure 17-39: Multi-Region Architecture

As another option for optimized high availability, you can even connect spokes to hubs in different regions. For example, as shown in Figure 17-40, region 2 and region 3 have a single hub. The spokes in those regions might connect both to their local hub and a hub in another region.

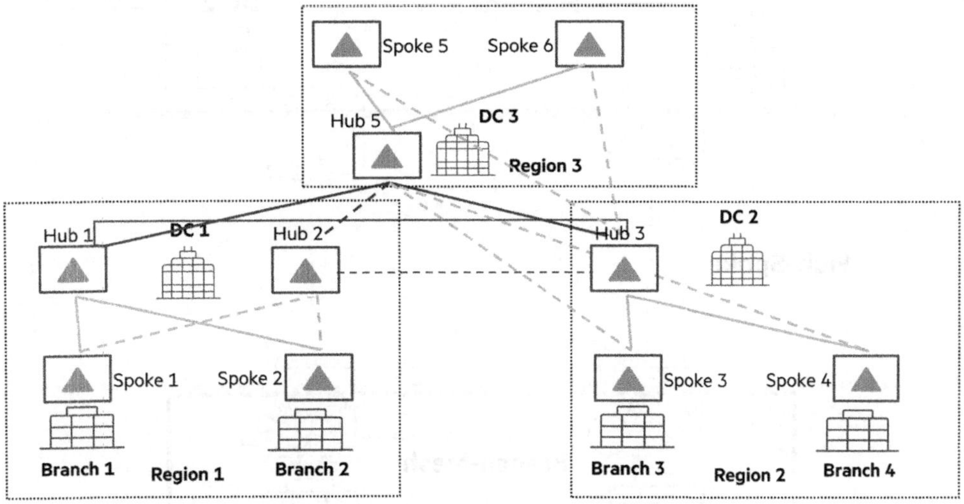

Figure 17-40: Multi-Region Redundant Architecture

Example EdgeConnect SD-Branch Architectures

EdgeConnect SD-Branch also uses the standard spoke and hub terminology with BGWs acting as spokes and VPNCs acting as hubs. In HPE Aruba Networking Central 2.5.3 and later, EdgeConnect SD-Branch supports multiple topology options, which you can combine together flexibly. These topologies are illustrated in Figure 17-41; as before, each line represents a secure IPSec tunnel or tunnels established over the internet, MPLS, or so on.

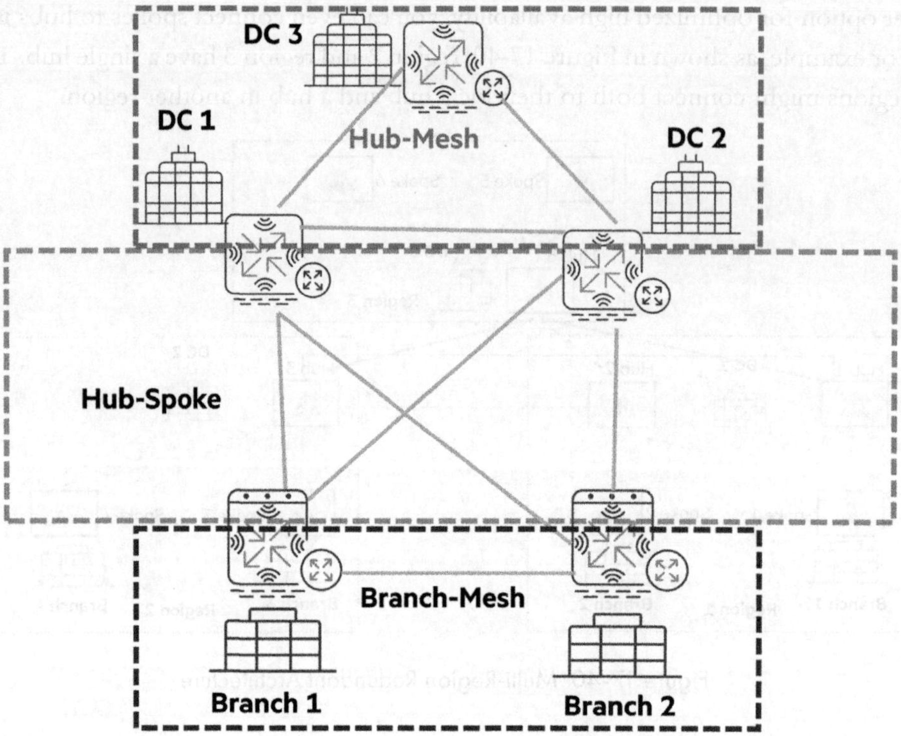

Figure 17-41: Example EdgeConnect SD-Branch Architectures

Hub-Spoke

EdgeConnect SD-Branch solutions are founded on hub-spoke topology. This topology consists of tunnels between BGWs and VPNCs. The hub-spoke topology enables branch clients to communicate with devices at a data center or main campus. It can also feature redundancy, as shown in the "hub-spoke" section of Figure 17-41. As you see, a BGW can connect to two VPNCs. However, if the architecture becomes too large or complex having only hub-spoke tunnels can limit scalability.

Hub-Mesh

You can place VPNCs in a hub-spoke architecture within one or more hub-mesh topologies, illustrated at the top of Figure 17-41. The VPNCs in the hub-mesh establish tunnels between each other. Then a BGW in branch 1, for example, can reach DC 3 through a connection to DC 1. A hub-mesh topology supports greater scalability for a company with multiple VPNs at multiple sites.

Branch-Mesh

For many companies branch clients only need to communicate with the internet and data center. However, some companies need to establish communications directly between some branches. These customers can add a branch-mesh topology to the solution, as shown at the bottom of Figure 17-41. A branch-mesh architecture requires each BGW to connect to at least one VPNC. You can then choose to connect BGWs to one or more BGWs in the same HPE Aruba Networking Central group or a different group. You can pick and choose the precise BGWs that connect together for great flexibility.

Refer to HPE Aruba Networking Central documentation for the most up-to-date limitations on the number of BGWs in a branch-mesh topology.

Centralized IPsec SA Establishment

Whichever EdgeConnect solution you are using its centralized orchestrator automates the establishment of secure IPsec tunnels across the architecture, as indicated in Figure 17-42.

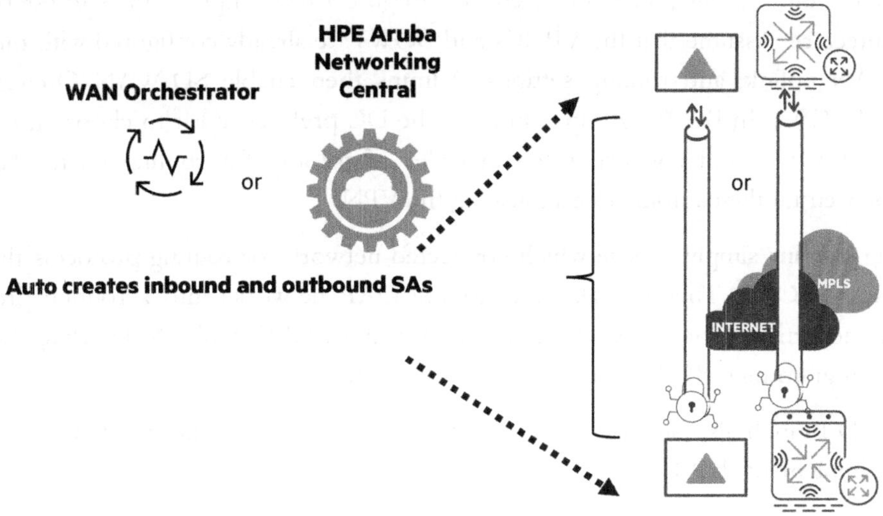

Figure 17-42: Centralized IPsec SA Establishment

EdgeConnect SD-WAN

This solution supports a choice of establishing IPsec tunnels with or without Internet Key Exchange (IKE), an industry standard for secure key agreement between IPsec tunnel endpoints. HPE Aruba Networking recommends IKE-based tunnels for organizations who require FIPS compliance and IKE-less tunnels (sometimes called IPsec UDP) for other organizations.

When using IKE-less tunnels, appliances securely derive shared keys using a nonce and ephemeral keying material sent by Orchestrator. The keys themselves are never transmitted over the lines.

EdgeConnect SD-Branch

HPE Aruba Networking Central's SD-WAN orchestrator automates the establishment of IPsec SAs without the need for IKE. For each IPsec tunnel, it creates an inbound and outbound ESP SA (direction from the viewpoint of the initiator). It sets the SPI for the SA and also chooses the encryption and authentication algorithms. It also generates the encryption and algorithm keys.

SD-WAN Orchestrator then distributes this information to both sides of the tunnel, also informing the gateways the IP address of their peer. The initiator, which is the BGW for hub-spoke tunnels, then contacts the peer, securing traffic with the keys provided by HPE Aruba Networking Central. The peer detects that the keys are correct for the SA with that peer, and the tunnel is established.

In addition to setting up the IPsec tunnels automatically, HPE Aruba Networking Central also orchestrates routing over those tunnels using Overlay Agent Protocol (OAP).

From admins' point of view, they only need to configure a few simple settings to establish a hub-spoke architecture. Assume that the VPNCs and BGWs are already configured with their base IP, VLAN, WAN uplink, and routing settings. Admins then enable SD-WAN Orchestration on VPNCs and BGWs. In BGWs' groups, they use the DC preference list to choose the VPNCs to which the BGWs should connect, listed in order of priority for routing traffic. HPE Aruba Networking Central then automates establishing the VPN.

For routing, admins simply choose which connected networks or routing protocols they want to redistribute into OAP. They can also redistribute OAP networks into a routing protocol (for example, to advertise branch networks in the data center). HPE Aruba Networking Central then automates programming BGWs' and VPNCs' routing tables.

Note that SD-Branch also supports NAT-T for cases when one or both gateways are behind a device that implements NAT.

For more information on setting up an EdgeConnect SD-Branch solution, refer to HPE Aruba Networking Central documentation.

Periodic Rekeying

Even though the centralized orchestrator distributes keys securely, best practices call for periodically rotating the keys. The more traffic that is encrypted by the same key, the more likely it is that the key could be cracked. By frequently changing the keys the solution maintains data security, as shown in Figure 17-43.

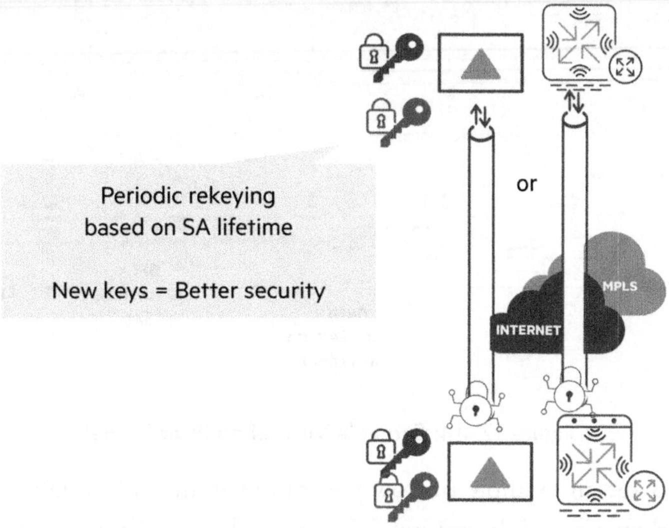

Figure 17-43: Periodic Rekeying

As of the release of this study guide, the default SA lifetime for EdgeConnect SD-Branch is 7200 seconds (2 hours). *Before* the old SA expires, HPE Aruba Networking Central creates a new SA and distributes the information and new keys for that SA to the tunnel endpoints. This process prevents traffic disruption.

For EdgeConnect SD-WAN, you can define both a rotation period, which defines how often the solution attempts to rotate keys, and a key lifetime, which defines the maximum length of time for keys. The lifetime must be at least three times the rotation period.

Viewing IPsec SAs on HPE Aruba Networking Central

When you use EdgeConnect SD-WAN or EdgeConnect SD-Branch, you do not need to configure granular IPsec settings. Instead, you simply specify which sites or gateways you want to establish tunnels. The centralized orchestration solution then handles the SA establishment.

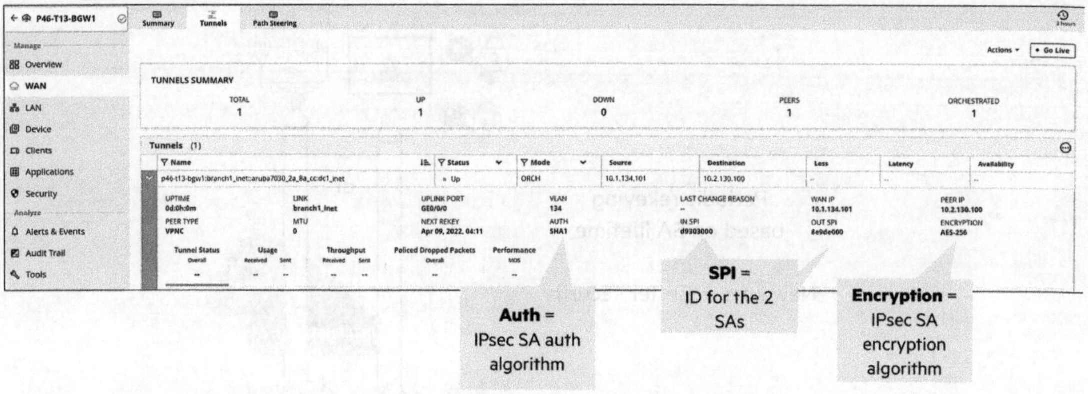

Figure 17-44: Example View of an IPsec Tunnel

In Figure 17-44, you see an example of how you can view the auto-established IPsec Sas within HPE Aruba Networking Central. In version 2.5.6, go to the specific gateway view and click **WAN**. Click the **Tunnels** tab. (As always, check the latest documentation.)

You will see a list of tunnels. You can expand each tunnel and see information about the SA such as the encryption and authentication algorithms in use. You can also see the SPIs that identify the two SAs for the tunnel.

End-to-End, Role-Based Segmentation

In addition to securing traffic flowing over the WAN, EdgeConnect SD-WAN and EdgeConnect SD-Branch implement the end-to-end, role-based segmentation required for zero trust networks. In other words, the solution does not treat traffic flowing over the WAN in the same way. Instead it adapts, enforcing the correct policies for each user and device.

As shown at the top of Figure 17-45, EdgeConnect SD-WAN uses VLANs implemented in the branch infrastructure to sort traffic into different segments. It maintains strict isolation between those segments. In this way, you can, for example, segment IoT traffic from traffic associated with sensitive applications. If you are using HPE Aruba Networking APs and switches at the branch site, they can receive roles for branch users and devices from HPE Aruba Networking ClearPass Policy Manager (CPPM). Those roles can help sort traffic into the correct VLANs associated with various segments on EdgeConnect SD-WAN.

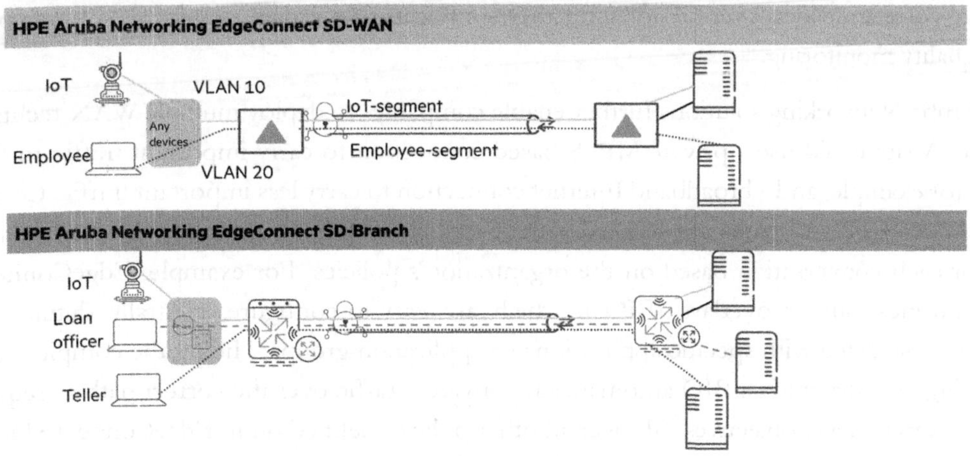

Figure 17-45: End-to-End, Role-Based Segmentation

As shown at the bottom of Figure 17-45, EdgeConnect SD-Branch even more tightly integrates with HPE Aruba Networking LAN solutions. As you have already examined in this study guide, AOS-CX switches can implement user-based tunneling (UBT) to BGWs, which then apply roles to that traffic. Similarly, APs can tunnel traffic to BGWs, which apply roles received from CPPM. (Or the APs can apply the roles locally and then the BGWs can apply roles again based on various rules configured on them.) In short, EdgeConnect SD-Branch extends the same role-based segmentation that you have studied throughout this study guide across the WAN.

Summary of EdgeConnect Benefits

In summary, EdgeConnect SD-WAN and EdgeConnect SD-Branch solutions use automation to simplify and optimize secure WAN connectivity.

An Internet-based WAN requires VPNs for security, but protocols like IPsec traditionally require extensive and complex setup. SD-WAN automates the establishment of IPsec tunnels over the internet. Then it continues to orchestrate the tunnels, making it easy for companies to deploy the solution and add new branches.

Throughout this study guide you have learned about the importance of end-to-end, role-based segmentation. Both EdgeConnect SD-WAN and EdgeConnect SD-Branch can enforce such segmentation for every user and device.

While this study guide focuses on security, you should understand other benefits of these solutions.

They optimize performance. For example, EdgeConnect SD-WAN technologies such as patch conditioning, WAN optimization, and SaaS optimization deliver premiere performance for even sensitive voice and video over broadband Ethernet. For its part, EdgeConnect SD-Branch provides patch quality monitoring.

HPE Aruba Networking solutions further enable companies to deploy multiple WAN technologies at a site. A site could use a private MPLS-based connection to carry important traffic to the data center, for example, and a broadband Internet connection to carry less important traffic. Companies can even use LTE. All paths are secured with IPsec. The solutions automatically choose the right path for each conversation based on the organization's policies. For example, EdgeConnect SD-WAN business intent overlays (BIOs), which are easy-to-configure, centrally defined logical networks associated with specific applications or application groups. Hiding the complexity of the underlying infrastructure, a BIO automatically forwards traffic over the correct path as required to meet the service level objectives (SLOs) and other policies defined on it. EdgeConnect SD-Branch uses dynamic path steering to select the right path for various types of traffic.

EdgeConnect SD-WAN appliances or EdgeConnect SD-Branch gateways can also output traffic destined to the Internet directly out the local Internet connection.

Both solutions offer centralized orchestration. EdgeConnect SD-WAN Orchestrator is deployed on-prem or accessed as a cloud service, and you can also integrate EdgeConnect SD-WAN with cloud-based HPE Aruba Networking Central. EdgeConnect SD-Branch is also managed from the cloud with HPE Aruba Networking Central. From that same cloud solution, admins can also manage any other HPE Aruba Networking components, such as switches and APs.

These benefits are summarized in Figure 17-46.

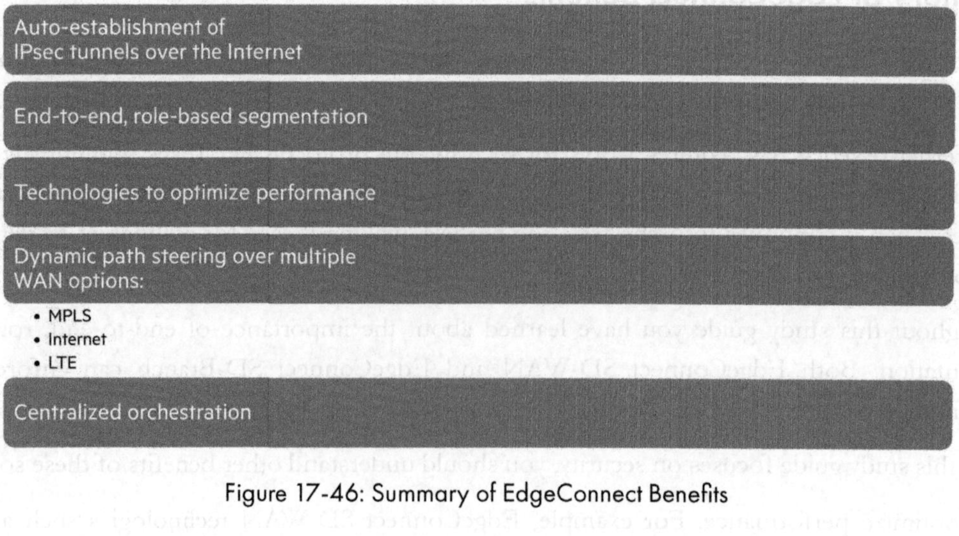

Figure 17-46: Summary of EdgeConnect Benefits

Summary

Congratulations! You have learned all about securing remote access with HPE Aruba Networking ZTNA, HPE Aruba Networking SWG, and HPE Aruba Networking CASB. You also learned how to secure branch access with HPE Aruba Networking EdgeConnect.

Learning checks

1. What is a key difference between HPE Aruba Networking ZTNA and a traditional client-to-site VPN?
   a. ZTNA uses the latest version of IPsec.
   b. ZTNA does not continuously monitor users' access.
   c. ZTNA uses on-prem firewalls rather than VPN concentrators.
   d. ZTNA grants access to individual apps rather than to the private network.
2. What is required to permit remote users access to your private apps through HPE Aruba Networking ZTNA?
   a. Deploying connectors that can reach those apps
   b. Installing agents on the application servers
   c. Registering your application servers' IP addresses with HPE Aruba Networking
   d. Integrating the applications with an IdP supported by HPE Aruba Networking ZTNA
3. Assume that a company has NOT deployed HPE Aruba Networking EdgeConnect. What is one requirement that applies to BOTH HPE Aruba Networking SWG and HPE Aruba Networking CASB?
   a. Clients have agents installed on them.
   b. Users authenticate to the built-in HPE Aruba Networking SSE IdP.
   c. A connector is deployed at the customer's private site.
   d. An external Web profile with Threat Intelligent Protection is specified in the policy rules.

4. Which type of EdgeConnect SD-Branch topology do you need to add if you want VPNCs to create tunnels to other VPNCs?
   a. Hub-spoke
   b. Branch-mesh
   c. Hub-mesh
   d. Spoke-mesh

Note

For answers to all learning checks, refer to the Appendix.

18 Configure HPE Aruba Networking Gateway IDS/IPS

EXAM OBJECTIVES

✓ Describe HPE Aruba Networking 9x00 Gateways

✓ Define and apply UTM policies

Assumed Knowledge

- Basics of HPE Aruba Networking solution architectures

Overview

HPE Aruba Networking 9x00 Gateways provide a Unified Threat Module (UTM), which helps them further protect your network from malware, ransomware, denial of service (DoS) attacks, and many other threats.

This chapter explains exactly what HPE Aruba Networking gateway IDS/IPS is, how it works, and how to configure it. After reviewing 9x00 Gateway security features, you will learn about the difference between an Intrusion Detection System (IDS) and an Intrusion Prevention System (IPS). You will also look at the gateway IDS/IPS ruleset and IDS/IPS policies. At the end the first section, you will take a high-level look at how gateway IDS/IPS works. Then you will spend the rest of the chapter learning how to configure gateway IDS/IPS to meet practical needs. You will explore considerations for choosing between IDS and IPS mode. You will also how to manage rules to maintain security while supporting productivity, as well as how to integrate gateway IDS/IPS with a third-party Security Information and Event Management (SIEM) solution.

HPE Aruba Networking Gateway IDS/IPS

In this section, you will learn about HPE Aruba Networking Gateway IDS/IPS.

HPE Aruba Networking 9x00 Gateway Security Features

As indicated in Figure 18-1, HPE Aruba Networking 9x00 Gateways support the AOS security features with which you are already familiar, including a role-based stateful firewall that provides deep packet inspection (DPI) and Web content control (WebCC). In this chapter, you will focus on the integrated Intrusion Detection and Prevention System (IDS/IPS) capabilities, which further protect the corporate WAN against ever-evolving threats.

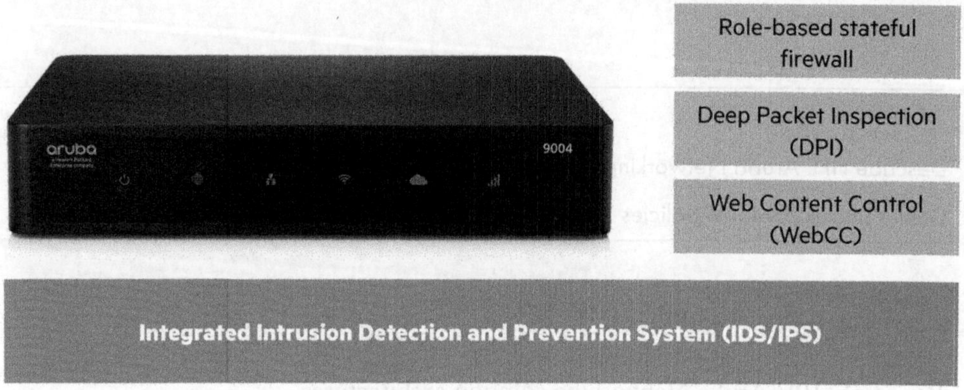

Figure 18-1: HPE Aruba Networking 9x00 Gateway Security Features

HPE Aruba Networking mobility solutions support Wireless IDS/IPS (WIDS/WIPS), which detects RF and 802.11-specific threats. This chapter focuses on gateway IDS/IPS, which detects a variety of different types of threats in traffic that flows through the gateway.

IDS and IPS

An IDS, as the name implies, focuses on detecting threats. In this chapter, you are focusing on network-based IDS. A network IDS monitors traffic that flows through it for signs of malicious activity. If it detects a threat, it generates a threat event, but does not take action to prevent the threat. It is up to admins to respond to the threat and determine how to mitigate it.

An network IPS also monitors traffic that flows through it, and, if it detects a threat, it generates a threat event. But it goes beyond detection to mitigate threats. In the case of HPE Aruba Networking gateways, the IPS mitigates threats by dropping packets associated with the threat.

View Figure 18-2 for a comparison of IDS vs IPS.

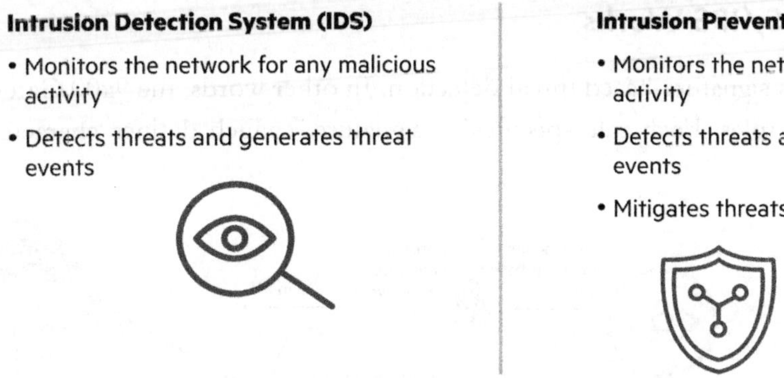

Figure 18-2: IDS and IPS

Licensing for Gateway IDS/IPS on 9x00 Gateways

To enable gateway IDS/IPS features on 9x00 Gateways, you must purchase one of these licenses:

- Foundation with Security
- Advanced with Security

These licenses provide the same features as either the Foundation or Advanced license, including SD-WAN. However, as shown in Figure 18-3, they add gateway IDS/IPS, antimalware, and the Security Dashboard. You choose between the Foundation and Advanced options based on your other requirements. For example, the Advanced license provides SaaS Express.

Figure 18-3: Licensing for Gateway IDS/IPS on 9x00 Gateways

How Gateway IDS/IPS Works

Gateway IDS/IPS uses signature-based threat detection. In other words, the 9x00 Gateway checks traffic against a set of rules. Each rule specifies a "signature," which defines characteristics of a particular threat.

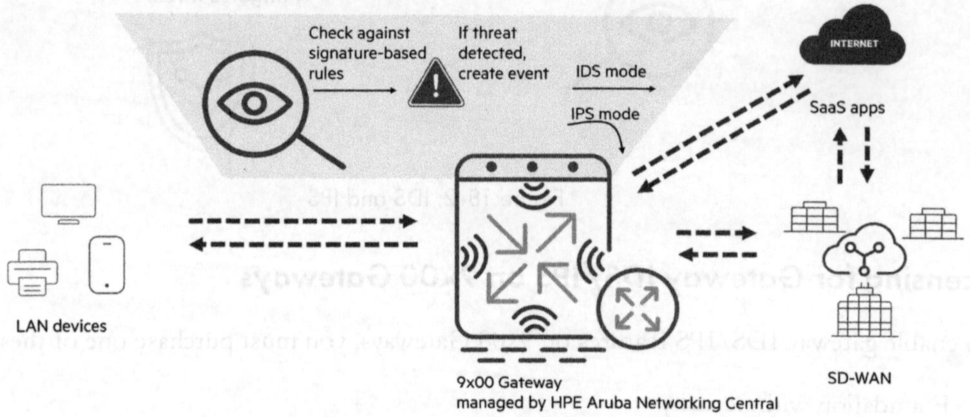

Figure 18-4: How Gateway IDS/IPS Works

As shown in Figure 18-4, the gateway inspects all traffic that flows through it, including traffic from internal clients, traffic received over the SD-WAN, and traffic from external devices on the Internet. When the gateway notices that a packet matches a signature in one of its rules, it takes action. If it is operating in IDS mode, it creates a threat event to inform admins of the detected threat. In IPS mode, the gateway creates a threat event and also drops the traffic.

Gateway IDS/IPS Ruleset

Gateway IDS/IPS uses a ruleset with tens of thousands of signature-based rules. HPE Aruba Networking also updates the rules frequently, and you can configure your gateways to automatically download the updates.

The gateways can detect threats in many categories. For example, they can detect traffic to and from compromised hosts, based on a private, reliable list of known compromised hosts that is updated daily. They can detect Bitcoin mining, exploit kit activity, malware, and mobile-device specific malware. They can detect signs of phishing or Trojans, as well as many other vulnerability exploits, attack types, and malware types.

For a complete list of threat categories, visit:

https://www.arubanetworks.com/techdocs/central/latest/content/sd-branch/idps/appendix/appendix.htm

Gateway IDS/IPS Policies and CVSS

Each of the threat categories includes many rules. You configure your gateway to use particular rules by selecting one of three simple policies for IDS or for IPS. The lenient policies detect or prevent only the most severe threats. The moderate policies detect or prevent threats of a moderate severity or higher. And the severe policies detect or prevent the broadest range of threats.

HPE Aruba Networking uses industry standards to rank the severity of each threat. As you may recall from Chapter 13, the security industry uses the Common Vulnerability Scoring System (CVSS) to assign known vulnerabilities or threats a score from 1.0 to 10.0, 1.0 being the least severe and 10.0 being the most severe. Each rule in the gateway IDS/IPS ruleset has a CVSS, based on the vulnerability or threat associated with that rule.

As shown in Figure 18-5, if you select the IDS Lenient or IPS Lenient policy, the gateway applies all rules from the current year and previous two years with a CVSS of 8.0 or higher. The IDS Moderate or IPS Moderate policy applies rules from the current year and previous two years with a CVS of 6.0 or higher. And applying the IDS Strict or IPS Strict policy causes the gateway to use rules from the current year and previous three years that have a CVSS of 4.0 or higher.

IDS/IPS Policy	Use rules	
	with this CVSS	from these years
IDS Lenient IPS Lenient	>8.0	Current and previous 2
IDS Moderate IPS Moderate	>6.0	Current and previous 2
IDS Strict IPS Strict	>4.0	Current and previous 3

Figure 18-5: Gateway IDS/IPS Policies and CVSS

Gateway IDS/IPS Configuration

This section explains how to configure Gateway IDS/IPS.

Overview of Gateway IDS/IPS Configuration

To configure IDS/IPS on 9x00 Gateways, first enable traffic inspection, which provides visibility into the packets that move through the gateways. Then update the gateway IDS/IPS rulesets to ensure that gateways use the latest signatures to detect threats. After that, choose between IDS or IPS for the inspection mode. You then act on that choice by configuring IDS or IPS.

These steps are summarized in Figure 18-6.

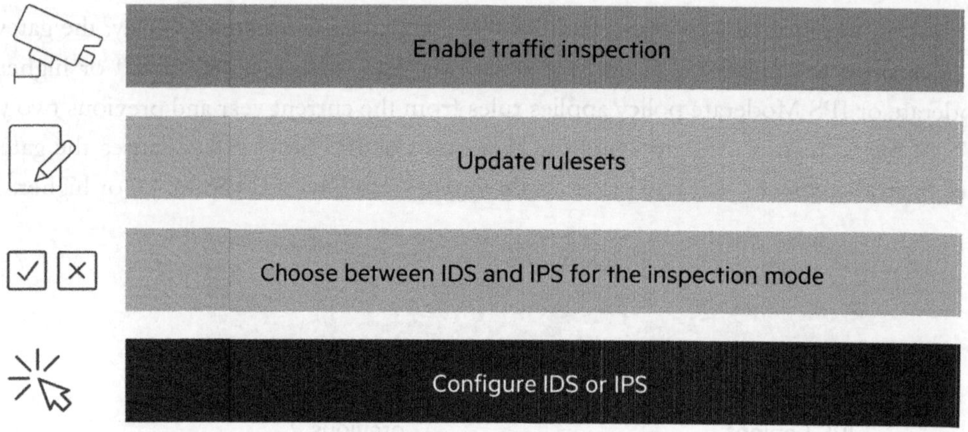

Figure 18-6: Overview of Gateway IDS/IPS Configuration

Enable Traffic Inspection

To enable traffic inspection on a gateway, you must first ensure that you have an active gateway subscription with the Security license discussed previously. And you must have successfully onboarded the gateway in HPE Aruba Networking Central.

As indicated in Figure 18-7, you can configure the gateway IDS/IPS settings at the group level or the individual device level, but the group level is typically preferred. In HPE Aruba Networking Central 2.5.5, you find these settings at **Security > Gateway IDS/IPS**. You then click the **Config** icon. In the **General** tab, select the **Enable traffic inspection** check box. (As always, check the latest documentation for up-to-date steps.)

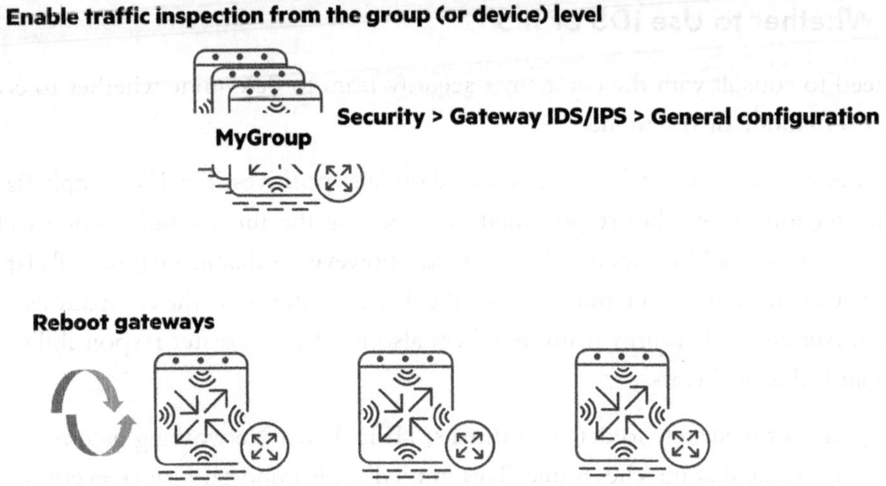

Figure 18-7: Enable Traffic Inspection

Check the current version to see whether you must reboot the gateway after you first enable traffic inspection; if you do need to, plan to set up IDS/IPS during a scheduled outage.

Update Rulesets

You can update the IDS/IPS ruleset to ensure that the gateways are using the latest signatures. As you see in Figure 18-8, you can perform this update from the **Security > Gateway IDS/IPS > General** settings.

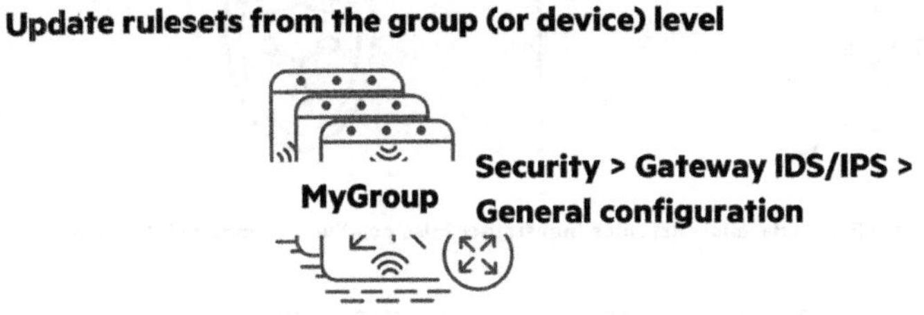

Figure 18-8: Update Rulesets

When you first enable gateway IDS/IPS, you should update the ruleset manually. Do so from the same page in which you enable traffic inspection.

You should generally also enable automatic updates to ensure that your gateway learns how to detect new threats. Select the option for automatically updating the ruleset. You then choose when the periodic update occurs by selecting **Week** or **Day**. Then choose the day of the week and time, for a weekly update, or the time of day, for a daily update, and save your settings.

Choose Whether to Use IDS or IPS

You will need to consult with the company's security team to determine whether to configure the gateways in IDS mode or IPS mode.

IDS mode minimizes the risks of disruption based on false positives. The IDS simply flags potential threats. The security team takes responsibility for viewing the threats and making well-informed decisions about how to address them. The downside, however, is that more time will elapse between the gateway detecting a threat and mitigation of the threat. In this time, the company can be exposed to ill effects. Admins and security team members also must take greater responsibility to monitor the system and address threats.

An IPS responds immediately to detected threats; HPE Aruba Networking gateways operating in IPS mode drop traffic that matches a rule. This immediacy is important, since every second counts when defusing an active threat. However, if the policy is too strict for the company's needs, IPS may react to valid traffic as if it is a threat, interfering with legitimate business operations. These advantages and disadvantages are summarized in Figure 18-9.

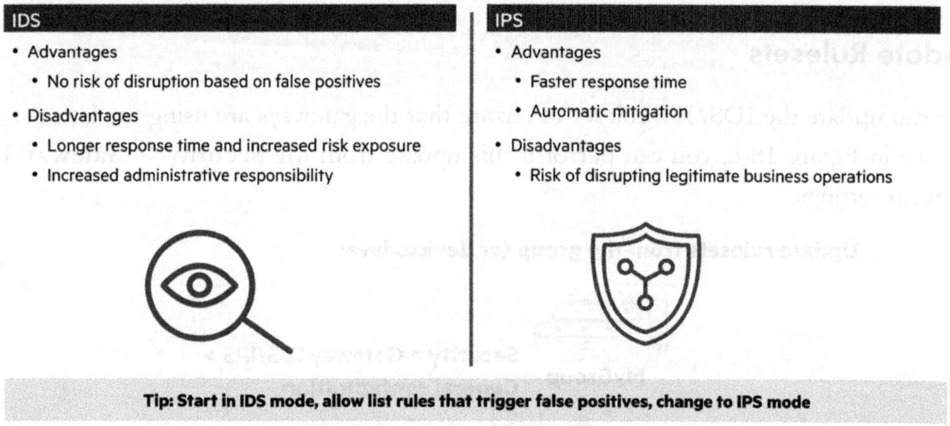

Figure 18-9: Choose Whether to Use IDS or IPS

Before you enable IPS mode, you might enable IDS mode for a period of time. Admins can then monitor what threats the gateways detects and determine which threats are real and which detect vulnerabilities that the company might want to ignore based on its own needs and requirements. You can then allow list the specific rules that are flagging false positives. When the security team is confident that all necessary traffic is being allowed, you can change the mode to IPS.

Configure IDS

You will now learn how to configure IDS, if you have determined that IDS is the best fit for a company.

Again you should be at the group view and the **Security > Gateway IDS/IPS** configuration settings. (Alternatively, you can configure the settings at the device level, but this is less typical.) Within the general settings, as shown in Figure 18-10, select IDS for the mode.

You must also select the IDS policy. Click the **Policies** tab. When IDS mode is enabled, you can choose from the three IDS policies (IDS Lenient, Moderate, and Strict), and the IPS policies are grayed out. As you learned earlier, the policy determine which rules the gateway IDS/IPS engine uses to inspect traffic. The IDS Lenient policy has the fewest rules, and the IDS Strict policy, the most.

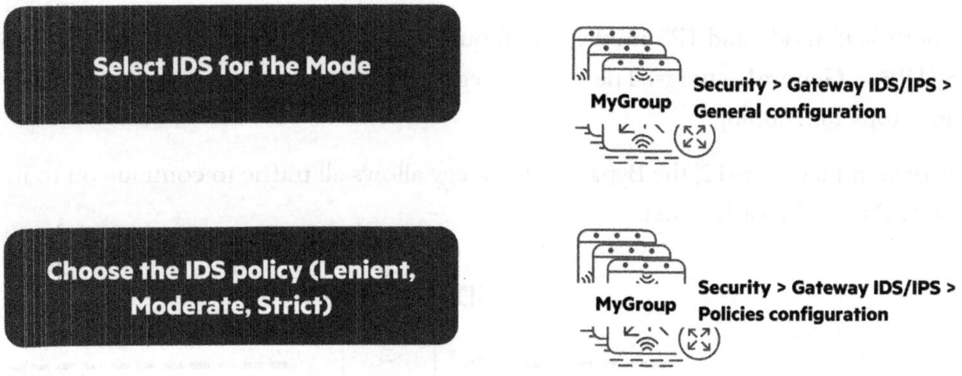

Figure 18-10: Configure IDS

Configure IPS

Configuring IPS is very similar to configuring IDS. As shown in Figure 18-11, you simply select IPS for the mode in the general settings. Then you click the **Policies** tab and select the IPS Lenient, IPS Moderate, or IPS Strict policy. Keep in mind that the Lenient, Moderate, and Strict policies have the same characteristics—concerning ruleset timeframes and CVSS scores—as the IDS policies.

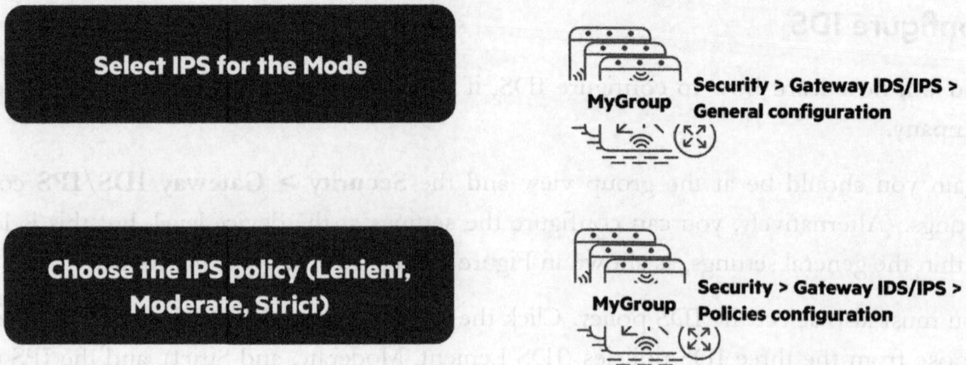

Figure 18-11: Configure IPS

Select the Gateway IDS/IPS Fail Strategy

For both IDS mode and IPS mode, you should also configure the Fail Strategy in the **Gateway IDS/IPS > General** settings. The Fail Strategy determines how the gateway behaves if its IDPS engine stops functioning.

As shown in Figure 18-12, the Bypass fail strategy allows all traffic to continue on to its destination in the LAN, WAN, or Internet.

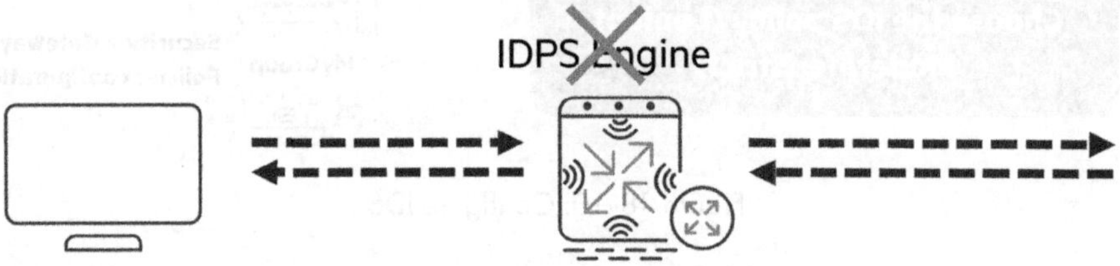

Figure 18-12: Bypass Fail Strategy

The Block strategy, on the other hand, prevents all traffic flow if the IDPS engine goes down (see Figure 18-13). This holds true even if the Inspection Mode is IDS.

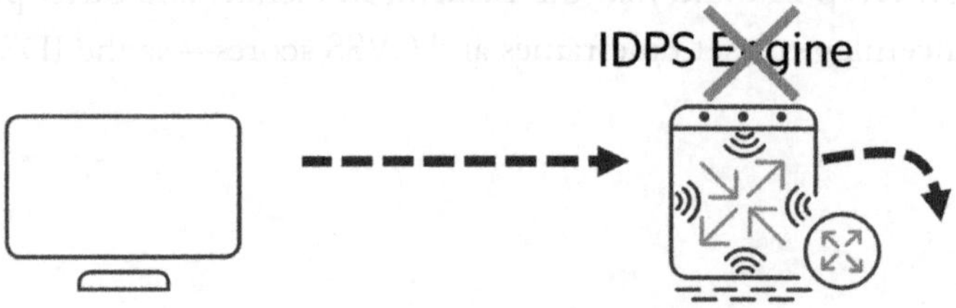

Figure 18-13: Block Fail Strategy

Most companies will prefer the Bypass strategy to prevent complete disruption if a single security system goes down. However, for some use cases that require the utmost security at all times, such as certain banking applications or classified government intelligence, the Block strategy is appropriately severe.

View Rules and Allow List Rules

Sometimes you want to allow list specific rules because your security team has investigated them and determined that the traffic flagged by these rules does not pose a significant risk to the company. When a rule is allow listed and the gateway detects a threat based on that rule, the gateway still creates a threat event, but the event indicates that the traffic is allow listed. If the gateway is in IPS mode, it does *not* drop the traffic.

Under the **Policies** tab, for both IDS and IPS, you can click on the name of the policy in effect (Lenient, Moderate, or Strict) to see a list of the rules that are included in that policy.

You can then select one or more rules and move them to the Allow List, as shown in Figure 18-14.

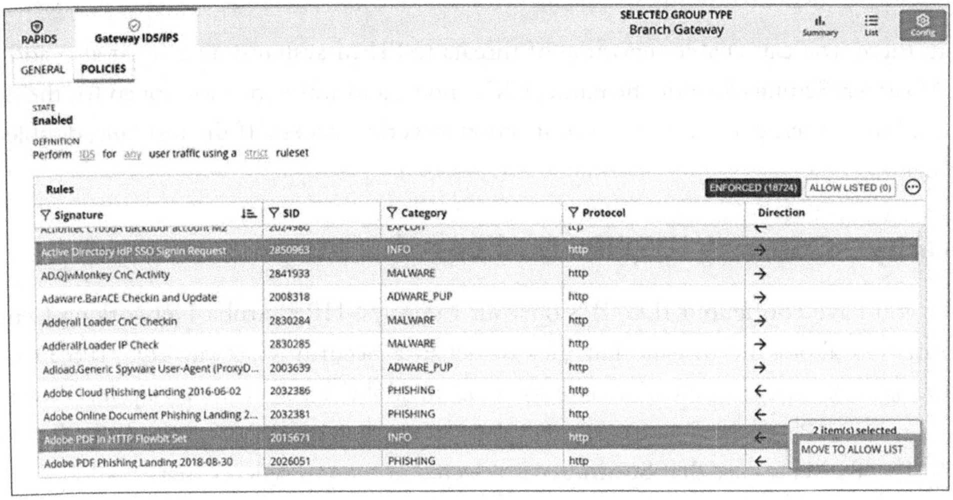

Figure 18-14: Example List of Gateway IDS/IPS Rules Associated with a Policy

Send Data to a SIEM

As shown in Figure 18-15, an 9x00 Gateway can send threat event data to a third-party Security Incident and Event Management (SIEM) server, such as Splunk. SIEM provides a holistic picture of the security posture of the organization by aggregating and correlating data from disparate sources in the network. Admins can then perform advanced analysis and generate reports. A SIEM often provides automated alerts and workflows. For example, it might generate help tickets that notify admins of issues and give them the information they need to start addressing the problem.

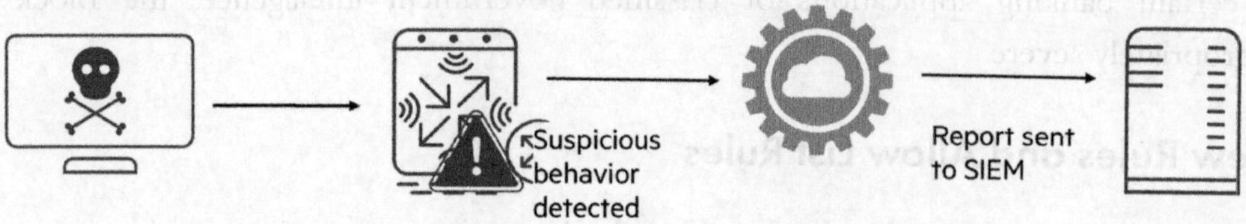

Figure 18-15: Send Data to a SIEM

Before configuring SIEM integration on the gateways, check to make sure you have an active subscription with a third-party SIEM provider. You will also need to know the server URL, an index number, and an authentication token.

You enable reporting to SIEM systems in the **Global** view under Gateway IDS/IPS configuration settings. In HPE Aruba Networking Central 2.5.5, set the filter to **Global**. Under **Manage**, navigate to **Security >Gateway IDS/IPS**. Click the **Config** icon to open the **Gateway IDS/IPS** configuration page. Go to the **SIEM** tab.

From there, you can enable reporting of threats to SIEM systems. You must also add at least one SIEM server. Settings include the name, URL, index, and authentication token for the server. While you add the server, you can test the connection to verify success. If the test fails, double-check your settings.

Security > Gateway IDS/IPS Dashboard

After you have configured IDS/IPS on your gateways, HPE Aruba Networking Central collects information about the threats that they detect in a Security Gateway IDS/IPS Dashboard (see Figure 18-16). You can view this dashboard at the global level, at the group level for a group that includes IDS/IPS-enabled gateways, or at the gateway level. Go to Security and select the Gateway IDS/IPS tab. Make sure that **Summary** is selected in the top right corner.

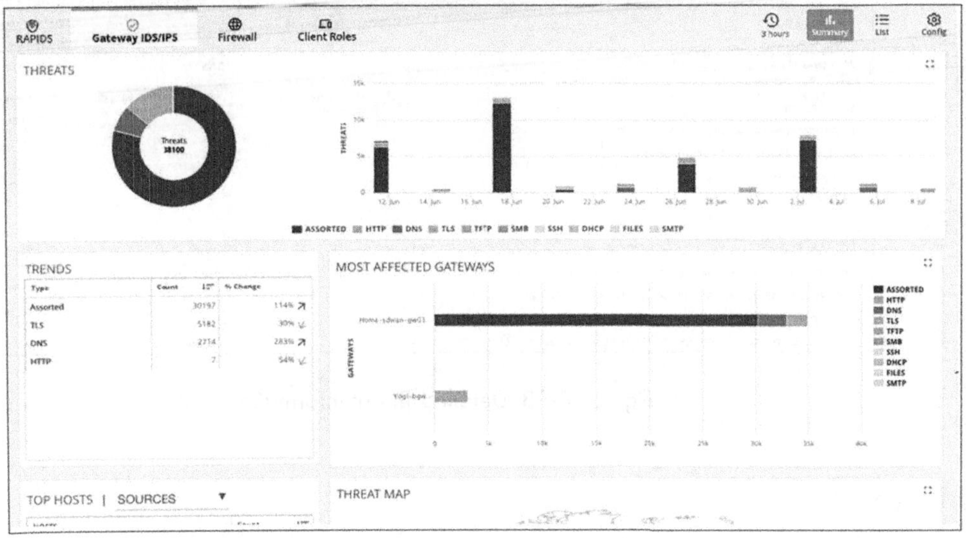

Figure 18-16: Security > Gateway IDS/IPS Dashboard

As you see, the dashboard provides an overview of the detected threats, breaking them down by type. It further presents trends, allowing you to see if the types of detected threats are changing over time. The **Most Affected Gateways** section lets you focus on the gateways finding the most threats while the **Top Hosts** section lets you see the top sources triggering the threats.

Example Threat Details

You can use the list view to see a list of detected threats. Then you can choose to see more details about a particular threat, as shown in Figure 18-17.

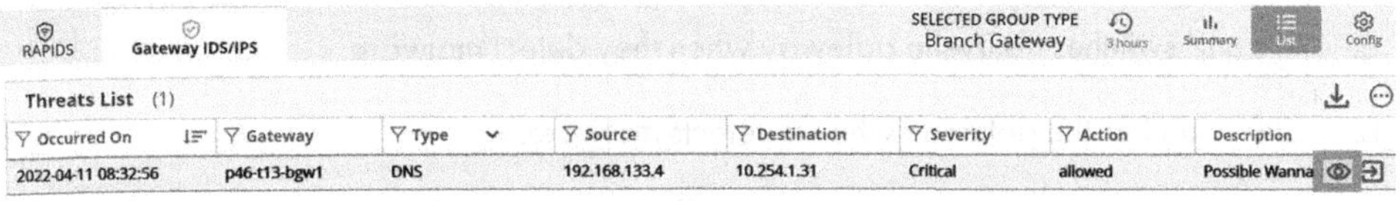

Figure 18-17: Threat List

In the detailed view, shown in Figure 18-18, you can see the detected signature and the relevant threat category for that signature. The page also displays the source IP address for the malicious traffic and its destination IP address, as well as a deeper dive into the exact contents of the traffic.

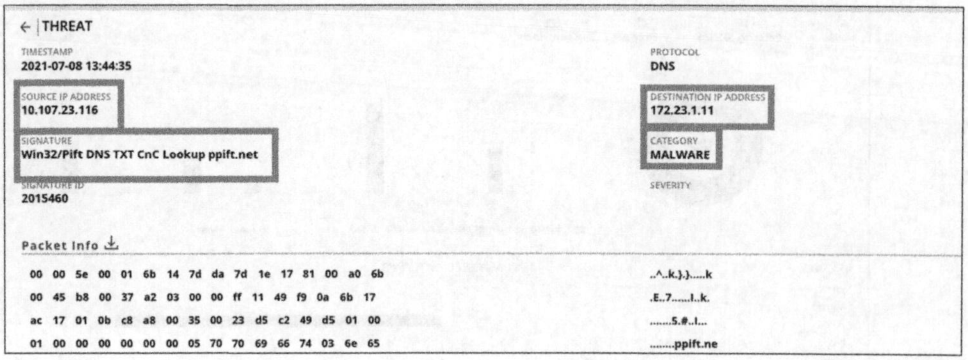

Figure 18-18: Detailed Threat Information

Summary

You began this chapter by learning about HPE Aruba Networking gateway IDS/IPS from a conceptual standpoint. You reviewed 9x00 Gateways, distinguished between IDS and IPS, and reviewed licensing for gateway IDS/IPS. You also learned how the gateway IDS/IPS engine uses rulesets and the actions that the engine can apply.

Then you learned how to configure IDS, how to configure IPS, how to manage specific rules, and how to send data to a SIEM.

Learning checks

1. How does HPE Aruba Networking gateway IDS/IPS detect malicious traffic?
   a. APs and switches notify the gateway when they detect malware.
   b. CPPM notifies the gateway when it detects malware.
   c. The gateway detects signatures in the traffic that indicate the presence of threats.
   d. The gateway detects anomalies in the traffic that differ from usual traffic patterns.
2. For which use cases should you recommend IDS, as opposed to IPS? (Consider your answer to this open-ended question.)

3. Why might a customer integrate gateway IDS/IPS with a SIEM server?

   a. To allow traffic to continue through a gateway if the IDPS engine goes down

   b. To prevent a gateway from forwarding traffic if the IDPS engine goes down

   c. To integrate gateway IDS/IPS with the company's automated security event management

   d. To ensure that gateways have the most up-to-date rules

Note

For answers to all learning checks, refer to the Appendix.

---

19 Use HPE Aruba Networking Central Alerts

EXAM OBJECTIVES

- ✓ Investigate HPE Aruba Networking Central alerts
- ✓ Recommend action based on the analysis of HPE Aruba Networking Central alerts

Overview

HPE Aruba Networking Central alerts give admins insights into the network from a centralized source. With timely notification of potential issues, admins can take action to secure the network.

This chapter provides an overview of the alerts supported in HPE Aruba Networking Central. You will learn about the notification options and how to configure the precise options that you want for each alert.

HPE Aruba Networking Central Security Alerts

This section explains how to configure and use HPE Aruba Networking Central security alerts.

HPE Aruba Networking Central Alert Overview

Throughout this study guide, you have learned about how to configure several features related to detecting and mitigating threats. However, admins are not constantly in network management interfaces, checking on functions. They need to be alerted to the most critical issues as those issues occur so that they can take timely action.

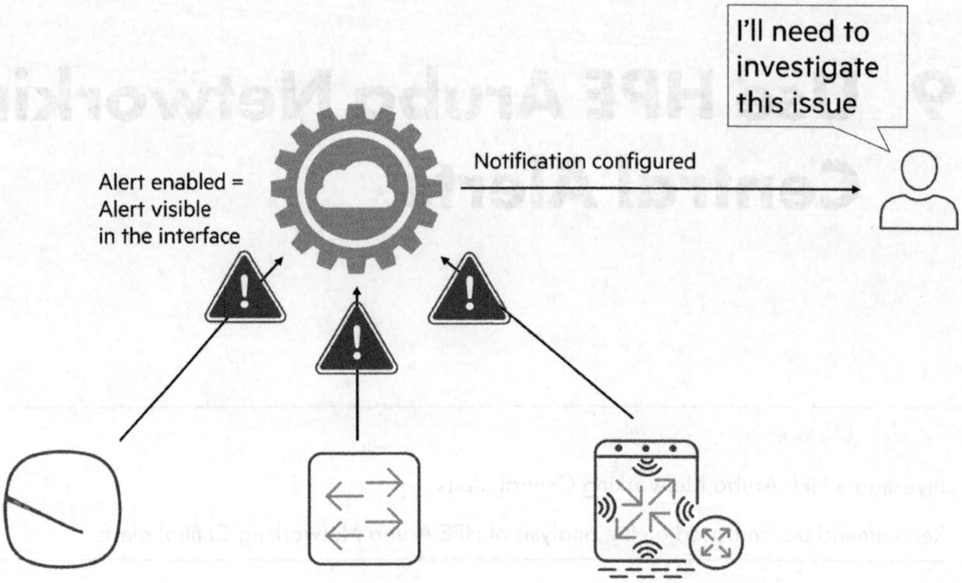

Figure 19-1: HPE Aruba Networking Central Alert Overview

HPE Aruba Networking Central collects alerts from the APs, switches, and gateways that it manages. Admins can look for the alerts in the cloud portal. But perhaps even more critically, HPE Aruba Networking Central can send notifications in response to alerts, calling the most important issues to admins' immediate attention, as shown in Figure 19-1. Using the alert as the starting point, admins can drill down into the details and begin an investigation.

HPE Aruba Networking Central alerts help admins to track a variety of issues, but this study guide will focus on the security-related ones.

HPE Aruba Networking Central Alert Types

HPE Aruba Networking Central groups alerts into seven categories:

- **User**—User alerts relate to admin accounts on HPE Aruba Networking Central.
- **Access Point**—HPE Aruba Networking Central gathers these alerts from managed APs. Some of these alerts deal with general connectivity while others, like Rogue AP Detected, refer to security events.
- **Switch**—HPE Aruba Networking Central gathers these alerts from managed switches. Again, the alerts deal with a variety of issues from connectivity issues to security ones.

- **Gateway**—HPE Aruba Networking Central gathers these alerts from managed gateways. They report on issues like gateway connectivity and WAN health, as well as IDS/IPS functions.
- **Connectivity**—These alerts report on client connectivity and authentication.
- **Audit**—These alerts trigger when configurations change.
- **Site**—Site alerts reduce the number of device disconnect alerts when a widespread issue such as a site outage occurs. For example, if site alerts are configured with **Severity** as Major, **Duration** as 10 minutes, and **Site** as site1, a single alert saying "Aggregated Device Disconnects" is raised on the user interface for every set of device belonging to "site1" that goes down within 10 minutes of the first DOWN event, limited to 100 devices per alert.

These categories are shown in Figure 19-2.

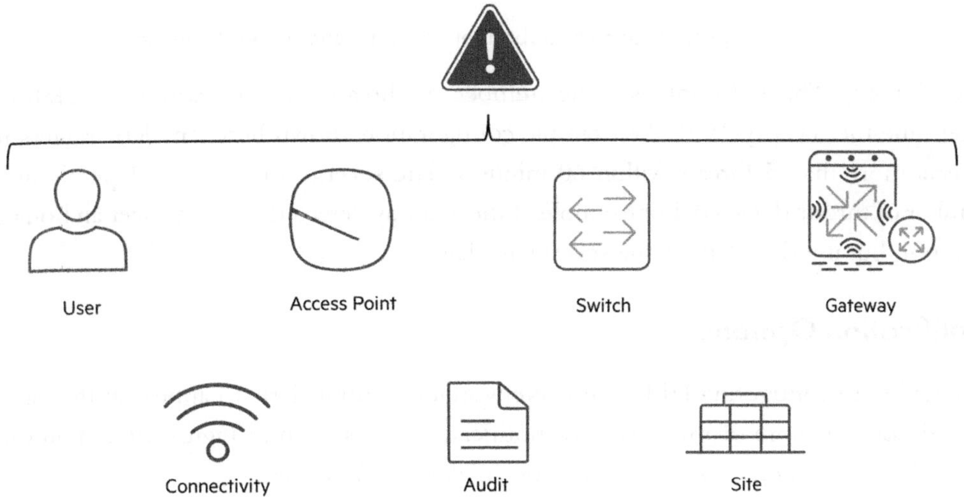

Figure 19-2: HPE Aruba Networking Central Alert Types

HPE Aruba Networking Central Alert Severities

Each alert has a severity level. From least to most severe, the four levels are: Warning, Minor, Major, and Critical.

Some alert types, such as Rogue AP detected, have a single severity assigned to any alert of that type, as shown on the left of Figure 19-3. For other alert types, such as Gateway Threat Count, you can configure multiple severities, as shown on the right of Figure 19-3. In that case, each severity is associated with a different condition.

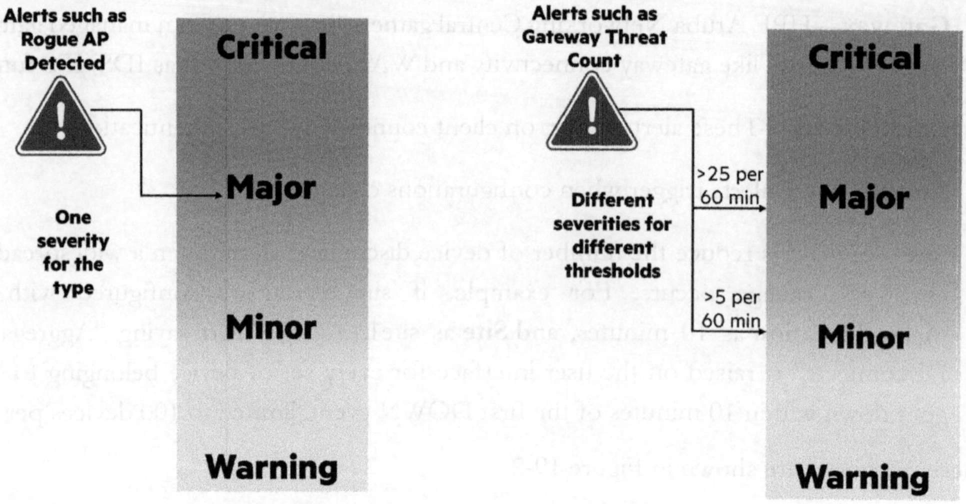

Figure 19-3: HPE Aruba Networking Central Alert Severities

The Gateway Threat Count uses the number of threats detected within a duration period to determine the severity. With the example configuration shown here, an alert triggers if a gateway detects more than 5 threats within 60 minutes. The severity of the alert depends on the precise number of threats detected. For example, if the gateway detected 6 threats over an hour, the severity is Minor, but if it detected 30, the severity is Major.

Notification Options

In addition to configuring HPE Aruba Networking Central alerts to appear in the dashboard, you can add one or more of three types of notifications, as shown in Figure 19-4. You can select the particular notification options to enable on an alert-by-alert basis.

Figure 19-4: Notification Options

Email notifications send an email to one or more recipients. You should set up email alerts for the most critical alerts to which admins need to respond immediately. For example, if an alert indicates an immediate threat to the company's services or data, admins need to know about it as soon as possible so that they can address the problem before the business is disrupted or data is stolen.

The Webhooks option enables you to integrate the HPE Aruba Networking Central alerts with third-party applications such as ServiceNow and Zapier, which are systems for automating IT ticketing and workflows. In this way, the company can apply the same processes that it uses to automate response to other events to the events occurring within the network.

The streaming notification option similarly enables integration with third party applications. However, this option provides more flexibility. Customers can program their choice of applications to receive data that is published to an API streaming interface.

Configuration of streaming or Webhooks is an advanced topic outside the scope of this study guide.

Configure HPE Aruba Networking Central Alerts

You configure all alerts settings at the Global view. As shown in Figure 19-5, those settings then apply to devices in all groups and at all sites.

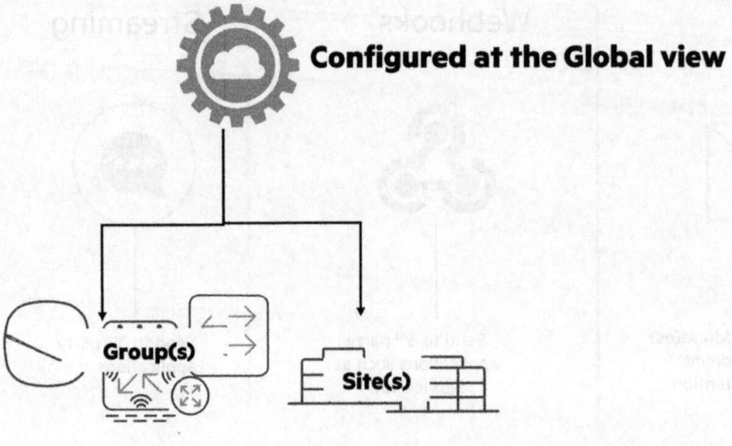

Figure 19-5: Configure HPE Aruba Networking Central Alerts

In HPE Aruba Networking Central 2.5.5, you go to the **Alerts & Events** page and click **Config** in the top right corner. You configure the settings in the **Alerts** tab. (As always, check the latest documentation for updates.)

Configure the Default Email Recipient

While you can configure different email addresses for different alert types, you often want to send many of the alerts to the same addresses. As shown in Figure 19-6, you can configure a default recipient or recipients, which you can then easily reference in individual alert settings. You add the default recipient from the Global Alert configuration settings. You can specify one or more email addresses.

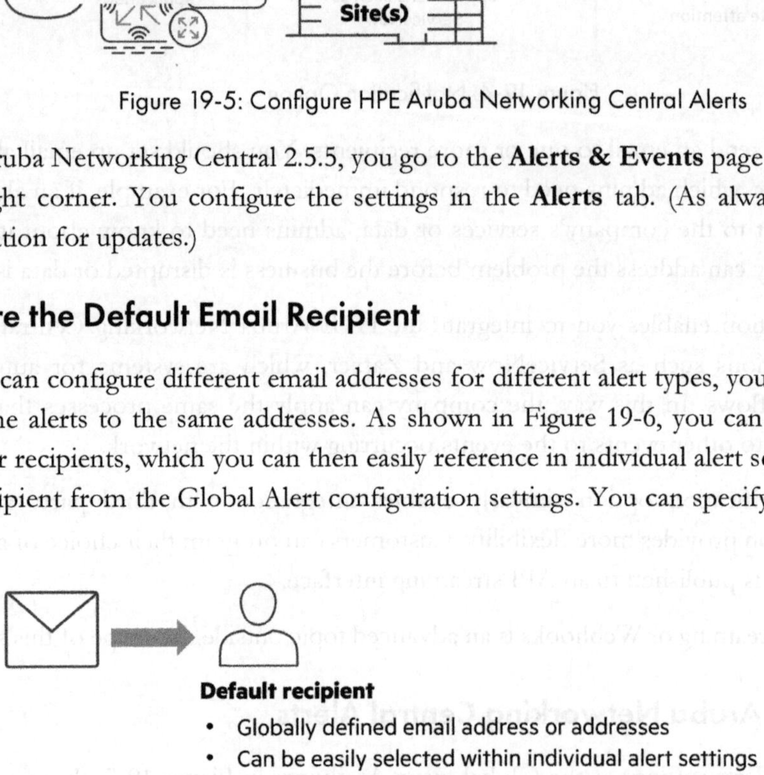

Figure 19-6: Configure the Default Email Recipient

Configuring Site-Specific Email Notifications

HPE Aruba Networking Central also lets you configure site-specific email addresses for notifying alerts. As shown in Figure 19-7, when an alert is generated for a specific site, HPE Aruba Networking Central will automatically send an email notification to the addresses configured for that site.

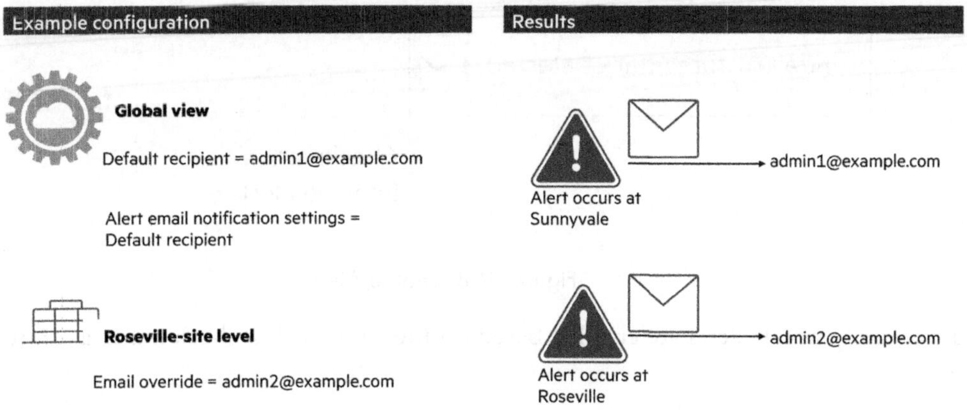

Figure 19-7: Configuring Site-Specific Email Notifications

You configure the site-specific notifications from the particular site's Alerts configuration settings. (In version 2.5.5, you go to the **Site** view. Go to the **Alerts & Events** and click the **Config** icon. Then select the **Alerts** tab.) You will see email configuration override settings, in which you can add email address or addresses for admins at this side.

The site-specific email addresses will override email destinations specified from global alert configuration pages. For example, you might set up an alert to send emails to the default recipient, which is globally set to admin1@example.com. You also set up an email override in the Roseville site to admin2@example.com If the alert occurs in most sites, an email is sent to admin1@example.com. But if the alert occurs at Roseville, an email is sent to admin2@example.com.

Enable Alerts

You enable alerts at the Global level. Select an alert category and click the + icon for each alert that you want to show in the HPE Aruba Networking Central interface. The alert is enabled with the default settings. At the default settings, when the alert is triggered, the interface shows it in the alerts list with the default severity for that alert. However, notifications are not enabled by the alert by default.

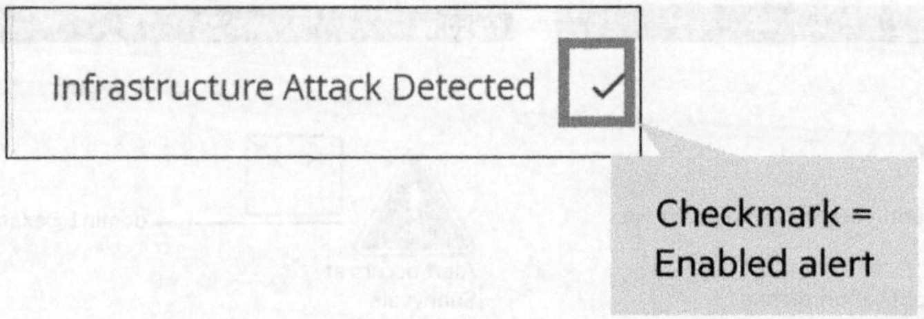

Figure 19-8: Enable Alerts

You can verify which alerts are enabled based on the alerts with a check mark, as shown in Figure 19-8.

If you want to change the alert settings, click the name of the alert.

Configure Alerts Settings

In the page that appears, you can change the severity for the alert. For some alerts, you can select multiple severities and set the threshold with which each severity is associated. The **Duration** field refers to the length of time over which the threshold is calculated. In the example shown in Figure 19-9, if the gateway detects more than 25 threats in 5 minutes, a major alert is triggered.

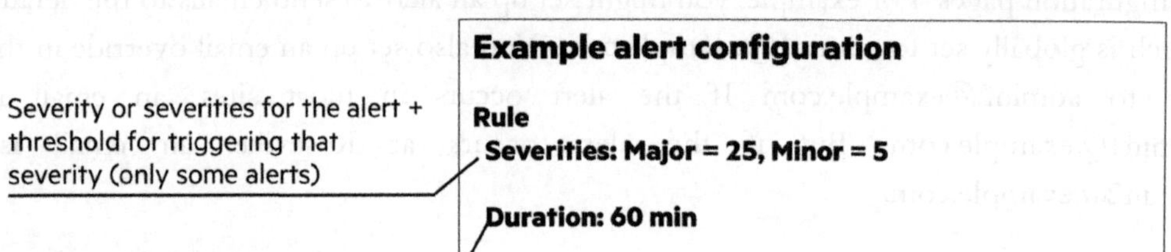

Figure 19-9: Configure Alerts Settings

By default, enabling the alert means that the alert triggers based on events on devices in any group, with any label, and at any site. You can use the device filter settings to change that behavior and, for example, enable an alert for some sites, but not others. As soon as you add one group, label, or site to the filter, then the alert can only be triggered by devices in that group, label, or site.

You can also configure one or more notification options. When you enable the email option, you must specify the recipients or, in other words, the destination email addresses. Selecting the default recipient sends an email to all the recipients list. You can add other recipients, separating their addresses with a comma. If you enable Webhook notifications, you must select a pre-configured Webhook, which defines the URL for the third-party solution.

Useful AP Security Alerts

Earlier in this study guide you learned about configuring WIDS/WIPS on HPE Aruba Networking wireless devices. When the APs are managed by HPE Aruba Networking Central, you can set up alerts and notifications to inform admins when APs detect any of the attacks enabled under IDS. These alerts fall into the three categories shown in Figure 19-10. When deciding for which alert categories to enable email notification, take into consideration how likely the alert is to indicate a serious threat. For example, you might enable an email notification for detecting a rogue AP, but not for all client attacks.

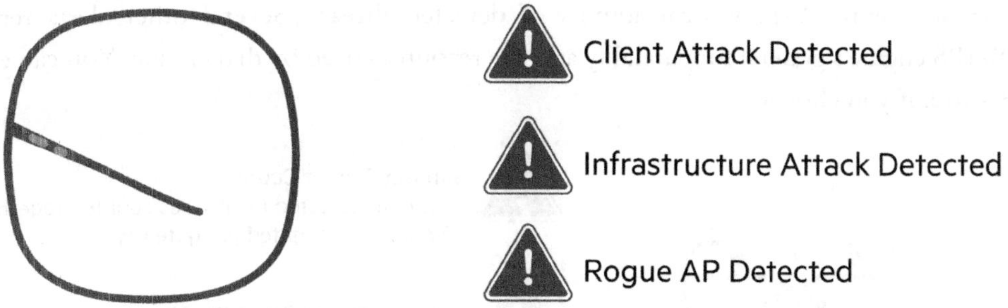

Figure 19-10: Useful AP Security Alerts

Note that these alerts are only available if the account supports WIDS/WIPS.

Follow Up on AP Security Alerts

When you receive an AP security alert, you should follow up on the issue.

To gain more information, go to the **Security > RAPIDS** page. There you can see the security events and expand for more information. Record some information about the attack. First, what is the threat type, and is this likely to be a true threat or a false positive. Earlier in this study guide, you received guidelines about which IDS settings to enable and which IDS settings are likely to cause false positives. Refer back to these guidelines as you assess whether a particular type of client or infrastructure attack poses a real threat.

Once you have determined that the event indicates a true threat, you should record information about the client or radio that triggered the alert. The event reports the station MAC address. Then take steps to locate and remove the device. In locating the device, you can consider the detecting devices. For example, if an AM on the north side of floor 1 detected the threat, you know that the offending device was in that general area. If you have uploaded floorplans into HPE Aruba Networking Central and placed your APs and AMs on them, then those floorplans can further help you find the location of the threat.

You should then visit that location and investigate further. Talk with employees and admins. It could be possible that users introduced a threat without understanding the issue (such as a rogue AP). Or if the threat truly comes from a hacker, the employees might have insight into who was in the location at the time the threat was detected.

Useful Gateway Security Alerts

Similarly, you learned about setting up gateway IDS/IPS earlier in this study guide. The four alerts shown in Figure 19-11 inform admins of detected threats. Several other alerts report on the IDS/IPS engine's general functionality and the resources used by that engine. You can enable those alerts too, if you choose.

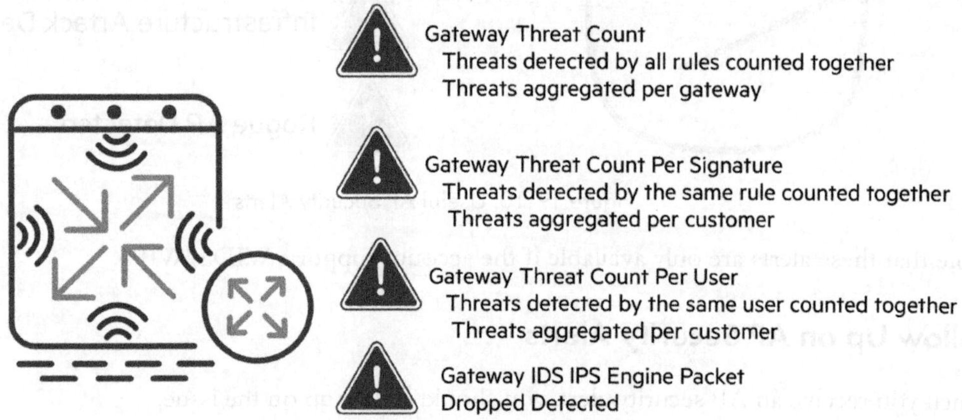

Figure 19-11: Useful Gateway Security Alerts

Note also that the severity configured for the alerts is different from the **Threat Lists** and **Threat Details** pages in the **Security > Gateway IDS/IPS** tab. The Severity value displayed in the Gateway IDS/IPS tab is specific to the threat signature.

Gateway Threat Count

This alert is triggered when a gateway detects more threats than the configured threshold (from 50 to 500 threats) in a given duration.

Aggregation for this alert occurs at the gateway level. For example, you might have set the duration to 30 minutes, the threshold to 10 for warning, and the threshold to 100 for major. If a company has three gateways, and one detects 12 threats over the hour, one detects 20 threats over the hour, and a third detects 80 threats over the hour, HPE Aruba Networking Central creates three warning alerts, one on each gateway.

Gateway Threat Count Per Signature

This alert also triggers based on the number of threats detected over a duration. However, it applies to threats of the same signature. It is aggregated at the customer level. For example, you might have set the duration to 30 minutes and the threshold to 10 for warning and 100 for major. If a company has three gateways, and one detects 12 threats *with the same signature* over the hour, one detects 20 threats *with the same signature* over the hour, and a third detects 80 threats *with the same signature* over the hour, HPE Aruba Networking Central creates one major alert.

Gateway Threat Count Per User

This alert also triggers based on the number of threats detected over a duration. However, it applies to threats detected by the same user. It is aggregated at the customer level.

Gateway IDS IPS Engine Packet Dropped Detected

If the gateways are using IPS mode, they drop packets in which they detect threats. Gateways generate a "IDS IPS Engine Packet Dropped Detected" alert when the percentage of packets over the duration dropped exceeds the configured threshold value. For example, you set the duration to 30 minutes and threshold to 1 for a warning alert. The warning alert triggers when a gateway is dropping, on average over 30 minutes, one out of 100 packets because those packets contain a threat.

Follow Up on Gateway Security Alerts

When you see one of the security-related gateway alerts, you should go to the **Security > Gateway IDS/IPS** page (global, group, or site). You can use the **Summary** view to look for trends in the Security Gateway IDS/IPS Dashboard. Go to the **List** view to see a list of threats.

From there you can record the particular threat type. You can refer to HPE Aruba Networking documentation for more information on each type. You should also be aware that allowlisted rules still create alerts. So check whether the alert indicates that the traffic is allow listed, in which case you or other admins have already determined that this traffic does not indicate a serious threat.

If you decide that you need to follow up on a potential security issue, you can click the threat in the Threats list and gather more information. For example, you can check the source of the threat. (Keep in mind, though, that the address might be spoofed.) You can also download the packet that triggered the threat.

Using that information, make your best effort to find the attack source and stop it from sending more attacks. If the gateway is not using IPS, mitigating the threat is particularly important. But even if the gateway is dropping the packets with IPS, you want to track down the attack source and prevent more attacks, as far as possible.

If the attack source is internal and you have CPDI and CPPM, you might mitigate the threat as follows. You could find the client in CPDI and gather information about it. What is the type? Which user or users have logged on with this client? You can apply a tag indicating that the client is compromised. If CPDI is integrated with CPPM, you should apply a tag that is used in an enforcement policy for clients of this type and associated with a quarantine or remediation enforcement profile.

You should investigate further as much as possible. Did the user on the device launch the attack deliberately? Or has the device been compromised? Security team members might need to take over the client and interview the user so that they can determine when and how the compromise occurred.

Useful Connectivity Alerts for Security Auditing

As indicated in Figure 19-12, the "Authentication Failure Detected" alert, under the Connectivity category, can also be useful for security auditing. This alert is generated when there are a high number of client authentication failures in the network. These failures include Wi-Fi security key-exchange failures, 802.1X authentication failures, MAC authentication failures, and captive portal authentication failures.

Figure 19-12: Useful Connectivity Alerts for Security Auditing

This alert triggers based on the authentication failures as a percentage of total authentication attempts over a specified time period.

Follow Up on Authentication Failures

Widespread authentication failures typically arise from configuration issues, rather than security breaches. If many users at a particular site fail to authenticate, perhaps admins have accidentally changed the RADIUS secret or other settings on a NAD at that site. Or perhaps Android devices do not trust the certificate used by your HPE Aruba Networking ClearPass servers. Certificate expiration on a ClearPass server could also cause widespread authentication failures.

Occasionally authentication failures could indicate that a hacker is attempting to gain access to a less secure WLAN that is using MAC-Auth or PSK. Hackers could also attempt to gain access to a network using 802.1X using stolen passwords. However, if you are using certificate-based authentication, you can be less concerned with such attempts.

AI Insights

HPE Aruba Networking Central also offers AI Insights as another way for admins to keep on top of what is happening within the network. With a more traditional approach, admins need to search through alerts and determine whether the alerts indicate troubling changes or unusual activity. AI Insights automatically detects anomalies and highlights them.

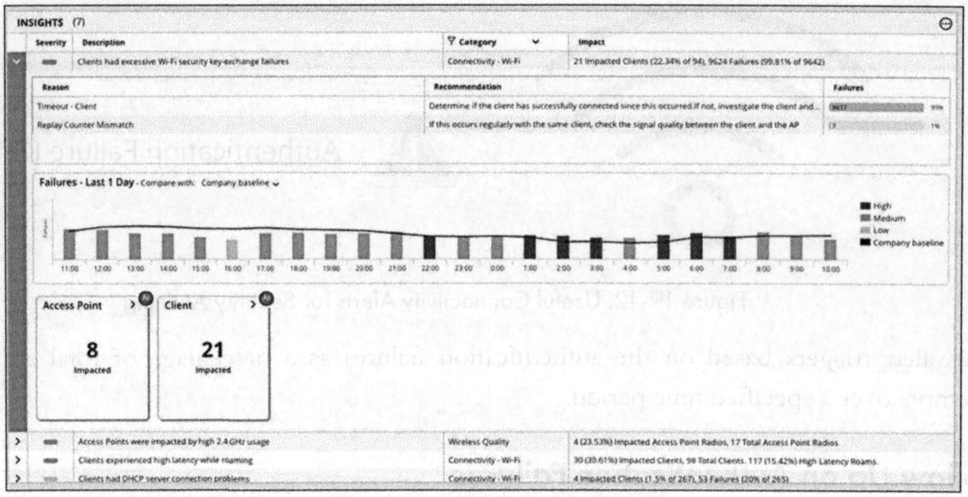

Figure 19-13: Example AI Insights

Admins can see AI Insights from the Global or Site view, as shown in Figure 19-13. (In version 2.5.5, at the **Overview** page, they should click the **AI Insights** tab.) By checking AI Insights periodically admins can more quickly detect potentially concerning issues.

In this example, AI Insights highlights that the number of clients failing to complete the Wi-Fi security key exchange is unusually high. It further shows the reason for this issue and recommends a next step.

Summary

Congratulations! You learned how to use HPE Aruba Networking Central's alerts. You learned how to configure the alert settings, including notification settings.

Learning checks

1. Match the configuration task with the appropriate HPE Aruba Networking Central scope. Note that you can assign more than one task to a scope.

   Configuration tasks

   a. Which alerts to enable

   b. Email overrides

   c. Alert settings

   Scope

   __ Site

   __ Global

2. Which of these alerts occurs when the percentage of packets dropped by gateway IPS exceeds the configured threshold?

   a. Infrastructure Attack Detected

   b. Authentication Failure Detected

   c. Gateway Threat Count per Signature

   d. Gateway IDS IPS Engine Packet Dropped Detected

Note

For answers to all learning checks, refer to the Appendix.

---

20 Practice Exam

Introduction

This exam validates candidates' knowledge, skills, and ability to implement Zero Trust Security (ZTS) using the HPE Aruba Networking portfolio. It covers how to configure HPE Aruba Networking network infrastructure and ClearPass solutions to authenticate and control both wired and wireless users, as well as remote users on a client-to-site VPN. The exam further tests candidates' ability to collect a variety of contextual information using ClearPass Policy Manager (CPPM) and implement advanced role mapping and enforcement policies. Candidates should also understand how to use HPE Aruba Networking ClearPass Device Insight to enhance visibility. They should be able to set up features such as the AOS-CX Network Analytics Engine (NAE), HPE Aruba Networking Wireless Intrusion Detection System/Intrusion Prevention System (WIDS/WIPS), and HPE Aruba Networking gateway IDS/IPS, as well as understand how to investigate alerts. They should also understand how to use HPE Aruba Networking unified SASE solutions.

Minimum Qualifications

The candidate should be a network engineer or help desk engineer working in either a customer or partner environment. The candidate should have about three years of experience working with wired and wireless technologies.

Exam Details

Below are details about the HPE7-A02 exam.

- Exam type: Proctored
- Types of exam items: multiple choice, single response, and multiple responses
- Exam duration: 1 hour 45 minutes
- Exam length: 70 questions
- Passing score: 67%

- Delivery languages: English, Japanese, Latin American Spanish

To register for the exam, you need an HPE Learner ID and a Pearson VUE login and password.

No reference material is allowed at the testing site. This exam may contain beta test items for experimental purposes, which are not taken into consideration for grading. During the exam, you can make comments about the exam items. We welcome these comments as part of our continuous improvement process.

Testing Objectives

This exam validates that you can successfully perform the objectives listed in Table 20-1. The objectives are grouped into three sections, and each section is given a weighting.

Table 20-1: Weightings and objectives

Weighting (% of exam)	Objectives
76%	**Protect and Defend** • Describe PKI dependencies • Mitigate threats by using ClearPass Device Insight (CPDI) to identify traffic flows and apply tags and using ClearPass Policy Manager (CPPM) to take actions based on tags • Explain methods and benefits of profiling • Explain Zero Trust Security in relation to HPE Aruba Networking solutions • Explain Wireless Intrusion Protection System (WIPS) and Wireless Intrusion Protection Detection (WIDS), as well as describe the HPE Aruba Networking 9x00 Series IDS/IPS • Describe log types and use the CPPM ingress event engine to integrate with third-party logging solutions • Explain unified SASE concepts, such as SD-WAN, ZTNA, SWG, and CASB

	• Set up secure authentication and authorization of network infrastructure managers (with a focus on advanced topics such as TACACS+ authorization and multi-factor authentication) • Secure L2 and L3 protocols, as well as other network protocols such as SFTP • Deploy Authentication, Authorization, and Accounting (AAA) for WLANs with CPPM • Define and apply advanced firewall policies (AppRF, PEF, WIPS, WebCC) • Set up integration between the HPE Aruba Networking infrastructure and CPPM, allowing CPPM to take action in response to events • Configure rogue AP detection and mitigation and other WIDS/WIPS features • Deploy AAA for wired devices with CPPM • Configure 802.1X authentication for APs • Deploy dynamic segmentation • Deploy certificate-based authentication for users and devices • Understand that HPE Aruba Networking EdgeConnect SD-WAN and SD-Branch automate VPN deployment for the WAN • Design and deploy HPE Aruba Networking SSE • Deploy and apply endpoint classification to the device • Define endpoint classification methodology using active and passive methods • Define, deploy, and integrate CPPM and CPDI
22.5%	**Analyze** • Investigate HPE Aruba Networking Central alerts

	• Interpret packet captures • Recommend action based on the analysis of HPE Aruba Networking Central alerts • Evaluate endpoint posture • Deploy and analyze Network Analytics Engine (NAE) scripts for monitoring and correlation • Perform packet capture on HPE Aruba Networking infrastructure locally and using HPE Aruba Networking Central • Analyze endpoint classification data to identify risk • Analyze endpoint classification data on CPDI
1.5%	**Investigate** • Explain CPDI capabilities for showing network conversations on supported HPE Aruba Networking devices

Advice to Help You Take This Exam

Complete the training and review all course materials and documents before you take the exam.

Exam items are based on expected knowledge acquired from job experience, an expected level of industry-standard knowledge, and possibly other prerequisites (events, supplemental materials, etc.). Successful completion of the course alone does not ensure you will pass the exam.

Read this HPE Exam Preparation Guide and follow its recommendations.

Visit HPE Press for additional reference materials, study guides, practice tests, and HPE books.

Practice Exam Items

1. An AOS-CX switch implements an access control list (ACL). You want to track trends in the number of hits against certain deny rules in that list and receive alerts if the hits pass a threshold. Which can you do to support this use case?

   a. Set up interim RADIUS accounting to ClearPass Policy Manager (CPPM) and enable Insight on CPPM.

b. Discover the switch as a NAS in ClearPass Device Insight (CPDI) and use CPDI's Network Activity views.

c. Deploy a Network Analytics Engine (NAE) agent to monitor the ACL hits.

d. Manage the switch in HPE Aruba Networking Central and use that solution's show command tool to view ACL hits.

2. Refer to the exhibit in Figure 20-1. You set up this EAP-TLS authentication method on CPPM. You then select that method as the only authentication method in a service. How will CPPM validate client certificates?

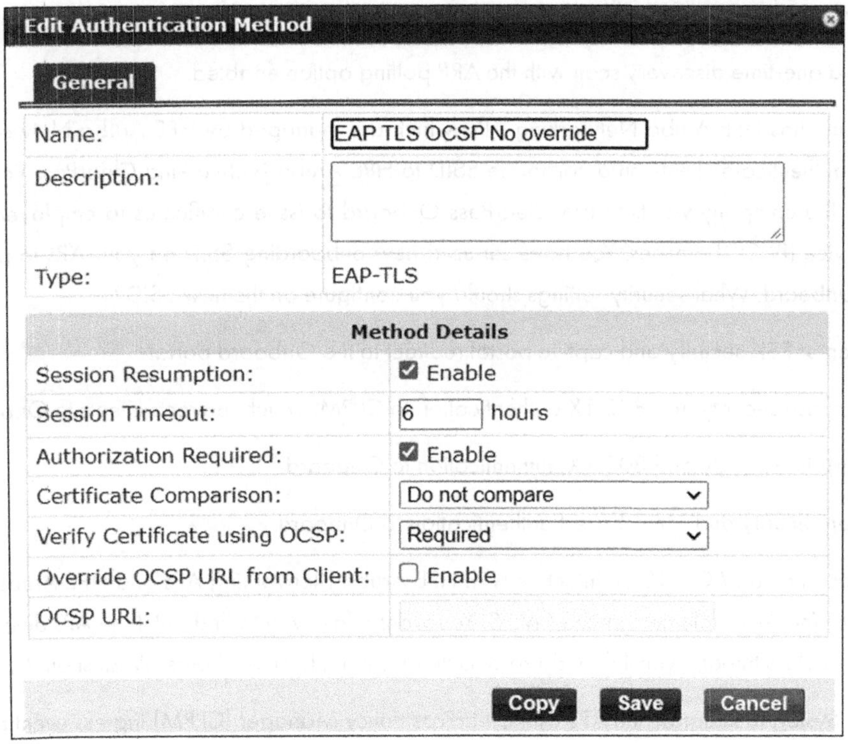

Figure 20-1: Item 2 Exhibit

a. Instead of checking the CA signature to validate the certificate, CPPM checks whether the certificate's serial number is in its OCSP database.

b. In addition to validating the certificate's signature, CPPM queries the OCSP URL in the certificate about the certificate's status.

c. CPPM will trust any certificate that is considered valid by CPPM's OCSP server, regardless of whether the certificate's CA is in CPPM's Trusted CA list.

d. In addition to validating the certificate's signature, CPPM checks whether the certificate serial number is on a CRL list, which it downloads using OCSP.

3. You intend to schedule periodic subnet scans on ClearPass Policy Manager (CPPM) to actively scan Windows domain clients. What should you do to ensure that CPPM can scan these devices correctly?

   a. On CPPM, map the domain clients' subnets to WMI accounts for accessing the clients.

   b. Configure the domain client firewalls to permit NetFlow traffic in and out of the devices.

   c. Enable periodic polling in the CPPM network device entries for the APs and switches to which the clients connect.

   d. Run a one-time discovery scan with the ARP polling option enabled.

4. A company has HPE Aruba Networking APs, which are managed by HPE Aruba Networking Central. The APs authenticate clients on a corporate SSID to HPE Aruba Networking ClearPass Policy Manager (CPPM). The company wants to use ClearPass Onboard to issue certificates to employees' Bring Your Own Device (BYOD) devices. You have set up a new onboarding SSID on your APs to permit users to access Onboard. What security settings should you configure on the new SSID?

   a. Open or PSK security and captive portal redirect to the Onboard portal

   b. Enterprise security and 802.1X authentication to CPPM, which redirects clients to Onboard

   c. Enterprise security and 802.1X authentication to Onboard

   d. Open security and MAC-based authentication to Onboard

5. A company has an AOS-10 architecture with HPE Aruba Networking APs and HPE Aruba Networking gateways. The APs implement WIDS/WIPS. You want a fast way to find out if the APs detect a rogue AP or other wireless threat, even if they do so outside of normal business hours. What should you do?

   a. Use Syslog to integrate the APs with ClearPass Policy Manager (CPPM) ingress event engine.

   b. Use Network Analytics Engine (NAE) to monitor for WIDS/WIPS events.

   c. Configure the APs to tunnel traffic to gateways, and set up alerts from the gateways' Threat Dashboard.

   d. Set up email notifications using HPE Aruba Networking Central's global alert settings.

6. Refer to the exhibit in Figure 20-2.

   A company's AOS-CX switches enforce manager authentication using TACACS+ to ClearPass Policy Manager (CPPM). One of the TACACS+ enforcement profiles to which CPPM assigns managers is named "SwitchAdmins1" and is shown in Figure 20-2.

   You have been instructed to configure a list of permitted and denied commands for managers assigned to this profile. However, you cannot find the Commands tab in the profile. What should you do?

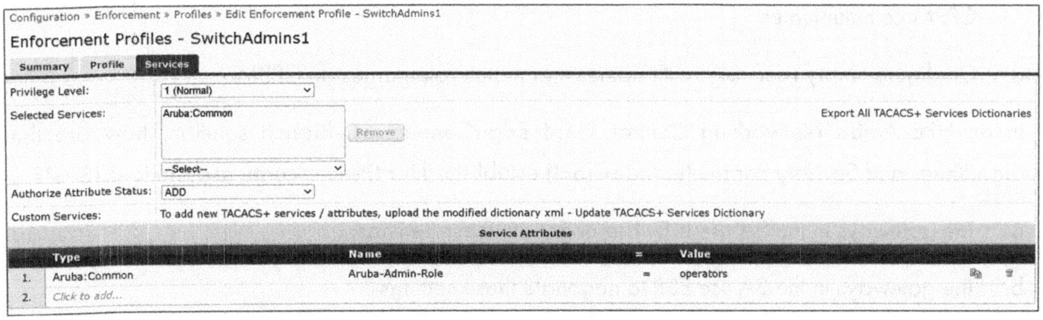

Figure 20-2: Item 6 Exhibit

   a. Enable TACACS+ command authorization in the CPPM server's TACACS+ settings.

   b. Change the SwitchAdmins1 profile Priv-Lvl to 15.

   c. Add the Shell service to the SwitchAdmins1 profile.

   d. Change the Aruba-Admin-Role VSA in the SwitchAdmins1 profile to "administrators."

7. You are setting up the Gateway IDS/IPS features on HPE Aruba Networking 9x00 gateways. The company wants to use IDS mode, and you need to set the IDS policy. What is the main difference between the moderate and strict policies?

   a. The moderate policy causes the gateway to detect threats but let all traffic pass, while the strict policy causes the gateway to detect threats and drop the threats.

   b. The moderate policy and strict policy both cause the gateway to detect threats, but let all traffic pass. But the strict policy matches traffic against a larger ruleset that includes less severe threats.

   c. The moderate policy and strict policy both cause the gateway to detect threats, but let all traffic pass. But the strict policy causes the gateway to drop traffic if the IDPS engine goes down.

   d. The moderate policy causes the gateway to detect threats in WAN and Internet traffic, while the strict policy causes the gateway to detect threats in WAN, Internet, and LAN traffic.

8. A company has AOS-CX switches and ClearPass Policy Manager (CPPM). The company wants switches to implement 802.1X authentication to CPPM and download user roles. What is one task that you must complete on the switches to support this use case?

   a. Install the root CA certificate for CPPM's HTTPS certificate in a TA profile.

   b. Specify CPPM as the RADIUS server with the exact CN in CPPM's RADIUS/EAP certificate.

   c. Specify a clearpass username and password that match credentials in the AD server with which CPPM communicates.

   d. Configure empty user-roles with names that match role names on CPPM.

9. In an HPE Aruba Networking Central 2.5.4 EdgeConnect SD-Branch solution, how are the keys, algorithms, and Security Parameter Index (SPI) established for IPsec security associations (SAs)?

   a. The gateways in the SA use IKEv1 to negotiate these settings.

   b. The gateways in the SA use ESP to negotiate these settings.

   c. Admins configure these settings manually and apply them to appropriate gateways or gateway groups in HPE Aruba Networking Central.

   d. HPE Aruba Networking Central SD-WAN Orchestrator establishes these settings and automatically pushes them to the correct gateways.

10. Which is a use case for using device auth-mode on an AOS-CX switch?

    a. To enable the switch to authenticate multiple downstream clients that connect to it through an unmanaged, non-EAP capable switch

    b. To enable the switch to authenticate a client and VoIP phone, which uses tagging, on the same port

    c. To enable the switch to authenticate an AP, but not the wireless clients connected to that AP

    d. To enable the switch to implement both MAC-Auth and 802.1X on the same port

11. Refer to the exhibit in Figure 20-3.

    What can you interpret from the packets that you see?

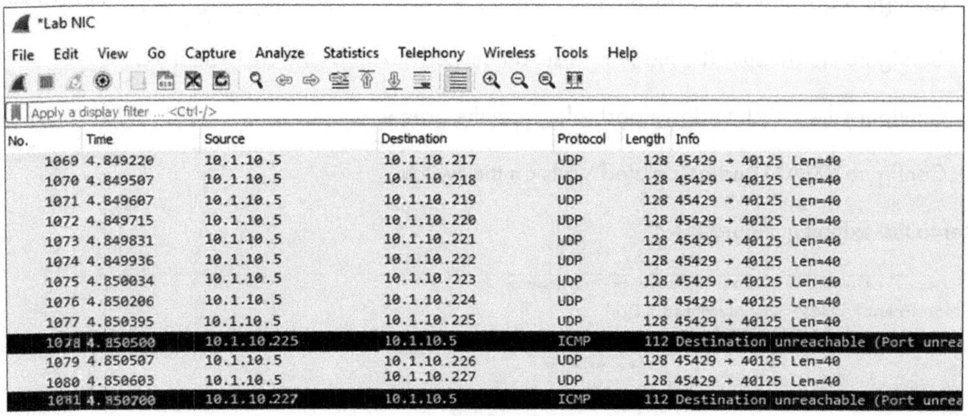

Figure 20-3: Item 11 Exhibit

   a. 10.1.10.5 is likely running a UDP sweep to detect active devices in the network.

   b. 10.1.10.5 is likely attempting to launch a denial of service (DoS) attack using UDP packets.

   c. 10.1.10.5 is likely attempting to implement a man-in-the-middle (MitM) attack against devices in the 10.1.10/24 subnet.

   d. 10.1.10.5 is likely attempting to gain unauthorized Ethernet access using UDP tunneling.

12. You are configuring the Device Insight Integration settings on ClearPass Policy Manager (CPPM). For which use case should you set the "Tag Updates Action" to "apply for selected tag updates?"

   a. You want to minimize traffic between CPDI and CPPM by configuring CPDI to inform CPPM only about specific tags.

   b. You use some Device Insight tags in enforcement policies, but other tags are only informational, and you want to minimize device disruption.

   c. You have a ClearPass cluster with servers at multiple sites; you want to minimize unnecessary traffic between the sites.

   d. You are using both CPPM and CPDI to collect device profile information; you want to control precisely which information the solutions exchange.

13. You are configuring an AOS-CX switch to implement user-based tunneling (UBT) to an HPE Aruba Networking gateway. Only untagged traffic is required to pass over the tunnels. Admins are configuring

VLAN assignments for wired clients in roles on the gateway. What should you configure on the switch to work with these requirements?

a. Configure UBT in VLAN extend mode.

b. Configure the default UBT mode and set an unused VLAN as the UBT client VLAN.

c. Configure the switch to download roles from the gateway.

d. Configure VXLAN and associated VNIs on the switch.

14. Refer to the exhibit in Figure 20-4.

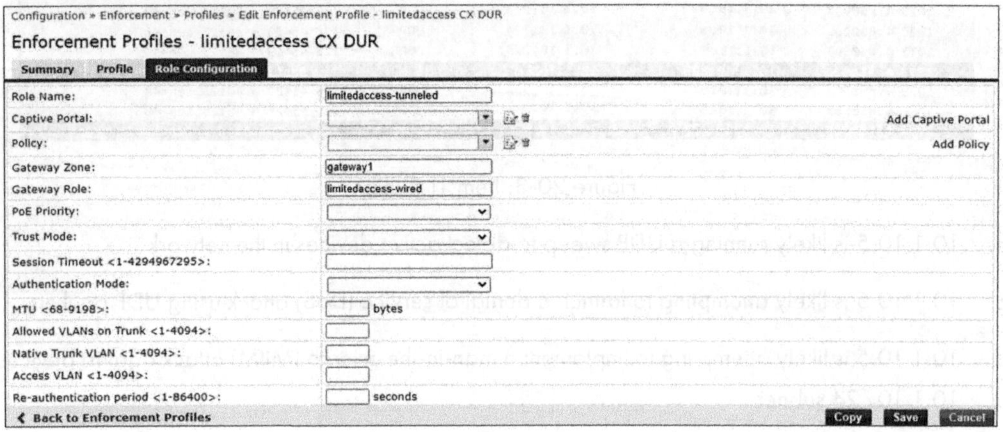

Figure 20-4: Item 14 Exhibit

You have configured an Aruba Downloadable Role enforcement profile on ClearPass Policy Manager (CPPM). As shown in Figure 20-4, this role implements tunneling between AOS-CX switches and HPE Aruba Networking gateways. You also configured an enforcement policy to output this profile; that policy is used by a service that authenticates clients on the AOS-CX switches. What is one step that you must complete on network infrastructure devices to ensure that clients assigned to this profile can forward traffic correctly?

a. Configure a user-role named "limitedaccess-tunneled" on the AOS-CX switches.

b. Configure a user-role named "limitedaccess-wired" on the AOS-CX switches.

c. Configure a role named "limitedaccess-wired" on the gateways.

d. Configure a gateway zone named "gateway1" on the gateways.

15. You need to obtain a Certification Authority (CA)-signed certificate for some HPE Aruba Networking APs, which are managed by HPE Aruba Networking Central.

You used this command to create a certificate signing request (CSR):

openssl req -new -out file1.pem -newkey rsa:3072 -keyout file2.pem

The intermediate CA returns the signed certificate in a file named file3.pem. What should you include in the PEM file that you upload to HPE Aruba Networking Central?

a. The contents of file 3 and the intermediate CA certificate only

b. The contents of file 1, file 2, and file 3

c. The contents of file2, file3, and the intermediate CA certificate

d. The contents of file 3 only

16. For which AOS-CX feature can you help prevent traffic disruption by saving IP-to-MAC bindings to external storage?

    a. User-based tunneling (UBT)

    b. Virtual network based tunneling (VNBT)

    c. MAC authentication

    d. ARP inspection

17. A company has HPE Aruba Networking APs that are controlled by HPE Aruba Networking Central and that implement WIDS. When you check WIDS events, you see a "detect ad hoc with valid SSID" event. What can you interpret from this event, and what steps should you take?

    a. Devices operating at default settings often create ad hoc networks, so this event is likely to be a false positive. You should turn off detection for this threat type.

    b. Hackers or users are likely trying to introduce an unauthorized device. You should use the detecting radio information and immediately track down the device that triggered the event.

    c. Clients are likely bridging traffic across multiple wireless NICs; you need to work with client admins to adjust these settings.

    d. The company likely has some third-party APs; you should add them to the authorized radios list manually.

18. The following AOS firewall role is configured on HPE Aruba Networking Central-managed APs:

```
wlan access-rule employees
  index 3
  rule any any match 17 67 67 permit
```

```
rule any any match any 53 53 permit
rule 10.2.2.0 255.255.255.240 match any any any deny
rule 10.2.2.0 255.255.255.0 match 6 80 80 permit
rule 10.2.2.0 255.255.255.0 match 6 443 443 permit
rule 10.2.0.0 255.255.0.0 match any any any deny
rule any any match any any any permit
```

A client has authenticated and been assigned to the "employees" role. The client has IP address 10.2.4.4. Which correctly describes behavior in this policy?

a. HTTPS traffic from 10.2.2.30 in an active session between 10.2.2.30 and 10.2.4.4 is denied.

b. HTTPS traffic from 10.2.4.4 to 10.2.2.6 is denied.

c. HTTPS traffic from 10.2.4.4 to 203.0.113.12 is denied.

d. HTTPS traffic from 198.51.100.12 in an active session between 10.2.4.4 and 198.51.100.12 is denied.

19. Refer to the exhibits in Figures 20-5, 20-6, and 20-7.

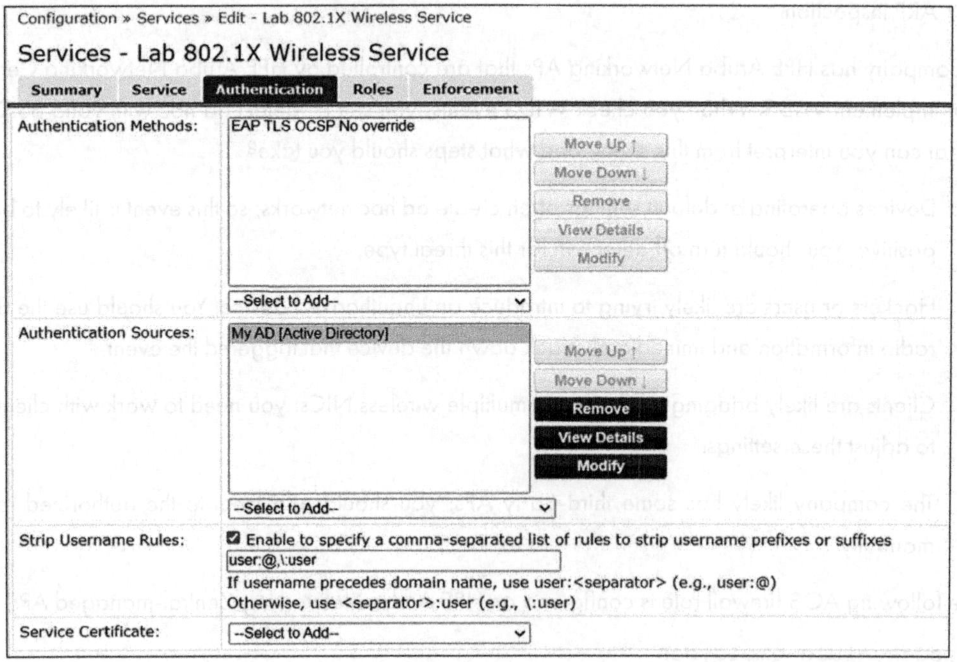

Figure 20-5: Item 19 Exhibit 1

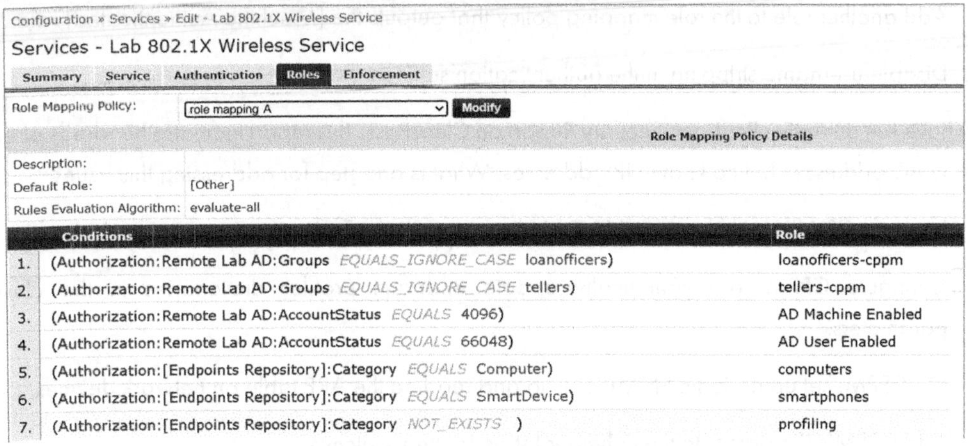

Figure 20-6: Item 19 Exhibit 2

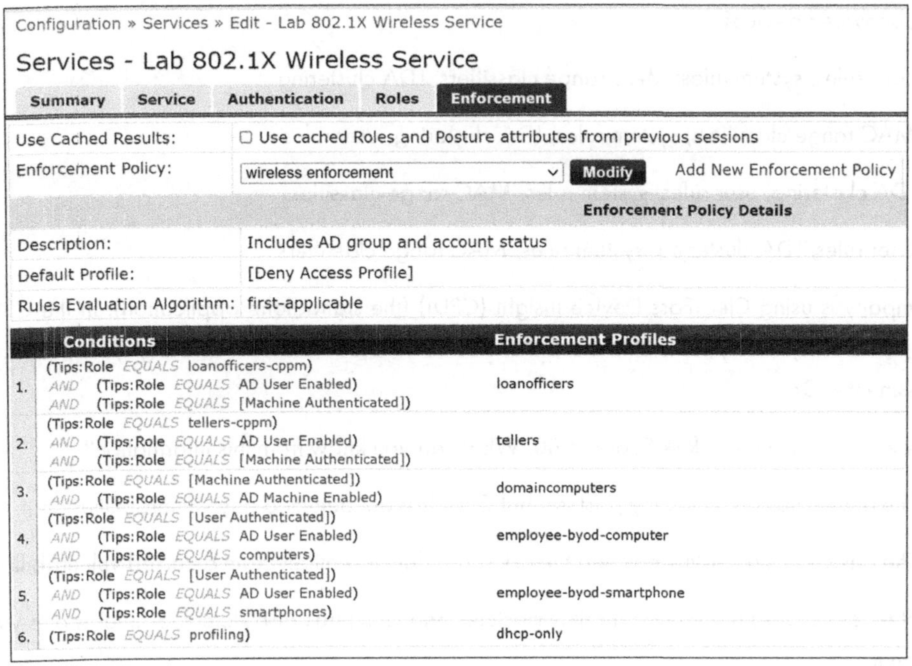

Figure 20-7: Item 19 Exhibit 3

You have configured a ClearPass Policy Manager (CPPM) service as shown in the exhibits. What additional step should you take to ensure that some clients are not left at the wrong access level?

a. Enable role and posture caching in the enforcement settings.

b. Enable the Profiler in the general settings and then set the CoA action for profile changes.

c. Add another rule to the role mapping policy that outputs the [Machine Authenticated] role.

d. Disable username stripping in the authentication settings.

20. You have run an Active Endpoint Security Report on ClearPass. This report indicates hundreds of endpoints with MAC addresses but no known IP addresses. What is one step for addressing this issue?

    a. Import a CSV file with the IP-to-MAC bindings to CPPM's Endpoints Repository.

    b. Configure CPPM to automatically update endpoint profiler dictionaries in the cluster-wide parameters.

    c. On CPPM, set up periodic SNMP polling and reading the ARP table on network devices.

    d. Add CPPM's IP address to the IP helper list on routing switches.

21. In which order does HPE Aruba Networking ClearPass Device Insight (CPDI) apply classification methods to a discovered device?

    a. User rules, system rules, MAC range classifiers, LDA clustering

    b. MAC range classifiers, system rules, LDA clustering, user rules

    c. LDA clustering, user rules, system rules, MAC range classifiers

    d. User rules, LDA clustering, system rules, MAC range classifiers

22. A company is using ClearPass Device Insight (CPDI) (the standalone application). In the CPDI security settings, Security Analysis is On, the Data Source is ClearPass Device Insight, and Enable Posture Assessment is On.

    You see that device has a Risk Score of 80. What can you know from this information?

    a. The device has an unhealthy posture, but CPDI has not detected any vulnerabilities on it.

    b. The device has a healthy posture; the risk score derives entirely from detected vulnerabilities.

    c. The device does not have a healthy posture, and CPDI has also detected at least one vulnerability on it.

    d. The device has an unknown posture, and CPDI has detected exactly 30 vulnerabilities on it.

23. A company has HPE Aruba Networking APs (AOS-10), which authenticate clients to ClearPass Policy Manager (CPPM). CPPM has some policies which apply [ArubaOS Wireless - Terminate Session] enforcement profiles. What else should you set up on CPPM to ensure that those profiles take effect?

    a. Set the CoA delay to 0 in the CPPM server's RADIUS server settings.

b. Enable interim accounting processing in CPPM's general server settings.

c. Enable RadSec in the network device entries for the APs.

d. Enable Dynamic Authorization in the network device entries for the APs.

24. A company has an AOS-8-based mobility architecture with mobility controllers (MCs) and HPE Aruba Networking Campus APs (CAPs). The security team wants you to capture traffic from a particular wireless client, which is connected to AP 1. This AP is controlled by MC 1 and connects to AOS-CX switch 1. You should direct this wireless client's traffic to the security team's machine for further analysis for potential security threats. What should you do to meet these requirements?

    a. Access the CLI for AP 1. Set up a remote mirroring session between its Ethernet port and the security team's machine.

    b. Access the CLI for switch 1. Set up a remote mirroring session between AP 1's port and the security team's machine.

    c. Access the CLI for MC 1. Set the datapath capture destination to the security team's machine; then start a datapath capture with the client's MAC address.

    d. Access the CLI for MC 1. Set the controlplane capture destination to the security team's machine; then start a controlplane capture with the client's IP address.

25. A company wants HPE Aruba Networking ClearPass Policy Manager (CPPM) to respond to Syslog messages from the company's Palo Alto Next Generation Firewall (NGFW). You have added the firewall as an event source and set up an event service. However, test Syslog messages are not triggering the expected actions. What is one CPPM setting that you should check?

    a. You have enabled all default device fingerprints on CPPM.

    b. The Palo Alto Extension is installed through ClearPass Guest.

    c. The Palo Alto NGFW is configured as a Syslog server on CPPM.

    d. Insight and ingress event processing are enabled on the CPPM server.

26. On HPE Aruba Networking ClearPass Policy Manager (CPPM), you have configured a posture policy that uses the ClearPass OnGuard agent. This policy has rules for assigning clients to the Healthy and Quarantine postures. How do you configure CPPM to respond to agents' reports and assign the correct postures to clients?

    a. Configure a Web-based Heath Check Service that references the posture policy.

    b. Configure a RADIUS Enforcement Policy that uses the Healthy and Quarantine postures in rules.

c. Install an OnGuard extension from the ClearPass Guest Administration page.

d. Enable caching roles and postures in the CPPM service to which clients authenticate.

27. You are setting up HPE Aruba Networking SSE for a company. The company wants the solution to control which actions, such as uploading and downloading files, remote employees can perform on services such Dropbox. What is one necessary step?

    a. Creating an application tag for software as a service (SaaS) applications

    b. Adding an integration for Dropbox to the HPE Aruba Networking SSE account

    c. Creating a self-hosted web application that refers to the Dropbox URL

    d. Creating an external web profile that enables SSL encryption

28. You have already deployed a virtual Data Collector on a VMware ESXi host for use with HPE Aruba Networking ClearPass Device Insight (CPDI). You now need this Data Collector to collect flow information about client's network behavior. You will configure core routing switches to mirror traffic to the ESXi host's NICs 3 and 4, which are owned by virtual switch 2. What should you tell virtual network admins to do?

    a. Connect the Data Collector's adapter with the third lowest MAC address to virtual switch 2 in VLAN 4095.

    b. Connect the Data Collector's adapter with the lowest MAC address to virtual switch 2 in VLAN 0.

    c. Connect the Data Collector's adapter with the highest port ID to virtual switch 2 in VLAN 4095.

    d. Connect the Data Collector's adapter with the lowest port ID to virtual switch 2 in VLAN 0.

29. Refer to Figures 20-8 and 20-9.

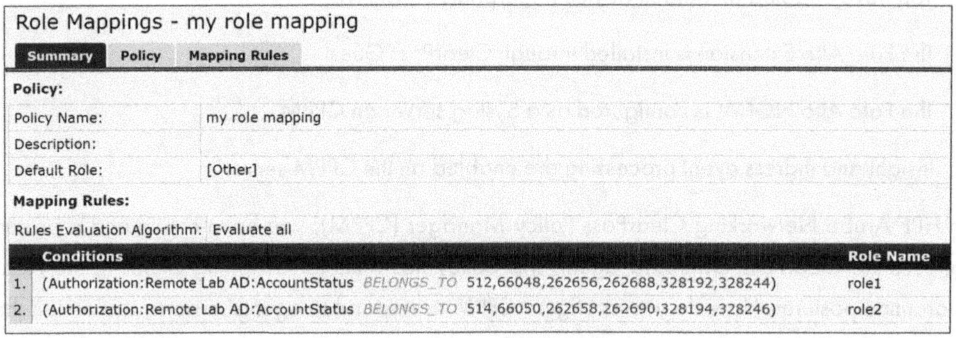

Figure 20-8: Item 29 Exhibit 1

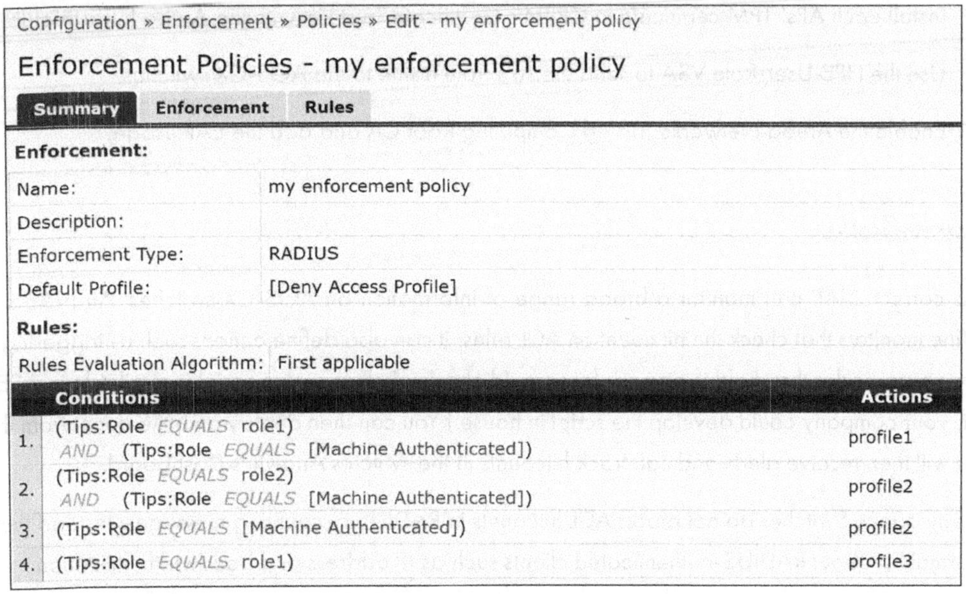

Figure 20-9: Item 29 Exhibit 2

ClearPass Policy Manager (CPPM) has a custom attribute for AD that uses AccountStatus alias for userAccountControl. Which enforcement profile does CPPM apply to a client that meets both these conditions:

- Succeeds in authenticating as a user, but to a disabled AD user account (userAccountControl value 66050)
- Succeeds at authenticating as a computer

a. profile1

b. profile2

c. profile3

d. Deny Access profile

30. A company has HPE Aruba Networking APs that connect to AOS-CX switches. The APs need to authenticate as 802.1X supplicants to ClearPass Policy Manager (CPPM). The company wants to use the Device (TPM) certificates for this purpose, as well as check that APs' MAC addresses are known in CPPM's Endpoints Repository. What is one recommended setting on CPPM?

a. Implement EAP-TEAP for the authentication method with both EAP-TLS and MAC-Auth as inner methods.

b. Install each APs' TPM certificate to CPPM's Certificate Trust List with the Aruba Infrastructure usage.

c. Use the HPE-User-Role VSA to send the APs' role name to the AOS-CX switches.

d. Enable the Aruba Networks Trusted Computing Root CA and add the EAP usage.

Explanations

1. **C** is correct. NAE can monitor a broad range of information on AOS-CX switches. An NAE script can define monitors that check the hit count on ACL rules; it can also define actions such as triggering an alert when a particular threshold is passed. (You could check whether such a script is available in ASE, and, if not, your company could develop the script in house.) You can then deploy an NAE agent from the script. You will then receive alerts and can track hitcounts in the switch's Analytics Dashboard.

   A is incorrect. Switches do not report ACL hitcounts in RADIUS accounting messages. Instead they report information about RADIUS-authenticated clients such as IP address and bandwidth utilization.

   B is incorrect. CPDI's Network Activity view for a client shows client's network conversations based on flow information, not ACL hits.

   D is incorrect. If you manage the switch in HPE Aruba Networking Central, you could use the show command tool to view ACL hits on the switch. However, this action does not meet all the criteria in the use case, including getting alerts when a threshold is passed.

   For more information on this topic, refer to Chapter 9.

2. **B** is correct. Enabling OCSP tells CPPM to use Online Certificate Status Protocol (OCSP) to query a server about the status of the certificate. The certificate includes an OCSP URL, which tells CPPM where to send this query. The OCSP check is *in addition to* all of the typical checks of certificate validity.

   A and **C** are incorrect. When using OCSP, CPPM still needs to validate that the certificate is signed by a CA that is in CPPM's trusted CA list with the EAP Usage. (In addition, CPPM does not have an OSCP database; it queries an OCSP server.)

   D is incorrect. Using OCSP is an alternative to using CRLs. Using CRLs requires a device to download an entire list and check that a certificate's serial number is not on that list. But with OCSP the client (such as CPPM) queries the server about just the serial number in question.

   For more information on this topic, refer to Chapters 2 and 3.

3. **A** is correct. To run an active scan properly, CPPM needs at least one external account mapped to the subnet or subnets being scanned. These accounts dictate which protocols and associated credentials to use during the scan. WMI is a good protocol for scanning Windows domain clients.

B is incorrect. CPPM does not use NetFlow for active scans.

C is incorrect. Enabling periodic polls in a CPPM network device entry lets CPPM poll that network device itself, not the clients connected to it.

D is incorrect. You can run a discovery scan before running subnet scans, but it is not required.

4. **A** is correct. HPE Aruba Networking generally recommends using a separate SSID for onboarding BYOD devices. The onboarding SSID should use open or PSK security; however, PSK is recommend for greater security. Then the devices can connect without needing to trust any ClearPass certificates in advance, and they can easily switch to a more secure WPA3-Enterprise SSID after being provisioned by Onboard. The onboarding SSID must redirect clients to the Onboard portal so that users can provision their devices. Note that the Onboard portal should enable Prevent CNA to bypass CNAs (configured in the Provisioning Profile Web Login settings).

   B is incorrect. CPPM does not receive clients' traffic; network devices such as APs must redirect clients. Also enterprise security is less recommended for the onboarding SSID. Users will move to a more secure SSID that uses enterprise (802.1X) security after onboarding.

   C and **D** are incorrect. Onboard is not an authentication server.

   For more information on this topic, refer to Chapter 3.

5. **D** is correct. AOS-10 APs (and AOS-10 gateways) are managed in HPE Aruba Networking Central. HPE Aruba Networking Central alerts include ones related to WIDS/WIPS events such as rogue AP detection, infrastructure threat detection, and client threat detection. You can enable these alerts in HPE Aruba Networking Central's global alert settings and set up email notifications for when these alerts trigger. These actions meet the scenario's requirements of informing you quickly when WIDS/WIPS events occur.

   A is incorrect. In theory, it is possible to configure APs to send Syslog messages to CPPM and to configure CPPM's ingress event engine to process those messages (however, doing so would require creating custom event dictionaries, which is an expert topic). However, these actions would not fulfill the requirements of this scenario, which is to make admins aware of threats more quickly.

   B is incorrect. NAE can help alert admins to events and issues on *AOS-CX switches*.

   C is incorrect for several reasons. The APs are responsible for running the WIDS/WIPS feature regardless of whether they bridge wireless traffic or tunnel it. The gateway's Threat Dashboard relates to gateway IDS/IPS, not WIDS/WIPS. (Also note that you do *not* set up gateway IDS/IPS alerts directly from the dashboard. Instead you can set those alerts up from HPE Aruba Networking Central's global alert settings, much like you do the AP alerts.)

   For more information on this topic, refer to Chapter 19.

6. **C** is correct. Adding the Shell service makes the **Commands** tab appear; you can then configure the permitted and denied commands in that tab.

   A is incorrect. You do not need to enable authorizing commands globally.

   B and **D** are incorrect. These actions relate to changing the group to which the switch assigns the manager. (In more detail, the Aruba-Admin-Role VSA takes precedence over the TACACS+ privilege level.) These actions do not add the **Commands** tab. You are allowed to configure commands for managers assigned to any group on the AOS-CX switches.

   For more information on this topic, refer to Chapter 6.

7. **B** is correct. When the gateway acts in IDS mode, it detects threats but lets all traffic pass. The difference between IDS policies lies in the number of rules against which the gateway matches traffic. The strict ruleset is larger and includes less severe threats (as well as more severe threats). In more detail, the strict policy includes threats with CVSS 4 and higher from the current and previous 3 years, while the moderate policy only includes threats with CVSS 6 and higher from the current and previous 2 years.

   A is incorrect. In IDS mode, the gateways detect threats, but let all traffic pass, regardless of policy.

   C is incorrect. It is the fail strategy that dictates whether the gateway drops traffic if its IDPS engine goes down. You can choose either the bypass or block fail strategy with any IDS policy.

   D is incorrect. The gateways detect threats in WAN, Internet, and LAN traffic when using any of the IDS (or IPS) policies.

   For more information on this topic, refer to Chapter 18.

8. **A** is correct. To download the downloadable user-roles (DURs), the switch must trust CPPM's *HTTPS* certificate. This trust includes having a TA profile with the root CA certificate for the CA that issued CPPM's HTTPS certificate.

   B is incorrect. The switch should specify CPPM as the RADIUS server with the exact name in the CN of CPPM's *HTTPS* certificate, not RADIUS/EAP certificate.

   C is incorrect. You must specify a clearpass username and password on the switch, but the credentials must match an admin account on CPPM. That admin account should have the Aruba User Role Download privilege level.

   D is incorrect. When the switch is downloading roles, it does not need user-roles configured locally in advance.

   For more information on this topic, refer to Chapter 7.

9. **D** is correct. HPE Aruba Networking Central SD-WAN Orchestrator is responsible for creating the SAs and securely distributing the information to the gateways in that SA.

   A is incorrect. IKE *is* a protocol for negotiating IPsec SAs. However, HPE Aruba Networking EdgeConnect SD-Branch gateways do not need to use IKE to negotiate the SAs. SD-WAN Orchestrator handles the SA establishment itself.

   B is incorrect. ESP does not provide negotiation for IPsec SAs.

   C is incorrect. With HPE Aruba Networking EdgeConnect SD-Branch, admins do not configure keys and SPIs for SAs manually.

   For more information on this topic, refer to Chapter 17.

10. **C** is correct. Device auth-mode tells the switch to authenticate the first MAC address on the port and then to permit all traffic, regardless of MAC address. This mode lets the switch authenticate the AP, but then permits traffic that the AP bridges from clients to flow freely. (Presumably, the AP has already authenticated those clients.)

    A and **B** are incorrect. In these scenarios, you want the switch to authenticate each MAC address on the port separately. Client auth-mode acts in this way.

    D is incorrect. The switch can implement both MAC-Auth and 802.1X in device auth-mode, but it can also do so in client auth-mode. The need to implement both methods is not a specific reason for enabling device auth-mode.

    For more information on this topic, refer to Chapter 7.

11. **A** is correct. These packets show several signs of a sweep. One IP address, here 10.1.10.5, is rapidly contacting many other IP addresses within a range. This indicates that 10.1.10.5 is trying to reconnoiter the network and discover which IP addresses are active. This particular sweep is called a UDP sweep because it uses UDP packets to an uncommonly used port. Active devices will send an ICMP unreachable packet because the port is closed.

    B and **C** are incorrect. While 10.1.10.5 might move on from reconnaissance to launching other attacks later, these particular packets are not associated with DoS or MitM attacks.

    D is incorrect. UDP tunneling is not a way to obtain unauthorized Ethernet access. This device already has access to the network, and its user is trying to discover more information.

    For more information on this topic, refer to Chapter 16.

12. **B** is correct. The "Tag Updates Action" setting tells CPPM whether to issue a CoA when a change occurs to a client's Device Insight tags. The "apply for selected tag updates" option lets you choose precisely

which tags' updates trigger the CoA. You can then select tags that you reference in enforcement policies to trigger CoAs, but leave informational only tags off the list. This approach minimizes device disruption.

A is incorrect. CPDI informs CPPM about all tags that apply to devices, regardless of the "Tag Updates Action."

C is incorrect. The tag information propagates across the cluster, regardless of the "Tag Updates Action."

D is incorrect. The "Tag Updates Action" does not control information exchanged between CPPM and CPDI. (Note that this exam focuses on CPPM 6.9, which disables profiling on CPPM when CPPM is integrated with CPDI. In ClearPass 6.10 and later, it is possible to enable profiling on both CPPM and CPDI. However, the "Tag Updates Action" continues to have nothing to do with an exchange of information.)

For more information on this topic, refer to Chapter 15.

13. **B** is correct. When an AOS-CX switch operates in reserved VLAN mode for UBT, you configure clients' VLAN assignments in the clients' role on the gateway. You do not have to configure matching assignments in the roles on the switch. To enable this mode, you simply configure a VLAN, not used for any other purpose, as the UBT client VLAN on the switch.

    A is incorrect. When you use VLAN extend mode, you must configure matching VLAN assignments for clients on both the gateway and the switches.

    C is incorrect. Switches download user-roles from HPE Aruba Networking ClearPass Policy Manager (CPPM), not from gateways.

    D is incorrect. AOS-CX switches use VXLAN with VNBT, not with UBT.

    For more information on this topic, refer to Chapter 8.

14. **C** is correct. The gateway role must match a role that is configured on the HPE Aruba Networking gateways that terminate the tunnels. The gateways use that role to apply settings to clients' traffic.

    A and **B** are incorrect. The enforcement profile shows a DUR, which the switch downloads from CPPM. The switches do not require any user-role settings pre-configured on it for them to download this DUR.

    D is incorrect. The *switches* need a gateway zone named "gateway1" configured on them. However, the gateways simply terminate the tunnel from switches that contact them, and you do not need to configure a zone on the gateways for this purpose.

    For more information on this topic, refer to Chapter 8.

15. **C** is correct. The APs need to use this certificate as their own, which means that they need the private key. The PEM file should include the private key associated with the certificate (file 2) and the signed certificate (file 3). It should also include the intermediate CA certificate; when APs submit their certificate to another entity, they need to include the intermediate CA certificate as part of the chain of trust.

    A and **D** are incorrect because the file must include the private key.

    B is incorrect because the file does not need to include the CSR and should include the intermediate CA certificate.

    For more information on this topic, refer to Chapter 2.

16. **D** is correct. When a switch implements ARP inspection, it checks ARP replies and only lets them pass if the MAC address that claims a particular IP address actually does own that IP address. The switch uses its IP-to-MAC binding table for this check, and typically it builds this table over time using DHCP snooping. If the switch reboots, it has not yet had time to rebuild the table, which can cause disruptions. Saving the table to external storage lets the switch obtain the table more quickly after a reboot (as often clients try to use the same addresses that they had before).

    A, **B**, and **C** are incorrect. These features do not rely on the IP-to-MAC binding table.

    For more information on this topic, refer to Chapter 6.

17. **B** is correct. If someone has set up an adhoc network that uses one of the company's SSIDs, that person is likely deliberately trying to hide the network or lure users to connect to it for malicious purposes.

    A is incorrect. It is true that some devices create ad hoc networks without users' awareness, so merely seeing that WIDS has detected an ad hoc network does not necessarily indicate a threat. However, this ad hoc network specifically has the company's SSID. That fact can indicate malicious intent, and you should follow up to learn more.

    C is incorrect. Bridging traffic across interfaces would not look like an ad hoc network.

    D is incorrect. Third-party APs would not advertise SSIDs in ad hoc mode.

    For more information on this topic, refer to Chapter 10.

18. **B** is correct. HTTPS traffic from 10.2.4.4 to 10.2.2.6 first hits rule 3 in the list (10.2.2.0 255.255.255.240 match any any any deny). For the source, AP firewall rules implicitly use the IP addresses of clients assigned to the role (wlan access-rule) in question. Therefore, 10.2.2.0 255.255.255.240 refers to the destination. The 10.2.2.6 IP address falls within the specific range (10.2.2.0-10.2.2.15). All traffic to these IP addresses is denied.

A is incorrect. The AOS firewall is stateful, so it always permits return traffic for active sessions. The client is permitted to send traffic to 10.2.2.30 by rule 5 (the first match), and 10.2.2.30 can also send traffic back. Note that this traffic does *not* match rule 3, which only applies to destinations 10.2.2.0-10.2.2.15.

C is incorrect. This traffic is permitted by the final rule in the list, which is the first match. Note that this traffic does not match rule 6 because 10.2.0.0 255.255.0.0 in that rule refers to the destination.

D is incorrect. The AOS firewall is stateful. The client is permitted to send traffic to 198.51.100.12 by the final rule in the list, and return traffic from 198.51.100.12 is permitted.

For more information on this topic, refer to Chapter 4. Note that the format of the rules might look a bit different based on whether you view them from the CLI or inside HPE Aruba Networking Central's UI. However, you are expected to be able to interpret rules in either case.

19. **B** is correct. This service is authenticating clients with 802.1X and specifically EAP-TLS. The role mapping and enforcement policy show that CPPM is using device category, as well as other context, to assign enforcement profiles to authenticated devices. When a device connects for the first time, CPPM will not know its device category, so it will assign the "profiling" role and, for BYOD devices, the "dhcp-only" enforcement profile. But after CPPM learns the device category, it should move the client to a new role and enforcement profile. The Profiler setting is required for this purpose. After enabling this setting, you must set the CoA action. Then CPPM can tell network devices to reauthenticate profiled clients, and the clients can receive the correct access level.

    A is incorrect. In this scenario, CPPM is not using posture in policy rules, and it does not need to cache posture or roles from earlier connections.

    C is incorrect. CPPM automatically assigns the [Machine Authenticated] role to any client that authenticates to a computer account on AD. You do not need to configure a role mapping rule for CPPM to do so.

    D is incorrect. Username stripping is often used with the EAP-TLS method.

    For more information, refer to Chapter 12 and also to Chapter 3.

20. **C** is correct. CPPM can use network devices' ARP tables to find out which IP address is associated with each MAC address. (It can also use interim RADIUS accounting messages for this purpose). To enable CPPM to discover this information, you can configure it with SNMP credentials for network devices, set up periodic SNMP polling of those network devices, and also enable reading the ARP table from the device.

    A is incorrect. You cannot import the IP addresses in a CSV file.

B is incorrect. Updating the endpoint profile dictionaries gives CPPM new device fingerprints so that it can better determine the device category, family, and name for each endpoint. However, these dictionaries do not help CPPM figure out the devices' IP addresses.

D is incorrect. Adding CPPM's IP address to the IP helper list on routing switches is very important for helping CPPM receive DHCP Discover messages and profile endpoints. However, it does not specifically help CPPM figure out the devices' IP addresses (as CPPM does not receive the DHCP Offers).

For more information on this topic, refer to Chapter 12.

21. **A** is correct. CPDI always applies user rules first. If it does not find a match, then it applies system rules. If it still does not find a match, it uses MAC range classifiers. Finally, it uses machine learning-based LDA clustering to group together the "generic" unclassified devices, as well as look for recommendations based on similarities to classified devices.

    B, **C**, and **D** are incorrect because they indicate the wrong order.

    For more information on this topic, refer to Chapters 13 and 14.

22. **C** is correct. The posture contributes up to 30 points to the risk score. Because the risk score is greater than 30, you know that CPDI has detected at least one vulnerability. The vulnerability contributes up to 70 points. With a risk score of 80, you know that the posture score is 10 or more. So the client might have an unknown posture (15 points toward the risk score) or an unhealthy posture (30 points toward the risk score), but not a healthy posture.

    A is incorrect. The posture contributes up to 30 points to the risk score. Because the risk score is greater than 30, you know that CPDI has detected at least one vulnerability.

    B is incorrect. The vulnerability contributes up to 70 points. With a risk score of 80, you know that the risk score contributed by the client posture is more than 0. So the client cannot be healthy.

    D is incorrect. The posture might be unknown (or unhealthy). But the vulnerability score does not come from the number of vulnerabilities detected. Instead it derives from the vulnerabilities' CVE scores.

    For more information on this topic, refer to Chapter 14.

23. **D** is correct. The [ArubaOS Wireless - Terminate Session] profile tells CPPM to issue a Change of Authorization (CoA) message. Dynamic Authorization must be enabled in the APs' network device entries for CPPM to send CoAs to the APs.

    A is incorrect. The CoA delay tells the CPPM server how long to wait before sending the CoA message; however, it will not send the CoA to a network device at all unless Dynamic Authorization is enabled for that device.

B is incorrect. Enabling interim accounting processing lets CPPM process interim accounting messages sent by network devices to it. Processing these messages is not required for CPPM to send CoAs.

C is also incorrect. Enabling RadSec between CPPM and a network device tells the devices to set up a secure TLS session for RADIUS messages; it does not affect whether or not CPPM can send CoAs to the network device.

For more information on this topic, refer to Chapter 1.

24. **C** is correct. You can capture a particular client's traffic as it passes through the MC datapath, based on the client MAC address. You can set the destination for the capture to a remote IP address, such as the security team's machine, as specified in this scenario.

    A is incorrect. The APs in this scenario are controlled by an MC, and you do not manage them through their CLI. Also capturing traffic at the AP's Ethernet port would not be the best approach because that would include a lot of packets non-related to the client in question—and the packets would still be encrypted.

    B is incorrect. Capturing the traffic at the switch port that connects to the AP would capture a lot of irrelevant traffic. Also the wireless client's packets would still be encrypted.

    D is incorrect. A controlplane capture on the MC captures the packets that are destined to the MC itself, not all the packets from clients that flow through it. (Also you cannot set the controlplane capture destination to a remote IP.)

    For more information on this topic, refer to Chapter 16.

25. **D** is correct. In addition to the steps outlined in the question, you must make sure ingress event processing is enabled on the CPPM server. You should also enable Insight to ensure that CPPM can properly process events and correlate an IP address of an attacker with a client MAC address.

    A is incorrect. Device fingerprint dictionaries are not associated with ingress event processing. *Event* dictionaries are, and you should enable the correct one for the Palo Alto NFGW's Syslog messages.

    B is incorrect. You do not need to install an extension for the ingress event processing capabilities to work.

    C is incorrect. The Palo Alto NGFW acts as a Syslog client, not server, for this scenario. To process the messages, CPPM must define the NGFW as an event source, have ingress event processing enabled, and have the correct event dictionary enabled.

    For more information on this topic, refer to Chapter 11.

26. **A** is correct. The Web-based Health Check Service processes SHV reports from OnGuard agents. It uses the plugin checks and rules in the referenced posture policy to assign the correct posture to clients.

B and **D** are incorrect. You might want to complete both of these steps to use postures assigned to clients to control their access. However, these steps do not help assign the posture to clients, which is the task about which the question asks.

C is incorrect. CPPM does not use an OnGuard extension to enable the OnGuard features.

For more information on this topic, refer to Chapter 5.

27. **D** is correct. The use case described in this item requires cloud access security broker (CASB), an HPE Aruba Networking SSE feature that analyzes users' traffic destined to SaaS applications and enforces configured restrictions. A policy that implements CASB must have an external web profile that enables SSL encryption.

    A is incorrect. You can optionally create tags to apply to applications. However, this action is not required.

    B is incorrect. HPE Aruba Networking SSE does not have a concept of SaaS integrations. Instead, it is pre-programmed to recognize hundreds of SaaS applications.

    C is incorrect. Self-hosted web applications are intended for private HTTP/HTTPS services hosted by a company for use in zero trust network access (ZTNA) policies. This use case requires a SaaS application.

    For more information on this topic, refer to Chapter 17.

28. **A** is correct. To collect the flow attributes and information about devices' network activity and network behavior information, the virtual Data Collector needs to receive information on its SPAN port. Virtual Data Collectors have two SPAN adapters; one has the third lowest (second highest) MAC address, and one has the fourth lowest (highest) MAC address. On ESXi hosts, VLAN 4095 passes all traffic, tagged and untagged, which is the best way to send all the traffic mirrored by the core routing switches.

    B and **D** are incorrect because VLAN 0 passes untagged traffic only. **B** is also incorrect because the adapter with the adapter with the lowest MAC address is the Mgmt port, which the Data Collector uses to connect to HPE Aruba Networking Central.

    C is incorrect because the port ID is not a reliable way to determine the virtual Data Collector adapter usage. HPE Aruba Networking recommends using the MAC address. (**D** is incorrect for this reason also.)

    For more information on this topic, refer to Chapter 14.

29. **B** is correct. Based on the role mapping policy, clients that authenticate to a disabled AD user account receive role2. (The AccountStatus values in rule 2 indicate disabled accounts.) Clients that authenticate as computers always receive the [Machine Authenticated] role. That means that this client has two CPPM roles: role2 and [Machine Authenticated]. The first rule that this client matches in the enforcement profile is rule 2, so CPPM applies profile2.

A and **B** are incorrect. The client does not have role 1, so these rules do not match. **D** is incorrect because the client does match a rule; therefore, the default enforcement profile does not apply.

For more information on this topic, refer to Chapter 3.

30. **D** is correct. Because the APs are authenticating with EAP-TLS (for certificate-based authentication), CPPM must validate the certificate signatures. The Aruba Networks Trusted Computing Root CA is the root CA for the APs' TPM certificates. So you must enable this CA and add the EAP usage.

    A is incorrect for several reasons. APs do not support EAP-TEAP as an authentication method. Also MAC-Auth is not a valid inner method for EAP-TEAP. Although the scenario did state that CPPM should check the APs' MAC addresses, as well as their certificates, MAC-Auth is not required for that purpose. Instead you can set up role mapping or enforcement policy rules to check that the APs are in the Endpoints Repository.

    B is incorrect. You do not need to install every AP TPM certificate in CPPM's Certificate Trust List. Enabling the root CA is sufficient. Also the correct usage is EAP for this use case, not Aruba Infrastructure.

    C is incorrect. AOS-CX switches receive the role name for authenticated clients in the Aruba-User-Role VSA.

    For more information on this topic, refer to Chapter 7 and also to Chapter 2.

Appendix A
Learning Check Answers

Chapter 1

1. Which correctly describes Zero Trust Security?

   a. External clients are untrusted, while internal clients are trusted.

   b. The perimeter is enhanced by a demilitarized zone (DMZ).

   c. **<u>All users and devices are identified and controlled with granular identity- and context-based policies.</u>**

   d. All users are considered equally untrustworthy and are given the same privileges regardless of identity.

2. Match the HPE Aruba Networking Zero Trust Security principles to the products that allow customers to act on those principles.

   ZTS principles

   a. Visibility

   b. Authentication

   c. Role-based access control, enforcement and response, continuous monitoring

   Solutions

   _c_ Unified Infrastructure, ClearPass Exchange

   _b_ ClearPass Onboard

   _a_ CPDI

3. What role do service rules play within a CPPM service?

   a. Determining whether a user is permitted access

   b. Selecting the preferred authentication method

   c. Matching the incoming request to the correct service

   d. Determining whether the NAD is authorized

4. What is a reason to use role mapping policies?

   a. Because role mapping policies simplify configuration by eliminating the need for enforcement profiles and policies

   b. Because you want to characterize clients with a global policy that you can use in multiple services

   c. Because you want to use more complex conditions which are not supported in enforcement policies

   d. Because you want to define role-based firewall policies directly within CPPM

Chapter 2

1. What is required for a client to trust a web server's certificate? (Select two.)

   a. The client has the certificate's public key installed on it in advance.

   b. The certificate was issued within the previous six months.

   c. The client can trace the certificate back to a trusted root.

   d. The certificate DNS SAN matches the hostname accessed by the client.

2. What is one requirement for joining a Subscriber to a ClearPass cluster?

   a. The Subscriber trusts the entire chain for the Publisher's RADIUS/EAP certificate.

   b. The Subscriber trusts the entire chain for the Publisher's HTTPS certificate.

   c. The Subscriber has an HTTPS certificate with its IP address listed as a DNS SAN.

   d. The Publisher has an HTTPS certificate with its IP address listed as a DNS SAN.

3. What correctly describes the Device Certificate on HPE Aruba Networking APs?

   a. It is signed by an HPE Aruba Networking CA and installed in a TPM.

   b. It is a self-signed certificate, which should be replaced in production.

c. It uses the AP's default IP address as its CN in the subject name.

d. It is signed by a CA of the company's choice and must be installed manually.

Chapter 3

1. A CPPM service uses the default [EAP-TLS] method for the authentication method and AD for the authentication source. What is required for a client to pass authentication? (Select two.)

   a. The client certificate is signed by a different CA from the one that signed CPPM's RADIUS/EAP certificate.

   b. The client certificate is signed by a CA in the CPPM trust list with the EAP usage.

   c. The serial number for the client certificate is listed on a CRL.

   d. The client username, derived from the client certificate, matches an account in AD.

2. What function does the Onboard Pre-Auth service provide in the onboarding solution?

   a. It authenticates users who are using 802.1X to authenticate with certificates issued by Onboard.

   b. It checks whether the user is attempting to onboard an authorized device type.

   c. It checks the credentials that the user submits to the Onboard portal.

   d. It authenticates users who are attempting to connect to an Onboard SSID.

3. A client has completed TEAP. It succeeded at TEAP Method 1 and failed at TEAP Method 2. What role does CPPM assign it?

   a. [Unauthenticated]

   b. [Partial Authenticated]

   c. [Machine Authenticated]

   d. [User Authenticated]

Chapter 4

1. An AOS device uses 802.1X to authenticate clients to ClearPass Policy Manager (CPPM). How does HPE Aruba Networking generally recommend you assign roles to clients that successfully authenticate?

   a. Using the authentication default method

b. Using server-derived rules

c. Using user-derived rules

d. **Using the Aruba-User-Role VSA sent by CPPM**

2. If a client attempts to access a particular subnet, the company wants to lock that client out of the network for a period. How can you enforce this action?

   a. Enable DPI and create an application rule that denies traffic sent to that subnet.

   b. Enable WebCC and create an application rule that denies traffic sent to that subnet.

   c. **Create a firewall rule that denies traffic sent to that subnet and also applies denylisting (blacklisting).**

   d. Create a firewall rule that denies traffic sent to that subnet and also applies NAT.

3. A customer has APs managed by HPE Aruba Networking Central. Based on Aruba-User-Role VSA sent by CPPM, an AP has assigned a client to a role named "loan-officers." Refer to Figure 4-20 to see the rules associated with this role. Then consider the following questions. Take notes on your answers.

   a. To which IP address are loan officers permitted to send DNS requests? **10.254.1.31**

   b. The domain controller is at 10.254.1.31. What services (list by name or protocol/port) are loan officers permitted to access on the domain controller? **DNS (UDP/TCP 53), HTTP, Kerberos (UDP/TCP 88) TCP 135 (RPC Endpoint Manager), Netbios SSN (TCP 139), HTTPS (TCP 443), SMB-TCP (TCP 445), TCP and UDP 389 (LDAP), and dynamic ports (TCP 1023-65434). Note that rules that apply to all of 10.254.1.0/24 also permit MSRPC (TCP and UDP) to 10.254.1.31.**

   c. Are loan officers permitted to access web services on 10.254.1.15? **No**

   d. Are loan officers permitted to access web services on 10.254.1.24? **Yes**

   e. Are loan officers permitted to access web services on 10.1.19.1? **No**

   f. Are loan officers permitted to access the Internet? **Yes**

```
allow udp on port 67 to all destinations
allow dns on server 10.254.1.31/255.255.255.255
allow tcp on port 53 on server 10.254.1.31/255.255.255.255
allow http on server 10.254.1.31/255.255.255.255
allow kerberos on server 10.254.1.31/255.255.255.255
allow tcp on port 88 on server 10.254.1.31/255.255.255.255
allow tcp on port 135 on server 10.254.1.31/255.255.255.255
allow netbios-ssn on server 10.254.1.31/255.255.255.255
allow https on server 10.254.1.31/255.255.255.255
allow smb-tcp on server 10.254.1.31/255.255.255.255
allow tcp on port 389 on server 10.254.1.31/255.255.255.255
allow udp on port 389 on server 10.254.1.31/255.255.255.255
allow tcp on port 1023-65434 on server 10.254.1.31/255.255.255.255
deny any to network 10.254.1.0/255.255.255.240
allow http to network 10.254.1.0/255.255.255.0
allow https to network 10.254.1.0/255.255.255.0
allow msrpc-tcp to network 10.254.1.0/255.255.255.0
allow msrpc-udp to network 10.254.1.0/255.255.255.0
allow https on server 10.1.19.11/255.255.255.255
allow tcp on port 6658 on server 10.1.19.11/255.255.255.255
deny any to network 10.0.0.0/255.0.0.0
allow any to all destinations
```

Figure 4-1: ACL Rules

Chapter 5

1. What is one difference between the OnGuard persistent agent and the dissolvable agent?
   a. The persistent agent only works for Windows systems while the dissolvable agent supports Windows and Linux.
   b. **The persistent agent supports ongoing monitoring while the dissolvable agent supports a one-time check.**
   c. The persistent agent integrates with Microsoft Intune while the dissolvable agent offers no integration.
   d. The persistent agent posture overrides the posture delivered by any other source while the dissolvable agent posture is overridden by other sources.

2. Match the Component with its Role.

<u>Component</u>

a. Posture policy

b. WebAuth enforcement policy

c. Web-based Health Check Only service

<u>Role</u>

<u>b Provides rules that assign enforcement profiles based on system posture</u>

<u>a Provides SHV plugin configuration and rules for assigning posture based on SHV checks</u>

<u>c Responds to requests from OnGuard agents</u>

3. You have added an Intune extension, and now you need to configure it. How do you determine what to specify for the "tenantId," "clientId," and "clientSecret?

a. You should always leave the default values for these fields.

b. You should consult with HPE Aruba Networking support to assess the right values for your environment.

c. <u>You should register ClearPass as an App in Azure AD and collect the values there.</u>

d. You should use XML files available in GitHub to look up the right values.

Chapter 6

1. What purpose does the "aaa authentication allow-fail-through" command serve on AOS-CX switches?

a. It lets the switch authenticate users against the secondary method only if the primary method times out.

b. It lets the switch actively test whether a TACACS+ server is available.

c. It lets the switch allow users operator level access if their authentication fails.

d. <u>It lets the switch authenticate users against the secondary method if the primary method rejects the user.</u>

2. How does ARP inspection prevent MITM attacks?

a. It discovers which IP and MAC addresses belong together.

b. It drops all ARP responses that arrive on untrusted ports.

c. **It drops ARP responses on untrusted ports that do not match the IP-to-MAC table.**

d. It stops clients from generating gratuitous ARP responses.

Chapter 7

Answers to Consideration Questions

You should have answered that CPPM should be set up to send desired roles in an Aruba-CPPM-Role VSA. To download the role, the switch must be set up for DURs, as you learned earlier.

The switch-port should use "guest" for the auth and critical roles. It should use "unauthenticated" for the fallback role. The reject role should be set to a role that denies all access in order to prevent the fallback role from applying to rejected clients. The "guest," "unauthenticated," and "denyall" roles need to be configured locally. The switch will dynamically download the roles for authenticated clients.

Answers to Learning Checks

1. CPPM already acts as the authentication server for wireless 802.1X. Now the company wants to add wired 802.1X.

   Which CPPM components does HPE Aruba Networking generally recommend should be the same for wired and wireless 802.1X? (More than one answer might be correct.)

   a. Service

   b. **Authentication source**

   c. **Role mapping policy**

   d. Enforcement policy

2. How does a switch receive the configuration for a DUR?

   a. CPPM uses SSH to log into the switch and configure the DUR on it.

   b. **The switch uses HTTPS to retrieve the role configuration from CPPM.**

   c. Admins configure the DUR settings on the switch, and CPPM directs the switch which DUR to apply.

   d. The switch uses SSH to contact an HPE Aruba Networking gateway and receive the DUR from it.

3. What is the significance of the following syntax on an AOS-CX switch?

```
AOS-CX(config)# class ip NEW-CLASS
AOS-CX(config-class-ip)# 10 match tcp any
10.101.12.0/24 eq ssh
```

a. The switch will drop all traffic from any IP address destined for an address in 10.101.12.0/24

b. The switch now has a class that matches all SSH traffic from devices within 10.101.12.0/24

c. The switch now will permit all IP traffic destined for SSH ports on 10.101.12/24

d. The switch now has a class that matches all SSH traffic destined to addresses in 10.101.12.0/24

Chapter 8

1. How do you apply UBT to a client's traffic?

a. Use the gateway zone setting in the role to which the client is assigned.

b. Enable the UBT (gateway) zone globally, which applies it to all clients' traffic.

c. Apply the UBT zone to the client's port.

d. Apply the UBT zone to the client's VLAN.

2. How can VNBT help a company secure clients' traffic?

a. By tunneling the clients' traffic to an HPE Aruba Networking gateway, which performs advanced security functions

b. By redirecting the clients' traffic to a captive portal server, which authenticates the clients

c. By redirecting the clients' traffic to a cloud security solution, which performs advanced security functions

d. By tunneling the clients' traffic to a switch in the data center, which passes the traffic to an application-level firewall

Chapter 9

1. What functions does an NAE agent perform? (Multiple answers can be correct.)

a. Collects data from the current state database and records it in a time-series database

b. Performs authorization control of management users

c. Allows you to virtualize legacy switch code within an AOS-CX switch

d. Triggers actions based on whether a monitored attribute meets a particular condition

2. Which correctly describes HPE Aruba Networking NAE scripts?

a. All HPE Aruba Networking-certified scripts are loaded on the AOS-CX switch at factory defaults.

b. You can find HPE Aruba Networking-certified scripts in ASE and Github.

c. HPE Aruba Networking provides support for all NAE scripts, as long as the scripts use Python code.

d. You are only allowed to load HPE Aruba Networking-certified scripts on AOS-CX switches.

Chapter 10

1. What distinguishes a Rogue AP from an Interfering AP?

a. A Rogue AP has launched a DoS attack.

b. A Rogue AP is connected to the corporate LAN.

c. A Rogue AP uses unauthorized security settings.

d. A Rogue AP is connected to Interfering clients.

2. What are some HPE Aruba Networking recommendations for IPS and IDS settings to enable? (Select two.)

a. Enable all IDS settings.

b. Enable all IPS settings.

c. Enable IDS settings that are unlikely to produce false positives.

d. If automatic containment is required, enable rogue AP containment with wireless deauthentication.

e. If automatic containment is required, enable rogue AP containment with wired containment.

Chapter 11

1. What is an event source on CPPM?

a. An MDM from which CPPM receives information about endpoint posture

b. A security appliance to which CPPM provides user information

c. A device from which CPPM will accept and process Syslog messages

d. An endpoint on which malware or another threat has been detected

2. What purpose does the Ingress Event Dictionary serve?

   a. It lists the valid event sources from which CPPM is allowed to process events.

   b. It transforms Syslog content into an event with attributes that can be referenced.

   c. It specifies the actions that CPPM should take in response to an event.

   d. It defines the Syslog message that CPPM should send in response to specific requests.

3. Match the definition to the component.

   Definitions

   a. Defines which attributes to update in the endpoint profile

   b. Defines the condition that triggers applying the post-authentication profile

   c. Processes events

   Components

   c Event service

   b Event enforcement policy

   a ClearPass Entity Update Enforcement profile

Chapter 12

Answers to Consideration Questions

1. What is the difference between passive and active profiling?

   Passive = CPPM discovers info about the endpoint without directly interacting with the endpoint (examples: DHCP, HTTP User-Agent, MAC OUI)

   Active = CPPM contacts the endpoint to collect info about it (examples: Nmap, WMI, SSH)

2. How can DHCP packets help CPPM determine the device type?

   Every OS has a unique way of ordering options in a DHCP Discovery packet, called a DHCP fingerprint. CPPM has a dictionary of these fingerprints, indicating which fingerprints go with which device types.

3. What useful information can CPPM find in clients' HTTP/HTTPS traffic?

   HTTP User-Agent can help CPPM distinguish between devices with similar DHCP fingerprints. (If traffic is HTTPS, CPPM must be the server involved.)

Answers to Learning checks

1. How do you enable CPPM to receive device fingerprint updates from HPE Aruba Networking?

   a. Integrate CPPM with CPDI.

   b. Integrate CPPM with HPE Aruba Networking Central.

   c. Enable auto endpoint profiler fingerprint updates in the cluster-wide parameters.

   d. Import an XML file from the Administration > Dictionaries > Device Fingerprints page.

2. What steps do you complete to scan endpoints with SSH? (Select two.)

   a. Define SSH credentials and map them to the endpoints' subnet.

   b. Run a Discovery scan on the network.

   c. Run a periodic poll on the endpoints' NADs.

   d. Run a Subnet scan on the endpoints.

3. How can you create a custom device fingerprint on CPPM?

   a. Import the new device fingerprint from an XML file.

   b. Create the new device fingerprint and then import rules from a discovered device.

   c. Create the new device fingerprint and then define rules for it manually.

   d. Import the new device fingerprint from a third-party solution.

4. You want to configure the profiling action for a CPPM service, but you do not see the Profiler tab. What should you do?

   a. Recreate the service, selecting a template that includes "Profiling" in the name.

   b. In the Service tab, change the template to one that includes "Profiling" in the name.

   c. In the Service tab, select the Profile Endpoints check box.

   d. Add a rule that uses the "IsProfiled" attribute to a role mapping or enforcement policy.

5. In which of these circumstances does CPPM set an endpoint's Conflict attribute to true?

   a. When the endpoint first connects before it is profiled

   b. When the endpoint category changes from unknown to "Computer"

   c. When the endpoint category changes from "Cameras" to "Computer"

   d. When the endpoint has fingerprints that match multiple device fingerprint entries

Chapter 13

1. When CPDI applies LDA clustering to generic devices, how does it produce recommendations for the cluster classification?

   a. It sends the cluster to HPE Aruba Networking experts who make the recommendation.

   b. It uses the MAC address ranges assigned to vendors by IEEE.

   c. It looks for classified devices that exhibit similar attributes.

   d. It uses system rules to create the recommendations.

Chapter 14

1. What port must be open between a CPDI Data Collector and HPE Aruba Networking Central?

   a. TCP 8080

   b. TCP 443

   c. UDP 22

   d. UDP 4422

2. What do the augmentation methods for a segment define?

   a. They provide expanded MAC address ranges to supplement classification.

   b. They define which additional user rules to apply to classify devices.

   c. They define which ML clustering methods are used to classify devices.

   d. They define the methods and credentials that the Data Collector uses to scan devices.

3. Which correctly describes a tag?

   a. A custom rule that reclassifies device types based on a variety of attributes

   b. A description that applies to devices based on various criteria

   c. A device attribute that is collected by a Data Collector

   d. A flow attribute that is collected by a Data Collector

4. What is a requirement for CPDI to collect vulnerability and posture information on Windows devices?

   a. The devices must have 32-bit operating systems.

   b. The devices must use Windows 7.

   c. CPDI agents must be installed on the devices.

   d. A WMI augmentation method must be attached to the segment for the devices.

Chapter 15

1. What must you obtain to enable Device Insight integration on CPPM?

   a. A valid admin email address

   b. An admin password

   c. A private key

   d. A registration token

2. CPPM is integrated with CPDI. What determines whether CPPM sends a CoA when a device's tag is updated by CPDI?

   a. The settings in the Profiler tab for the service that authenticated the device

   b. The settings in the ClearPass Device Integration page

   c. The cluster-wide General settings

   d. The network device settings for the NAD to which the device is connected

Chapter 16

1. What is one difference between Wireshark capture filters and display filters?

   a. Only capture filters can be removed and changed during an active capture.

   b. Only capture filters can filter packets based on protocol.

   c. **Only display filters can be removed and changed during an active capture.**

   d. Only display filters can filter packets based on protocol.

2. What is a sign of a TCP Syn scan? (Multiple answers can be correct.)

   a. **A device that immediately sends TCP RST packets in response to Syn + Ack messages**

   b. Any TCP packet that has the Syn flag set without the Ack flag

   c. Any TCP packet that has the Syn flag set with the Ack flag

   d. **Many TCP Syn packets from the same source to different destinations and ports**

3. Below are listed several different instructions that the security team might give you. For each one consider what steps you would take next.

   a. "We have detected suspicious activity on a wireless client at IP address 10.1.55.36. Please capture its traffic for 15 minutes and give us the results in a PCAP file." (This environment uses an AOS-8 architecture with MCs and CAPs.)

      Use the Mobility Conductor (MCR) interface to find the client in question. Find its MAC address and the MC to which it is connected. Go to the CLI for the MC. Set the packet-capture destination to local filesystem. Initiate a datapath capture for the client MAC address. After 15 minutes, stop the packet capture. Save out the file at /var/log/oslog/datapath.pcap.

   b. "We have detected suspicious activity on a wireless client at IP address 10.5.32.17. Please capture its traffic for 30 minutes and give us the results in a PCAP file." (This environment uses an AOS-10 architecture with APs and HPE Aruba Networking Central.)

      Find the client in HPE Aruba Networking Central and open its client details page. Go to **Analyze > Live Events**. Stop the troubleshooting session. Enable **Packet Capture** and re-start the troubleshooting session. After 15 minutes, download the file and start another troubleshooting session. After 15 minutes, download the second file.

c. "We think that an intrusion might have occurred in Building 1. We want to capture all traffic coming between the access switch in that building and the data center or Internet. Please send the traffic to 10.2.6.8, where we are running an analyzer." (This switch belongs to the CX 6300 series.)

Log into the access switch CLI. Create a mirror session. Set the source interface to the switch's uplink(s) and **both** (for rx and tx traffic). Set the destination to tunnel with remote IP, 10.2.6.8. Enable the session.

Chapter 17

1. What is a key difference between HPE Aruba Networking ZTNA and a traditional client-to-site VPN?

   a. ZTNA uses the latest version of IPsec.

   b. ZTNA does not continuously monitor users' access.

   c. ZTNA uses on-prem firewalls rather than VPN concentrators.

   d. ZTNA grants access to individual apps rather than to the private network.

2. What is required to permit remote users access to your private apps through HPE Aruba Networking ZTNA?

   a. Deploying connectors that can reach those apps

   b. Installing agents on the application servers

   c. Registering your application servers' IP addresses with HPE Aruba Networking

   d. Integrating the applications with an IdP supported by HPE Aruba Networking ZTNA

3. Assume that a company has NOT deployed HPE Aruba Networking EdgeConnect. What is one requirement that applies to BOTH HPE Aruba Networking SWG and HPE Aruba Networking CASB?

   a. Clients have agents installed on them.

   b. Users authenticate to the built-in HPE Aruba Networking SSE IdP.

   c. A connector is deployed at the customer's private site.

   d. An external Web profile with Threat Intelligent Protection is specified in the policy rules.

4. Which type of EdgeConnect SD-Branch topology do you need to add if you want VPNCs to create tunnels to other VPNCs?

   a. Hub-spoke

   b. Branch-mesh

c. **<u>Hub-mesh</u>**

d. Spoke-mesh

Chapter 18

1. How does HPE Aruba Networking gateway IDS/IPS detect malicious traffic?

   a. APs and switches notify the gateway when they detect malware.

   b. CPPM notifies the gateway when it detects malware.

   c. **<u>The gateway detects signatures in the traffic that indicate the presence of threats.</u>**

   d. The gateway detects anomalies in the traffic that differ from usual traffic patterns.

2. For which use cases should you recommend IDS, as opposed to IPS?

   <u>Ideas include: Preventing disruption of business based on false positives.</u>

3. Why might a customer integrate HPE Aruba Networking gateway IDS/IPS with a SIEM server?

   a. To allow traffic to continue through a gateway if the IDPS engine goes down

   b. To prevent a gateway from forwarding traffic if the IDPS engine goes down

   c. **<u>To integrate gateway IDS/IPS with the company's automated security event management</u>**

   d. To ensure that gateways have the most up-to-date rules

Chapter 19

1. Match the Config with the HPE Aruba Networking Central Scope.

   a. Global

   b. Site

   <u>b Email overrides</u>

   <u>a Which alerts to enable</u>

   <u>a Alert settings</u>

2. Which of these alerts occurs when the percentage of packets dropped by gateway IPS exceeds the configured threshold?

   a. Infrastructure Attack Detected

   b. Authentication Failure Detected

   c. Gateway Threat Count per Signature

   d. Gateway IDS IPS Engine Packet Dropped Detected